Michael Breuer
Franz Durst
Christoph Zenger
Editors

High Performance Scientific and Engineering Computing

Proceedings of the 3rd International FORTWIHR Conference on HPSEC, Erlangen, March 12–14, 2001

With 265 Figures

Springer

Editors

Michael Breuer
Franz Durst

Lehrstuhl für Strömungsmechanik (LSTM)
Friedrich-Alexander-Universität Erlangen-Nürnberg
Cauerstraße 4
91058 Erlangen, Germany
e-mail: breuer@lstm.uni-erlangen.de
durst@lstm.uni-erlangen.de

Christoph Zenger

Institut für Informatik
Technische Universität München
80290 München, Germany
e-mail: zenger@in.tum.de

Front cover figure: designed from an original idea by Michael Breuer

Cataloging-in-Publication Data applied for

Die Deutsche Bibliothek - CIP-Einheitsaufnahme
High performance scientific and engineering computing :
proceedings of the 3rd International FORTWIHR Conference on HPSEC,
Erlangen, March 12 - 14, 2001 / Michael Breuer ... ed.. - Berlin ; Heidelberg ; New York ;
Barcelona ; Hong Kong ; London ; Milan ; Paris ; Tokyo : Springer, 2002
(Lecture notes in computational science and engineering ; Vol. 21)
ISBN 3-540-42946-8

Mathematics Subject Classification (2000):
35-XX, 37-XX, 49-XX, 65-XX, 68-XX, 70-XX, 74-XX, 76-XX, 80-XX, 93-XX

ISSN 1439-7358
ISBN 3-540-42946-8 Springer-Verlag Berlin Heidelberg New York

Springer-Verlag Berlin Heidelberg New York
a member of BertelsmannSpringer Science + Business Media GmbH
http://www.springer.de

Printed in Germany

Cover Design: Friedhelm Steinen-Broo, Estudio Calamar, Spain
Cover production: *design & production*
Typeset by the authors using a Springer TEX macro package

Printed on acid-free paper SPIN: 10653059 46/3142/LK - 5 4 3 2 1 0

Preface

In Douglas Adams' book *'Hitchhiker's Guide to the Galaxy'*, hyper–intelligent beings reached a point in their existence where they wanted to understand the purpose of their own existence and the universe. They built a supercomputer, called Deep Thought, and upon completion, they asked it for the answer to the ultimate question of life, the universe and everything else. The computer worked for several millennia on the answers to all these questions. When the day arrived for hyper–intelligent beings the to receive the answer, they were stunned, shocked and disappointed to hear that the answer was simply

42.

The still open questions to scientists and engineers are typically much simpler and consequently the answers are more reasonable. Furthermore, because human beings are too impatient and not ready to wait for such a long period, high–performance computing techniques have been developed, leading to much faster answers. Based on these developments in the last two decades, scientific and engineering computing has evolved to a key technology which plays an important role in determining, or at least shaping, future research and development activities in many branches of industry. Development work has been going on all over the world resulting in numerical methods that are now available for simulations that were not foreseeable some years ago. However, these days the availability of supercomputers with Teraflop performance supports extensive computations with technical relevance. A new age of engineering has started.

The 3rd International FORTWIHR Conference, which took place at the University of Erlangen–Nürnberg on March 12–14, 2001, brought together developers and users of computer codes that have been set up to carry out high-performance computations to solve scientific and/or engineering problems on supercomputers (FORTWIHR is the Bavarian Consortium for High–Performance Scientific Computing). Researchers from all over the world presented their results and discussed the state of the art in high–performance computing. Furthermore, invited lectures presented major developments in the field and provided an outlook into the future. The main topics included the development of advanced numerical methods, parallel computing techniques, grid generation, and visualization. Applications of these techniques are directed to fluid dynamics, material sciences, chemical engineering, combustion, dynamic systems, optimal control, and electronic system design.

The 1st FORTWIHR Conference took place in Munich in June 1993. The two–day event was attended by more than 300 scientists from industry and from the academic field. Focusing on FORTWIHR's main research topics, leading experts in applied and numerical mathematics, informatics, science, and engineering discussed recent results and new trends in the field of numerical simulation on high-performance architectures. The conference proceedings were published by Vieweg–Verlag as Vol. 48 of its 'Notes on Numerical Fluid Mechanics' series. The 2nd FORTWIHR Conference, held in Munich in March 1998, was a three–day international event with nearly 200 participants. The proceedings with 38 contributions were published by Springer–Verlag in a preceding edition of 'Lecture Notes in Computational Science and Engineering' (Vol. 8).

The present volume of LNCSE is the result of the 3rd FORTWIHR Conference. In 14 sessions, about sixty papers including five invited lectures were presented, covering a broad range of high-performance computing: computational fluid dynamics, combustion and porous media related flows, computational structural dynamics, dynamic systems and optimal control, crystal growth and materials, optimization of electronic circuits, and mathematical methods. After a reviewing process, 44 papers were selected for this volume.

The proceedings represent the final act of the Bavarian research initiative FORTWIHR, launched in April 1992. The objectives of this initiative – major progress in selected areas of applications of high–performance scientific computing (HPSC) by a strictly interdisciplinary approach in close cooperation between university and industry, rapid transfer of all results to the industrial partners (especially small– and medium–sized companies), education of qualified and experienced experts for industry, strengthening of HPSC in the degree programs of participating universities – have definitely been reached. FORTWIHR expired in March 2001; however, the activities in this highly interesting field of research and development are certainly continuing in all the participating university research groups and in the cooperating companies.

The editors would like to express their thanks to several persons and institutions having contributed in different ways to the conference and to these proceedings. First, we thank our host, the Friedrich–Alexander–University Erlangen–Nürnberg for generously providing the perfect conference facilities, including the lecture halls.

Next, we thank the entire FORTWIHR crew, especially Dr. G. Brenner, Prof. Dr. H.-J. Bungartz, S. Enger, N. Gergova, M. Glück, J. Grasser, Dr. L. Kadinski, F. Kaschak, P. Lammers, Dr. D. Melling, I.V. Paulus, M. Selder, V. Vassilev, Th. Zeiser, and Dr. St. Zimmer, for all kinds of support before and during the conference. Furthermore, during the compilation of this book the support of M. Ihme and M. Glück was highly valuable.

Throughout the preparation and realization of this volume, cooperation with Dr. Martin Peters, Thanh–Ha Le Thi, and Leonie Kunz of Springer–Verlag was fruitful and pleasant and will not be forgotten.

Finally, we express our gratitude to the *'Bayerische Forschungsstiftung'* and especially its former president, Prof. Dr. N. Fiebiger, furthermore to the *'Bayerische Staatsministerium für Wissenschaft, Forschung und Kunst'* represented by Ministerialdirigent J. Großkreutz, and last but not least FORTWIHR's numerous industrial partners and sponsors. Neither the conference including these proceedings and their procursors nor an important part of the results reported in many of the contributions would have been possible without their continuous financial support over a period of more than eight years. This support is gratefully acknowledged.

Erlangen, September 2001

Michael Breuer
Franz Durst
Christoph Zenger

Contents

Part IV. Dynamic Systems and Optimal Control

Part V. Optimization of Electronic Circuits

Part I

Fluid Flow

Large-Scale Fluid-Structure Interaction Simulations Using Parallel Computers

R. Löhner[1], J.D. Baum[2], Ch. Charman[3], and D. Pelessone[4]

[1] School of Computational Sciences and Informatics
M.S. 4C7, George Mason University, Fairfax, VA 22030-4444, USA
[2] Applied Physics Operations
Science Applications International Corp., McLean, VA 22102, USA
[3] General Atomics, San Diego, CA 92121, USA
[4] Engineering Software System Solutions, Solana Beach, CA 92075, USA

Abstract. A methodology to simulate large-scale fluid-structure interaction problems on parallel machines has been developed. Particular emphasis was placed on shock-structure interaction problems. For the fluid, a high-resolution FEM-FCT solver based on unstructured, moving, body fitted grids is used. For the structure, a Lagrangean large-deformation finite element code is employed. The coupled system is solved using a loose coupling algorithm, with position and velocity interpolation and force projection. Several examples, run on parallel machines, demonstrate the range of applicability of the proposed methodology.

1 Introduction

The threat of intentional (Lockerbie (1988) [1], World Trade Center (1993) [2], Khobar Towers (1996), Nairobi and Dar Es Salaam (1998), etc.) or accidental (Seveso (1976), oilrigs, oil refineries, etc.) explosions is a reality that planners have to take seriously when designing buildings, particularly those that will have a large concentration of people (airports, embassies, etc.). Any comprehensive methodology that seeks to predict or reproduce these events must account for the relevant physical phenomena, which include:

- HE ignition and detonation;
- Shock wave propagation through fluid and solid media;
- Load and position/velocity transfer between fluid and structure;
- Possible catastrophic failure of the structure with consequent changes to the flowfield domain/topology.

At the same time, the simulations required must be run as fast as possible in order to have an impact on design or understanding. This implies that

- Integrated pre- and post-processing for fluid and structural dynamics;
- Fully automatic grid generation; and
- Scalability on parallel computers

should be an integral part of any such simulation methodology.
This paper reports on recent developments towards and simulations with the envisioned methodology.

2 HE Ignition and Detonation

Most high explosives are well modelled by the Jones-Wilkins-Lee (JWL) equation of state, given by:

$$p = A\left(1 - \frac{\omega}{R_1 v}\right)e^{-R_1 v} + B\left(1 - \frac{\omega}{R_2 v}\right)e^{-R_2 v} + \omega\rho e \quad , \tag{1}$$

where v denotes the relative volume of the gas:

$$v = \frac{V}{V_0} = \frac{\rho_0}{\rho} \quad . \tag{2}$$

Afterburning is modelled by adding energy via a burn coefficient λ that is obtained from

$$\lambda_{,t} = a p^{\frac{1}{6}} \sqrt{1-\lambda} \quad , \tag{3}$$

where $\lambda = 0$ for the unburned state, and $\lambda = 1$ for the fully burned material. After updating λ, the energy released is added as follows:

$$(\rho e)|^{n+1} = (\rho e)|^{n} + \rho Q(\lambda^{n+1} - \lambda^{n}) \quad , \tag{4}$$

where Q is the afterburn energy. Compared to the five unknowns required for the Euler equations with an ideal air equation of state, we require an additional two: the burn fraction b to determine which part of the material has ignited, and the afterburn coefficient λ. Observe that in the expanded state ($v \rightarrow \infty$), the JWL equation of state reduces to

$$p = \omega\rho e = (\gamma - 1)\rho e \quad , \tag{5}$$

where the correlation of ω and γ becomes apparent. The transition to air is made by comparing the density of air to the density of the high explosive. Given that $A >> B$, the decay of the first term in Eqn.(1) with increasing v is much faster. This implies that as v increases, we have

$$p \rightarrow B e^{-R_2 v} + \omega\rho e \quad . \tag{6}$$

The mixture of high explosive and air is considered as air when the effect of the B-term may be neglected, i.e.

$$\frac{p}{p_{cj}} = \epsilon = B e^{-R_2 v} \quad , \tag{7}$$

where p_{cj} denotes the Chapman-Jouget pressure and $\epsilon = O(10^{-3})$. For a typical explosion simulation, the major portion of CPU time is required to simulate the burning material. This is because the pressures are very high, and so are the velocities of the fluid particles. Once the material has burned out, one observes a drastic reduction of pressures and velocities, which implies a dramatic increase in the allowable timestep. Even though shocks travel much larger distances, this post-burn 'diffraction phase' takes less CPU-time than the burn phase. In order to speed up the simulation, the portions of the grid outside the detonation region are deactivated. The detonation velocity provides a natural speed beyond which no information can travel. Given that the major loops in an unstructured-grid flow solver are processed in groups (elements, faces, edges, etc.) for vectorization, it seemed natural to deactivate not the individual edge, but the edge-group. In this way all inner loops can be left untouched, and the test for deactivation is carried out at the group-level. The number of elements in each edge-, face-, or element-group is kept reasonably small (O(128)) in order to obtain the highest percentage of deactivated edges without compromising performance. The points are renumbered according to their ignition time assuming a constant detonation velocity. In this way, the work required for point-loops is minimized as much as possible. The edges and points are checked every 5-10 timesteps and activated accordingly. This deactivation technique leads to considerable savings in CPU at the beginning of a run, where the timestep is very small and the zone affected by the explosion only comprises a small percentage of the mesh.

3 Shock Wave Propagation Through Fluid and Solid Media

The simulation of shock wave propagation through the fluid and solid media is carried out using explicit timestepping schemes.

For the fluid, a body-fitted, arbitrary Lagrangean-Eulerian (ALE) unstructured grid solver is employed. The formulation is based on finite elements, and uses flux-corrected transport to propagate accurately shocks [11]. The motion of the surfaces attached to deforming fluids is imposed from the structural dynamics solver. The mesh motion is smoothed using a nonlinear Laplacian [16] in order to reduce the amount of global and local remeshing required. At the end of each fluid timestep, the loads exerted on the 'wetted surfaces' belonging to the structural dynamics solver are computed and transferred back.

For the solid, a Lagrangean finite element code based on the large displacement formulation is used. Truss-, beam-, triangular and quadrilateral shells, as well as hexahedral solid elements have been incorporated. Innovative features of this structural dynamics code include adaptive refinement [23,24], advanced contact algorithms [25] and the ability to fail elements or open cracks based on physics (i.e. not predetermined). At the beginning of each structural dynamics timestep, the loads from the fluid are updated. At the end of each structural dynamics timestep, the positions and velocities of the 'wetted surfaces' are extracted and transferred back to the flow solver.

4 Load and Position/Velocity Transfer

Optimal discretizations for the structure and the fluid are, in all probability, not going to be the same. As an example, consider an explosion in a room. For an accurate fluid solution, element sizes below 1cm have proven reliable. For the structure, shell or brick elements of 10cm are more than sufficient. There may be situations where fluid and structural surfaces are represented by abstractions of different dimensionality. A typical example here would be a commercial aircraft wing undergoing aeroelastic loads. For an accurate fluid solution using the Euler equations, a very precise surface representation with 60-120 points in the chordwise direction will be required. For the structure, a 20×40 mesh of plate elements may be more than sufficient. Any general fluid-structure coupling strategy must be able to handle efficiently the information transfer between different surface representations. This is not only a matter of fast interpolation techniques [14,22], but also of accuracy, load conservation [5,6], geometrical fidelity [6], and temporal synchronization [10,6].

4.1 Load Transfer and Conservation

When considering different mesh sizes for the CSD and CFD surface representation, the enforcement of accuracy in the sense of:

$$\sigma_s(\mathbf{x}) \approx \sigma_f(\mathbf{x}) \tag{8}$$

and conservation in the sense of:

$$\mathbf{f} = \int \sigma_s \mathbf{n} d\Gamma = \int \sigma_f \mathbf{n} d\Gamma \tag{9}$$

proves to be non-trivial. The best way to date to handle this problem for similar surface representations is via an adaptive Gaussian quadrature [5]. In some cases, the conservation of forces may not be as important as conservation of other physically meaningful quantities, e.g. work or torque. It is important to note that conservation of only one of these: force, torque or work, can be guaranteed. The reliable and accuate treatment different levels of abstraction or dimensionality between fluid and structures represents an open question to date.

4.2 Geometrical Fidelity

In many instances, the structural model will either be coarse as compared to the fluid model, or may even be on a different modeling or abstraction level. On the other hand, it is the structural model that dictates the deformation of the fluid surface. It is not difficult to see that an improper transfer of structural deformation to the fluid surface mesh can quickly lead to loss of geometrical fidelity. For the cases considered here, the fluid surface was 'glued' to the surface of the structure. For 'non-glued' fluid and structural surfaces a number of recovery techniques have been proposed to date [8,15,6]. However, we consider the proper treatment of surface deformation for 'non-glued' fluid and structural surfaces, as well as some form of error indicator to warn the unsuspecting user, to be areas deserving further study.

5 Topology Change

Suppose that due to cracking, failure, spalation, etc., the 'wetted surface' of the structure has been changed. This new surface, given by a list of points and faces, has to be matched with a corresponding fluid surface. The fluid surface data typically consists of surface segments defined by analytical functions that do not change in time (such as exterior walls, farfield boundaries, etc.), and surface segments defined by triangulations (i.e. discrete data) that change in time. These triangulations are obtained from the 'wetted structure surface' at every timestep. When a change in topology is detected, the new surface definition is recovered from the discrete data, and joined to the surfaces defined analytically, as indicated in Fig. 1.

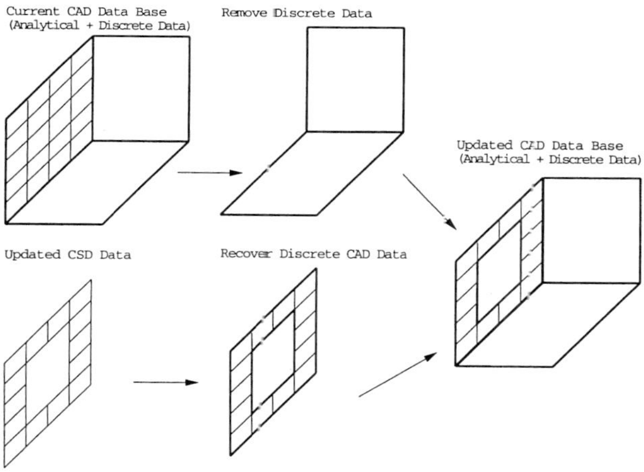

Fig. 1. Automatic surface reconstruction

The discrete surface is defined by a support triangulation, with lines and end-points to delimit its boundaries. In this sense, the only difference with analytically defined surfaces is the (discrete) support triangulation. The patches, lines and end-points of the 'wetted structure surface' are identified by comparing the unit surface normals of adjacent faces. If the scalar product of them lies below a certain tolerance, a ridge is defined. **Corners** are defined as points that are attached to:

- Only one ridge;
- More than two ridges; or
- Two ridges with considerable deviation of unit side-vector.

Between corners, the ridges form **discrete lines**. These discrete lines either separate or are embedded completely (i.e. used twice) in **discrete surface patches**. Figure 2 sketches the recovery of surface features and the definition of discrete surface patches for a simple configuration. For more information, see [17].

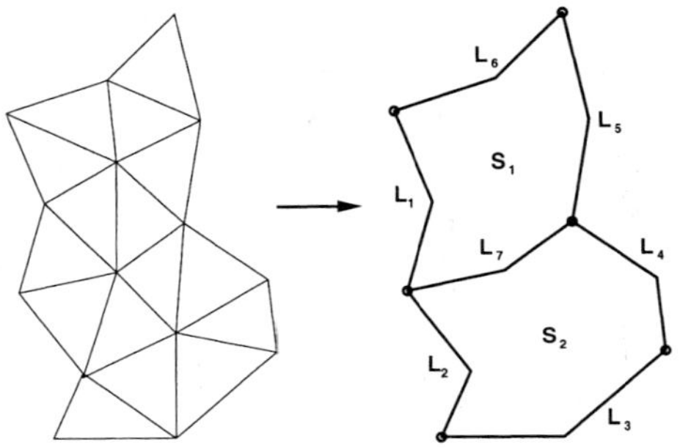

Fig. 2. Discrete surface recovery

For the old surface definition data set, the surface patches attached to wetted structure surfaces are identified and all information associated with them is discarded. The remaining data is then joined to the new wetted structure surface data, producing the updated surface definition data set. This data set is then used to generate the new surface and volume grids.
The surface reconstruction procedure may be summarized as follows:

- For the Updated Discrete Data, Obtain:
  - Surface Patches + B.C.
  - Lines
  - End-Points
  - Sources
- For the Old Analytical+Discrete Data:
  - Remove Discrete Data
  - Reorder Arrays
- Merge:
  - Old Analytical Data
  - Updated Discrete Data

A topology change as described above will imply a complete surface mesh regeneration, with subsequent flowfield mesh regeneration. If topology change occurs frequently, this can represent a considerable portion of the total computing cost. A way that has proven effective in reducing this cost is to identify which surfaces have actually changed, and to regenerate locally only the regions where changes have occured. This local mesh generation procedure may be summarized as follows:

- For the Updated and Old Surface Data:
  - Compare Surface Patches
  - Compare Lines
  - Compare End-Points
- Remove from the Old Surface Data:
  - Surfaces that Have Disappeared;
  - Lines that Have Disappeared;
  - Points that Have Disappeared;
- Remove from the Current Mesh:
  - Faces of Surfaces that Have Disappeared;
  - Elements Close to These Faces;
- Obtain Surface Mesh for New Surfaces;
- Locally Remesh Close to New Surfaces.

The procedure is shown schematically in Fig. 3.

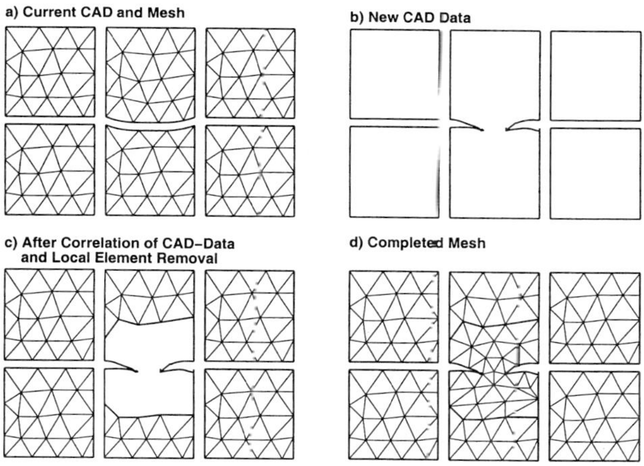

Fig. 3. Topology change with local remeshing

6 Scalabiliy on Parallel Computers

Despite the striking successes reported to date, only the simplest of all solvers: explicit timestepping or implicit iterative schemes, perhaps with multigrid added on, have been ported without major changes and/or problems to massively parallel machines with distributed memory. Many code options that are essential for realistic simulations are not easy to parallelize on this type of machine. Among these, we mention local and global remeshing [12,19], repeated h-refinement [13], such as required for transient problems [13], contact detection and force evaluation [9], some preconditioners [26], applications where particles, flow, and chemistry interact, fluid-structure interaction with topology change [20] and, in general, applications with rapidly varying load imbalances. Even if 99% of all operations required by these codes can be parallelized, the maximum achievable gain will be restricted to 1:100. If we accept as a fact that for most large-scale codes we may not be able to parallelize more than 99% of all operations, the shared memory paradigm, discarded for a while as non-scalable, will make a comeback. It is far easier to parallelize some of the more complex algorithms, as well as cases with large load imbalance, on a shared memory machine. And it is within present technological reach to achieve a 100 processor, shared memory machine.

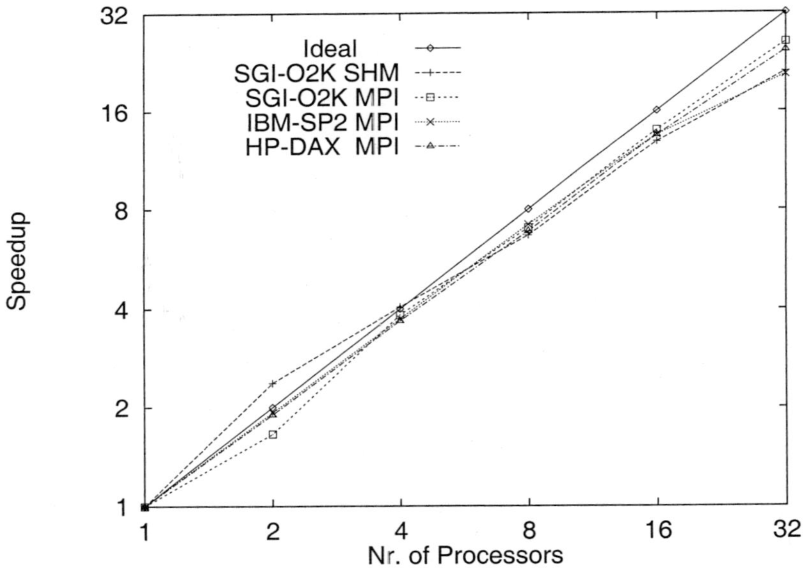

Fig. 4. Performance of FEFLO on different platforms

Figure 4 shows the performance of FEFLO, the fluid code used in the present work, on a variety of common US-DOD-HPC platforms. One can see that the speedup obtained using shared and distributed memory approaches is similar. Figure 5 shows grid generation timings for a pilot ejection simulation, a typical problem with moving bodies that requires several remeshings during the course of a simulation. The outline of the domain is shown in Fig. 5(a).

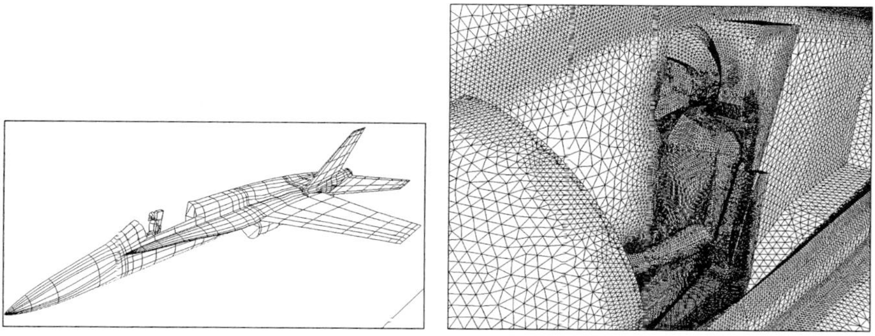

Fig. 5. **(a)-(b)** F18: Outline of domain and surface mesh (closeup)

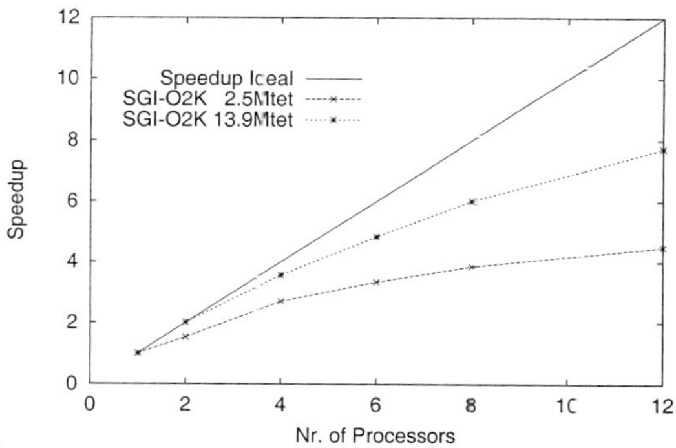

Fig. 5. **(c)** F18: Speedups obtained

The surface triangulation of the final mesh, which had approximately 14 million tetrahedra, is shown in Fig. 5(b). The smallest and largest specified element side lengths were 0.65 cm and 250.00 cm respectively, i.e. an edge-length ratio of approximately $1 : 4 \cdot 10^2$ and a volume ratio of $1 : 5.6 \cdot 10^7$. The spatial variation of element size was specified via approximately 110 sources [18]. The speedup obtained for two different grid sizes on the SGI Origin 2000 using the shared memory paradigm is displayed in Fig. 5(c). One can see that the grid generator scales well with the number of processors and that scalability improves with the amount of work required. Details of the parallel gridding technique have been described in [21].

7 Examples

We show three cases to illustrate the range of problems covered by the capabilities developed. The coupling complexity increases with each example. For the first example, the bodies move freely, based on the loads of the structure. The last two examples show structural failure, rupture and topology change. All cases were run on SGI Origin 2000 machines in shared memory mode, using up to 64 processors.

7.1 Weapon Fragmentation

As the first example, we consider a weapon fragmentation experiment. The weapon used consisted of a thick wall cylinder, serrated into 32 rows of fragments, 16 fragments per row, for a total of 512 'small' fragments, and thick nose and tail plates. While the cutting pattern was identical for all rows, the pattern was rotated between rows by angles that produced no symmetric pattern. Thus, the whole weapon had to be modeled. Each fragment weighed about 380 grams. The serrated weapon had tabs that kept the minimum distance between the fragments to 0.5 mm, while the average size of each fragment was a $1-4$ cm per side. In contrast, the room size was several meters. This huge disparity in dimensions required the use of sources attached to each flying body in order to ensure a uniform, high-resolution mesh about each fragment.
Figure 6(a) shows the computational domain as well as snapshots of the surface grid employed. Each of the fragments is treated as a separate, freely flying rigid body whose velocity and trajectory were determined by integrating the six ordinary differential equations describing the balance of forces and moments. Figures 6(b)-(i) show a sequence of snapshots of pressure, detonation products velocity, and fragment and mesh velocity, at several times. The pressure results are shown for a planar cut. Note that this is not a plane of symmetry. The figures show the detonation wave, the accelerating fragments, and the sharp capturing of the shock escaping through the opening gaps. The results indicate that acceleration to the final velocity takes about 120 microseconds. The predicted velocities fall within the 10% error band of the experimental measurements. Given the complexity of the physical phenomena being modelled, such a correlation with experimental data is surprisingly good. Typical meshes for this simulation were of the order of 3-4 Mtets. The simulation was carried out on an 8 processor SGI Origin 2000, and took approximately 2 weeks. For more details, see [4].

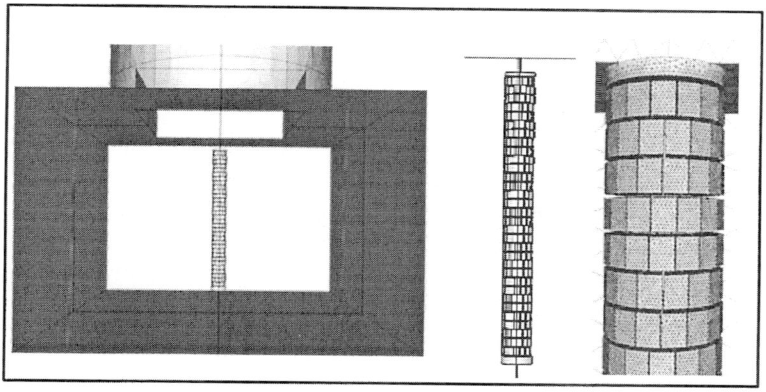

Fig. 6. (a) Computational domain

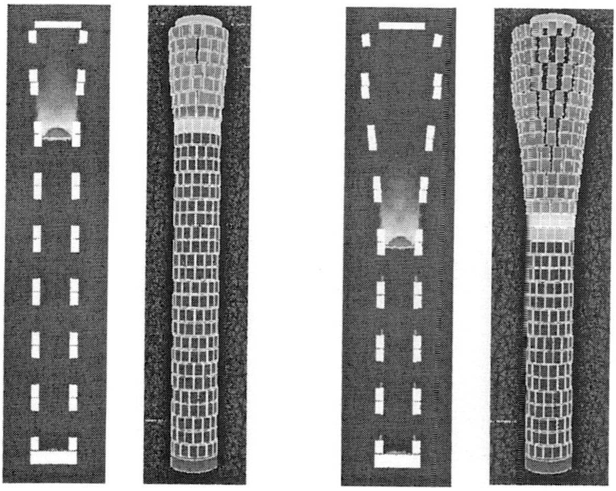

Fig. 6. (b)-(c) Pressure contours and fragment velocities at $t = 72\ \mu sec$ and $t = 139.7\ \mu sec$

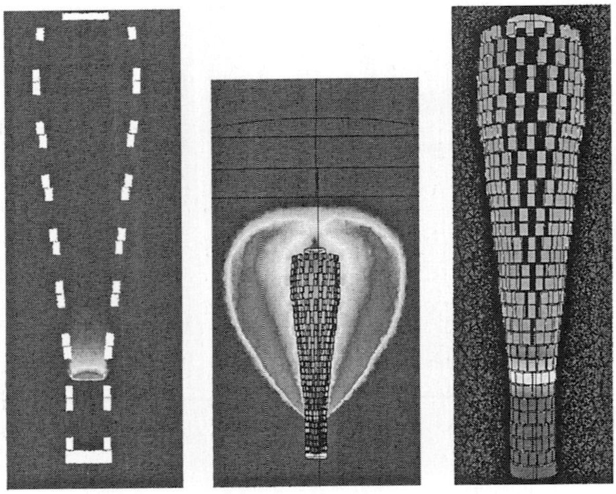

Fig. 6. (**d**)-(**e**) Pressure contours, detonation products, and fragment velocity at $t = 225.7$ μsec

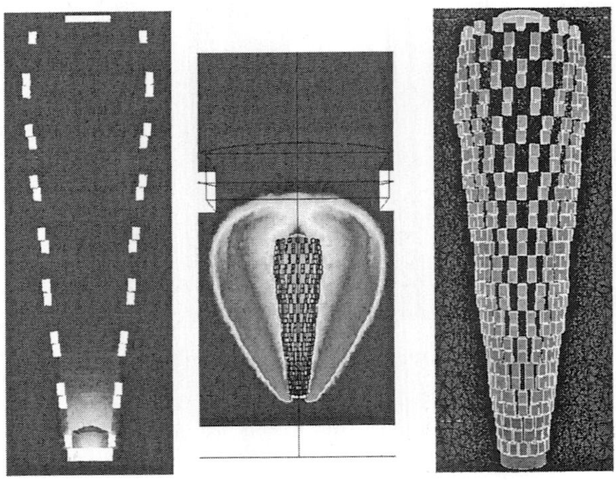

Fig. 6. (**f**)-(**g**) Pressure contours, detonation products, and fragment velocity at $t = 297.3$ μsec

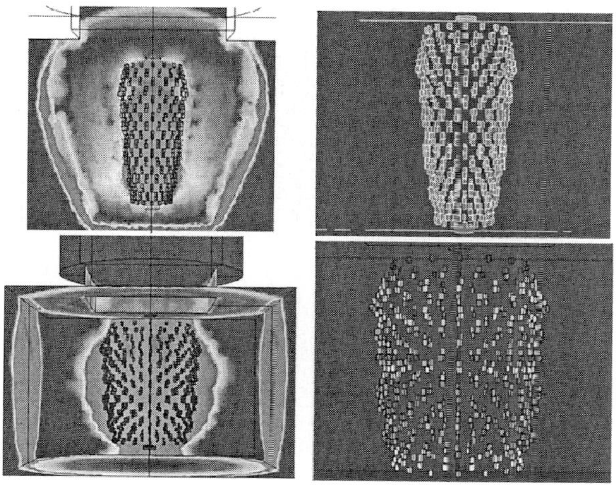

Fig. 6. **(h)-(i)** Pressure contours and fragment velocities at $t = 501\ \mu sec$ and $t = 900\ \mu sec$

7.2 Generic Weapon Fragmentation

The second example shows a fully coupled CFD/CSD run. The structural elements were assumed to fail once the average strain in an element exceeded 60%. At the beginning, the fluid domain consisted of two separate regions. These regions connected as soon as fragmentation started. In order to handle narrow gaps during the break-up process, the failed structural elements were shrunk by a fraction of their size. This alleviated the timestep constraints imposed by small elements without affecting the overall accuracy. The final breakup led to approximately 1200 objects in the flowfield. Figure 7 shows the fluid pressure and surface velocity of the structure at three different times during the simulation. Typical meshes for this simulation were of the order of 8.0 Mtets, and the simulations required of the order of 50 hours on the SGI Origin 2000 running on 32 processors.

7.3 Shock Interaction with a Ship Hull

The third example shows another fully coupled CFD/CSD run. In this case, an explosion close to a generic ship hull was considered. The structure was modeled with quadrilateral shell elements. The structural elements were assumed to fail once the average strain in an element exceeded 60%. As the shell elements fail, the fluid domain underwent topological changes. Local and global remeshing options were used. Figure 8 shows the pressure contours on the surface, as well as the velocity of the structure, at several times during

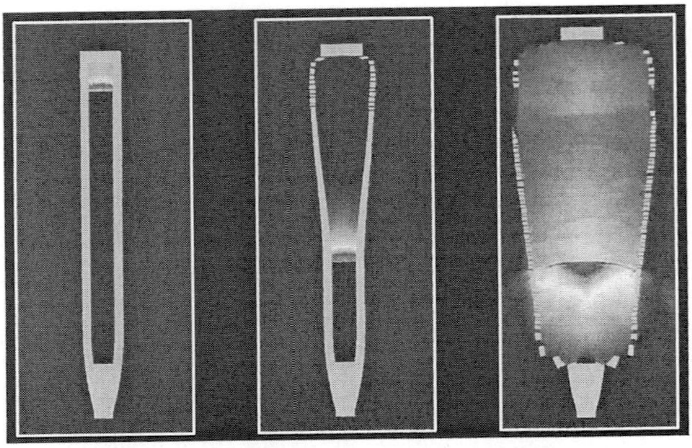

Fig. 7. (**a**) Pressure at different times

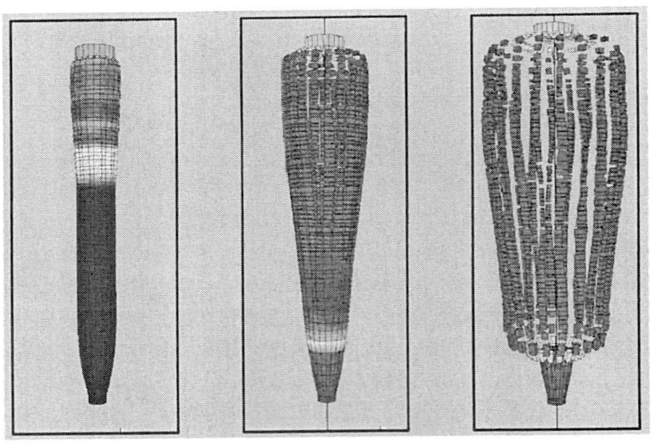

Fig. 7. (**b**) Surface velocity at different times

the run. The influence of walls can clearly be discerned for the velocity. Note also the failure of the structure, and the invasion of high pressure into the chamber.

8 Conclusions

A methodology to simulate large-scale fluid-structure interaction problems on parallel machines has been developed. Particular emphasis was placed on

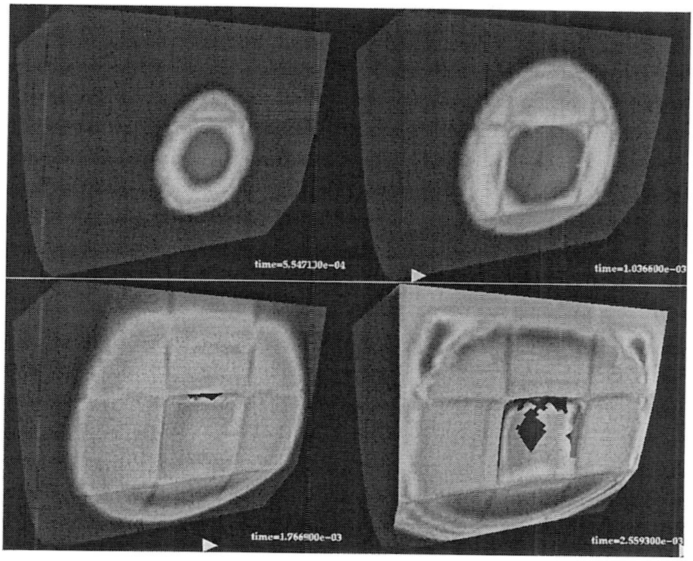

Fig. 8. (a) Surface pressure at different times

shock-structure interaction problems. For the fluid, a high-resolution FEM-FCT solver based on unstructured, moving, body fitted grids is used. For the structure, a Lagrangean large-deformation finite element code is employed. The coupled system is solved using a loose coupling algorithm, with position and velocity interpolation and force projection. The described methodology has been maturing rapidly over the last years, and is used on a routine basis to predict the complex phenomena encountered in shock-structure interaction problems.

Areas of current research include:

- Time-accurate implicit solvers for the drag phase;
- RANS gridding for arbitrary body shapes;
- Automatic element size reduction for gaps and narrow openings during topology change;
- Parallel grid generation for large number of processors;
- Failure criteria for concrete based in first principle physics;
- 'Hard' contact algorithms that guarantee no penetration; and
- Further parallelization of fluid, structure and coupling libraries.

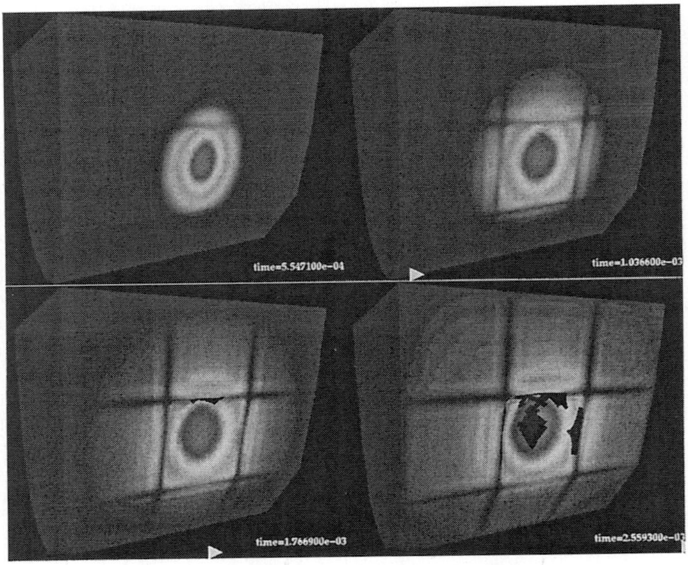

Fig. 8. (**b**) Surface velocity at different times

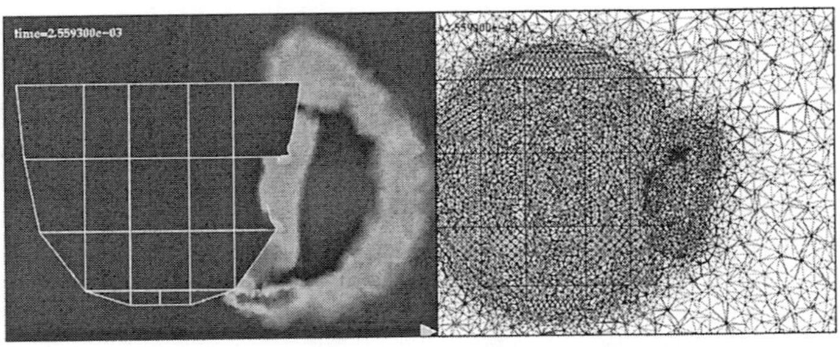

Fig. 8. (**c**) Plane cut at T=0.000255 (pressure, grid)

Acknowledgements

This work was partially supported by DTRA, with Dr. Michael Giltrud as the technical monitor, as well as AFOSR, with Dr. Leonidas Sakell as the technical monitor.

References

1. J.D. Baum. H. Luo and R. Löhner - Numerical Simulation of a Blast Inside a Boeing 747; *AIAA*-93-3091 (1993).
2. J.D. Baum, H. Luo and R. Löhner - Numerical Simulation of Blast in the World Trade Center; *AIAA*-95-0085 (1995).
3. J.D. Baum, H. Luo, R. Löhner, C. Yang, D. Pelessone and C. Charman - A Coupled Fluid/Structure Modeling of Shock Interaction with a Truck; *AIAA*-96-0795 (1996).
4. J.D. Baum, H. Luo and R. Löhner - The Numerical Simulation of Strongly Unsteady Flows With Hundreds of Moving Bodies; *AIAA*-98-0788 (1998).
5. J.R. Cebral and R. Löhner - Conservative Load Projection and Tracking for Fluid-Structure Problems; *AIAA J.* 35, 4, 687-692 (1997a).
6. J.R. Cebral and R. Löhner - Fluid-Structure Coupling: Extensions and Improvements; *AIAA*-97-0858 (1997b).
7. J.R. Cebral and R. Löhner - Interactive On-Line Visualization and Collaboration for Parallel Unstructured Multidisciplinary Applications; *AIAA*-98-0077 (1998).
8. G.P. Guruswamy and C. Byun - Fluid-Structural Interactions Using Navier-Stokes Flow Equations Coupled with Shell Finite Element Structures; *AIAA*-93-3087 (1993).
9. E. Haug, H. Charlier, J. Clinckemaillie, E. DiPasquale, O. Fort, D. Lasry, G. Milcent, X. Ni, A.K. Pickett and R. Hoffmann - Recent Trends and Developments of Crashworthiness Simulation Methodologies and their Integration into the Industrial Vehicle Design Cycle; *Proc. Third European Cars/Trucks Simulation Symposium (ASIMUTH)*, Oct. 28-30 (1991).
10. M. Lesoinne and Ch. Farhat - Geometric Conservation Laws for Flow Problems With Moving Boundaries and Deformable Meshes, and Their Impact on Aeroelastic Computations; *Comp. Meth. Appl. Mech. Eng.* 134, 71-90 (1996).
11. R. Löhner, K. Morgan, J. Peraire and M. Vahdati - Finite Element Flux-Corrected Transport (FEM-FCT) for the Euler and Navier-Stokes Equations; ICASE Rep. 87-4, *Int. J. Num. Meth. Fluids* 7, 1093-1109 (1987).
12. R. Löhner - Three-Dimensional Fluid-Structure Interaction Using a Finite Element Solver and Adaptive Remeshing; *Computer Systems in Engineering* 1, 2-4, 257-272 (1990).
13. R. Löhner and J.D. Baum - Adaptive H-Refinement on 3-D Unstructured Grids for Transient Problems; *Int. J. Num. Meth. Fluids* 14, 1407-1419 (1992).
14. R. Löhner - Robust, Vectorized Search Algorithms for Interpolation on Unstructured Grids; *J. Comp. Phys.* 118, 380-387 (1995a).
15. R. Löhner, C. Yang, J. Cebral, J.D. Baum, H. Luo, D. Pelessone and C. Charman - Fluid-Structure Interaction Using a Loose Coupling Algorithm and Adaptive Unstructured Grids; *AIAA*-95-2259 [Invited] (1995b).

16. R. Löhner and Chi Yang - Improved ALE Mesh Velocities for Moving Bodies; *Comm. Num. Meth. Eng.* 12, 599-608 (1996a).
17. R. Löhner - Regridding Surface Triangulations; *J. Comp. Phys.* 126, 1-10 (1996b).
18. R. Löhner - Extensions and Improvements of the Advancing Front Grid Generation Technique; *Comm. Num. Meth. Eng.* 12, 683-702 (1996c).
19. R. Löhner, Chi Yang, J.D. Baum, H. Luo, D. Pelessone and C. Charman - The Numerical Simulation of Strongly Unsteady Flows With Hundreds of Moving Bodies; *Int. J. Num. Meth. Fluids* 31, 113-120 (1999a).
20. R. Löhner, C. Yang, J. Cebral, J.D. Baum, H. Luo, E. Mestreau, D. Pelessone and C. Charman - Fluid-Structure Interaction Algorithms for Rupture and Topology Change; *Proc. 1999 JSME Computational Mechanics Division Meeting*, Matsuyama, Japan, November (1999b).
21. R. Löhner - A Parallel Advancing Front Grid Generation Scheme; *AIAA*-00-1005 (2000).
22. N. Maman and C. Farhat - Matching Fluid and Structure Meshes for Aeroelastic Computations: A Parallel Approach; *Computers and Structures* 54, 4, 779-785 (1995).
23. D. Pelessone and C.M. Charman - Adaptive Finite Element Procedure for Non-Linear Structural Analysis; *1995 ASME/JSME Pressure Vessels and Piping Conference*, Honolulu, Hawaii, July (1995).
24. D. Pelessone and C.M. Charman - An Adaptive Finite Element Procedure for Structural Analysis of Solids; *1997 ASME Pressure Vessels and Piping Conference*, Orlando, Florida, July (1997).
25. D. Pelessone and C.M. Charman - A General Formulation of a Contact Algorithm with Node/Face and Edge/Edge Contacts; *1998 ASME Pressure Vessels and Piping Conference*, San Diego, California, July (1998).
26. R. Ramamurti and R. Löhner - Simulation of Flow Past Complex Geometries Using a Parallel Implicit Incompressible Flow Solver; pp. 1049,1050 in *Proc. 11th AIAA CFD Conf.* , Orlando, FL, July (1993).

MEGAFLOW - An Industrial Flow Simulation Tool for Aircraft Applications

N. Kroll

German Aerospace Center (DLR), Institute of Aerodynamics and Fluid Mechanics, D-38108 Braunschweig, Germany

Abstract. Some years ago, the national CFD project MEGAFLOW was initiated in Germany which combines many of the CFD development activities from DLR, universities and aircraft industry. Its goal is the development and validation of a dependable and efficient numerical tool for the aerodynamic simulation of complete aircraft. Today, the MEGAFLOW software system has reached a high level of quality and efficiency. As a consequence, the MEGAFLOW system is being intensively used by the German aerospace industry in the design processes for new aircraft. This paper highlights recent enhancements of the software and it´s capability to simulate viscous flows around complex industrial applications

1 Introduction

During the last decade, considerable progress has been made in the development and validation of numerical simulation tools for aerodynamic applications. However, even despite recent advances, CFD still suffers from deficiencies in accuracy, robustness and efficiency for complex applications, such as complete aircraft flow predictions. From the aircraft industry's point of view, numerical simulation tools are expected to deliver detailed viscous flow analysis for complete configurations at realistic Reynolds numbers, prediction of aerodynamic data with assured high accuracy and known error bands, fast response time per flow case at acceptable total costs as well as aerodynamic optimization of the main aircraft components. In order to meet these requirements in earlier and earlier stages of the design process, considerable improvements of the CFD methods currently available in industry are necessary. Within the framework of the German aerospace research program, the national CFD project MEGAFLOW was initiated under the leadership of DLR with the objective of enhancing the capabilities of current CFD methods and supporting the establishment of numerical simulation as an effective tool in the industrial design process [1]. The goal of the project is to produce a dependable, efficient and quality controlled program system for the aerodynamic simulation of complete transport aircraft in cruise as well as take-off and landing configuration. The MEGAFLOW software is designed to meet the requirements of industrial implementation. The extensive development and validation work has been carried out in concentrated cooperation with aircraft industry, DLR and several universities.

2 Software Tools

The basic components of the MEGAFLOW software are the block-structured flow solver FLOWer and the unstructured hybrid flow solver TAU.

The FLOWer code [2] solves the compressible, three-dimensional Reynolds - averaged Navier-Stokes equations for rigid bodies in arbitrary motion. It uses a finite-volume formulation on block-structured meshes. The baseline method employs either central space discretization combined with artificial viscosity or upwind discretization. Integration in time is done using explicit multistage time-stepping schemes. For steady calculations convergence is accelerated by implicit residual smoothing and multigrid. Preconditioning is used for low-speed flows. For time accurate calculations an implicit time integration according to the dual time stepping approach is employed. The code is highly optimized for vector computers and parallelization based on MPI is realized through the high level communication library CLIC-3D. Using the NEC SX5 computer, typically the FLOWer code achieves 2 GFLOPS on a single processor and its parallel efficiency is about 80 %. For turbulent flow calculations advanced linear and nonlinear one- and two-equation turbulence models were implemented. An implicit integration of the turbulence equations ensures efficent calculations on highly stretched cells as they appear in high Reynolds number flows. Figure 1 shows the convergence history of FLOWer for the calculation of viscous flow around the RAE 2822 airfoil at different Reynolds numbers using the implicit treatment of the turbulence transport equations. The advantage of the implicit method for high Reynolds numbers is evident. In order to enhance the flexibility of the block-structured approach, the Chimera technique was implemented in the FLOWer code. This approach allows the grid blocks to overlap each other. The Chimera technique combined with flexible meshes considerably simplifies the simulation of bodies in relative motion.

The Navier-Stokes code TAU [3] makes use of the advantages of unstructured grids. The mesh may consist of a combination of prismatic and tetrahedral cells and therefore combine the advantages of regular grids for the accurate resolution of viscous shear layers in the vicinity of walls and the flexibility of grid generation techniques based on unstructured meshes. The use of a dual mesh makes the solver independent of the type of cells that the initial grid is composed of. Various discretization schemes were implemented, including a central scheme with artificial dissipation and several upwind methods. For the simulation of turbulent flows, classical one- and two-equation turbulence models were implemented. In order to accelerate convergence, a multigrid procedure was developed based on the agglomeration of the control volumes of the dual grid for coarse grid computations. In order to efficiently detect detailed flow features, a grid adaptation algorithm for hybrid meshes based on local grid refinement was implemented. The TAU code is optimized for vector and parallel computers. Message passing parallelization is realized using the MPI-library. On the NEC SX5 computer the TAU code achieves a parallel efficiency of about 95% with 1 GFLOPS for a single processor. With respect to unsteady calculations, the TAU code was extended to simulate a rigid body in arbitrary motion and to allow grid deformation. In order to bypass the severe time-step restriction associated with explicit schemes, the implicit method based on the dual time stepping approach was implemented. For the calculation of low-speed flows, preconditioning of the compressible flow equations similar to the FLOWer code was realized. Figure 2 demonstrates the convergence improvement for low Mach numbers using preconditioning for a viscous flow simulation around a 3-element airfoil. The convergence rate of both, the residual and the drag coefficient, is independent of the Mach number.

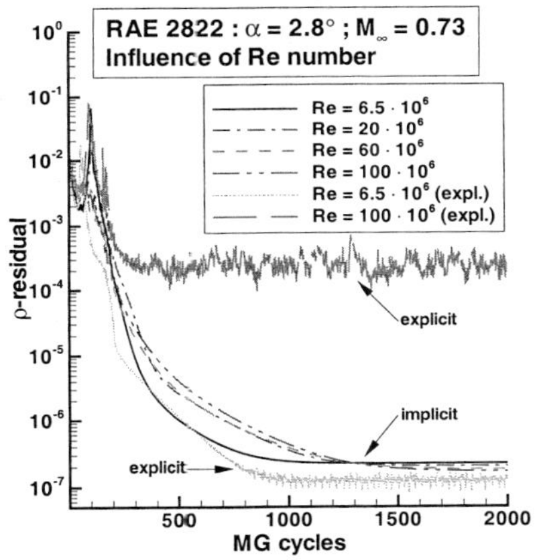

Fig. 1. Effect on Reynolds number on convergence, FLOWer code

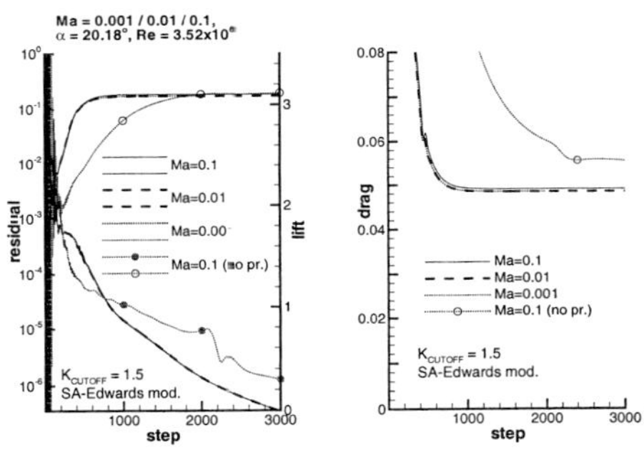

Fig. 2. Improvement of convergence for low speed flow, TAU code, viscous flow around 3-element airfoil

3 Applications

3.1 Cruise Flight and High Lift Test Cases

One of the validation objectives of the FLOWer code is the accurate prediction of transonic cruise drag for wing/body/engine/pylon configurations with an accuracy of 3-5 % with respect to experimental data. The test case selected is the DLR-F6 configuration shown in Fig. 3. Calculations have been performed with the block-structured Navier-Stokes code FLOWer using both the algebraic Baldwin/Lomax turbulence model and the standard two-equation $k-\omega$ model of Wilcox [4]. The grid was generated with the DLR grid generation software MegaCads. The grid consists of 45 blocks and 3.8 million grid points. Figure 3 shows the isobars for the flow conditions $M_\infty = 0.75$,

$\alpha = 0.98^o$ and $Re = 3 \cdot 10^6$. The lift and drag polars are also given in Fig. 3. It can be seen that the lift predicted with the Baldwin/Lomax model is consistently too high. In contrast, the Wilcox k-ω computations predict both lift and drag that agree well within 5 % of the measurements. The differences are in part a result of the modeling of the nearly separated flow at the trailing edge of the wing, which is more accurately represented by the Wilcox k-ω model. This in turn has an expectedly strong effect on the shock position and lift coefficient.

Concerning 3d high lift applications, the MEGAFLOW software has been validated based on the DLR-ALVAST wing/fuselage configuration with deployed high-lift system consisting of a wing with nose slat and inboard and outboard flaps. Figure 4 presents results of viscous simulations carried out with both the block-structured FLOWer code and the hybrid TAU code for $M_\infty = 0.22$, $Re = 2 \cdot 10^6$ [5]. The structured grid generated with MegaCads consists of about 50 blocks with 9.2 million grid points . The unstructured grid was generated using the commercial software package CENTAUR. It contains a quasi-structured prismatic cell layer around the geometry surface, in order to accurately resolve the viscous effects in the boundary layers. The outer domain of the flow field is covered with tetrahedral cells. After adaptation the grid consists of about 5.8 million points. In Fig. 4 the lift coefficient versus the angle of attack and versus the drag coefficient are plotted. The overall agreement between the predicted and measured data is quite good. The lift coefficients computed with the hybrid TAU code are consistently overpredicted by about 5 % in the linear range of the $C_l(\alpha)$-curve. However, the angle of attack for maximum lift agrees well with measured data. The aerodynamic coefficients computed with the block-structured FLOWer code show a somewhat different behaviour. Whereas the agreement of the structured computations with experimental data is excellent for $\alpha < 15^o$, the lift coefficients are underpredicted by 5 % between $15^o < \alpha < 22^o$. The differences between the structured and unstructured calculations may be partly explained by the different grid densities and the different turbulence models used in the calculations. For FLOWer computations the two-equation k-ω model has been used, whereas the TAU calculations have been performed with the one-equation model of Spalart-Allmaras. With the unstructured approach the simulation turn around time is about 2 weeks including grid generation, flow computation and evaluation of results.

3.2 Wake Vortex

To avoid risks for an aircraft flying in the wake vortex field of a preceding aircraft during take-off and landing, strict rules for minimum required separation distance have been established. These air traffic control separation standards are merely based on the maximum take-off weights of both aircraft. Due to the increasing number of congested airports and the development of

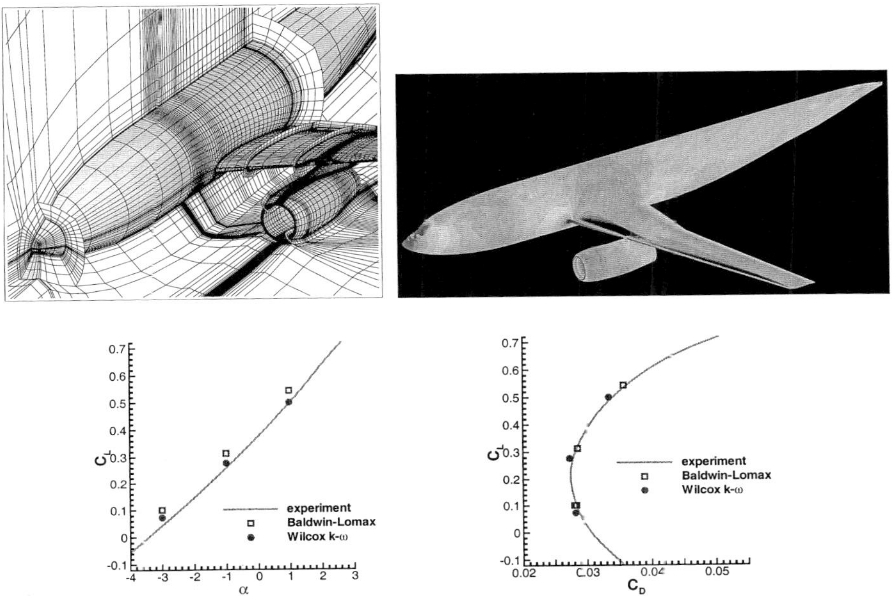

Fig. 3. Viscous simulation of DLR-F6 configuration with the FLOWer code, $M_\infty = 0.75$, $Re = 3 \cdot 10^6$; grid, surface pressure, lift and drag polars

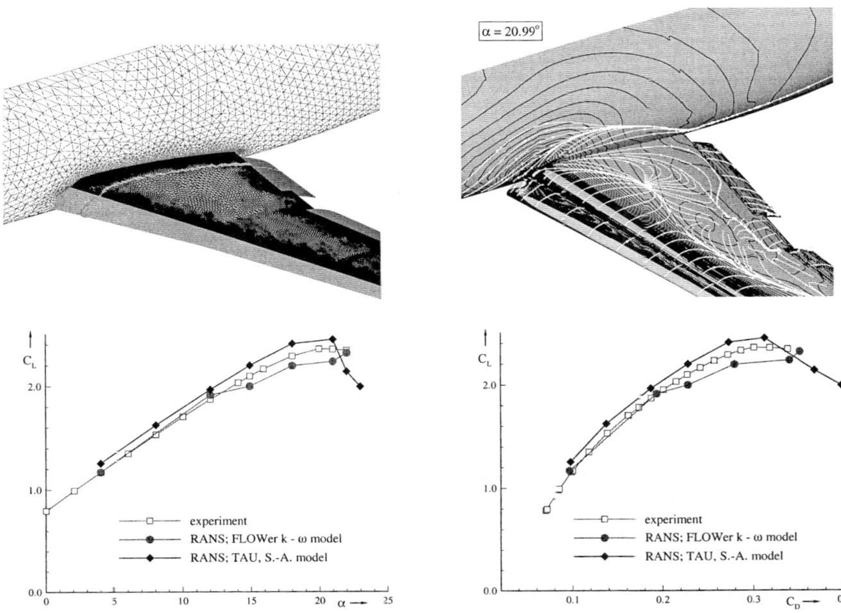

Fig. 4. Viscous simulation of ALVAST high lift configuration with FLOWer and TAU code, $M_\infty = 0.22$, $Re = 2 \cdot 10^5$; hybrid grid, surface streamlines, lift and drag polars

very high capacity transport aircraft, in the past few years considerable research effort has been put on the wake vortex problem aiming at a reduction of today's separation standards.

One of the research areas is to enhance the knowledge of the formation and development of the nearfield wake vortex of transport aircraft in take-off and landing configuration. At DLR investigations have been carried out to assess the predictive capability of an Euler code using unstructured grids [6]. The aircraft model used for these studies is the DLR ALVAST configuration. The results have been obtained with the TAU code. In order to resolve the wake vortex, grid adaptation has been used. Since Euler calculations over-predict lift considerably, the angle of attack is adjusted in order to meet the measured lift coefficient. The left picture of Fig. 5 shows the formation of the nearfield wake vortex for the landing configuration. A grid with roughly 6 million points and 10 adaptation steps has been used. The right picture of Fig. 5 shows a comparison of the predicted and measured wake structure at a plane 0.5 wing spans downstream of the wing. The five-hole-probe measurements do not cover the complete cross section and are not continuous due to mechanical restrictions of the rake positioning system. The figure demonstrates that the computational results match the experimental data remarkably well. These results show that Euler calculations on unstructured, adapted grids with adjusted lift coefficents are capable to simulate the formation and development of nearfield wake vortices.

Within the European project WAVENC, investigations for wake vortex encounter have been carried out. A small generic aircraft model has been placed downstream into the vortical flow-field of a preceding aircraft. The experimental data base contains forces, moments and surface pressure distributions at several model positions with respect to the vortices generated by the preceding aircraft. Inviscid computations with the Chimera option of the FLOWer code have been performed [7]. Component meshes have been generated separately around the fuselage, the wings and the horizontal and vertical tails of the generic aircraft model (see Fig. 6). These component meshes are embedded in a Cartesian background grid, which is well suited for vortex dominated flows. The incoming vortices of the preceding aircraft have been prescribed at the inflow boundary of the computational domain. In order to avoid the diffusion of the vortices, special Cartesian meshes (vortex transport grids) have been incorporated into the Chimera grid structure to transport the incoming vortices to the following aircraft. Computations have been carried out for five positions of the aircraft. For all positions the comparison of the predicted and measured rolling moment coefficient is quite good. Furthermore, the surface pressure distribution at the mid-span position of the port-side wing compares well with experimental data. These investigations show that the Chimera technique is well suited for the simulation of a wake vortex encounter situation. The main advantages of this approach are that a mutual interaction of vortices and airplane is captured in the simulation and that a parametric study of the position of the aircraft relative to the vortices is possible without generating new meshes.

4 Conclusions

The main objective of the MEGAFLOW initiative is the development of a dependable, effective and quality controlled program system for the aerodynamic simulation of complete aircraft. Due to it's high level of maturity, the MEGAFLOW software system is being used extensively throughout Germany for solving complex aerodynamic problems. However, since industry is still demanding more accurate and faster simulation tools, furher development is aimed at improvement of physical modelling, significant reduction of problem turn-around time for large scale computations and efficient integration into an interdisciplinary simulation system.

Acknowledgement

The author would like to thank all colleagues of the MEGAFLOW project who carried out the developments and obtained the results shown in this paper.

References

1. Kroll, N., Rossow, C. C., Becker, K., Thiele, F. (2000) The MEGAFLOW Project. Aerosp. Sci. Technol **4**, 223–237
2. Kroll, N., Radespiel, R., Rossow, C.-C. (1995) Accurate and Efficient Flow Solvers for 3D Applications on Structured Meshes. AGARD Report **R-807**, 4.1–4.59
3. Gerhold T., Friedrich, O., Evans J., Galle, M. (1997) Calculation of Complex Three-Dimensional Configurations Employing the DLR-TAU Code. AIAA 97-0167
4. Monsen E., Franke M., Rung T., Aumann P., Ronzheimer A. (1999) Assessment of Advanced Transport-Equation Turbulence Models for Aircraft Aerodynamic Performance Prediction. AIAA 99-3701
5. Melber, S., Rudnik, R., Ronzheimer, A. (2000) 3D RANS Structured and Unstructured Numerical Simulation in High-Lift Aerodynamics. Proceedings of Workshop on EU-Research on Aerodynamic Engine/Airframe Integration for Transport Aircraft, 13.1 - 13.10.
6. Stumpf, E., Rudnik, R., Ronzheimer, A. (2000) Euler computations of the near wake vortex of an aircraft in take-off configuration. Aerosp. Sci. Technol **4**, 535–543
7. Heinrich, R. (2000) Numerical Simulation of Wake-Vortex Encounters using the Chimera-Technique. Notes on Numerical Fluid Dynamics, Vieweg Verlag Braunschweig, to be published.

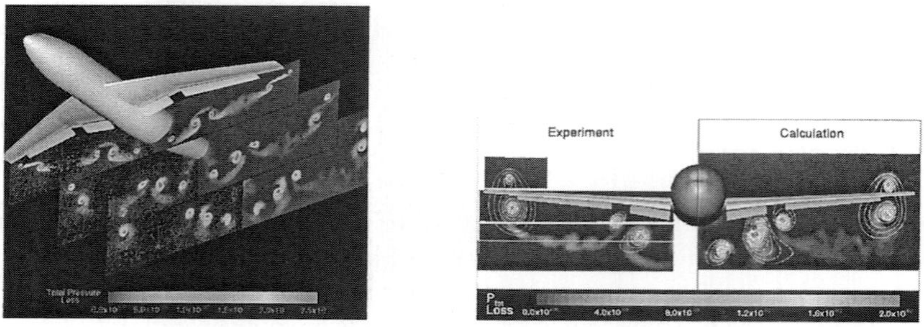

Fig. 5. Wake development calculated with TAU, DLR ALVAST landing configuration, $M_\infty = 0.22$, comparison of vortex structure between experiment and Euler calculation at a downstream plane (0.5 wing span)

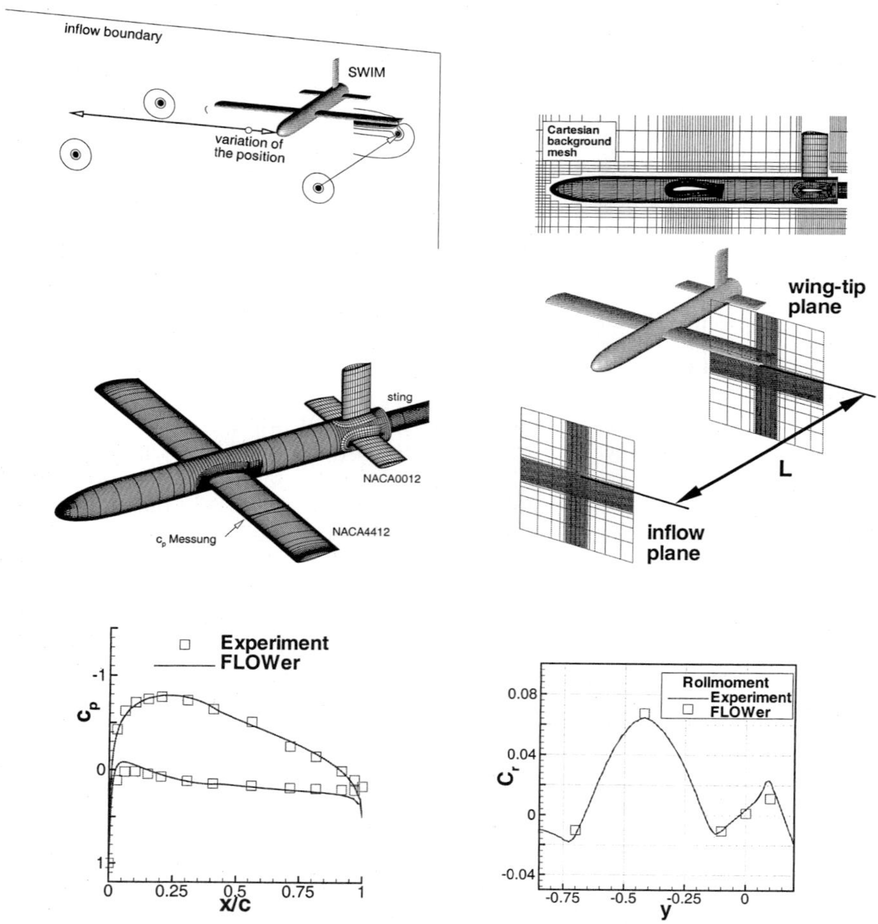

Fig. 6. Simulation of vortex encounter with the Chimera option of the FLOWer code

Development of a Parallel FVM Based Groundwater Flow Model

B. Witte, R. Hinkelmann, and R. Helmig

Institute of Hydraulic Engineering, University of Stuttgart, Germany

Abstract. Although parallel computing is gaining in importance in environmental engineering, there has been a lack of graduate education in this field. There are not enough models using parallel algorithms, which are simple enough to be understood with only the educational background of a student. The authors want to close this gap by developing a parallel FVM based groundwater flow model.

1 Motivation

Today, there is an urgent need for parallel computing for many practical engineering applications because the performance of single-processor computers is not sufficient for complex realtime simulations. Most of the users are not computer scientists but engineers or natural scientists mostly with little knowledge of programming and even less of parallel computing. There are now some lectures at university level that deal with parallel computing, but the practical applications are much too complex to be understood by non-specialists, while examples of lecture notes (e.g. [1]) or technical documentations of communication standards such as MPI [4] are too simple to show effects of speedup and what influence the network type or bandwidth will have.
It was demonstrated that a complete parallel code can be developed by students that have no high level programming skills in a rather limited time. Such a simple parallel code can show the performance of parallisation and parallel computers, e.g. by observing the running time reduction or speed up, respectivly. The parallel groundwater flow model serves as a basis for further efficeint programming and advanced numerical techniques.

2 Groundwater Flow Equation

The continuity equation for stationary groundwater flow is given by:

$$\int_{\Omega} \underbrace{div\{\rho * \underline{v}\}}_{Laplace\ term} \, d\Omega - \int_{\Omega} \underbrace{\rho * q}_{source/sink\ term} \, d\Omega = 0 \tag{1}$$

In this equation
$\underline{v}$ denotes the Darcy-velocity of the fluid (water) $[m/s]$,
q denotes the source / sink term $[m^3/(m^3 * s)]$ and
ρ denotes the density of the fluid (water) $[kg/m^3]$
Ω denotes the observed area.
The momentum equation is simplified yielding the well-known *Darcy's Law*:

$$\underline{v} = -\underline{\underline{k}}_f * grad\ h \tag{2}$$

In this equation
$\underline{\underline{k}}_f$ denotes the hydraulic conductivity $[m/s]$
h denotes the piezometric head $[m]$
If (2) is inserted into (1), this leads to

$$\int_\Omega div\{-\underline{\underline{k}}_f * grad\ h\}\ d\Omega - \int_\Omega q\ d\Omega = 0 \tag{3}$$

If the *Green Gauß* theorem is applied, the volume integral is replaced by a boundary integral:

$$\oint_\Gamma \left\{-\underline{\underline{k}}_f * grad\ h * \underline{n}\right\} d\Gamma - \int_\Omega q\ d\Omega = 0 \tag{4}$$

with
$\underline{n}$: a normal vector which can be ignored because of the rectangular grid to be used in the computations,
Γ : the boundary of the observed area.

3 Discretisation

The two-dimensional computational domain is meshed by rectangular elements / cells of the same size and geometry i.e. a topologically uniform grid. For the grid generation, two different possibilities, which are called element- and mesh-orientated, exist (see Fig. 1).

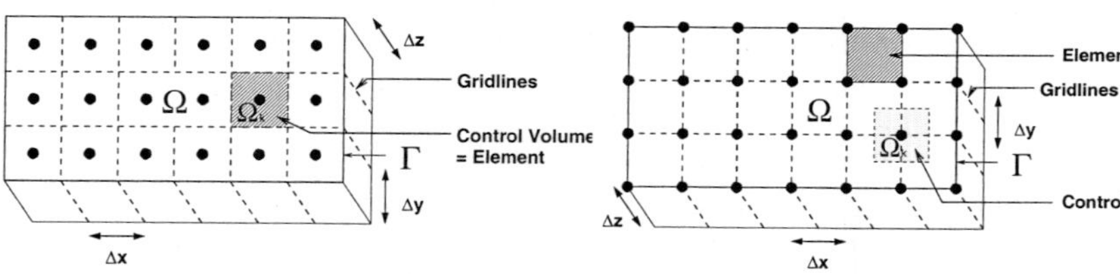

Fig. 1. Element orientated (left) and mesh orientated (right) grid

The application of a central difference approach for the gradient ($grad\ h$) leads the translation of the boundary integral in (4) to (shown here for the

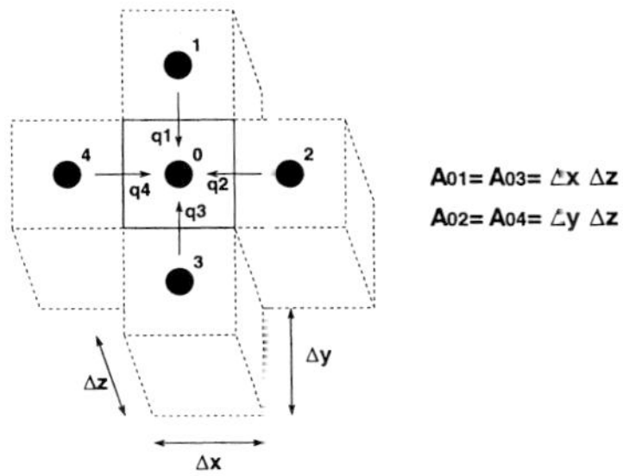

Fig. 2. Schematic model for a single cell using the FVM approach

x-direction):

$$\sum_{i=1}^{4}\left\{A_{0i} * k_{f_i} * \frac{h_0 - h_i}{\Delta x}\right\} = (\Delta x \Delta y \Delta z) q \tag{5}$$

Here, A_{0i} stands for the interface of cells 0 and i.
After some transformations, we obtain for cell 0 (see Fig. 2)

$$\begin{aligned} & h_0\left[k_1(\Delta x)^2 + k_2(\Delta y)^2 + k_3(\Delta x)^2 + k_4(\Delta y)^2\right] \\ - \quad & h_1 * \left[k_1(\Delta x)^2\right] - h_2 * \left[k_2(\Delta y)^2\right] \\ - \quad & h_3 * \left[k_3(\Delta x)^2\right] - h_4 * \left[k_4(\Delta y)^2\right] = (\Delta x \Delta y)^2 * q_0 \end{aligned} \tag{6}$$

If this integration is carried out for all cells, a sparse linear and symmetric system of equations is solved with a *preconditioned conjugate gradient (PCG)* method for the unkown piezometric head. Finally, the flow velocity is computed using Darcy's Law (2).
For the work described here, the mesh orientated variant was chosen, because it allows us to treat 'interprocessor boundaries" (see next section) in the same way as outer boundaries. Special considerations are required for the nodes on the edges and corners (see Fig. 3).

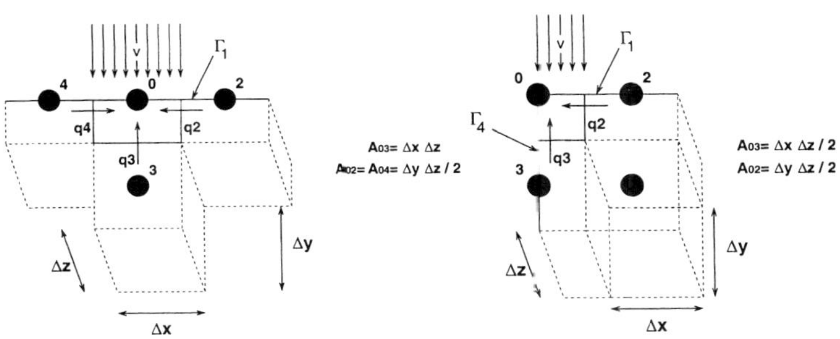

Fig. 3. Node on an edge boundary (left), node on a corner boundary (right)

4 Parallelisation Techniques

The parallel groundwater flow model was developed for a *Multiple-Instruction Multiple-Data (MIMD)* architecture. An algebraic parallelisation strategy (see [2,3]) was chosen. The grid must be divided into subdomains which are then assigned to the processors, the data must be distributed to the processors. This is done by *recursive coordinate bisection*. Along the larger coordinate direction, the mesh is divided; this procedere is then repeated recursively up to a given level. The number n of dividing steps is defined by the number p of processors: $n = log_2\, p$. Figure 4 shows how it works for four processors.

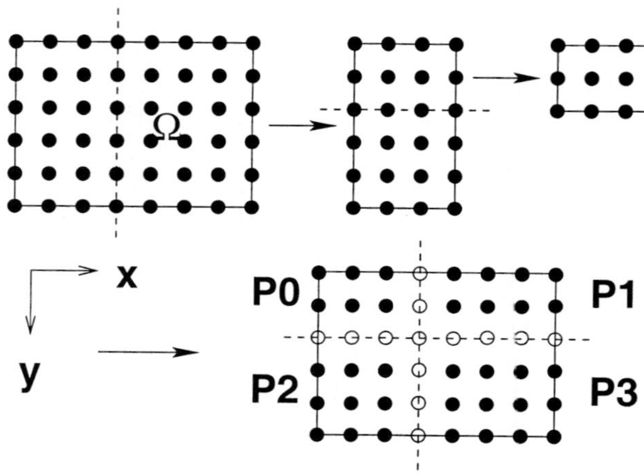

Fig. 4. Load balancing for four processing units (PU)

For the communication, the MPI standard was chosen. The use of highly developed communication algorithms was rejected because of their complexity. So the only subroutines used are:

- `MPI_Ssend(...)` for synchronous and blocking sending of data
- `MPI_Recv(...)` for receiving data from another processor
- `MPI_Allreduce(...)` for data collection from / distribution to all processors

5 Verification and Practical Application

For the verification of the numerical result, two different cases were simulated:

5.1 Parallel Flow

The simplest example is a parallel flow in a completely homogeneous and isotropic porous medium because the analytical solution is easy to calculate and therefore the result easy to check.
The system length is 1000m, the width 500m. A difference in the hydraulic head of 1.00m os prescribed as the Dirichlet boundary condition between $y = 0m$ and $y = 1000m$. For a given hydraulic conductivity of $k_f = 10^{-4}\frac{m}{s}$, a Darcy velocity of $v_y = 10^{-4} * \frac{1}{1000} = 10^{-7}\frac{m}{s}$ is obtained. Figure 5 shows the result of the parallel flow which is identical to the analytical solution.

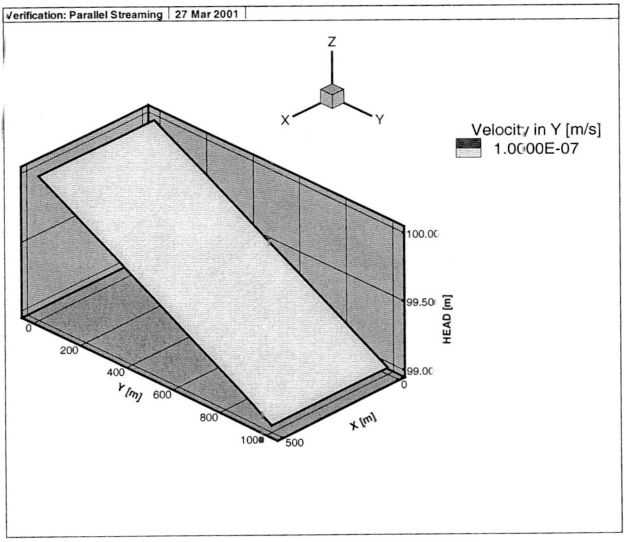

Fig. 5. Parallel flow: velocity in expected flow direction

5.2 Empirical Single-Well Formula by *Sichardt*

This example serves as a further verification of the parallel groundwater flow model by comparing the results with the empirical solution of *Sichardt*. The empirical formula assumes an infinite groundwater system, the computational area is given by a finite quadratic area. Therefore, the water-level drop is forced to have a quadradic shape, while this shape is circular in the empirical solution. But, with a small discretisation ($\Delta x = \Delta y = 1/800$ of $\Delta X = \Delta Y$), the results reasonably agree (see [5, p. 40f]).

5.3 Practical Application

The verified parallel groundwater flow model was applied to an engineering practical example in order to demonstrate its suitability for such problem classes.

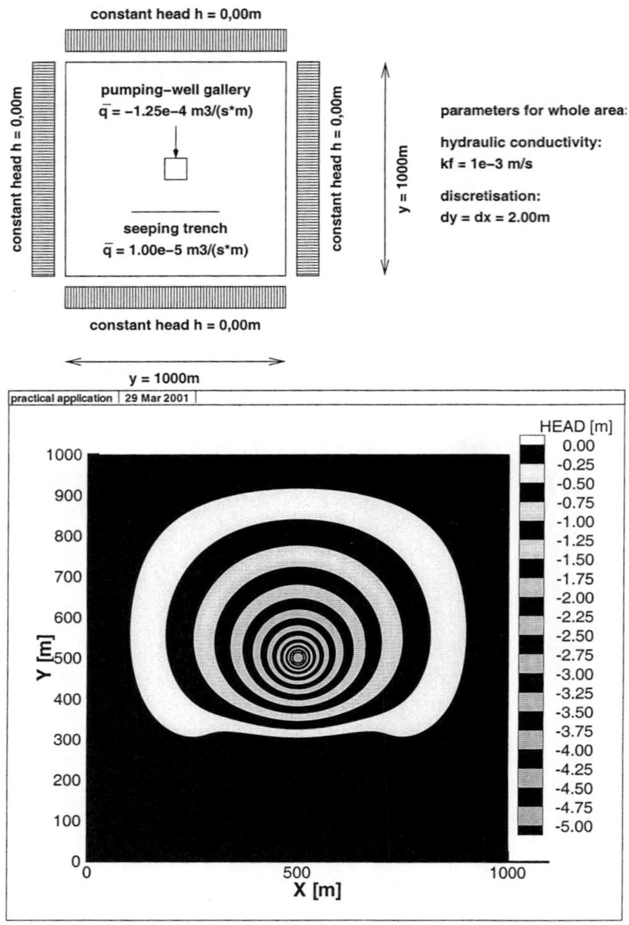

Fig. 6. System set-up and results of practical application

A building pit is to be established in a quadratic area. On one hand, the groundwater must be kept out of the pit, so a gallery of pumping wells is set up around the pit. On the other hand, it is necessary to keep the water level constant on one side of the pit, so a seeping trench was set up. The parameters are given in the system set-up, being shown in Fig. 6.

6 Running-Time Analysis

The simulations were carried out on two different kinds of parallel architectures, a massively parallel CRAY T3E and a cluster of standard PCs. The most important hardware characteristics are given in Table 1.

Table 1. Hardware characteristics of parallel computers used

Parameter	T3E	Cluster
CPU Type	Alpha EV5	Intel Pentium III
Clock Rate	450 MHz	450 MHz
1st Level Cache	2 MByte	256 KByte
Network Topology	Torus by CRAY	Hub with additional traffic
Bandwidth, theoretically	500 MByte/sec	up to 12.5 MByte/sec
Bandwidth, measured	166 MByte/sec	up to 9 MByte/sec

For the benchmark, the parallel flow of the verification problem was simulated in two different variants:

(1) $\Delta x = 1.0m, \Delta y = 2.0m \rightarrow 501 * 501 = 251001$ nodes
(2) $\Delta x = 2.0m, \Delta y = 4.0m \rightarrow 251 * 251 = 63001$ nodes

Figure 7 shows the paralle efficiency measured. The parallel efficiency is the ratio of the running time on one processor to the product of running time on p processors with p. A significant running time reduction was obtained with the parallel groundwater flow model when compared to sequential code.

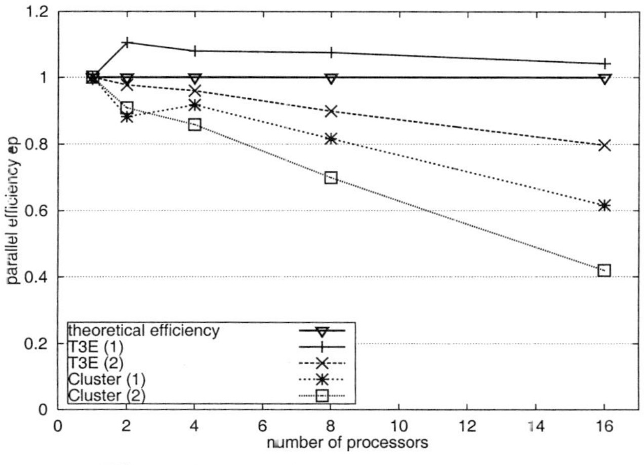

Fig. 7. Parallel efficiency for two variants

The results for more than 4 processors are as expected:
For a constant problem size the parallel overhead increases with increasing number of processors, which causes the efficiency to decrease. For a constant number of processors, the parallel overhead with increasing problem size, so the efficiency increases, too. Due to the higher network bandwidth, the efficiency is higher on the T3E when compared to the cluster.
Efficiency higher than 1.0 which is obtained on the T3E in variant *(1)* is caused by cache effects.

7 Limitations and Conclusions

Up to now, the model has been rather simple which is also shown by its limitations:

- only two types of boundaries are available:
  - constant head (DIRICHLET)
  - constant flow (NEUMANN)
- simulation of stationary flow only,
- simulation area must be of rectangular geometry,
- no blank areas inside are supported, a penalty approach is used instead (low hydraulic conductivity)
- no boundaries inside the simulation area
- no transport is possible

But the simple approach and easy implementation will give students a much better chance to understand the idea of parallel computing. In the contest of groundwater flow modelling with the FVM, the parallel program allows to experience the effect of speedup and to investigate changes in the amount of memory used on each node.

For the future the limitations will be reduced, e.g. leakage boundaries implemented and blank areas as well as inner boundaries used.

References

1. Bastian, P. (1997) Parallele Numerische Methoden für partielle Differentialgleichungen, Lecture Notes, University of Stuttgart
2. Bastian, P. and Helmig, R. (1999) Efficient Fully-Coupled Solution Techniques for Two Phase Flow in Porous Media. Parrallel Multigrid Solution and Large Scale Computations. Published by Advances in Water Resources, 23(3), 199-216
3. Hinkelmann, R. (1997) Parallelisierung eines Lagrange-Euler-Verfahrens für Strömungs- und Stofftransportprozesse in Oberflächengewässern, PhD-Thesis, Bericht Nr. 51, Institut für Strömungsmechanik und Elektronisches Rechnen im Bauwesen, Universität Hannover
4. Ohio Supercomputer Center (1996) MPI Primer / Developing With LAM, The Ohio State University, `http://www.osc.edu/lam.html`
5. Witte, B. (2000) Parallelisierung eines auf der Integralen-Finite-Differenzen-Methode basierenden Grundwasserströmungsmodells, Diploma Work, TU Braunschweig

Adaptive Hybrid Mixed Finite Element Discretization of Instationary Variably Saturated Flow in Porous Media

P. Knabner and E. Schneid

University of Erlangen-Nürnberg, Institute of Applied Mathematics,
Martensstr. 3, D-91058 Erlangen, Germany

Abstract. For the Richards equation, a nonlinear elliptic-parabolic partial differential equation modelling saturated-unsaturated flow in porous media, we present a hybrid mixed finite element discretization. The efficiency of the algorithm is improved by local time step and grid adaption. The adaption algorithms are based on rigorous error estimators, whose derivation is indicated. Examples elucidate the performance of the algorithm.

1 Introduction

The availability of numerical models to predict the spreading of contaminants or the influence of selective measures within remediation strategies relies on the detailed description of water movement in the subsurface. The complete modelling of large subsurface areas, necessary for such reliable predictions in real cases, is still a task that most often fails due to insufficient computing performance and inefficient numerical techniques. An appropriate model for the water movement, taking into account unsaturated subregions near the surface, is the Richards equation, which in the saturated subregions degenerates from parabolic to elliptic type. Thus the model is much more complex than a linear model for the saturated movement in an aquifer and similar in character to a general multiphase flow problem. Concerning the spatial discretization, the picture is as follows: Taking for granted that the complex interior structuring of a field site requires a discretization method based on an unstructured grid, there are still several possilities: Besides the well-known conformal finite element method the related node-centered finite volume method is popular with the engineering community. More recent approaches are the mixed finite element method and the related cell-centered finite volume method. There are two qualitative requirements for the method, namely

- local mass conservation
- an intrinsic and accurate approximation of the flux.

The last requirement is in particular important, as the flux is the driving force of the forced convection in the solute transport problems mentioned above. The only discretization method satisfying these two requirements is the mixed finite element method, but to the expense that even for linear problems the emerging discrete systems of equations are difficult to solve due to being the solution of a saddle point problem. For linear (stationary) problems the remedy of hybridization and static condensation is well-known and will be extended here to the nonlinear (and degenerate) case. A further building stone of an efficient algorithm is an appropriate nonlinear solver, where we rely on Newton's method and the multigrid method for the linear subproblem in the Newton iteration. Finally the aim of our work is to improve the efficiency of the numerical simulation by local adaption of discretization parameters. This development is based on error indicators, derived for the hybrid mixed finite element discretization of the Richards equation in its pressure based formulation. We consider both error estimation in time and in space and corresponding adaptation algorithms. The adaptation in space, i.e. the refinement and coarsening of the grid, defines the hierarchy of grids being the basis of the multigrid method. A detailed representation of the following methods and results are contained in [5].

2 The Model

The Richards equation results from the requirement of mass conservation (in form of volume conservation assuming the incompressibility of water):

$$\partial_t \Theta(\psi) + \nabla \cdot \boldsymbol{q} = 0 \tag{1}$$

and Darcy's law

$$\boldsymbol{q} = -K(\psi)\nabla(\psi + z) \,. \tag{2}$$

Here t denotes time, $\boldsymbol{x}$ space with the coordinate z against the direction of gravity, $\psi = \psi(\boldsymbol{x}, t)$ the hydraulic pressure, Θ the water content (porosity $\cdot$ saturation) and K the hydraulic conductivity. There are two coefficient functions: the soil-water retention $\Theta(\psi)$, relating the saturation and the pressure, and the unsaturated hydraulic conductivity $K(\psi)$, relating the conductivity and the pressure. A common representation of these material properties is given by the models of van Genuchten and Mualem [3].

By inserting (2) into (1) we reach at

$$\partial_t \Theta(\psi) - \nabla \cdot K(\psi)\nabla(\psi + z) = 0 \,. \tag{3}$$

For negative pressure values, corresponding to unsaturated subregions, the nonlinearities are monotone increasing, therefore (1) is a (strongly) nonlinear parabolic differential equation there, but positive pressure values lead to the constant value of maximum saturation and represent the region below the groundwater table, where the pressure obeys an elliptic equation. Thus in total the model (3) (or (1), (2)) represents a nonlinear elliptic-parabolic degenerate partial differential equation. This equation supplemented by initial and boundary conditions on a spatial domain Ω and a time interval $[0,T]$ constitutes the model to be discretized. Concerning the boundary conditions we assume a subdivision of the boundary: $\partial\Omega = \Gamma_{\mathrm{D}} \cup \Gamma_{\mathrm{F}}$ such that

$$\begin{aligned} \psi &= g_{\mathrm{D}} \quad \text{for } \boldsymbol{x} \in \Gamma_{\mathrm{D}}, \ t \in (0,T)\,, \\ \boldsymbol{q}\cdot\boldsymbol{n} &= g_{\mathrm{F}} \quad \text{for } \boldsymbol{x} \in \Gamma_{\mathrm{F}}, \ t \in (0,T) \end{aligned}$$

for given functions g_{D} and g_{F}, where $\boldsymbol{n}$ denotes the outward normal on $\partial\Omega$.

3 Discretization

To discretize the Richards equation, we use the implicit Euler method for the time and mixed finite elements for the spatial dependence, applied to the form (1), (2). Thus we ensure elementwise mass conservation with continuity of fluxes across edges between elements. This method prevents artificial source terms within subsequent transport simulation as if based on a fluid flow failing to be locally mass conservative like linear finite elements. This means that in one time step of stepsize τ the mixed finite element method is applied to

$$\Theta(\psi) + \tau\boldsymbol{\nabla}\cdot\boldsymbol{q} = \Theta(\psi^0)\,, \tag{4}$$

$$K^{-1}(\psi)\boldsymbol{q} + \nabla z = -\boldsymbol{\nabla}\psi\,. \tag{5}$$

The resulting discrete system of equations is further enlarged by hybridization, introducing Lagrange multipliers on edges. Afterwards the flux variables are explicitly eliminated.

Thus the discrete set of equations attains the following form, using the lowest order Raviart–Thomas ansatz on triangles:

$$\Theta(\psi_T) + \frac{\tau b_T}{|T|}K(\psi_T)\sum_{E\subset T}(\psi_T - \lambda_E) = \Theta(\psi_T^o) \quad \forall\, T \in \mathcal{T}\,, \tag{6}$$

$$\sum_{T\supset E} K(\psi_T)\sum_{E'\subset T} B_{TEE'}^{-1}(\psi_T - \lambda_{E'} - z_{TE'}) = \int_E g_{\mathrm{F}}\, ds \quad \forall\, E \in \mathcal{E}, E \notin \Gamma_{\mathrm{D}}\,. \tag{7}$$

Here $\mathcal{T}$ and $\mathcal{E}$ denote the triangulation and its edges, T and E are indices for elements and edges, ψ_T and ψ_T^o denote the element pressures at the new and the old time level, λ_E the Lagrange multiplier on an edge. Furthermore $|T|$ is the area of an element, B_T the local contribution to the stiffness matrix, $B_{TEE'} := \int_T \boldsymbol{w}_{TE}\cdot\boldsymbol{w}_{TE'}\,dx$, where $\boldsymbol{w}_{TE}$ is the local shape function from the Raviart–Thomas space for the edge E, $b_T := \sum_E B_{TEE}^{-1}$, and z_{TE} stems from the gravitational term in Darcy's law.

A second reduction is realized by an implicit elimination: The mass conservation equation (6) on each element represents an implicit expression for the pressure value on this element. Expressing formally these values as functions of the Lagrange multipliers, (7) is the nonlinear system of equations that has to be solved for each time step.

The procedure described above is not only formal, but can be made rigorous by showing the equivalency between the mixed finite element formulation, the hybrid mixed finite element formulation and finally the reduced problem being only a system of equations for the Lagrange multipliers (see [4]).

The nonlinear system of equations (7) and also the local equations (6) are solved by a variably damped Newton's method. The linear subproblems within the Newton iterations for (7) are solved by a multigrid method. Here the proper definition of the intergrid transfer operators is decisive for an efficient performance of the multigrid method. We can take advantage of the relations of a discretization by mixed finite elements with the lowest order Raviart–Thomas space and a nonconforming finite element discretization with Crouzeix–Raviart elements, for which such transfer operators have been developed (see [2] for the linear case and [4]).

4 Adaptivity

In general it is not possible to seperate errors produced by time and by space discretization. Anyhow we start from such a point of view, but aim at a subsequent combination of the obtained results. This strategy is mainly motivated by the structure of the mixed formulation, separating mass conservation from the Darcy law, but also by [1].

The time step size is adapted based on the following considerations: The classical technique of the Richardson extrapolation would lead to an error estimator in time, but to the expense of two additional global problems to be solved (for the two time steps with step size $\tau/2$). The local equations (6) represent the discrete time evolution. By the technique of the Richardson extrapolation but approximating the discrete evolutions without loss in the approximation order, we can derive the following error indicator:

$$\eta_\tau = \frac{1}{|\Omega|} \sum_{T \in \mathcal{T}} \left\{ \tau \frac{N}{4} b_T \frac{K(\psi_T^\circ)}{\Theta'(\psi_T^\circ)} f_T(\psi_T^\circ, \Lambda_T^\circ, \Lambda_T, \tau) \right\} . \tag{8}$$

Here $N(= 3$ here$)$ denotes the number of edges of an element and $\Lambda_T := \sum_{E \subset T} \lambda_E$. The function f_T is defined by

$$\begin{aligned} &f_T(\psi_T^\circ, \Lambda_T^\circ, \Lambda_T, \tau) \\ &= \left| \Theta'(\psi_T^\circ) \tfrac{\Lambda_T - \Lambda_T^\circ}{\tau N} + \tfrac{b_T}{|T|} \left(K(\psi_T^\circ) + K'(\psi_T^\circ) \left(\psi_T^\circ - \tfrac{\Lambda_T^\circ}{N} \right) \right) (N\psi_T^\circ - \Lambda_T^\circ) \right| \end{aligned}$$

and has a similarity with the residual of (6). This indicator (8) can be evaluated after the computation of one time step without further equations to be solved. After the computation of one time step, the current error indicator is compared to a given control parameter: If it exceeds this parameter, the computation is repeated with half the time step size. Otherwise the result is accepted and if in addition the error indicator is less than a predefined ratio of the control parameter, the time step size is doubled.

Comparison (of the error in a numerical convergence test) between adaptive time stepping and constant step sizes shows the superiority of the adaptive algorithm. Figure 1 shows the results and adaptively generated time step sizes for the simulation of water infiltration into a double layered soil column. The bottom layer has a 100 times lower permeability than the upper layer. The time step sizes follow the predefined (by selection of geometry and material) time scale of water redistribution, that changes substantially to the bottom layer at about $t = 1000$.

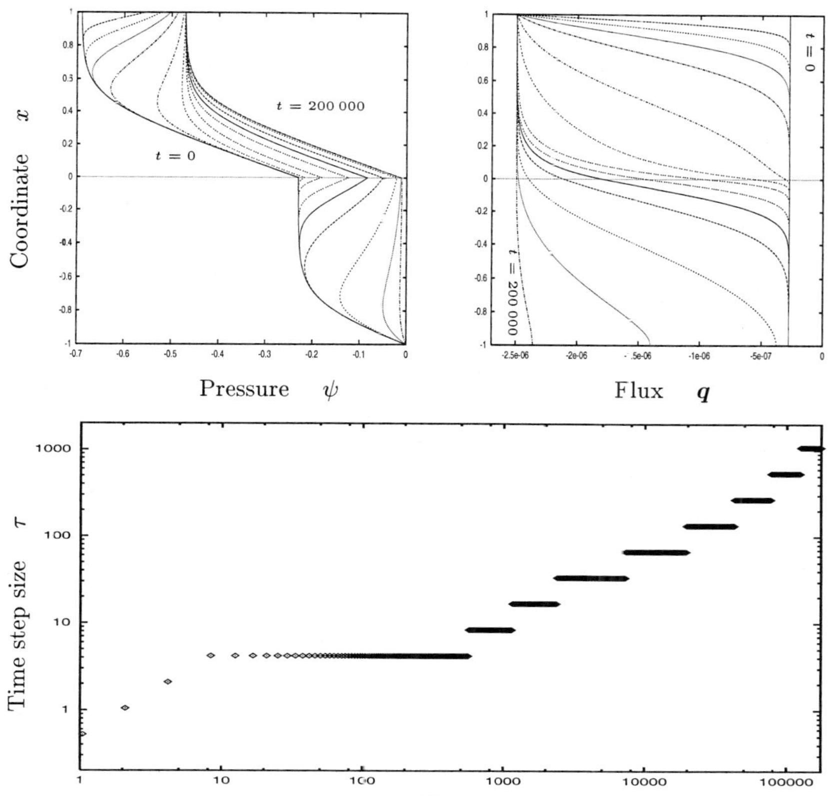

Fig. 1. Pressure and flux profiles and time step sizes for infiltration into a layered soil column

The error indicator for the grid adaption is derived from the residual of the weak formulation of Darcy's law, which in the mixed variational formulation (compare (5)) reads as

$$D(\psi, \boldsymbol{q}) = 0 \ ,$$

where $D : \mathrm{L}^2(\Omega) \times \mathrm{H}(\mathrm{div}; \Omega) \longrightarrow (\mathrm{H}_{0,\mathrm{F}}(\mathrm{div}; \Omega))^*$ is defined by

$$\langle D(\psi, \boldsymbol{q}), \boldsymbol{v} \rangle := \int_\Omega \left(\frac{\boldsymbol{q}}{K(\psi)} \cdot \boldsymbol{v} - \psi \boldsymbol{\nabla} \cdot \boldsymbol{v} + \boldsymbol{\nabla} z \cdot \boldsymbol{v} \right) d\boldsymbol{x} + \langle g_\mathrm{D}, \boldsymbol{v} \cdot \boldsymbol{n} \rangle_{\Gamma_\mathrm{D}} \tag{9}$$

for all $\boldsymbol{v} \in \mathrm{H}_{0,\mathrm{F}}(\mathrm{div}; \Omega)$. Here

$$\mathrm{H}_{0,\mathrm{F}}(\mathrm{div}; \Omega) := \left\{ \boldsymbol{v} \in (\mathrm{L}^2(\Omega))^2 \;\middle|\; \boldsymbol{\nabla} \cdot \boldsymbol{v} \in \mathrm{L}^2(\Omega) \, , \ \boldsymbol{v} \cdot \boldsymbol{n} = 0 \text{ on } \Gamma_\mathrm{F} \right\} \ .$$

Therefore

$$r(\psi, \boldsymbol{q}) := \| D(\psi, \boldsymbol{q}) \|_{(\mathrm{H}_{0,\mathrm{F}}(\mathrm{div};\Omega))^*} = \sup_{\boldsymbol{v} \in \mathrm{H}_{0,\mathrm{F}}(\mathrm{div};\Omega)} \frac{|\langle D(\psi, \boldsymbol{q}), \boldsymbol{v} \rangle|}{\|\boldsymbol{v}\|_{\mathrm{div};\Omega}}$$

seems to be an appropriate error estimator, which can also be made rigorous. Therefore it is desirable to find sufficiently sharp estimations of r, which are easily computable. Such an estimation of the residual gives the error indicator

$$\eta_h = \sum_{T \in \mathcal{T}_h} \left\| \tilde{\psi}_h - \psi_h \right\|_{0,T}^2 \tag{10}$$

as predominant part, where $\tilde{\psi}_h$ is a pressure, postprocessed from available discrete data.

Within the simulation of a time dependent process, the grid is once adapted (local refinement and coarsening) according to the error indicator after the computation of each time step, without subsequent control and recurrence.

As in the case of adaptive time stepping, comparison between the grid adapting algorithm and homogeneous refinement of the triangulation shows the superiority of the adaptive algorithm. Figure 2 shows adaptively generated grids for a groundwater recharge scenario. The adaptation algorithm identifies refinement and coarsening areas during the process. The picture on the upper left corner shows the coarsest grid at the beginning, which is also the base level grid for the multigrid algorithm.

5 Conclusions

Two fundamental ingredients of an adaptive algorithm for the simulation of instationary variably saturated flow in porous media are introduced: The error indicators for space and time discretization. In addition to the development of codes running efficiently on high performance computers, the technique of local adaptation of discretization parameters contributes to provide

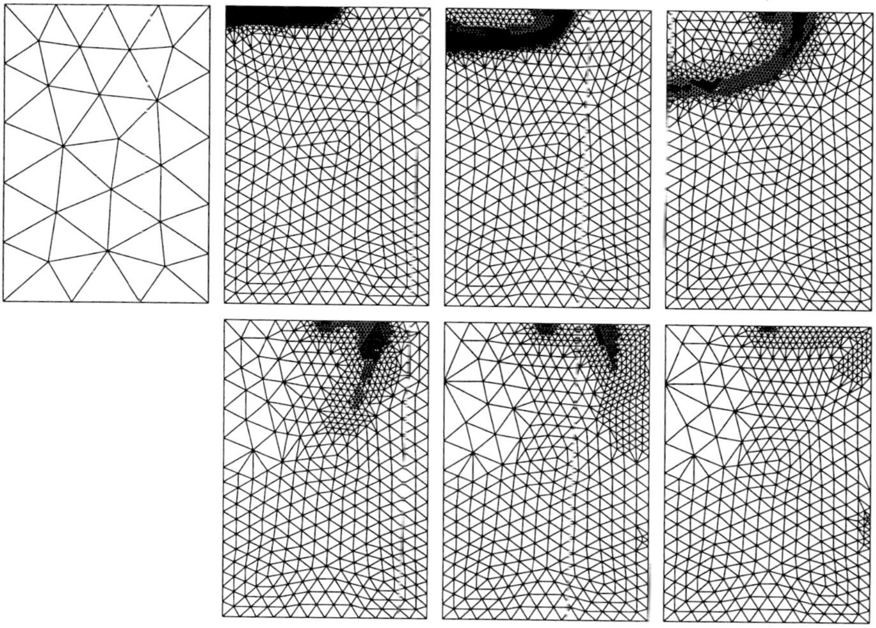

Fig. 2. Adaptively generated grids for a groundwater recharge scenario

a simulation tool capable of predicting processes of high complexity, as it is necessary for applications in groundwater risk assessment and the development of remediation strategies.

The future task to bring the adaptive algorithms for space and time together depends on the following: The projection of the pressure between differently refined grids produces errors, the time step size adaptation is responding to. In addition the interaction of the error indicators and the corresponding refinement/coarsening has to be treated.

References

1. Bornemann, F.A. (1991) An Adaptive Multilevel Approach to Parabolic Equations in Two Space Dimensions. PhD thesis, Konrad-Zuse-Zentrum für Informationstechnik, Berlin
2. Braess, D. and Verfürth, R. (1990) Multigrid methods for nonconforming finite elelemnt methods. SIAM J. Numer. Anal., 27 **4**, 979–986
3. Genuchten van, M.T. (1989) A closed-form equation for predicting the hydraulic conductivity of unsaturated soils. Soil Sci. Soc. Am. J., **44**, 892–898
4. Knabner, P. and Summ, G. Efficient realization of the mixed finite element discretization for nonlinear problems, in preparation.

5. Schneid, E. (2000) Hybrid-Gemischte Finite Elemente Diskretisierung der Richards-Gleichung. PhD thesis. Universität Erlangen, Institut für Angewandte Mathematik.

Simulation of High Pressure Liquid Chromatography (HPLC) Columns with CFD

H. Boysen[1], G. Wozny[1], M. Lisso[1], W. Arlt[2], and T. Laiblin[2]

[1] Chair of Process- and Plant Technology, KWT 9, Technical University Berlin, Straße des 17. Juni 135, 10623 Germany
[2] Chair of Thermodynamics and Thermal Separation, TK 7, Technical University Berlin, Straße des 17. Juni 135, 10623 Germany

Abstract. Steep concentration gradients of the chromatographical separation process lead to discretisation errors. Different approaches to reduce these inaccuracies are applied and compared. The following methods have been investigated:

- Local grid refinement
- Application of different discretisation schemes
- Moving mesh
- Relation of the discretisation with respect to time and space
- Weighting of the differential equation system

Finally the CFD model is validated by comparing simulated results with experiments.

1 Simulation of HPLC (High Performance Liquid Chromatography) Columns with CFD

The model consists of coupled differential equations which include derivatives with respect to time and space. In order to solve these equations the commercial CFD code Fluent 4.5 is in use.

Principle of a Preparative HPLC Column (Fig. 1)

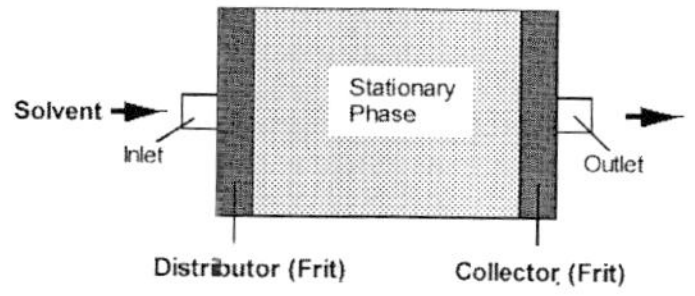

Fig. 1. Principle drawing of a preparative HPLC column

A solvent (eluent) flows through the inlet into the HPLC column. The frit (distributor) distributes this stream equally over its cross section. A small grained stationary phase is used to provide a large surface area.

A pulse of a mixture containing compounds which are difficult to separate (i.e. optical isomers) is fed into the column [4]. The eluent carries these compounds to the outlet. On the way the substances interact with the stationary phase. The most important of these interactions is adsorption. Adsorption is a temporary bond of single species to the stationary phase depending on their concentration. The greater the affinity of a substance to the stationary phase the later this compound will arrive at the outlet. The species leave the column separated in time.

2 Chromatography from a Numerical Point of View

Steep concentration gradients in axial direction lie within the nature of this separation technique. The discretisation with respect to time and space decreases these gradients while the concentration pulse passes the column. The effect of these inaccuracies is comparable with physical dispersion. Modern stationary phases cause very small physical dispersion. Therefore it is necessary to simulate HPLC columns with a high grid density as well as small time steps. This leads to a large demand of calculation time. Numerical dispersion often can not be decreased to a neglectable range. Therefore it has to be considered by modeling physical dispersion. Hence a simple input of the physical dispersion coefficient into the simulation is not possible. By knowing about the numerical dispersion of a specific CFD simulation the missing partial amount of the physical dispersion can be added as the dispersion coefficient into the simulation. However, to approach a model setup along this line, the numerical dispersion needs to be below the physical dispersion of a real HPLC column. To serve this demand, a high computational time is necessary. Therefore the most suitable configuration in means of a small discretisation error is desired and has been investigated.
During the study of the behaviour of numerical dispersion, the following approaches were considered:

- The concentration gradient in a real HPLC column becomes more shallow as the peak travels through the fixed bed due to physical dispersion. A high grid density at the inlet continuously getting less dense to the outlet was applied.
- Numerical dispersion is due to discretisation errors (truncation errors) [2]. Different discretisation schemes were applied to find the best one with respect to the smallest numerical dispersion.
- The species peak with steep concentration gradients in a HPLC column is present only in a part of the fixed bed. A moving mesh which travels with the peak was implemented to reduce the necessary number of cells [3].
- Numerical dispersion is affected by the grid density and the time step size. The general behaviour is shown based on numerical experiments. From these data an optimal *Courant Number* is drawn to maximize the numerical accuracy for constant calculation time.
- Good weighting of the differential equation system can be done by changing the time unit in the calculation to an appropriate range. This has a

significant influence on numerical dispersion. The study of the differential equation system leads to rules, which allow to change the time unit, even if the CFD code does not support such intention.

2.1 Local Grid Refinement

For every functional group (inlet, distributor, collector, and outlet) a separate grid segment is defined (Fig. 1). The approach to cluster the grid segment of the stationary phase more dense at the boundary to the distributor and to use grid lines more sparingly at the collector shows a shallow minimum of the discretisation error at a grid weighting factor[1] of 1.5 (Table 1). For every simulation similar grid configurations were used for the inlet, distributor, collector, and outlet. The number of cells of the segment for the stationary phase was kept constant. Only the distribution of the grid spacing in the segment has been altered.

Table 1. Numerical dispersion coefficient in $[10^{-6}\frac{m^2}{s}]$ using different grid weighting factors

	Grid weighting factor			
	1	1.5	2	3
num. D	1.59	1.57	1.59	1.65

2.2 Discretisation Schemes

Different discretisation schemes for the compound balance (Power Law, 2nd Order Upwind and Quick) were applied to the simulation of the chromatography process to study their impacts to numerical dispersion in the CFD model. It could be observed that the linear discretisation shows bad performance, so a higher order scheme should be used. The choice between Quick or Second Order Upwind is of neglectable importance (Table 2).

Table 2. Numerical dispersion coefficient in $[10^{-6}\frac{m^2}{s}]$ using different discretisation schemes

	Discretisation Schemes		
	Power Law	2nd Ord. Upwind	Quick
num. D	4.90	1.59	1.45

[1] Using the internal grid generator of Fluent 4.5 it is possible to modify the grid distribution within a segment via weighting factors to increase or decrease the grid density at the segment endpoint [1]. A weighting factor of 1 causes an uniform grid density for the respective segment.

2.3 Moving Mesh

In a moving mesh an area of high grid density moves with the traveling peak (steep concentration gradients) through the column. To distinguish the beginning and the end of the peak the criterion (1) was applied.

$$\tilde{c} \geq \frac{c_{Cell}}{c_{max(allCells)}} \cdot 100 \tag{1}$$

$\tilde{c}$	=	Defined concentration of the beginning and the end of the peak
c_{Cell}	=	Maximum concentration of all cells at a specific axial position
$c_{max(allCells)}$	=	Maximum concentration of all cells in the domain

The area of high grid density moves downstream (Fig. 1), if (1) becomes not true.

The main drawback of this approach is the fact that after every re-meshing the stationary velocity and temperature fields have to be converted again. These equations are coupled with the concentration field, which converts quicker if the steady state flow and temperature distribution is already converted. Therefore a simulation with moving mesh needs more iterations for each time step. Hence a part of the saved computational time due to a smaller number of cells is used up for re-meshing the grid and the larger number of iterations needed. For example the reduction of cells by 50% saves 30% of the computational time. Unfortunately a moving mesh with 50% of the cells is not able to resolve the peak fully. Therefore the numerical dispersion increases by 23% [3]. At this stage it was possible to make out the magnitude of the possible reduction of the computational time with moving mesh. Since strong conversion problems were occurring at some mesh configurations, further efforts to implement moving meshes were stopped.

2.4 Relation of Discretisation with Respect to Time and Space

The relation of the discretisation with respect to time and space can be expressed with the Courant Number $Cou = \frac{w \times \Delta t}{\Delta L}$.

Fig. 2 shows the numerical dispersion decreasing with a decreasing time step and an increasing number of cells. At constant time step size the numerical dispersion can only be reduced to a certain value just by increasing the grid density. An analog behaviour can be observed for constant grid density and reduced time step size. So it is possible to conclude that a minimum value of the numerical dispersion for a given computational effort must exist. To derive this value the data of Fig. 2 were used to approximate a three dimensional function.

This function is presented as a double logarithmic plot. It shows the numerical dispersion versus computational expenses with the Courant Number

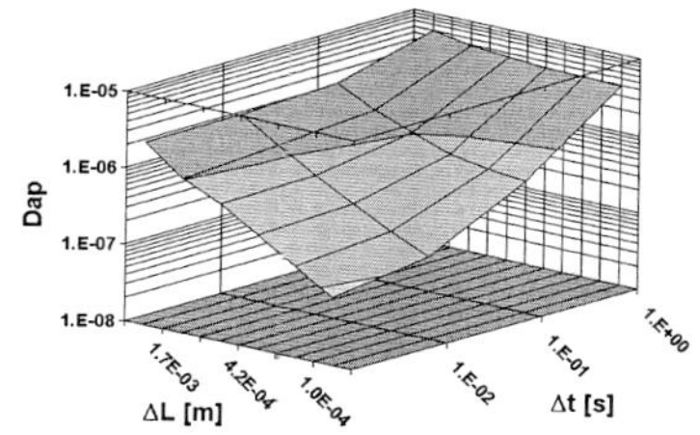

Fig. 2. The numerical dispersion coefficient of a modeled tube with a length of 5cm and an average velocity of 0.003 m/s. It is dependent on the grid density and the time step size

as parameter (Fig. 3). The computational effort can be expressed by the relation of the number of cells to the time step size.

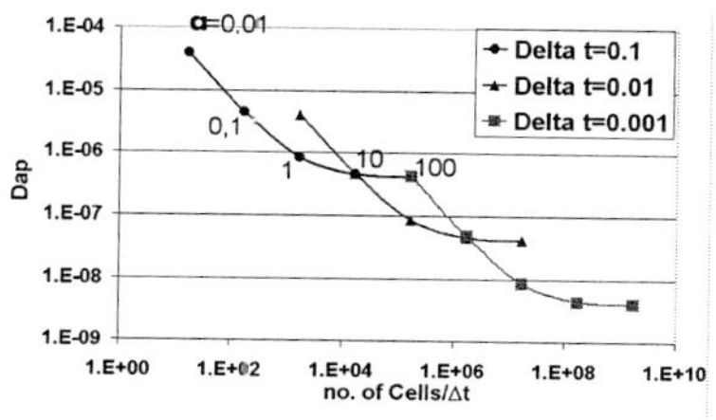

Fig. 3. The numerical dispersion coefficient versus the computational effort

The best simulation configuration is as far as possible in the lower left corner of the diagram. This is achieved best with a Courant Number of one for all plotted time step sizes.

2.5 Weighting the Applied Differential Equations

The simplified differential equations can be expressed as follows:

1. Navier Stokes Equation:

$$\varrho(\boldsymbol{v}\nabla)\boldsymbol{v} = -\nabla p + \eta(\frac{\boldsymbol{v}}{\alpha}) \tag{2}$$

2. Equation of continuity:

$$\nabla \cdot \boldsymbol{v} = 0 \tag{3}$$

3. Conservation of species:

$$\frac{\partial c_i}{\partial t} + F\frac{\partial q_i}{\partial t} = -\boldsymbol{v}(\nabla c_i) + \nabla(D_{ij}(\nabla c_i)) \tag{4}$$

4. Energy conservation:

$$\varrho cp(\frac{\partial T}{\partial t} + \boldsymbol{v}(\nabla T)) = \lambda(\nabla^2 T) \tag{5}$$

Introducing

$v = v^\star \cdot a$	$D_{ij} = {D_{ij}}^\star \cdot a$	$t = \frac{t^\star}{a}$
$\eta = \eta^\star \cdot a$	$\lambda = \lambda^\star \cdot a$	$p = p^\star \cdot a^2$

leads to:

$$\varrho(v^\star\nabla)v^\star = -\nabla p^\star + \eta^\star \cdot (\frac{v^\star}{\alpha}) \tag{6}$$

$$\nabla \cdot v^\star = 0 \tag{7}$$

$$\frac{\partial c_i}{\partial t^\star} + \frac{1-\epsilon}{\epsilon} \cdot \frac{\partial q_i}{\partial t^\star} = -v^\star(\nabla c_i) + \nabla(D^\star_{ij}(\nabla c_i)) \tag{8}$$

$$\varrho cp(\frac{\partial T}{\partial t^\star} + v^\star(\nabla T)) = \lambda^\star(\nabla^2 T) \tag{9}$$

v	velocity vector	p	pressure
η	viscosity	c_i	concentration of species i
t	time	ϵ	void fraction
q_i	load on stationary phase of species i	D_{ij}	dispersion coefficient of species i and j
ϱ	density	cp	thermal capacity factor
T	temperature	λ	thermal conductivity
α	permeability of porous media	a	constant weighting factor to time unit
$\star$	weighted value (superscript)		

Equations (6), (7), (8), and (9) show that it is possible to choose any constant a to re-scale the time unit of the differential equation system to achieve a better weighting.

In the CFD model all parameters with a unit including time have to be modified accordingly.

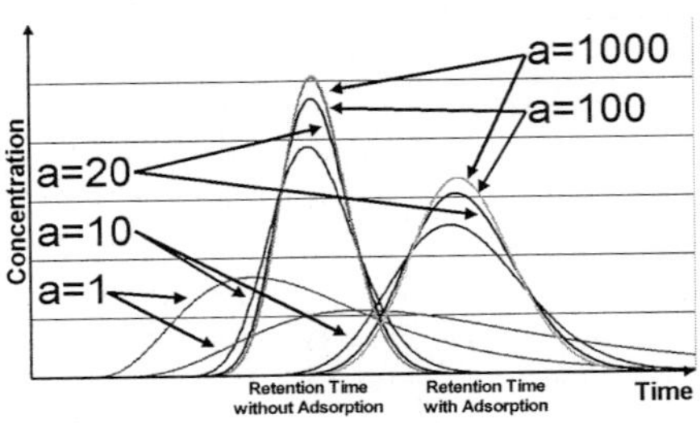

Fig. 4. The concentration at the outlet versus time modeled with a raising $'a'$ and a constant number of iterations. It shows that it converges to the same profile (a=100, 1000)

Fig. 4 shows the impact of the variation of a on the simulation results. Every simulation is done using the same grid density and number of iterations.

The step size is reduced constantly according to the increasing a. Keeping Fig. 2 in mind, it shows the positive influence of an suitable weighted differential equation system in matters of calculation time. The weighting of the involved equations show how it is possible to intervene into the equation system, even if the CFD code does not provide features for such intentions[2].

3 Validation of Results

To validate the CFD model, simulated results were compared with experimental data of a real chromatography column. More specific, the temperature and concentration field inside of the column, as well as the concentration versus time at the outlet of the column are used to show the accuracy of the CFD model.

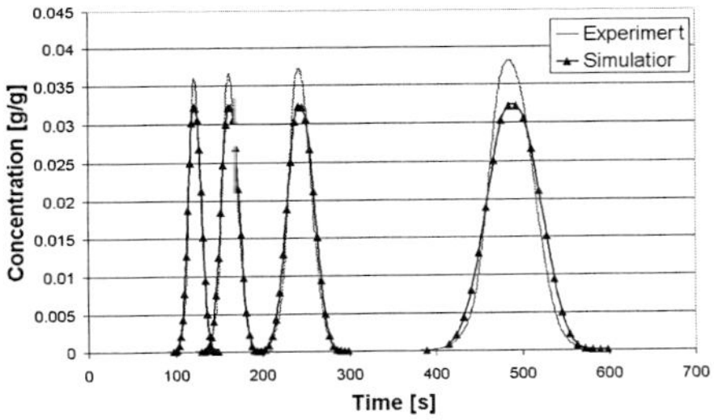

Fig. 5. Concentration at the outlet versus time modeled for different volume flows (D=100mm, 1, 2, 3 and 4 $\frac{ml}{min}$) in comparison to the experiment

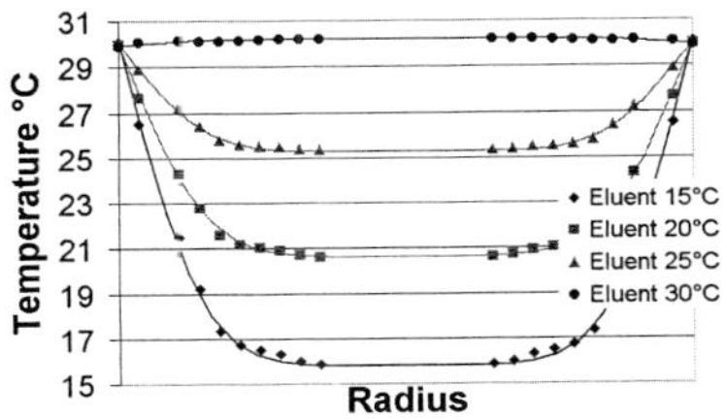

Fig. 6. Temperature profile in the column (D=100mm, 300 $\frac{ml}{min}$) near to the collector frit for different solvent temperatures in comparison to the experiment [5]

The grid validation had been performed in radial direction. Only a small change of the numerical dispersion coefficient can be observed by using different numbers of cells. Also the overall mass balance was checked.

It was observed that the error for the adsorbing and the non adsorbing peak is below 0.2% (Table 3).

4 Conclusions

The general approach to accept a certain inaccuracy due to discretisation and considering this error as a part of the physical dispersion coefficient in

[2] The simple change of the time unit in FLUENT does not have the desired effect since FLUENT calculated the input back to S.I. (or British) units before it starts the calculation of the differential equation system.

Table 3. Comparison of the mass injected with the mass leaving the HPLC column

ad. compound			Error
m_{inj}	$1,7453E-8$	kg	0.14 %
m_{outlet}	$1,7428E-8$	kg	

non ad. compound			Error
m_{inj}	$1,7453E-8$	kg	0.15 %
m_{outlet}	$1,7427E-8$	kg	

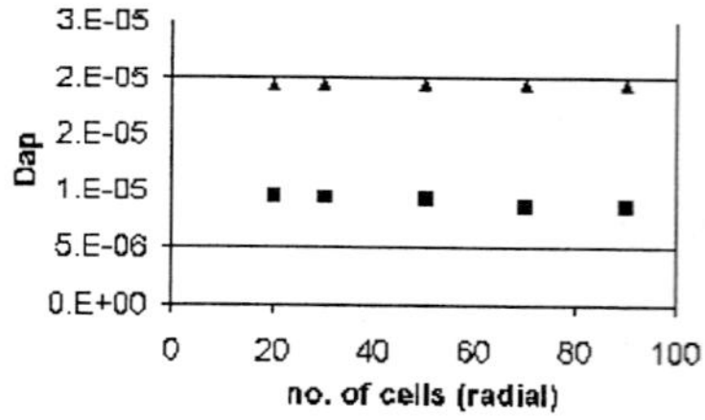

Fig. 7. Grid validation by showing constant numerical dispersion with an increasing number of cells in radial direction. ▲ denotes the adsorbing and ■ denotes the non adsorbing peak

the differential equations can be validated by comparing various simulation results with experimental data. However, due to the steep concentration gradients of the process numerical dispersion requires a high grid density and a small time step size which leads to a large computational demand for each simulation. To reduce this demand different approaches to reduce the discretisation error are applied and compared. It could be shown that, most of all it is necessary to achieve a good weighted differential system. That can be done by choosing an appropriate time unit (Fig. 4). An other important factor is an appropriate relation of the discretisation with respect to time and space. This relation can be expressed with the *Courant Number* which should be in the range of one (Fig. 3). Also the grid weighting and a higher order discretisation scheme helps to reduce the discretisation error. The influences are rather small in comparison to the approaches mentioned above. A moving mesh could be distinguished as not appropriate for the simulation of chromatographical columns, since the steady state temperature and velocity field has to be converted after every re-meshing. This consumes the saved calculation time due to the smaller number of cells.

References

1. (Aug. 1996) FLUENT Users's Guide. Fluent Inc., London
2. Seidel-Morgenstern, A. (1995) Mathematische Modellierung der präparativen Flüssigchromatographie, Deutscher Universitätsverlag DUV, Wiesbaden, ISBN 3-8244-2064-3
3. Klein, A. (1999) Parameterstudien zur hydrodynamischen Optimierung von chromatographischen Säulen, Studienarbeit TU-Berlin, 35–43
4. Coulson, J. M., Richerson J. F. (1991) Chemical Engineering, Butterworth-Heinemann, Oxfort, ISBN 0-7506-2942-8
5. Laiblin T., Arlt W., Wozny G., Lisso M., Boysen H., (2000) Conference Proceedings, Spica, Zürich

CFD Calculations of Flow, Dispersion and Chemical Reactions in Fixed Bed Tubular Reactors Using the Lattice Boltzmann Method

Th. Zeiser*,[1], H. Freund[2], J. Bernsdorf[3], G. Brenner[1], and F. Durst[1]

[1] Lehrstuhl für Strömungsmechanik, Universität Erlangen-Nürnberg, Cauerstraße 4, D-91058 Erlangen, Germany
[2] Lehrstuhl für Technische Chemie I, Universität Erlangen-Nürnberg, Egerlandstraße 3, D-91058 Erlangen, Germany
[3] C&C Research Laboratories, NEC Europe Ltd., Rathausallee 10, D-53757 Sankt Augustin, Germany

Abstract. Detailed 3-D simulations are required to correctly predict the coupled behaviour of flow and mass transport in chemical reactors with Θ-asymmetry. These simulations can be of great technical and economical interest as the reactor performance and safety during operation are significantly influenced by the local flow behaviour. One particular interesting example are randomly filled tubular fixed bed reactors with small tube-to-particle diameter ratios. New numerical methods such as the lattice Boltzmann approach together with the increasing computational power can provide detailed insight into the flow and transport processes in complex geometries. The present paper demonstrates the applicability of this approach by investigating complex flows and showing first results of the inhomogeneities in the 3-D flow of reacting species in a tubular fixed bed reactor.

1 Introduction

In contrast to other industrial sectors, CFD methods are not yet widely used in chemical engineering. The typical approach in this field is still to set up an experiment, to measure, to extract some empirical correlations and finally use them in semi-empirical equations which are in most cases of lowdimensional order. The reasons for this approach are quite easy to understand: The geometries typically used are complex. They cannot be described analytically and CAD data are, in contrast to applications in the automobile industry, often not available. A second point is, that the relevant phenomena are strongly non-linear and highly coupled. The geometrical structure determines the flow field. Both, the flow field and also the geometrical structure influence the mass transport. Therefore, the effective coefficients of the mass transport (e.g. dispersion coefficients) are not material properties but strongly depend on the actual set-up. Chemical reactions, especially in the case of heterogeneous surface-catalytic reactions, depend on the interaction

* e-mail: thzeiser@lstm.uni-erlangen.de

of all mentioned phenomena. If energy conversion and transport are relevant, it is becoming even more complicated than bulk transport, because the boundary layers and the conduction inside the solid have to be considered in addition. For simplicity, energy was not taken into account in the calculations presented here.

As a first conclusion, we can summarize: complex and coupled phenomena are typical for the field of chemical engineering. If we want to apply CFD methods in this area, they must be able to deal with these complex flows.

As has been shown in the last decade, lattice Boltzmann (LB) methods are an efficient numerical simulation approach especially in the case of complex geometrical structures such as porous media. The extension of the LB method to the needs of chemical engineering therefore seems to be very promising.

In Sect. 2, we define one particular example, tubular fixed bed reactors with small tube-to-particle diameter ratios, which clearly show the requirement of detailed simulations. The basic ideas of the lattice Boltzmann method are briefly summarized in Sect. 3. With some rather simple test-cases, the method and its implementation are quantitatively validated in Sect. 4. In Sect. 5, we discuss details of the flow field and its consequences on the mass transport and the surface reaction in a tubular fixed bed reactor. High-performance computing aspects are given in Sect. 6. Finally, the main points are summarized in the conclusions (Sect. 7).

2 Requirement for Detailed Simulations in Chemical Engineering: Tubular Fixed Bed Reactors with Small Tube-to-Particle Diameter Ratios

Tubular fixed bed reactors are of particular significance in the chemical process industries. They are, among others, used as catalytic reactors, adsorption columns and separation towers. Therefore, they are perhaps even the most widely used type of reactor at all. The typical structure consists of two major parts: a confining tube which is filled with particles either randomly or strongly ordered. In the simplest but often used case, the particles are monodisperse spheres. For some applications, e.g. in the case of strongly exothermic or endothermic reactions, very thin tubes must be used to remove or provide the heat. In order to nevertheless achieve the required throughput, up to several thousand tubes are arranged in parallel. In the calculations shown later, we focus on one single tube with the detailed particle structure inside.

The typically applied modelling approaches used in chemical engineering are based on empirical correlations derived from local averaging. The averaging can be performed on different levels, ranging from 0-D models with "plug-flow" assumption (i.e. homogeneous voidage distribution all over the packing, resulting in a constant axial velocity and no radial velocity at all) to sophisticated 2-D "non-parallel flow" models. The latter only assume symmetry in the circumferential (Θ) direction. The averaged velocity can vary in the axial and radial direction. All of these modelling approaches do not take the complete geometrical structure into account. They therefore rely on effective transport parameters, e.g. permeability coefficients for the calculation of the pressure drop, or dispersion coefficients for the determination of the mass transport with the convection-diffusion equation. In general, these values cannot be determined a priori. They must therefore be measured experimentally or determined form empirical correlations of similar set-ups.

In the case of tubular fixed bed reactors with small tube-to-particle diameter rations (i.e. aspect ratios lower than 10), a further complication occurs. The solid distribution is not homogeneous at all. Therefore even the "non-parallel flow" models which assume Θ-symmetry cannot predict the correct behaviour in detail. In the case of tubular fixed bed reactors with low aspect ratios, detailed 3-D simulations must be applied which consider the complete geometrical structure and the resulting velocity field and coupled transport processes to achieve reliable results.

The exact 3-D flow patterns and structures are of great interest as local inhomogeneities can cause channelling, hot-spots and thus reduce the conversion rate. They can also cause local catalyst deactivation leading to reduced selectivities. This demonstrates the importance of detailed information for the safe operation of the reactors at low costs.

In oder to perform detailed 3-D simulations, some requirements must be fulfilled: A first requirement is the three-dimensional description of the geometrical structure. In general, several methods can be applied to obtain the geometry. Simple or regular structures can be described analytically. Random packings, for example, can be generated synthetically with Monte Carlo methods [9] which imitate the technical filling process Another approach is based on computer tomography which can provide non-overlapping 3-D representations of arbitrary complex structures of many materials. Last but not least, CAD data sets are an interesting source, if available. But unfortunately, detailed CAD drawings of the internals of chemical reactors are often lacking. Anyway, there are, as described above, several methods which can provide the complex geometrical internals of complex apparatus. This brings us to the second requirement, efficient numerical methods. The result of some of the mentioned discretization schemes (e.g. the computer tomography) are Cartesian and equidistant grids. Therefore, the numerical methods should have low computational requirements per grid node and thus be able to handle large grids efficiently. The lattice Boltzmann method, briefly summarized in the next section, is an ideal candidate. The computational requirements per grid point are low, it is typically used on simple Cartesian grids, and it can successfully be used on high-performance vector-parallel computers.

3 Brief Description of the Lattice Boltzmann Approach

The fundamental idea of the lattice Boltzmann methods [5] is to look at the physical background of fluid mechanics from a *statistical point of view.* Within typical Navier-Stokes CFD solvers the conservation equations for mass and momentum (Navier-Stokes equations), based on the macroscopic quantities (pressure, density, flow velocity), are solved. In contrast, the underlying equation for the lattice Boltzmann technique is a space, momentum and time discrete formulation of the Boltzmann equation, describing the interaction of single fluid particles or an ensemble-averaged particle density distribution function $N_i(t_*, \boldsymbol{r}_*)$.

The current version of our lattice Boltzmann code is based on the 3-D 19-speed (D3Q19) lattice Boltzmann automata model with the so called single time Bhatnagar-Gross-Krook relaxation collision operator Δ_i^{Boltz} proposed by Qian [11]:

$$N_i(t_* + 1, \boldsymbol{r}_* + \boldsymbol{c_i}) = N_i(t_*, \boldsymbol{r}_*) + \overset{Boltz}{\underset{i}{\Delta}} , \tag{1}$$

$$\overset{Boltz}{\underset{i}{\Delta}} = \omega\left(N_i^{eq} - N_i\right), \tag{2}$$

where the first equation is the advection term meaning that the particle density distribution $N_i(t_*, \boldsymbol{r}_*)$ with discrete momentum index i located at the discrete position $\boldsymbol{r}_*$ at the discrete time t_* is shifted within one iteration (t_*+1) in direction $\boldsymbol{c_i}$ to its next neighbouring lattice node located at $\boldsymbol{r}_* + \boldsymbol{c_i}$.

The second equation describes the "collision" of the particles, realized by a local redistribution of the density distribution function N_i while locally conserving mass and momentum. This is implemented as a single time relaxation approach with a local equilibrium distribution function N_i^{eq}. The local equilibrium distribution function N_i^{eq} (3) is chosen in such a way, as to recover the time-dependent Navier-Stokes equation [11] in the low Mach number limit.

$$N_i^{eq} = t_p \rho \left\{ 1 + \frac{c_{i\alpha} u_\alpha}{c_s^2} + \frac{u_\alpha u_\beta}{c_s^2} \left(\frac{c_{i\alpha} c_{i\beta}}{c_s^2} - \delta_{ij} \right) \right\} \tag{3}$$

All macroscopic flow quantities can be simply calculated from the particle distribution function $N_i(t_*, \boldsymbol{r}_*)$ by the following relations:

density	$\rho = \sum_{i=0}^{n} N_i$,
flow velocity	$\boldsymbol{u} = \sum_{i=0}^{n} \boldsymbol{c_i} N_i / \rho$,
pressure	$p = c_s^2 \rho$, with: $c_s = 1/\sqrt{3}$,
viscosity	$\nu = \frac{1}{6}\left\{\frac{2}{\omega} - 1\right\}$.

In chemical engineering, it is essential to include multi-species transport. In the case of passive scalars, i.e. "tracers" which are convected by the flow and spread due to molecular diffusion but do not change the flow field, several approaches are possible within the lattice Boltzmann framework.

First of all, the same ideas, methods and equations as described above can be applied for each species in addition to the carrier fluid flow. But as the convection-diffusion equation only contains linear dependencies on the velocity and does not depend on the isotropy of the 4th order tensor, rigorous simplifications are possible for the species representation still recovering the convection-diffusion equation [7]. The 2nd order terms of the velocity can be neglected in the equilibrium distribution function (analogous to 3) for the species representation. But more important, it is sufficient to consider a velocity space discretisation with only six vectors along the Cartesian coordinates instead of 19 directions in the case of the carrier fluid. The drawbacks of these simplifications are slightly reduced stability and some limitations on the magnitude of the macroscopic velocity.

For the current work, we applied this simplified model for the multi-species transport combined with the D3Q19 model for the carrier fluid.

4 Validation

4.1 Pressure Drop in Randomly Packed Cubes

The pressure drop in porous media and packed beds is of great technical interest. Therefore some investigations of the pressure drop in randomly packed cubes are presented to show the quantitative agreement with experimental data. The plot in Fig. 1 [3] shows the pressure drop in different random packings in terms of the non-dimensional friction factor for a wide range of Reynolds numbers. For the flow in this type of porous media, it can be shown both, theoretically and experimentally, that the friction factor is constant for low Reynolds numbers and increases non-linearly at higher Reynolds numbers. For comparison, an experimental correlation for the pressure drop is also plotted. The numerical simulations quantitatively predict the pressure drop in these complex porous structures.

4.2 Taylor-Aris Dispersion Between Parallel Plates

Diffusion and dispersion in packed beds is one of the most important parameters for the design of many chemical reactors. As the simplest but nevertheless interesting test-case, the so-called Taylor-Aris dispersion between two parallel plates, for which an analytical solution is known, is shown here for validation purpose.

If a concentration pulse is injected into an axial plane of the fully developed laminar Poiseuille flow, the solute pulse gets dispersed as it moves

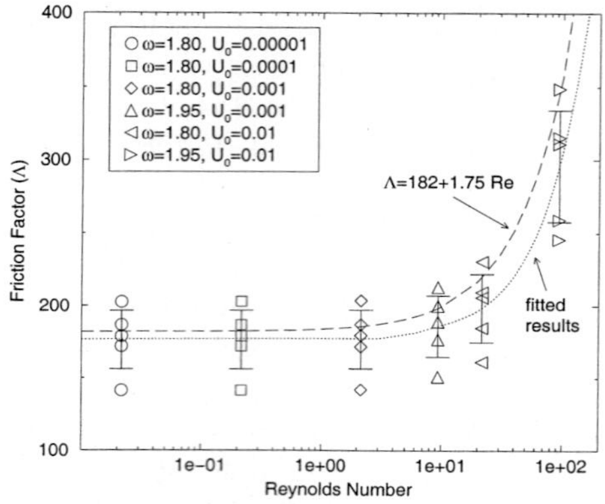

Fig. 1. Dimensionless friction factor (pressure drop) in a random packing of cubes as a function of the Reynolds number [3]

through the tube. The set-up and the physical principles are sketched in Fig. 2 [6]. The pulse is initially sharp (a), but the laminar flow quickly distorts the pulse as in (b). If there is no diffusion, the distortion continues unabated and the pulse is widely spread. If, instead, there is rapid diffusion, material in the centre of the tube tends to diffuse outward, into a region of solvent that is moving more slowly. Simultaneously, material that is left behind near the tube walls tends to diffuse towards the centre, into a region of faster flow. This radial diffusion inhibits the dispersion induced by axial convection. Thus, in the Taylor-Aris dispersion, fast diffusion unexpectedly produces little dispersion, and vice versa. The lattice Boltzmann simulations can correctly predict the analytical solution even on coarse grids as shown in Fig. 3.

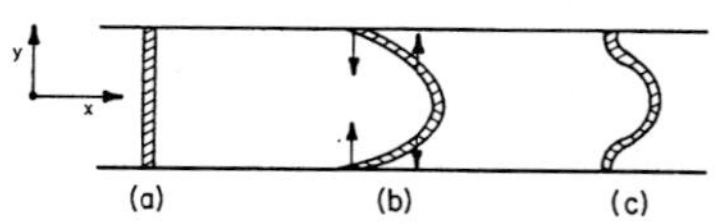

Fig. 2. Schematic plot skething the physics of the Taylor-Aris dispersion [6]

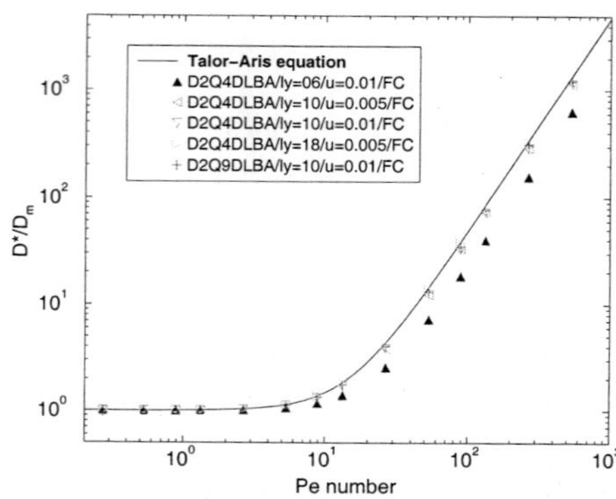

Fig. 3. Magnitude of the dispersion as a function of the Reynolds number for Taylor-Aris dispersion between parallel plates

5 Results of the Detailed Simulations of Randomly Packed Bed Reactors

A typical simplifying assumption of conventional models for tubular fixed bed reactors is Θ-symmetry. In the case of low aspect ratios this assumption is not valid as already outlined in Sect. 2. Figure 4 shows a packing with a tube-to-particle ratio of 5 and a packing length of 7.5 particle diameters. It can clearly be seen in Figs. 5–7 that large local inhomogeneities occur.

The local flow velocity and consequently also the reactant and product concentrations oscillate according to the geometric structure of the packed bed in the different cross sections. Large variations occur in both, the radial and the circumferential direction.

Owing to the large changes in the flow patterns behind the packing, it seems to be questionable whether experimental data, measured e.g. with hot-wire anemometry behind the packing, can be extrapolated to the inside of the packing [4,1,2]. The flow at a tenth of the particle diameter behind the packing (position +0.1) has already changed remarkably. This effect can especially be observed in the concentration distributions (Fig. 7).

The increased voidage near the wall leads to generally higher velocities in this region. As a result, the reactant concentration is rarely consumed in the outer region. Therefore, the reactant concentration in the wall region is still high at the end of the catalytic packing. This behaviour is commonly called "breakthrough". In adsorption processes, this effect can be crucial. Generally, in the case of chemical processes, local maxima of the product concentration or minima of the reactant indicate zones of high reactive activity. For exothermic reactions, this may result in hot spots and consequently lead to the deactivation of the catalyst or reduced selectivity.

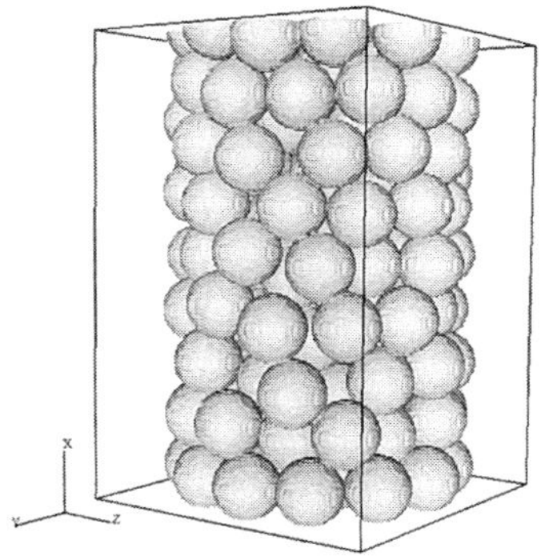

Fig. 4. Randomly packed tubular fixed bed reactor [10]

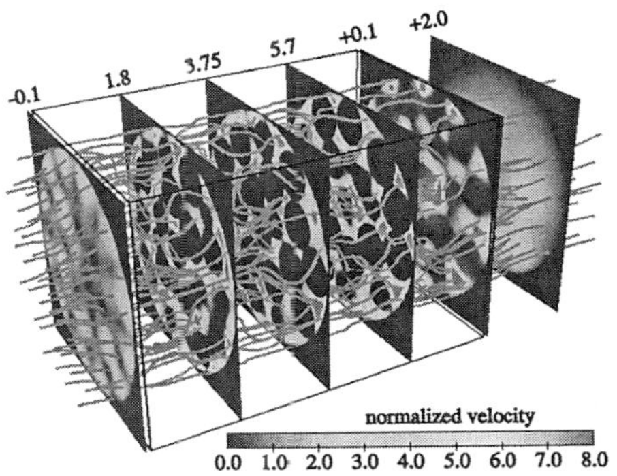

Fig. 5. Velocity contour plots in different cross-sections of the tubular reactor [10]

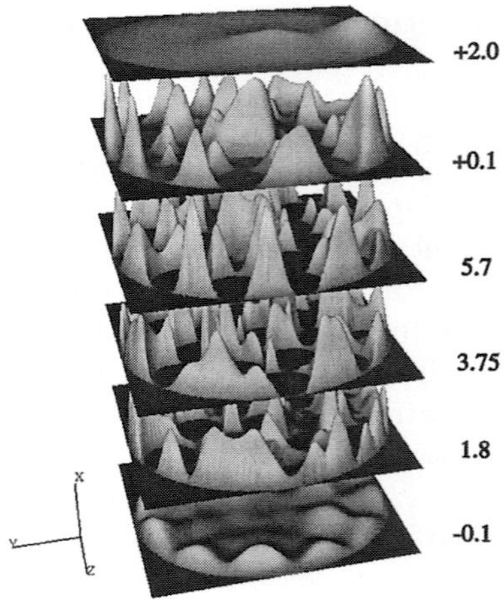

Fig. 6. Elevated surface (proportional to the velocity magnitude) coloured by the axial velocity [8]

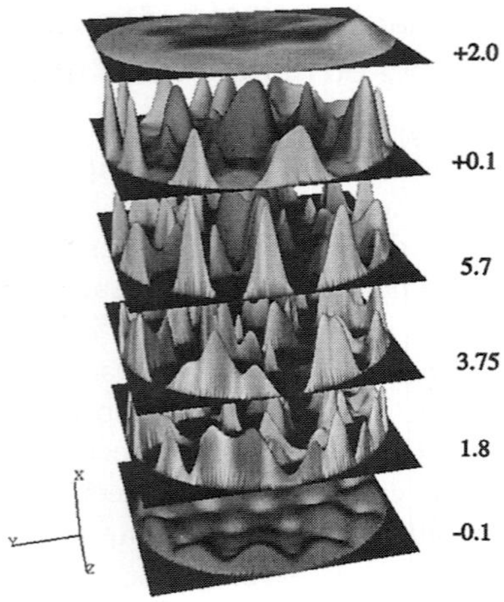

Fig. 7. Elevated surface (proportional to the velocity magnitude) coloured by the reactant concentration [8]

6 Vectorisation, Parallelisation, Performance and Resource Usage

Our lattice Boltzmann implementation is vectorized throughout all major parts. In typical production runs, about 2.55 GFlops are achieved per CPU on a NEC-SX5e (i.e. almost 65% of the peak performance) if only the flow field is calculated. In a combined calculation of the flow field, the mass transport and chemical surface reaction, still more than 2.0 GFlops (i.e. 50% of the peak performance) are typically sustained.

Due to the local character of the lattice Boltzmann algorithm, parallelisation is rather simple (as long as full matrices are used for storage) and, moreover, likely to be efficient. For the parallelisation, simple domain decomposition methods together with message passing are straightforward. But also shared memory parallelisation can give very good results. It is often claimed that the efficiency of shared memory parallelisation is quite low compared to explicit message passing. Figure 8 shows the results of extensive tests [12] with rather small test-cases (about 10% of typical production runs) on the NEC SX5e, where the number of processors was gradually increased keeping the total work (i.e. the total grid size) constant. Up to 15 processors, almost linear speedups and efficiencies higher than 90% are achieved. Less systematical tests with larger test-cases confirm this behaviour. The concurrent MFlops rate is about 15 GFlops on 6 processors of a NEC SX5e (calculation of the flow field only) and about 16 GFlops on 8 processors (calculation of both the flow field and the mass transport with chemical surface reactions) for production runs.

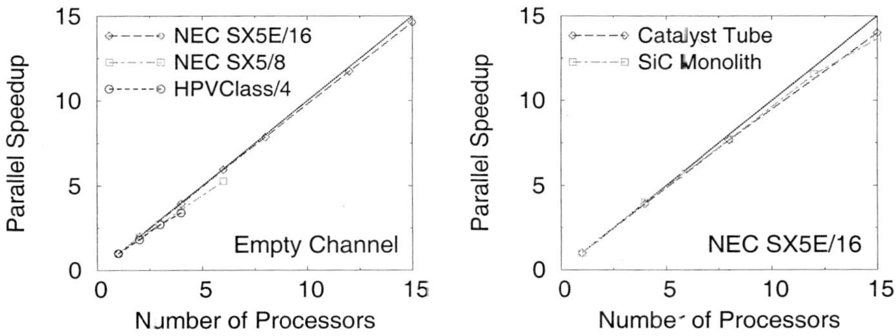

Fig. 8. Parallel speedup for shared memory parallelisation on the NEC SX5e with rather small test-cases and fixed total work in the case of simple (left) and complex (right) geometries [12]

7 Conclusions

In the first part, we pointed out that detailed 3-D simulations of flow, mass transport and reaction are necessary in certain cases as local inhomogeneities can dramatically reduce the performance and safety in operation of chemical reactors. One example are tubular fixed bed reactors with small tube-to-particle diameter ratios. Typical averaging approaches cannot successfully describe this type of reactor exactly.

The lattice Boltzmann method is an appropriate numerical method for the complex "flows" appearing in chemical engineering. Lattice Boltzmann simulations can predict quantitatively correct results in arbitrary complex geometric structures. The representation of the geometry on huge Cartesian grids and the lack of body-fitted grids (which are not of particular interest for the considered applications) are no serious problem as the computational requirements per lattice node are rather low and good performance can be obtained on high-performance computers.

Ongoing quantitative analysis of the data sets for fixed bed tubular reactors of low aspect ratios and comparisons with traditional methods will show the practical importance of these detailed simulations.

Acknowledgements

This work is supported by the German Research Foundation (DFG) under grant For262/2. The Monte Carlo simulations of the randomly packed bed are based on a program developed by Yong-Wang Li in the framework of an Alexander von Humboldt (AvH) foundation fellowship at the authors' institute. The calculations were mainly carried out on the machines of the

Leibniz Rechenzentrum in Munich and the High-Performance Computing-Center in Stuttgart, Germany.

References

1. R. Adler. Stand der Simulation von heterogen-gaskatalytischen Reaktionsabläufen in Festbettreaktoren - Teil 1. *Chem.-Ing.-Technik*, 72(6):555–564, 2000.
2. R. Adler. Stand der Simulation von heterogen-gaskatalytischen Reaktionsabläufen in Festbettreaktoren - Teil 2. *Chem.-Ing.-Technik*, 72(7):688–699, 2000.
3. J. Bernsdorf, G. Brenner, T. Zeiser, P. Lammers, and F. Durst. Numerical analysis of the pressure drop in porous media flow using the lattice Boltzmann computational technique. *???*, (in press). 1st International Conference on CFD (ICCFD), Kyoto, Japan, 2000.
4. O. Bey and G. Eigenberger. Fluid flow through catalyst filled tubes. *Chem. Engng. Sci.*, 57(8):1365–1376, 1997.
5. S. Chen and G. D. Doolen. Lattice Boltzmann method for fluid flows. *Annu. Rev. Fluid Mech.*, 30:329–364, 1998.
6. E. L. Cussler. *Diffusion: Mass Transfer in Fluid Systems.* Cambridge University Press, 2 edition, 1997.
7. E. G. Flekkøy. Lattice Bhatnagar-Gross-Krook models for miscible fluids. *Phys. Rev. E*, 47(6):4247–4257, 1993.
8. H. Freund, E. Klemm, G. Emig, T. Zeiser, G. Brenner, and F. Durst. Detailed 3D-simulations of single phase reacting flow in randomly packed beds with low aspect ratios. In *Proceedings of 3rd European Congress of Chemical Engineering, Nuremberg, 26-28 June 2001*, 2001. published on conference CD.
9. Y.-W. Li, T. Zeiser, P. Lammers, G. Brenner, E. Klemm, G. Emig, and F. Durst. Direct simulation of the structure and consequential flow field in a packed bed. *AIChE Journal*, submitted, 2000.
10. Y.-W. Li, T. Zeiser, P. Lammers, G. Brenner, E. Klemm, G. Emig, and F. Durst. Direct numerical simulations of the sphere structure in packed beds and the flow with chemical reactions. submitted to NATURE.
11. Y. H. Qian, D. d'Humières, and P. Lallemand. Lattice BGK models for Navier-Stokes equation. *Europhys. Lett.*, 17(6):479–484, Jan. 1992.
12. T. Zeiser, G. Brenner, P. Lammers, and J. Berndsorf. Performance aspects of lattice Boltzmann methods for application in chemical engineering. In C. Jenssen, T. Kvamdal, H. Andersson, B. Pettersen, A. Ecer, J. Periaux, N. Satofuka, and P. Fox, editors, *Parallel Computational Fluid Dynamics 2000, Proceedings of the Parallel CFD 2000 Conference, May 22-25, Trondheim, Norway*, pages 407–414. Elsevier, 2001.

Computational Engineering for Wind-Exposed Thin-Walled Structures

A. Halfmann[1], E. Rank[1], M. Glück[2], M. Breuer[2], F. Durst[2], J. Bellmann[3], and C. Katz[3]

[1] Lehrstuhl für Bauinformatik, Technische Universität München, 80290 München, Germany
[2] Lehrstuhl für Strömungsmechanik, Universität Erlangen-Nürnberg, 91058 Erlangen, Germany
[3] SOFiSTiK AG, 81541 München, Germany

Abstract. In this paper a computer-aided simulation approach for fluid-structure interaction of wind-exposed structures is presented. In the center of our software architecture is the geometric model of the structure, from which a finite element mesh for the structure and a finite volume mesh for the fluid are derived. The attention is concentrated to thin-walled structures like membranes and thin shells composed of light and flexible materials. The interaction of fluid and structure is effected by wind-induced vibrations and causes large elastic deformations of the structure. A powerful simulation tool is provided by coupling two codes developed for flow simulation and structural dynamics by a fully implicit coupling algorithm. Results of three-dimensional simulations of fluid-structure interaction for several test cases as well as for a real-life example will be presented.

1 Introduction

During the last few years much progress has been achieved to integrate analysis and design in civil engineering [9]. The computer-aided design process starts by setting up a geometric model being used for the definition of all structural properties as well as for the loads on the construction. From this geometric model a computational model is derived as the basis for the following dimensioning of the construction. This computer-aided engineering process has to support efficiently the primary goals of any civil engineering design, i.e. the development of constructions satisfying the requirements of safety during their whole lifetime, resisting all expected influences they might be exposed to. Considering for example light textile constructions like the convertible shade roof for a large court of the Prophet's Mosque in Madinah consisting of several

Fig. 1. Convertible shade roof [10]

large umbrellas (Fig. 1 [10]), the most demanding engineering task is to dimension the construction so that it will resist wind loads. Following the Davenport-Wind-Load-Chain an assumed local wind profile for a given terrain can be established according to the European Design Code EC 1 [5]. This wind distribution causes a pressure field on the structure. If simplified, conventional load assumptions for this pressure distribution are made, the structures may be strongly over-designed. Yet, also an under-design with the danger of a collapse may result from this simplified procedure, as the mutual influence of neighbouring constructions could change the wind pressure field dramatically. Therefore, todays standard of practice are time-consuming and costly experiments in wind tunnels. A model of the construction is considered and wind pressure fields on its surface are measured to be used for later numerical structural analysis. One of the major drawbacks of this mixed computational and experimental approach is the difficulty to quickly change the geometry of the model construction for the wind tunnel experiment. Therefore, it would be highly desirable to have a 'numerical wind tunnel' available, being directly integrated in the civil engineering design and analysis process, and enabling to investigate a structure also under the influence of neighbouring constructions. Several steps in this direction have been performed by the authors in a joint research project during the past two years and will be presented in this contribution.

2 Software Architecture

The software architecture shown in Fig. 2 is influenced by the idea of a loose coupling strategy based on highly specialized and well evaluated simulation codes, each developed for the special field of the interacting system. In a first step, the structural model is defined, using a classical CAD environment on a PC or a workstation. To consider different stress distributions in a membrane, it is possible to compute the geometric shape of pure membrane structures in a formfinding process [2]. Thereby the deformed shape of the structure under dead load is computed by a finite element analysis starting from an inital geometry. Using special membrane elements and assuming an initial stiffness caused by isotropic prestressing and neglecting any bending stiffness, the deformed shape will lead to the soap film form with its constant isotropic stress distribution. All geometric information is stored in a database describing a b-rep (boundary repre-

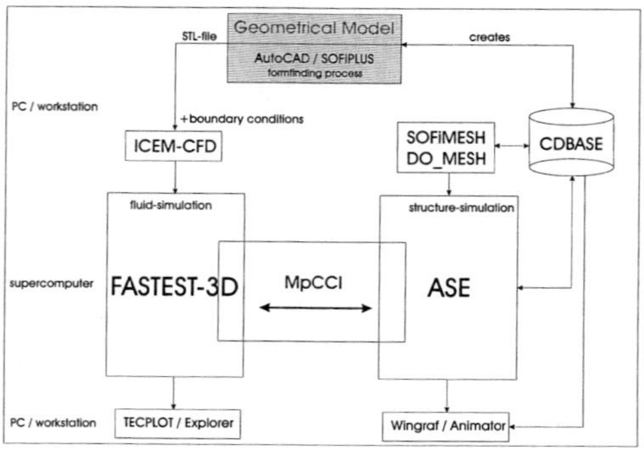

Fig. 2. Software structure

sentation) model completed by information concerning material properties and boundary conditions [9]. For the structural part of the analysis system this project database represents a central module. All programs used for the structural simulation have access to the information stored in the database. The geometric model for the fluid dynamic part is also derived from the CAD model completed by boundary conditions and discretized as a block-structured three-dimensional grid being adequate to the finite volume technique. Due to extremely high computational requirements, the numerical simulation of the dynamic fluid-structure interaction is performed on a high-performance computer [8] in the next step. Thereby the data transfer between the two simulation codes, each running on different nodes/processors, is performed via a common geometric model using a suitable coupling interface [1]. Finally, the results can be evaluated and visualized by powerful postprocessors on a workstation or on a PC.

3 Coupling Algorithm

The fluid-structure interaction is described by the structural deformations as response to wind forces, resulting in a modification of the fluid flow domain. The coupling is performed by a partitioned solution approach [3]. The Computational Fluid Dynamics (CFD) simulations are performed by a finite volume based code [4] solving the Reynolds-Averaged-Navier-Stokes (RANS) equations using a two-equation k-ϵ model. It is adapted to moving grids by an Arbitrary Lagrangian Eulerian (ALE) formulation. For the Computational Structure Dynamics (CSD) the dynamic non-linear structural response (large displacements/small deformations) is described by the equations of motion based on a finite element approach and an implicit time-stepping procedure [11]. For more details concerning the simulation codes we refer to [6]. The time-dependent simulation process is controlled by the coupling algorithm shown in Fig. 3. The solution is based on an iteration procedure between the CFD and CSD simulation until convergence is reached within each time-step. Thereby the nodal loads as input for the CSD simulation are computed from the results of the CFD simulation (pressure and wall shear stresses). The updated boundary geometry is based on the structural displacements as a result of the CSD simulation. In cases of large structural deformations an

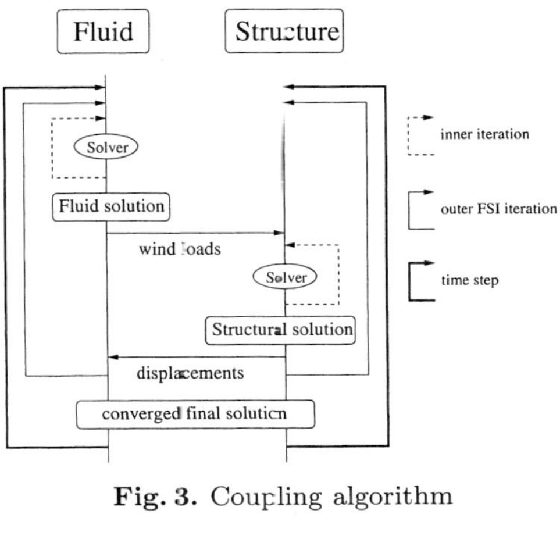

Fig. 3. Coupling algorithm

under-relaxation of the update of the boundary geometry was found to improve the convergence significantly. A reduction of the number of iterations within each time-step can be obtained by a predictor-corrector scheme [6]. Informations concerning the data transfer between the different numerical grids can be found in [7].

4 Numerical Examples

The coupling procedure presented in the previous section was applied to several test cases. In the following the results of two elementary systems will be discussed with respect to different parameters for spatial and temporal discretization. Based on this, the coupled application was used to simulate the fluid-structure interaction of a complex membrane structure.

4.1 Vertical Plate in a Resting Fluid

The first example describes a temporarily loaded rectangular plate clamped at the lower boundary in a closed cavity of a resting fluid. The dimensions of the flexible plate pictured in Fig. 4 are length/width/thickness = 1.0/0.4/0.06 m. For the material properties a polyester with a modulus of elasticity of $E = 2.5$ MPa, a Poisson's ratio of $\nu = 0.35$ and a density of $\rho_S = 2550$ kg/m^3 is assumed. The density and dynamic viscosity of the fluid are $\rho_F = 1$ kg/m^3 and $\mu_F = 0.2$ Pa. Using these parameters, a laminar flow is expected. During the first five time-steps a constant load is applied in x-direction. After removing the load the plate executes oscillations induced by the initial deflection damped by the surrounding fluid flow.

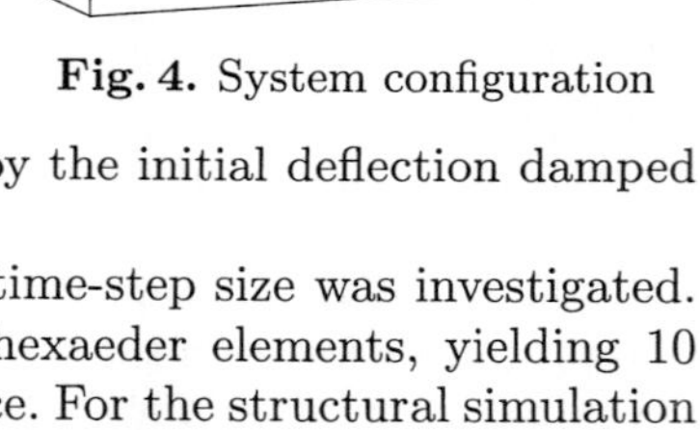

Fig. 4. System configuration

In a first simulation the influence of the time-step size was investigated. The fluid domain was discretized by 1650 hexaeder elements, yielding 10 quadrilateral elements to describe the interface. For the structural simulation a surface mesh of 8 elements was chosen. Both surface discretizations are shown in Fig. 5. Due to the boundary conditions for the fluid simulation the results of the coupled three-dimensional simulation shows a quasi two-dimensional behavior. Fig. 6 points out the computed displacement of the upper boundary of the plate for simulations with three different time-step sizes. The amplitude of the plate oscillation is damped by the ambient fluid. For the different time-step sizes only small deviations could be identified. Therefore, $\Delta t_1 = 0.025$ s was chosen for the following simulations.

The dependence of the displacement on different spatial discretizations is depicted in Fig. 8. Starting again from a discretization of 1650 hexaeder elements for the fluid domain (10 quadrilaterals on the interface, Fig. 7) a

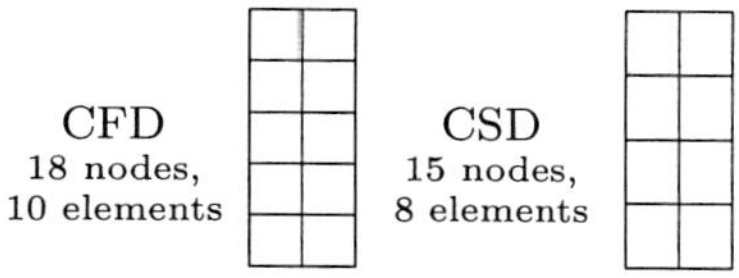

Fig. 5. Discretization of the interface

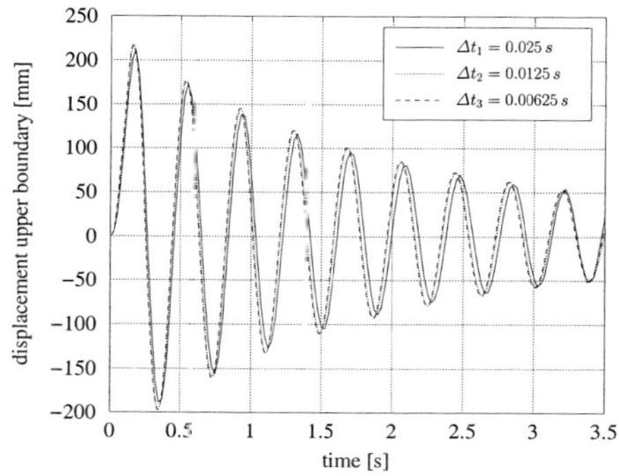

Fig. 6. Displacement for different Δt

uniform refinement was performed in two levels which leads to 13,200 and 105,600 hexaeder elements and 40 or 160 quadrilaterals for the interface, respectively. The coarsest grid for the CSD simulation consisting of only 2 elements is also shown in Fig. 7. Uniform refinement yields surface meshes of 8 or 32 quadrilateral elements for the structure simulation, respectively.

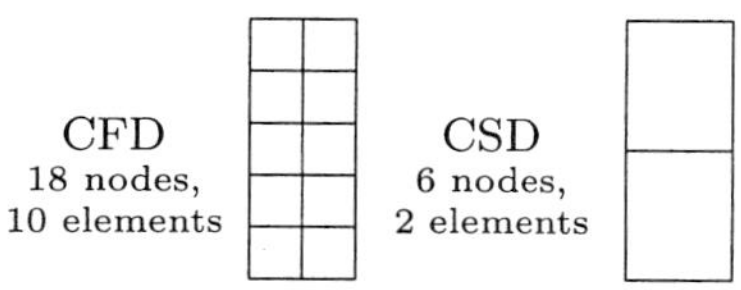

Fig. 7. Initial interface discretization

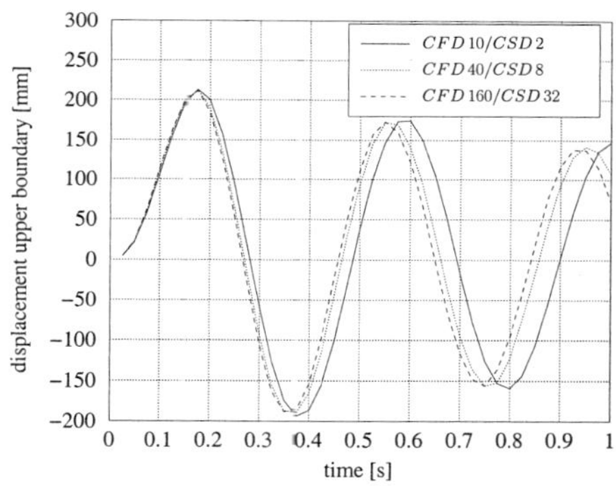

Fig. 8. Displacement for different interface discretizations

Fig. 8 points out the insufficient accuracy of the results corresponding to the initial interface discretizations and the good agreement of the transient displacements achieved with the refined meshes. In particular the CSD mesh consisting only of 2 elements for the y-direction is too coarse to obtain useful results.

4.2 Vertical Plate under a Sudden Wind Gust

The geometry and material properties of the second test case are the same as before. Fig. 9 shows a cutting plane through the fluid domain which is open with an inlet (left boundary), an outlet (right boundary), and symmetric boundary conditions at the top as well as in z-direction. At the beginning of the coupled simulation the fluid suddenly accelerates and adopts immediately a constant inflow velocity of $U_\infty = 10$ m/s being equivalent to a Reynolds number of $Re_L = 50$. As a consequence the structural oscillations are induced by the saltus of the fluid velocity and reach a stationary deformation state after a certain time. Fig. 10 shows again the displacement of the upper boundary of the plate for various time-step sizes. The interface discretizations comply to the grids pictured in Fig. 5. Only minor distinctions could be identified for the different time discretizations. Furthermore, the influence of the spatial interface discretization was investigated using the grids shown in Fig.11 and performing an uniform refinement of two levels for the CSD grid whereas the mesh for the fluid domain remains unchanged. It is again obvious that the bending oscillations of the plate cannot be represented sufficiently accurate by a CSD mesh with only two elements in y-direction, see Fig. 12.

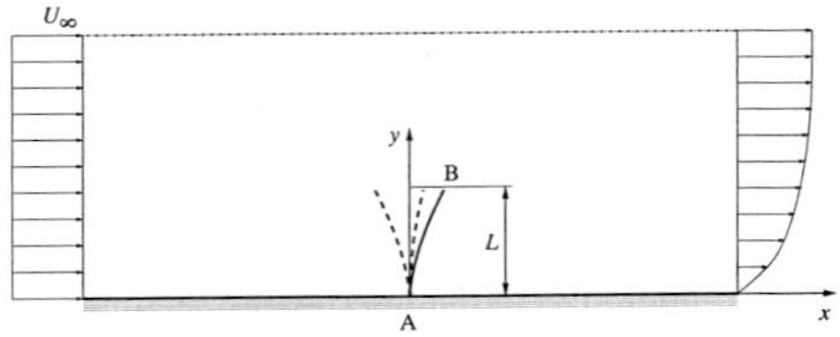

Fig. 9. System configuration

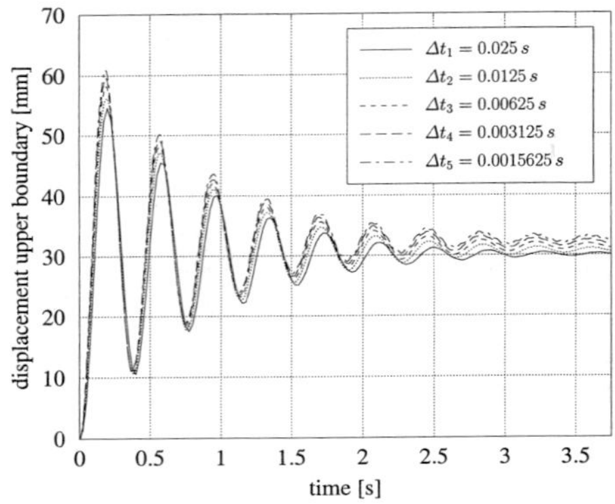

Fig. 10. Displacement for different Δt

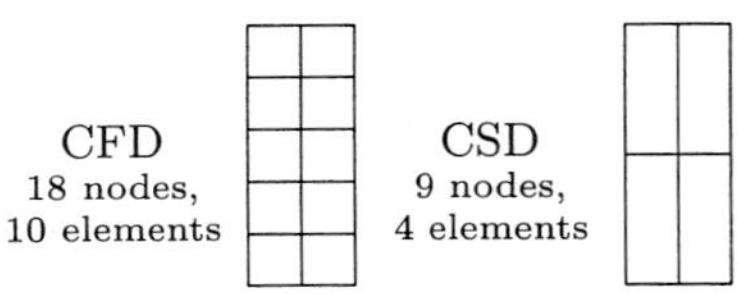

Fig. 11. Initial interface discretization

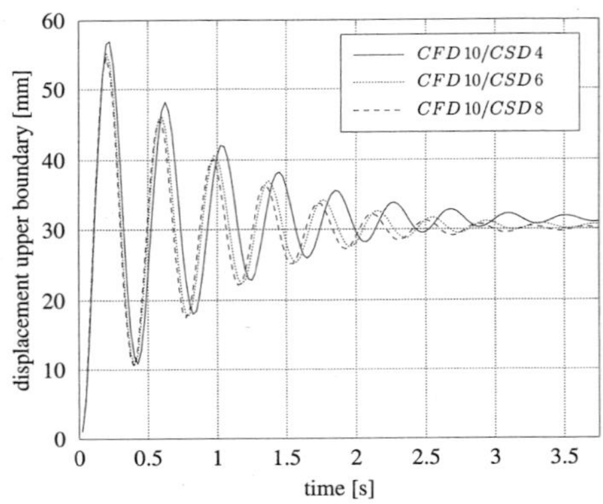

Fig. 12. Displacement for different CSD interface discretizations

4.3 Tent Roof

Fig. 13 shows a screenshot of the CAD-environment representing the geometric model of a real-life structure as a starting point for the simulation.

The shape of the structure is a result of the formfinding process sketched in Section 2 and specifies a textile roof being positioned in front of an office building. A material of glass-fibre synthetics with a thickness of 1.5 mm, a modulus of elasticity of $E = 3000$ MPa, and a shear modulus of $G = 1500$ MPa is assumed. The structure is 24 m long and shows a width between 3 m and 8.5 m. The boundary of the block-structured three-dimensional CFD grid on the interface and the CSD surface grid are shown in Fig. 14 and 15. Due to the fact that the k-ϵ model produces non-physical results for such complex geometries the CFD simulations were performed neglecting turbulence in this first step. The velocity distribution of the assumed wind gust and the displacement as structural response are pictured for four different time-steps in Fig. 16. The gust leads to a maximal displacement in z-direction of $z_{max} = 31$ cm.

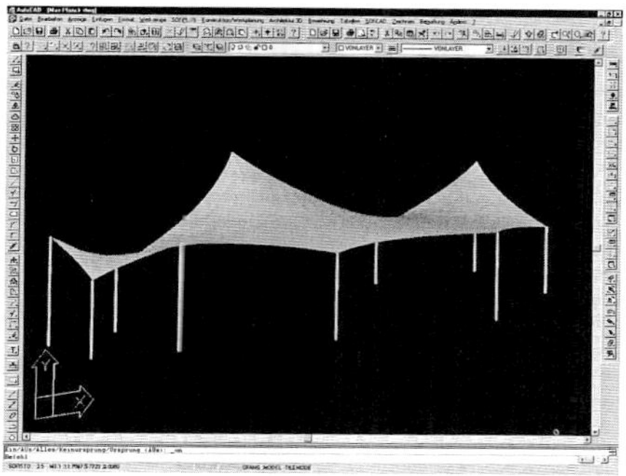

Fig. 13. Geometry of the tent roof

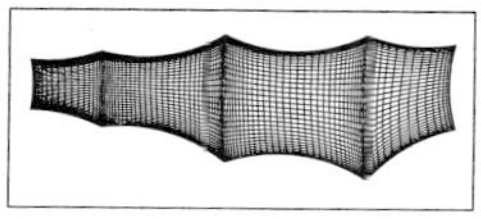

3536 nodes,
3400 elements

Fig. 14. CFD interface discretization

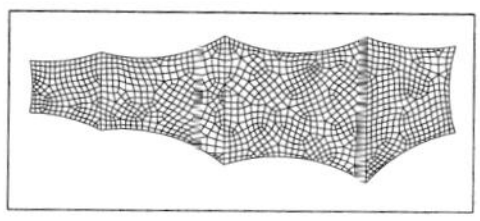

1409 nodes,
1311 elements

Fig. 15. CSD interface discretization

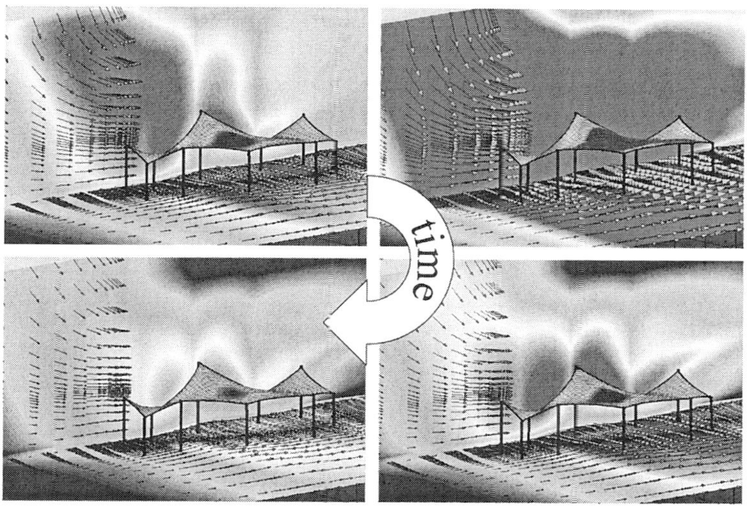

Fig. 16. Velocity distributions and diplacements of four different time-steps ($t_1 = 2.0$ s, $t_2 = 3.0$ s, $t_3 = 4.36$ s and $t_4 = 4.46$ s)

5 Conclusions

The partitioned solution approach has been investigated for the simulation of fluid-structure interaction of light civil engineering constructions under wind load. Before the outlined concept could be a practically used procedure, it will be necessary to incorporate more elaborate turbulence modeling such as large-eddy simulation, perform more numerical tests and compare the numerical results with verified experimental data. If thus sufficient confidence is gained in this simulation technique, it can be an application of high-performance computing with significant impact on engineering practice.

Acknowledgements

The work was financed by the Bayerische Forschungsstiftung in the Bavarian Consortium of High-Performance Scientific Computing, FORTWIHR (III). The simulations were partially carried out on the Hitachi SR8000-F1 at the LRZ München [8]. This support is gratefully acknowledged.

References

1. Ahrem, R., Hackenberg, M.G., Post, P., Redler, R., Roggenbuck, J. (2000) MpCCI – Mesh Based Parallel Code Coupling Interface. Institute for Algorithms and Scientific Computing (SCAI), GMD, http://www.mpcci.org/
2. Bellmann, J. (1998) Membrantragwerke und Seifenhaut – Unterschiede in der Formfindung. Bauingenieur, **3/98**, 118–123
3. Cebral, J.R. (1996) Loose Coupling Algorithms for Fluid-Structure Interaction. Ph.D.-Thesis, George Mason University Fairfax, Virginia
4. Durst, F., Schäfer, M. (1996) A Parallel Block–Structured Multigrid Method for the Prediction of Incompressible Flows. Int. J. Num. Methods Fluids **22**, 549–565
5. EUROCODE 1 (1995) Basis of Design and Actions on Structures; Part 2-4: Actions on Structures - Wind Actions. European Committee for Standardization Ref. No. ENV 1991-2-4
6. Glück, M., Breuer, M., Durst, F., Halfmann, A., Rank, E. (2000) Computation of Fluid-Structure Interaction on Lightweight Structures. Proceedings of Fourth International Colloquium on Bluff Body Aerodynamics & Applicastions, Bochum, To appear in J. of Wind Engineering and Industrial Aerodynamics
7. Halfmann, A., Rank, E., Glück, M., Breuer, M., Durst, F. (2000) A Partitioned Solution Approach for the Fluid-Structure Interaction of Wind and Thin-Walled Structures. Proceedings of IKM 2000, Weimar
8. http://www.lrz-muenchen.de/services/compute/hlrb/.
9. Rank, E., Halfmann, A., Rücker, M., Katz, C., Gebhard, S. (2000) Integrierte Modellierungs- und Berechnungssoftware für den konstruktiven Ingenieurbau: Systemarchitektur und Netzgenerierung. Bauingenieur, Februar 2000, 60–66
10. Sonderkonstruktionen und Leichtbau GmbH (1993) The Work of SL. Leinfelden-Oberaichen
11. SOFiSTiK AG (2000) ASE–Handbuch. München

Numerical Simulation of Wind Loads on Antenna Structures

R. Sieber, P. Droll, and M. Schäfer

Department of Numerical Methods in Mechanical Engineering
Technical University Darmstadt
Petersenstr. 30
64347 Darmstadt, Germany

Abstract. This contribution considers the modeling and simulation of the wind loads applied to a 35m deep space antenna. The performance of such large structures can be heavily affected by deformations due to wind effects. The aim of the investigation is the prediction of the deformation of the antenna structure caused by wind loads. The presented work constitutes an example for the use of numerical simulation techniques to complex practical multi–physics problems by suitably coupling the numerical methodologies available for the fluid mechanics and structural mechanics subtasks.

1 Introduction

For telemetry and telecommand operations of deep space and high elliptical orbit missions, in particular for Rosetta mission, the European Space Agency (ESA) is procuring a Deep Space S/X Band Antenna (DSA), presently under construction at the New Norcia site, nearby Perth. The antenna diameter for fulfilling the project requirements is 35 meters. The antenna, a dual reflector Cassegrain system, is of elevation over azimuth type, symmetric, with joke support by two elevation bearings. The mount is placed, with a large scale azimuth bearing, on top of a concrete tower. The baseline antenna radio frequency (RF) subsystem is including S/X–band receiving and transmitting capabilities. The antenna mechanics and structure is already designed and prepared for the future optional implementation of Ka–band receiving capabilities.

The pointing requirements of large reflector antennas operating at X/Ka–band frequencies are very demanding. This is due to the small beam width at these frequencies and the requirement to operate with full performance under environmental (wind load) conditions. By 'pointing requirements' is meant the accuracy with which the antenna RF–axis is aligned to the commanded, instantaneous pointing vector to the spacecraft. The examined variable is here the 'pointing error', which should not rise above two millidegree. The surface accuracy of the various reflectors is also a very important parameter, influencing both the phase error and the optical aberrations of the antenna aperture, which may result if not properly controlled in a dramatic deterioration of the antenna performance. By surface accuracy is meant e.g. for the main reflector the deviation of the real surface from the ideally shaped parabolic surface optimized upon RF requirements. Here the RMS–value is the determining value, which shouldn't be higher than 2 millimeters.

All kind of forces acting on the antenna have an effect on the pointing and on the surface accuracy. Therefore, the wind loads under which the antenna is operated play a very decisive role. The objective of the paper is to investigate the capabilities of computational fluid dynamics (CFD) to predict the wind load and the resulting RMS–value and pointing error with sufficient accuracy. Till nowadays it is common practice in antenna design, to use experimental data of simplified, scaled wind tunnel models to investigate effects resulting from aerodynamic flow around the antenna. Often it is not possible to ensure flow conditions in the experiments, which accurately reflect the real conditions like e.g. the Reynolds number. In addition, it is quite difficult and expensive to reproduce the geographical topology of the site of an antenna in the wind tunnel. Due to recent advances in CFD, numerical simulation can help to overcome these problems. However, due to the very complex geometrical structure of an antenna and the turbulent character of the flow, the numerical simulations are very challenging.

Once the pressure distribution is known by fluid analysis or experimental data, the deformation of the antenna can be obtained by proceeding structural analysis, i.e. by solving Cauchy equations, for the full antenna structure with the pressure loads as boundary conditions. From the deformations finally the pointing and the surface accuracies can be determined. Figure 1 illustrates the presented scheme. The discussion of the results also includes comparisons between numerical and experimental findings with regard to turbulence modeling and the applied numerical scheme.

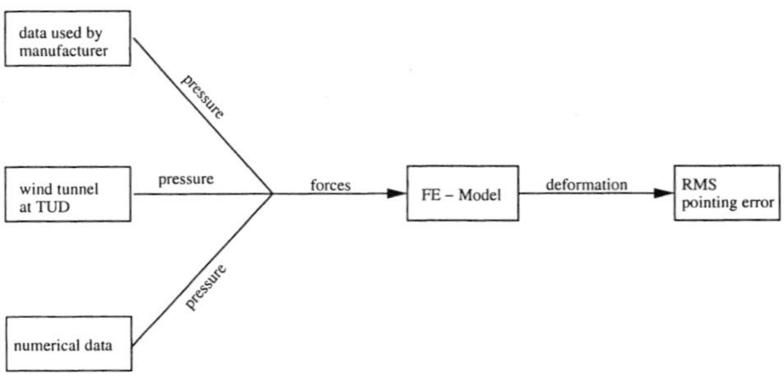

Fig. 1. Flowchart of the scheme computing RMS–value and pointing error

2 Numerical Methods

The structural and the flow calculations are done using different numerical methods. A single commercial code allowing the efficient calculation of structural mechanical and fluid mechanical problems is not available. The requirements for covering the different physical effects lead to two numerical models: one numerical model with the related assumptions for the structural analysis, and one for the fluid analysis.

2.1 Structural Analysis

The multi-purpose finite-element program ANSYS [1] is constituted to calculate the displacements and deviations of the antenna structure resulting from the various wind loads. Due to linear material properties and expected small displacements, all problems lead to a linear system of equations, which is solved by a preconditioned CG solver.

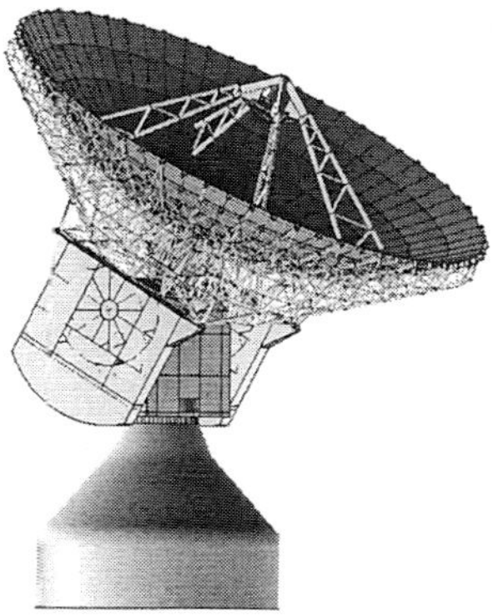

(a) Front side of DSA

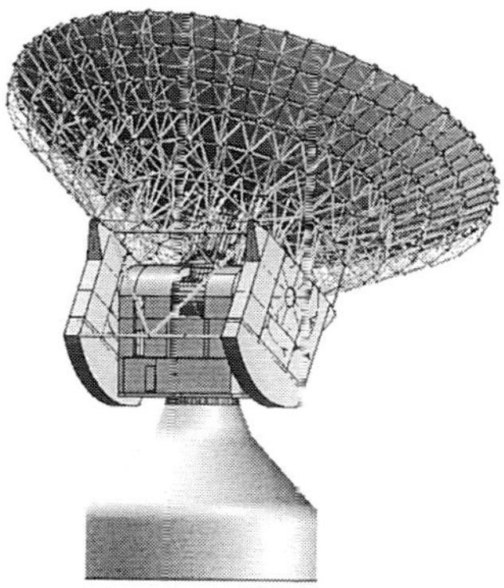

(b) Back side of DSA

Fig. 2. Structural FE–model with approx. 22 000 elements

The antenna is made out of steel. The tubes of the main and the sub reflector supporting structure are modeled by beam elements. The modeling of joints and T–branches with beam elements overpredicts the moment of inertia. To compensate for this, the inertial moment has been reduced to 20% of its theoretical value. This value was found by comparing T–branches of tubes with a typical length modeled with beam and shell elements. The panels of the main reflector are parasitic elements used to apply the pressure distribution resulting from the fluid analysis in an easy way. The rest of the antenna is built-up by shell and brick elements. The boundary conditions for the displacements are defined according to the function of the drive system, bearings and foundation of the antenna.

2.2 Fluid Analysis

The basic equations for this flow analysis are the Reynolds averaged Navier – Stokes equations (RANS) for incompressible fluids. The modeling of the

Reynolds stress tensor in this equations is done with an eddy–viscosity approach. For solving the closure problem various two–equation turbulence models are applied: the standard k–ϵ turbulence model [4], the k–ϵ model with Kato–Launder extension [3] and the RNG–model [10].

The applied CFD–code is the program FASTEST of Invent Computing GmbH [2]. The spatial discretisation of the above mentioned equations is done by means of a fully conservative second–order finite–volume method with colocated arrangement of the variables [7]. The convective terms are approximated once with a first–order upwind scheme (UDS), once with a second–order central differencing scheme (CDS) and with 2 hybrid schemes using 50% UDS and 50% CDS as well as 20% UDS and 80% CDS. The applied numerical grids are non–orthogonal boundary–fitted block–structured grids. They are varying in the range of 2×10^5 and 3×10^5 control volumes. For linearization and coupling of the equations an iterative pressure correction scheme based on the SIMPLE algorithm [6] is used, where a selective interpolation technique for avoiding oscillatory effects is applied. The arising linear systems of equations are solved with the strongly implicit iterative ILU solver [8] for all variables.

Fig. 3. Experimental antenna model

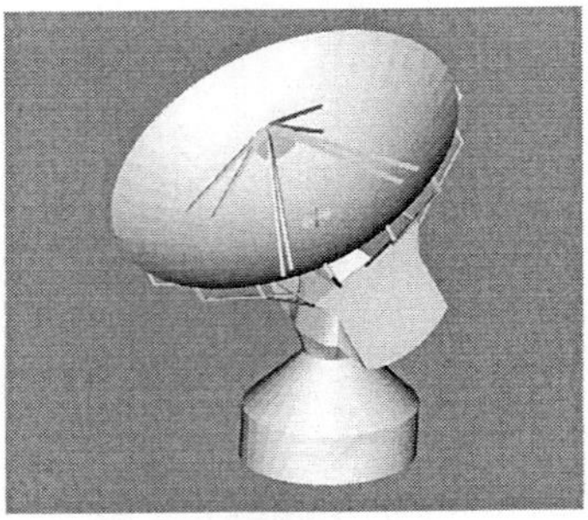

Fig. 4. Numerical antenna model

The verification of the CFD calculations is based on a comparison of experimental data ascertained in the wind tunnel of the Department of Fluid Mechanics and Aerodynamics of the TU Darmstadt on a simplified antenna model [9], and numerical data computed with the same simplified numerical model. Figures 3 and 4 display the wind tunnel model and the corresponding numerical model. The applied Reynolds number is Re=1.45×10^5 and consequently the flow is highly turbulent.

The resulting experimental and numerical pressure distributions, as well as the pressure distribution used by the manufacturer are applied on the FEM–model as it is shown in Fig. 1. The structural analysis with different pressure loads on the main reflector are performed and the RMS – values and pointing errors are computed.

3 Results

For the comparison three different load cases are investigated: case one with an elevation angle of 0° of the main reflector and front wind, case two with elevation 60° and front wind and case three with elevation 90° and side wind.

The presentation of results is divided into three parts relating to the investigations of the sensitivity with regard to turbulence modeling, the applied discretisation scheme and the spatial discretisation.

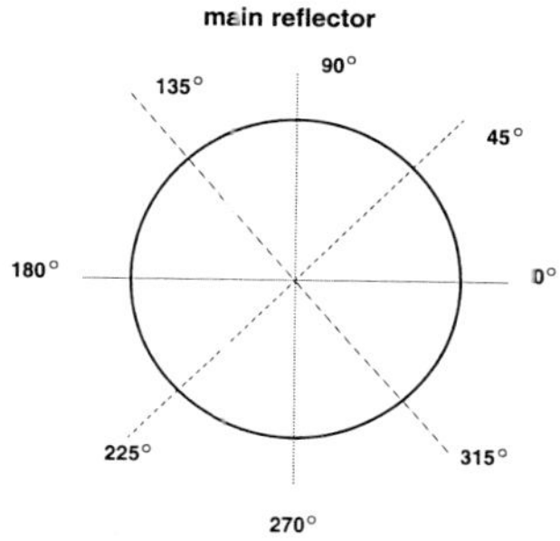

Fig. 5. Extracted profiles on the main reflector

The considered values are the above mentioned RMS–value and pointing error as integral values and several cutting lines through the diameter of the main reflector, where the dimensionless pressure difference Δc_p will be plotted as local value. Figure 5 indicates the location of different extracted profiles presented in this section.

3.1 Turbulence Modeling

Figures 6 and 7 show two exemplary ΔC_p profiles for case one. The applied turbulence models are the standard k–ϵ model with wall functions (KE w. WF) and without wall functions (KE wo. WF) as well as the RNG model (RNG).

In this case it was not possible to achieve a stable solution with the k–ϵ model with Kato–Launder extension (KL). The comparison for the different appropriate models with regard to Δc_p indicates that the RNG model represented by the solid line matches best the experimental data pictured as black dots. Especially on the edge of the main reflector the discrepancies between the turbulence models are relatively big. The RNG model catches most suitably the turbulent effects at the boundary. This results also in a difference in RMS and pointing error between the RNG model and the remaining turbulence models as given in Table 1. Indicated are the RMS–values in millimeter as well as the pointing errors around and perpendicular to the elevation axis (REL and RXEL) in millidegree for the particular turbulence model and the corresponding result of the experimental pressure distribution.

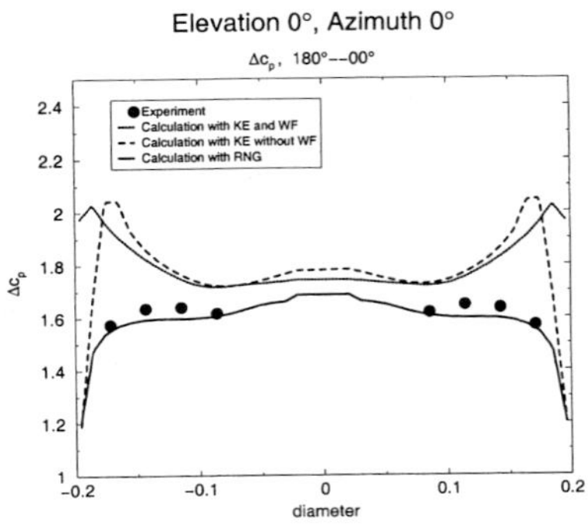

Fig. 6. Δc_p along horizontal line for different turbulence models

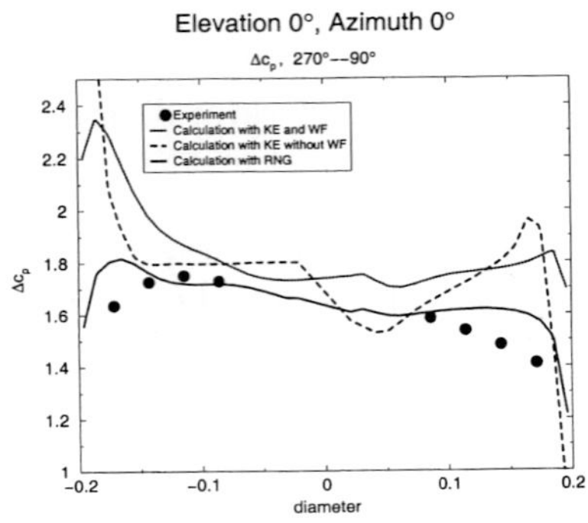

Fig. 7. Δc_p along vertical line for different turbulence models

Also the other cases show a big discrepancy among the different models. However, the RNG models provides as a whole the best results.

Table 1. Integral values for the applied turbulence models and experimental value

load case	value	KE w. WF	KE wo. WF	RNG	KL	exp.
0° elev.	RMS [mm]	0.058	0.059	0.056	–	0.035
	REL [mdeg]	-0.186	-0.034	-0.125	–	-0.013
	RXEL [mdeg]	-0.178	-0.174	-0.157	–	-0.140
60° elev.	RMS [mm]	0.164	–	0.473	0.194	0.309
	REL [mdeg]	-0.044	–	-0.343	-0.089	-0.260
	RXEL [mdeg]	-0.044	–	-0.024	-0.045	-0.060
90° elev.	RMS [mm]	0.265	0.277	0.253	0.366	0.156
	REL [mdeg]	0.020	0.031	0.013	0.017	0.010
	RXEL [mdeg]	-0.133	-0.107	-0.134	-0.093	-0.210

3.2 Discretisation Scheme

After determination of a suitable turbulence model the influence of the discretisation scheme will be investigated. The applied schemes are pure upwind (UDS), a hybrid scheme with 50% central differences (0.5 CDS), a hybrid scheme with 80% central differences (0.8 CDS) and a pure central differencing scheme (CDS). The variations of schemes is done for all solved equations. Exemplary results can be seen in Figs. 8 and 9 displaying Δc_p–profiles for the case with elevation 90°. It can be observed that the particular Δc_p profiles of the different numerical schemes do not differ significantly.

The integral values in Table 2 show that the RMS–values varying in a range of 10% and the pointing errors in a range of 50% among the different discretisation schemes. The percentage numbers are oriented on the maximum appearing value of RMS and pointing. However,with a rising quota of

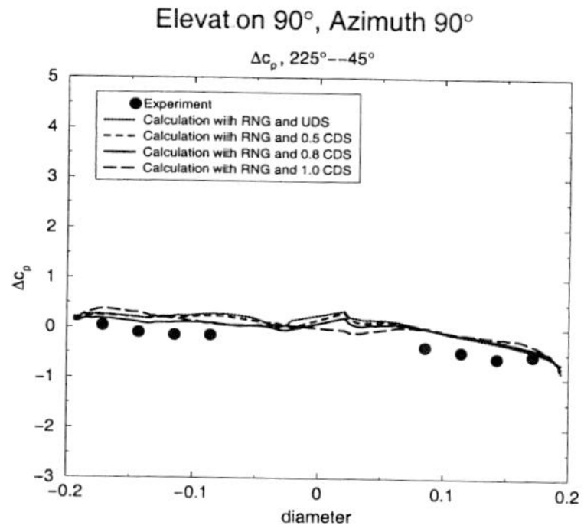

Fig. 8. Δc_p along 225°-45° diagonal line for different numerical schemes

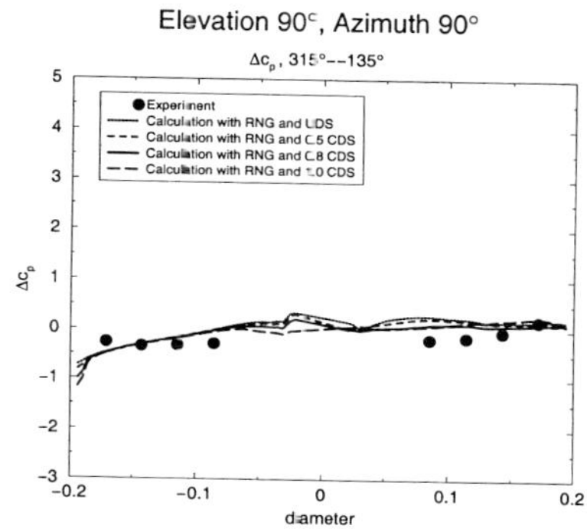

Fig. 9. Δc_p along 315°-135° diagonal line for different numerical schemes

central differencing a rising instability of the numerical scheme was identifiable. This behaviour results in divergence of the 60° case for 0.8 CDS and pure CDS. The flow around the antenna with this Reynolds number produces an unsteady vortex shedding. The numerical diffusion of the UDS scheme is damping the translative effects and makes it easier to find a steady state solution. However, the CDS–scheme predicts the vortex shedding better so that it is in some cases not possible to find a steady solution, what results in divergence of the computations.

Table 2. Integral values for the applied numerical schemes and experimental values

load case	value	UDS	0.5 CDS	0.8 CDS	CDS	exp.
60° elev.	RMS [mm]	0.490	0.543	–	–	0.309
	REL [mdeg]	-0.207	-0.320	–	–	-0.260
	RXEL [mdeg]	-0.031	-0.037	–	–	-0.060
90° elev.	RMS [mm]	0.253	0.251	0.253	0.230	0.156
	REL [mdeg]	0.013	0.022	0.026	0.043	0.010
	RXEL [mdeg]	-0.133	-0.150	-0.165	-0.184	-0.210

3.3 Grid Dependency

Because of convergence problems with the central differencing schemes it is needful to check the sensitivity of the results to a refinement of the numerical grid. For this investigation a two times in every logical direction refined grid with regard to the above mentioned original grid is computed.

The comparison of the Δc_p–profiles for the 60° case shows no significant dependency of the results on the spatial discretisation.

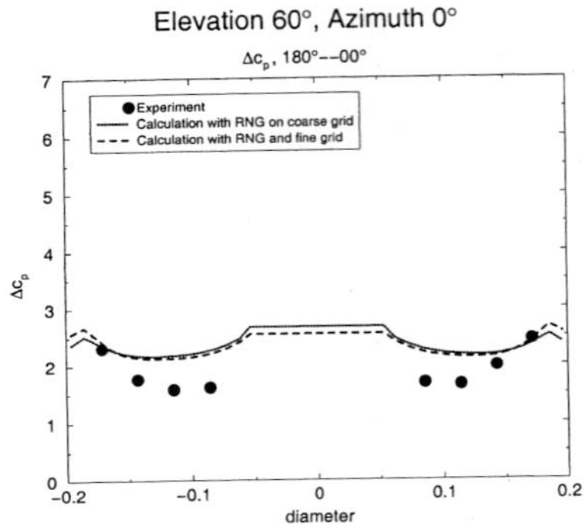

Fig. 10. Δc_p along horizontal line for different spatial discretisations

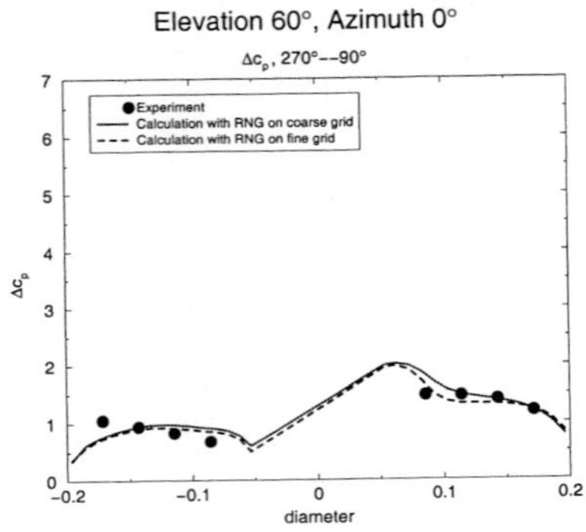

Fig. 11. Δc_p along vertical line for different spatial discretisations

Beside the RMS–values and the pointing errors Table 3 contains also the number of control volumes (cv) for the actual grid and the corresponding approximate computing time. The calculations with the original grid were done on a Pentium III/600 processor, whereas the refined grid was computed on a Power 2 processor. The performances of both chip sets are similar and consequently comparable. It is visible in Table 3, that the results between the original grid and the refined grid differ in a range of 10%, however the computing time rises tremendously.

Table 3. Integral values for the different spatial discretisation

load case	value	original grid	refined grid
60° elev.	number of cv	287491	2303528
	computing time	≈ 1d	≈ 12d
	RMS [mm]	0.473	0.498
	REL [mdeg]	-0.343	-0.314
	RXEL [mdeg]	-0.024	-0.025
90° elev.	number of cv	215381	1723048
	computing time	≈ 1d	≈ 10d
	RMS [mm]	0.251	0.243
	REL [mdeg]	0.022	0.020
	RXEL [mdeg]	-0.150	-0.181

This leads to the conclusion, that the improvement of accuracy on the finer grid is rather small compared to the additional computational effort.

3.4 Comparison

After having examined the sensitivity of the results with regard to different parameters we finally compare our results with the data used by the manufacturer. Since the computation of the integral values for all datasets rests upon the same FE–model (Fig. 1, the discrepancies among the results are only due to differences among the applied pressure distributions.

Table 4. Integral values for numerical, experimental and manufacturer data

load case	value	numerics	experiments	others
0° elev.	RMS [mm]	0.056	0.035	0.069 (lit.)
	REL [mdeg]	-0.190	-0.013	-0.340 (lit.)
	RXEL [mdeg]	0.180	0.140	0.230 (lit.)
60° elev.	RMS [mm]	0.473	0.309	0.370 (manu.), 0.570 (lit.)
	REL [mdeg]	-0.340	-0.260	-0.720 (manu.), -0.020 (lit.)
	RXEL [mdeg]	-0.020	-0.060	-0.020 (manu.), 0.000 (lit.)
90° elev.	RMS [mm]	0.230	0.156	
	REL [mdeg]	0.040	0.010	
	RXEL [mdeg]	-0.180	-0.210	

This comparison is displayed in Table 4. Besides the available manufacturer pressure distribution, some structural computations with pressure distributions out of literature [5] (lit.) were performed. The results indicate that the predicted values for RMS and pointing error agree fairly good with the other values, especially when taking into account the uncertainties with which these values are afflicted. The difference between the numerical model and the wind tunnel model is up to 50%. However, the numerical results are much closer to experomental data than those obtained using alternative assumptions about pressure distribution. The accuracy of the predicted RMS and pointing errors is sufficient enough to estimate the mechanical performance of the antenna.

4 Conclusion

We have presented investigations concerning the capabilities of computational fluid dynamics to estimate the wind load on a deep space antenna. For the validation additional wind tunnel experiments were considered and compared with the numerical calculation. The comparison was done by means of the dimensionless pressure difference between front and backside of the main reflector, as well as on the base of the relative surface accuracy (RMS) and the relative pointing accuracy computed by structural analysis. Different turbulence models were applied on three load cases to examine their accuracy. It turned out that the RNG model predicts the turbulent behaviour of the wind field better than the other turbulence models employed. The sensitivity of the results with regard to numerical parameters like discretisation scheme or grid refinement was also investigated. Because of the marginal difference in the results of the first–order upwind scheme and the hybrid schemes with higher quota of second–order central differences and the increasing instability problems of the numerical scheme in finding a steady state solution it is sensible to use the upwind scheme due to its damping of the transitive effects of the flow and the resulting higher stability. The tremendous rise of the computational effort and the small difference in the results allow the conclusion that the original grids with approximately 250000 control volumes are fine enough for the computations.

The comparison of the computed RMS–value and pointing error with the data used by the manufacturer shows that CFD is able to estimate the wind load on the antenna reasonably good. In combination with a structural model it is possible to predict if an antenna will fail under certain wind conditions. In case of the DSA with a maximum RMS–value of 2 millimeter and a maximum pointing of 2 millidegree it is recognizable that the antenna will work properly under the assumed conditions. The future steps will be the appraisal of the boundary conditions at the site of DSA and the performance of real case calculations including the real wind profile and the geographical topology in Australia.

References

1. ANSYS. ANSYS/Multiphysics product. *ANSYS Inc., http://www.ansys.com*, 1999.
2. Invent Computing GmbH. Fastest Manual. *Invent Computing, http://www.invent-computing.de*, 1999.
3. M. Kato and B. Launder, editors. *The Modeling of Turbulent Flows around Stationary Vibrating Cylinders*, 1993. 9th Symposium on Turbulent Shear Flows.
4. B. Launder and D. Spalding. The numerical computation of turbulent flows. *Methods in Applied Mechanical Engineer*, 3:269–289, 1974.
5. R. Levi. *Structural Engineering of Microwave Antennas*. IEEE PRESS, 1996.
6. S. V. Patanker and D.B. Spalding. A calculation procedure for heat, mass and momentum transfer in three dimensional parabolic flows. *International Journal of Heat and Mass Transfer*, 15:1787–1806, 1972.
7. M. Perić. *A Finite Volume Method for the Prediction of three-dimensional Fluid Flow in complex ducts*. PhD thesis, University of London, 1985.
8. H. L. Stone. Iterative solution of implicit approximations of multidimensional partial differential equations. *SIAM Journal for Numerical Analysis*, 5:530–557, 1968.
9. C. Tropea and K. Hufnagel, editors. *Private Communication*, 2000. Dep. of Fluid Mechanics and Aerodynamics.
10. V. Yakhot and S.A. Orszag. Renormalization Group Analysis of Turbulence, I. Basic Theory. *Journal of Scientific Computing*, 1:1–51, 1986.

Numerical Calculation of Turbulent Premixed Flames with an Efficient Turbulent Flame Speed Closure Model

F. Dinkelacker

Lehrstuhl für Technische Thermodynamik, Universität Erlangen,
Am Weichselgarten 8, 91058 Erlangen, Germany

Abstract. A numerical model for premixed gaseous turbulent combustion is investigated, where the combustion process is modelled in terms of a single transport equation for a reaction progress variable c. The reaction term of the progress variable is modelled with an efficient turbulent flame speed closure approach. The model is checked by comparing numerical results with experimental data from a turbulent premixed V-shaped flame, where the conditions of the approaching turbulent flow and of the chemical processes have been varied separately. Regarding the simple structure of this model, it is found to predict the flame shape and flame width well. Further extension of the model account for local variations of stoichiometry, allowing to consider the entrainment of additional air. With that, it was possible to calculate turbulent Bunsen flames of significantly different size (Reynolds number 7000 and 44000) and a free standing turbulent premixed low swirl stabilized flame for different flow conditions.

1 Approach

For numerical calculations of turbulent premixed flames a simple and efficient numerical model is not straightforward, since turbulent flow and reaction of the flame have strong interactions. While the fluctuating turbulent flow wrinkles and modifies the reaction zone and thus modifies the mean reaction rate, the burning process induces a large decrease of the gas density, modifying the flow. For the numerical calculation of turbulent non-reacting flows, typically the coupled partial differential equations are solved for mass, momentum, and turbulence, leading to a description of the mean properties of the flow. For reacting flows, additional equations for the chemical species and energy are necessary and are in general known. However, due to strong non-linearities especially in the reaction terms, a formal approach with the averaged Arrhenius type equations would lead to incorrect results, unless higher order correlation and modelling terms are handled with great care (e.g. [1,2]). An essential simplification is possible for turbulent premixed flames. Here, recent experiments show that for many operating conditions, the instantaneous reaction process is occurring within a thin flame front (thickness typically below 1 mm), dividing essentially the unburnt from the burnt gases (e.g. [3]). Although in turbulent flows this thin reacting front can have a strongly wrinkled shape, and the instantaneous flame location can fluctuate significantly within the turbulent flame brush, this leads to a strong correlation between temperature and all major species, as long as adiabatic flame conditions are assumed (neglecting heat losses due to radiation). Therefore, the main information of the reactive and thermal processes of turbulent premixed flames might be described in terms of a single reaction progress variable, from which the spatial distribution of the average heat release rate, temperature, density, and the major species concentrations can be deduced. Note that the thin flame approximation is not valid for some slow chemical reactions like the NOx formation. But since these minor species have low concentrations and only a minor influence on heat formation, their influence on the flow field is small, and will be neglected in the present investigation.

Based on the correlation between the major species and temperature, the numerical computation of turbulent premixed flames can be simplified essentially, by describing the main reactive and thermal processes within only one additional transport equation of the density weighted mean reaction progress variable $\tilde{c}$ ($\tilde{c} = 0$ in the unburnt mixture and $\tilde{c} = 1$ in the products), with the following structure [4]:

$$\frac{\partial}{\partial t}(\overline{\rho}\tilde{c}) + \frac{\partial}{\partial x_k}(\overline{\rho}\tilde{u}_k\tilde{c}) = \frac{\partial}{\partial x_k}(\overline{\rho}\frac{\nu_t}{Sc}\frac{\partial \tilde{c}}{\partial x_k}) + \overline{\omega}_c \tag{1}$$

where t is time, x_k and u_k are the coordinates and the flow components, ρ is the gas density and $\overline{\omega}_c$ is the source term. Density weighted (Favre averaged) quantities are defined with $\tilde{\varphi} = \overline{\rho\varphi}/\overline{\rho}$, having the advantage, that the common set of equations for mass and momentum of incompressible flows can be used for the calculation of reacting flows with density variations [2]. Thus, available computational fluid mechanics codes may be used directly for calculation. However, for the comparison with experimental data, a transformation to common Reynolds averages has to be regarded, being described later.

The first term on the right hand side of (1) might be interpreted as a description of the turbulent diffusion of $\tilde{c}$. Here, it is assumed that the products (and c) are behaving as a passive scalar, being mixed by the turbulent diffusion $D_t = \nu_t/Sc$ (where the Schmidt number $Sc = \nu/D$ connects the kinematic viscosity ν with the diffusion coefficient both for molecular and turbulent transport). In the present study, further details of the dynamic of the local flame front or counter-gradient diffusion effects are neglected in this part of the model.

One of the main problems is concerned with the modelling of the source term in the transport equation, describing the mean reaction rate. In the present approach the source term of (1) is closed with

$$\overline{\omega}_c = \rho_u \hat{s}_T |\nabla \tilde{c}| \tag{2}$$

Here, ρ_u is the density of the unburnt gas, and the term $|\nabla\tilde{c}|$ is related to the probability of finding the reaction zone, indicating that the mean location of reaction and of the mean gradient of the products (and of c) are connected. A field variable $\hat{s}_T(x)$ is introduced, being related to the turbulent burning velocity s_T of the premixed flame. While the turbulent burning velocity s_T is a property of the entire flame front, and commonly is defined as the normal velocity just ahead of the flame front, the variable $\hat{s}_T(x)$ is calculated at every position x inside the flame as a function of the physico-chemical characteristics of the combustible mixture and the local turbulence parameters. The closure is based on the intuitive idea that the reaction term is directly related to the turbulent flame propagation speed s_T, which includes the interaction between turbulence and physico-chemical reaction processes as a function of the local turbulence parameters [4]. For sufficiently high turbulence intensity, it is described by an algebraic equation

$$\frac{\hat{s}_T}{s_L} = A \cdot Pr^{1/4} \cdot Re_t^{1/4} \cdot \left(\frac{u'}{s_L}\right)^{1/2} \tag{3}$$

with a constant $A = 0.52$, the Prandtl number $Pr = \nu/a = 0.71$, the kinematic viscosity ν, the thermal diffusivity a, the turbulent Reynolds number $Re_t = u' \cdot l_x/\nu$, the integral length scale l_x, the root-mean-square velocity fluctuation u', and the laminar burning velocity s_L [4]. The turbulence quantities are calculated from the computed average turbulent kinetic energy $\tilde{k}$ and turbulent dissipation $\tilde{\varepsilon}$, using the standard definitions of the $k-\varepsilon$ model:

$$u' = \sqrt{\frac{2}{3}\tilde{k}}, \qquad l_x = c_\mu^{3/4}\frac{\tilde{k}^{3/2}}{\tilde{\varepsilon}} \tag{4}$$

with $c_\mu = 0.09$. The turbulent viscosity in the diffusion term is calculated as usual in the $k - \varepsilon$ model with:

$$\nu_t = c_\mu^{3/4}\frac{\tilde{k}^2}{\tilde{\varepsilon}} \tag{5}$$

and a Schmidt number of $Sc = 0.7$ is assumed. The values for the laminar burning velocity s_T depend on the stoichiometric ratio of the fuel-air mixture [5], calculated with a simple fit relation. For the computation of the gradient of in the source term, a second order central difference scheme has been implemented into the subroutines.

The reaction has a strong influence on the flow field due to the density change in the flame. With the thin flame assumption, the following relation can be shown to be valid for the calculation of the mean density [6]

$$\frac{1}{\bar{\rho}(\tilde{c})} = \frac{1-\tilde{c}}{\rho_u} + \frac{\tilde{c}}{\rho_b} \tag{6}$$

where the index u and b denote the unburnt and burnt values.

The equations described above have been implemented as subroutines into a commercial finite volume based fluid dynamics code [7], where the Favre-averaged Navier-Stokes equations are solved. The pressure-velocity coupling is based on the SIMPLE scheme, while for the modelling of turbulence the standard formulation of the $k-\varepsilon$ model [7] is used. The heat release sub-model couples with the flow field via the density $\bar{\rho}$ (6).

After finishing the calculation of the fields of flow, turbulence and reaction progress variable, the calculated Favre averaged progress variable $\tilde{c}$ can be transformed to the Reynolds averaged progress variable $\bar{c}$ with

$$\bar{c} = \frac{(1+\tau)\tilde{c}}{1+\tau\tilde{c}} \tag{7}$$

with the heat release parameter $\tau = (\rho_u/\rho_b - 1)$, in order to compare with experimental data in Reynolds averaged form [8].

2 Comparison with Experimental Data

In order to check the model, numerical results are compared with experimental data from several turbulent premixed flames. Firstly, this is a turbulent premixed V-shaped flame, which is stabilized on a 2 mm diameter wire, situated 10 mm above the burner exit (width 40 mm, Fig. 1). The conditions of the approaching turbulent flow and of the chemical processes have been varied separately and systematically with exit velocities between 1.8 and 3.1 m/s, turbulent Reynolds numbers between 45 and 87 and lean methane-air mixtures with stoichiometric ratios between 0.5 and 0.7 [6,9,10]. Note, that the decrease of the density in the reaction zone strongly influences the streamlines of the mean flow field. Therefore the flame location of the V-shaped flame is very sensitive to this effect. This is a challenge to correct flame modelling, and constant density models would fail to predict the flame angle. Additionally, the width of the flame zone increases with height, and the influence on this parameter can be investigated. Using the above written flame speed relation (2) the comparison between calculation and experiment shows that the calculated flame location and flame width fit well with the experimental data for the different flow rates and equivalence ratios, without tuning on fitting constants.

In an additional study [6] the influence of the formal structure of the reaction term (gradient approach $\overline{\omega}_c \sim |\nabla\tilde{c}|$) is compared with that of the common parabolic approach ($\overline{\omega}_c \sim \tilde{c}\cdot(1-\tilde{c})/L_y$, proposed e.g. in the Bray-Moss-Libby model [8] and in other approaches). While for the gradient approach, the calculated flame shape agrees with the experimentally found straight lines, for the parabolic approach a concave bounded flame shape is found for constant length scale L_y [6]. This can be understood by analyzing the reaction rate integral across the flame brush. In the first case this integral

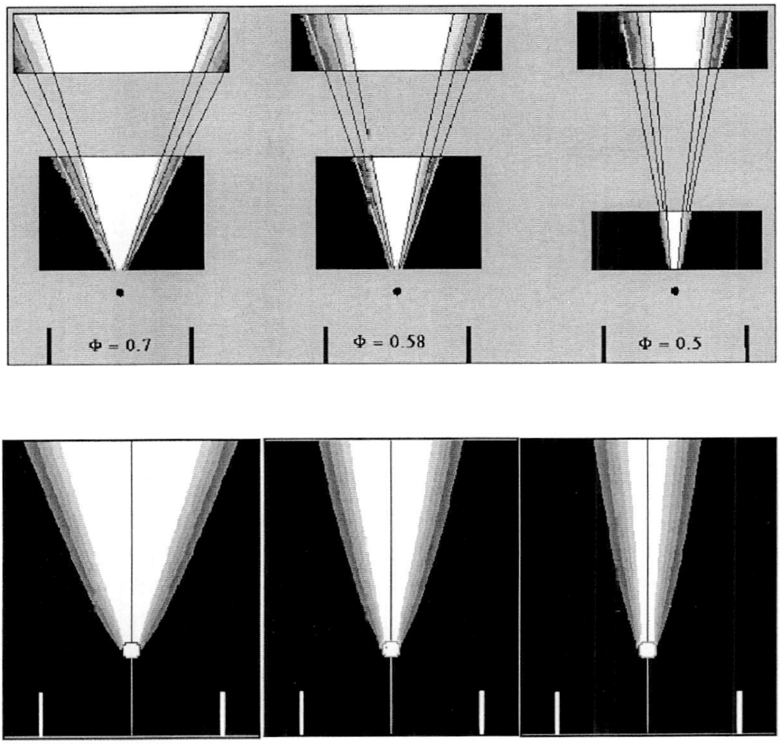

Fig. 1. Mean reaction progress variable $\bar{c}$ of turbulent premixed V-shaped flames in the plane perpendicular to the flame holder for different stoichiometric ratios Φ. Upper line, experimental data, determined with planar laser induced Rayleigh scattering ($\bar{c}$ = probability to find burnt gas). Lower line, calculated flames

is width-independent. In the second case the reaction rate integral is proportional to the width of the turbulent flame brush instead, which does not make sense for a flame with this kind of topology.

Additionally, the model has been extended to account for local variations of the stoichiometry with one additional transport equation for the fuel-air versus air mixture field. With that, the model has been applied for the calculation of a premixed Bunsen-type flame (exit diameter 30 mm, grid generated turbulence, Reynolds number based on exit diameter $Re = 7000$, heat release of about 8 kW), measured by Trimis [11]. Both, the reaction zone location and the mixing with ambient air are calculated in reasonable agreement to the experimental data (Fig. 2). For that, the relation 3 was slightly modified to avoid the turbulent flame speed to be below the laminar flame speed for regions with low turbulence intensity. Also a much larger turbulent premixed

Bunsen flame (exit diameter 80 mm, $Re = 44000$, heat release 130 kW [3]) could be calculated with sufficient accuracy (not shown here).

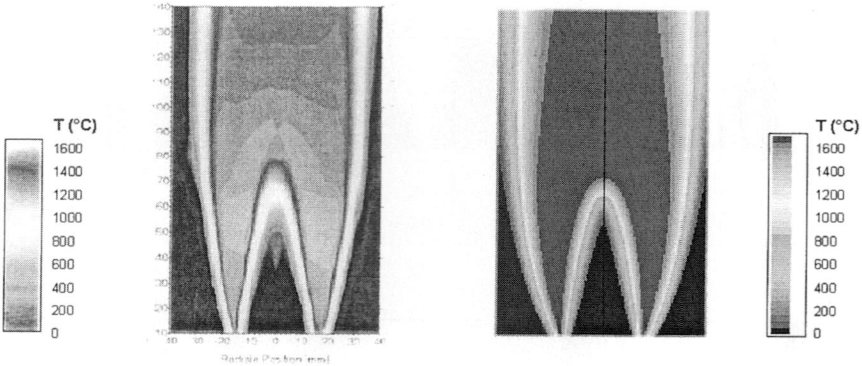

Fig. 2. Measured (left) and calculated (right) temperature field of turbulent premixed methane-air Bunsen-flame of Trimis 1995. Besides the reaction zone (cone in the middle) also the mixing zone with ambient air (outside) is calculated well

Furthermore, this model is applied for the calculation of the flame location and flame zone width of a free standing low swirl flame with turbulent Reynolds numbers up to 3000 (exit Reynolds number 40000). Here, a flat flame is stabilized due to a low swirl component leading to a diverging flow field (Fig. 3) [12,13]. The computational model is able to calculate this flame sufficiently well, leading to a turbulent flame brush thickness comparable with the experimental situation for both calculated burner loads. This shows also, that flames without fixed flame holder can be calculated numerically.

3 Conclusion

A rather simple combustion model is investigated for gaseous premixed flames, allowing also for locally leaned conditions due to additional air. The closure of the reaction term is based on a turbulent flame speed relation which allows a very effective and numerically robust code realization, working efficiently in connection with a commercial CFD code. This approach has been tested to calculate the location of the average heat release zone and the related density, temperature, main species and velocity fields of premixed flames of different size, Reynolds number and geometry, showing a wide range of applicability. Further adaption to account for different operating pressure, radiative heat transfer and for slow chemical processes like thermal NOx formation are in progress.

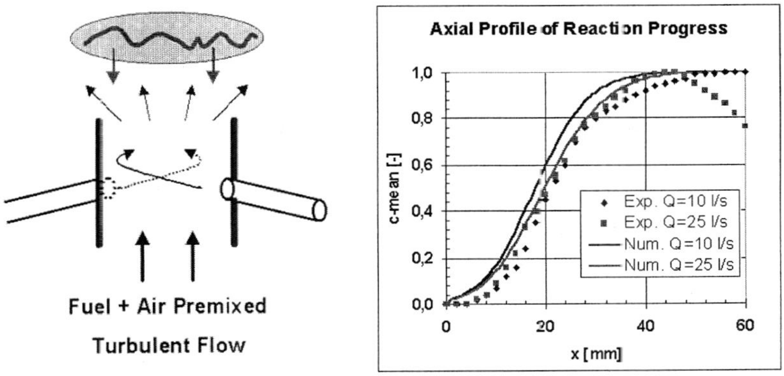

Fig. 3. Free standing turbulent premixed low swirl flame. Downward propagation of flame equal to axial component of upward flow. Schematics (left). Measured and calculated axial profiles of the Reynolds averaged reaction progress for two different burner loads (right)

References

1. Bray, K.N.C. (1980) Turbulent Flows with Premixed Reactants. In: P.A. Libby and F.A. Williams (Eds.) Turbulent Reacting Flows, Springer, Berlin, 115-183
2. Jones, W.P., Whitelaw, J.H. (1982) Calculation methods for reacting turbulent flows: a review. Comb. Flame 48, 1-26
3. Buschmann, A., Dinkelacker, F., Schäfer, T., Schäfer, M., Wolfrum, J. (1996) Measurement of the instantaneous detailed flame structure in turbulent premixed combustion. Proceedings of the Combustion Institute, 26, 437-445
4. Zimont, V.L., Lipatnikov, A.N. (1995) A numerical model of premixed turbulent combustion of gases. Chem. Phys. Reports 14, 993-1025
5. Egolfopoulos, F.N., Cho, P., Law, C.K. (1989) Laminar flame speeds of methane-air mixtures under reduced and evaluated pressures. Comb. Flame 76, 375-391
6. Dinkelacker, F., Hölzler, S. (2000) Investigation of Turbulent Flame Speed Closure Approaches for Premixed Flame Calculation. Combust. Sci. and Tech. 158, 321-340
7. Fluent (1996) Computational Fluid Dynamics Software. Fluent Incorporated, Lebanon, NH, USA
8. Bray, K.N.C. (1990) Studies of the turbulent burning velocity. Proc. R. Soc. Lond. A 431, 315-335
9. Dinkelacker, F., Hölzler, S., Helbig, K.-P., Leipertz, A., Polifke, W. (1998) Tests of a TFC model for premixed turbulent flame calculations at low Reynolds number flames of different stoichiometry. Fluent User Meeting, Darmstadt 21.-22.9.98
10. Soika, A. (1996) Untersuchungen zur Wechselwirkung zwischen Turbulenz und Verbrennung mittels Laser-Streulicht-Techniken. Diplomarbeit, Lehrstuhl für Technische Thermodynamik, Universität Erlangen

11. Trimis, D. (1995) Verbrennungsvorgänge in porösen inerten Medien. Dissertation, Lehrstuhl für Strömungsmechanik, Universität Erlangen
12. Bedat, B., Cheng, R.K. (1995) Experimental study of premixed flames in intense isotropic turbulence. Comb. Flame 100, 485-494
13. Plessing, T. (2000) Laseroptische Untersuchungen turbulenter Transport- und Reaktionsvorgänge in Vormischflammen. Dissertation, Institut für Technische Mechanik, RWTH Aachen

Monte Carlo Simulations of Radiative Heat Transfer with Parallel Computer Architectures

J.G. Marakis, J. Chamiço, G. Brenner, and F. Durst

Institute of Fluid Mechanics, University of Erlangen-Nürnberg Cauerstraße 4, D-91058 Erlangen, Germany

Abstract. This work presents a parallel Monte Carlo algorithm for the calculation of combined radiative and conductive heat transfer. The proposed formulation effectively separates the time-consuming ray-tracing part of the Monte Carlo method from the energy computations required in the iterative solution of the energy equation. The method is applied for a simple combined radiative and conductive heat transfer problem and excellent agreement with the benchmark results is found. The ray-tracing part of the algorithm is parallelised and applied in two configurations, which represent the opposite ends of the currently available parallel computer architectures; a PC cluster and the Hitachi SR8000-F1 supercomputer. For sufficiently large sampling sets, the computation experiments show an almost ideal speed-up.

1 Introduction

The solution of the energy equation in combined radiative and conductive/convective heat transfer problems is a time consuming task. The usual practice in solving this type of problems is to employ an approximate and fast model for radiation and thus to avoid the significant overloading of the iterative solution of the energy equation. The alternative option, the use of an accurate model, appears rarely in the literature. The reasons for that are generally associated with the lack of sufficient computational resources. The present paper exploits the straightforward method to circumvent this deficiency; namely it investigates the parallelisation of an accurate radiative heat transfer model, such as the Monte Carlo method. Descriptions of the serial version of this method can be found in textbooks [1,2], or monographs [3,4]. Recently, [5] reviewed its numerous applications in the field of radiative heat transfer. The parallelisation of this method has been addressed by [6,7].

In the Monte Carlo method, radiative heat exchange is simulated by stochastically tracing an amount of discrete packets emitted by each surface or volume element. All the possible radiative events (emission, transmission, absorption, scattering, reflection) are described by probability density functions. Macroscopic radiative quantities, such as radiative fluxes or source terms, are formed as averages over a sufficiently large statistical sample. The problem is that each time the temperature field changes, as it happens during the iterative solution of the energy equation, a new sample of energy packets has to be traced. Since even the tracing of a single set of energy packets is time consuming, this repetitive application makes the method unpractical for combined heat transfer problems of engineering interest. An alternative method is to reuse all, or part, of the previous sample. This approach has been proposed by [8] and further elaborated by [9]. A method to consistently reuse previous samplings is to introduce, as an intermediate step in the algorithm, the calculation of the direct exchange factors. This approach originates from the zonal method of [10] and it will be described in the next section.

The specific problem investigated in this work is the determination of the temperature field between two infinite parallel plates. The medium between the plates is absorbing, emitting and conducting. This simple configuration, apart from the interest it has in many applications, such as modelling of melting/solidification and heat transfer in semitransparent crystals, it also has the advantage of an accurate semi-analytic solution. It may therefore serve as a benchmark solution for the validation of the combined Monte Carlo and finite volume code used in this study. Another focal point is the evaluation of the efficiency of the parallel Monte Carlo implementation. This issue is investigated by running the ray-tracing part of the algorithm in the two platforms which represent the lowest and the highest ends of the current parallel computing architectures, namely a 10-node PC cluster and the Hitachi SR-8000 F1 supercomputer of the Leibniz Supercomputing Center in Munich.

2 Formulation

The steady state energy equation for a gray radiating and conducting medium reads

$$\frac{\partial}{\partial x_i}\left(k\frac{\partial T}{\partial x_i}\right) = \frac{\partial q_{r,i}}{\partial x_i} = \kappa\left(4\sigma T^4 - \int_{4\pi} I d\Omega\right) \tag{1}$$

where k is the thermal conductivity, q_r is the net raditive flux, κ is the absorption coefficient, σ the Stefan-Boltzmann constant, I the radiative intensity and Ω a solid angle. A semi-analytic solution of (1) based on the exponential integral functions can be found in standard textbooks, e.g. [1,2]. The method adopted in this study is the usual iterative solution of (1); based on a assumed temperature distribution, the radiative source term, $\frac{\partial q_{r,i}}{\partial x_i}$, is first calculated and then (1) is solved to obtain an improved estimation of the temperature field. In this iterative procedure, the Monte Carlo method is used to calculate the radiative source term. Details for the implementation of this method, the ray-tracing algorithm on which it is based, as well as a description of a parallelisation strategy can be found in [7] and therefore will not be repeated here. A schematic picture is shown only for illustration purposes in Fig. 1.

The previous description implies that the Monte Carlo algorithm has to be repeated once the new temperature field has changed. A very significant

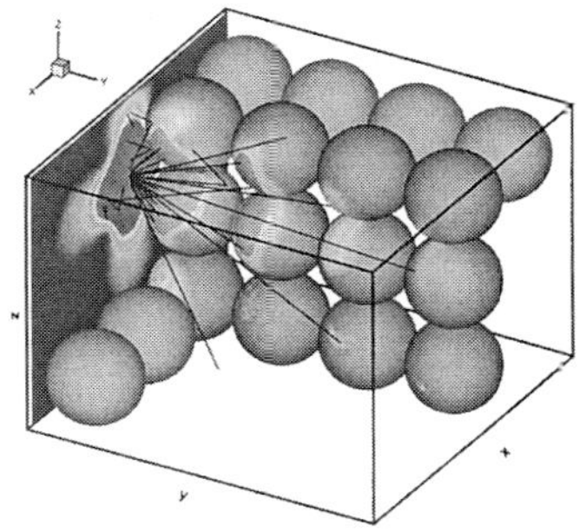

Fig. 1. An illustration of the direct exchange factors isolines on the surface of spheres on a regular packed bed

reduction in the computational time is achieved if the problem is expressed in terms of the direct exchange factors. [10,4]. This formulation allows splitting the problem in two parts. In the first part, the ray-tracing part of the Monte Carlo method is used to compute the direct exchange factors. These factors are defined as

$$R_{ij} = \frac{n_j}{n_i} \tag{2}$$

where n_i is the number of packets emitted by the i-th element and n_j the number of them that is absorbed in the j-th element ($\sum_j n_j = n_i$). This part of the algorithm is "energy-free", i.e. it is considered that the packets do not carry energy. In the second part, the direct exchange factors are used to quantify the energy transfer. This is done by computing the total amount of energy absorbed by the j-th element as

$$Q_{abs,j} = \sum_i R_{ij} Q_{emit,i} \tag{3}$$

and $Q_{emit,i}$ is either

$$Q_{emit,i} = \varepsilon_i \sigma T_i^4 \triangle A_i \tag{4a}$$

$$Q_{emit,i} = 4\kappa_i \sigma T_i^4 \triangle V_i, \tag{4b}$$

depending on whether i is a surface or volume element. In Eqs. (4a) and (4b), ε is the emissivity, κ is the absorption coefficient and $\triangle A_i$, or $\triangle V_i$, is the size of the i-th element.

The advantage of this formulation is that the time consuming ray-tracing algorithm is performed only once. Therefore it has the potential to reduce the computational time by a factor in the order of the number of iterations necessary for the solution of (1). Due to the very strong non-linearity of this equation, a large number of iterations are needed and therefore this saving is expected to be significant. The restrictions for the application of the direct exchange formulation are, first, that the spatial distribution of the radiative properties of the medium should remain unchanged with iterations, and second, that these properties are also temperature independent. Both restrictions are met in a wide range of applications, especially those involving a solid matrix embedded in a gas phase, as it happens for example in porous burners. In the cases where the aforementioned restrictions are not met, the repetition of the ray-tracing algorithm is necessary to recalculate the new set of direct exchange factors. In these cases, the direct exchange formulation may turn to represent a slight computational overhead, instead of saving, when compared with typical Monte Carlo implementations.

Another improvement reported in this work is the introduction of a set of cutoff criteria. This improvement is based on the observation that only a small group of the direct exchange factors represent almost completely the radiative heat exchange. It is therefore desirable to identify which factors belong in this group and subsequently to perform the energy calculations using only this small subset. The benefit is significant because, first the database for storing the direct exchange factors scales with n^2, the number of the discretisation elements, and second, the size of the loop implied by (3) also scales with the number of the direct exchange factors. The reduction of the database is achieved by sorting the factors in descending order and storing them until any of the cutoff criteria is reached. Two criteria have been implemented, the first referring to the value of a single factor and the second on their cumulative value which is calculated as they being sorted. The advantage of these threshold parameters is that they offer a consistent method to manage the accuracy vs. the computational requirements of the Monte Carlo algorithm.

3 Results

The problem considered in this section is to determine the temperature distribution in an absorbing, emitting and conducting medium confined between two infinite parallel, isothermal, gray, emitting and diffusely reflecting plates. The aim is twofold; first to validate the combination of the Monte Carlo algorithm and a finite volume code used to solve the energy equation, and second to demonstrate the feasibility and the benefits of the cutoff criteria.

The temperature distribution between the plates calculated by the semi-analytic method presented in [2] is shown in Fig. 2 for three values of the conduction-to radiation parameter N. The latter is defined as

$$N = \frac{k\kappa}{4\sigma T_{\max}^3} \tag{5}$$

and it is a measure of the relative importance of conductive and radiatve fluxes for optically thick media. The curves shown in Fig. 2 correspond to a

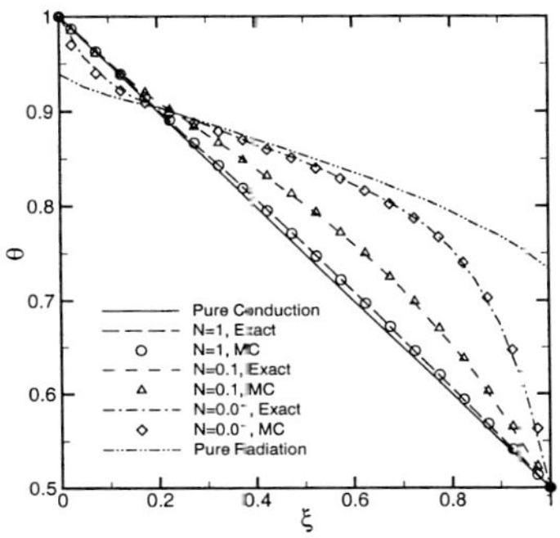

Fig. 2. Comparison between Monte Carlo results and benchmark data [2] for three values of the conduction-to-radiation parameter

slab with optical thickness $\tau_L = 1$, black walls ($\varepsilon_1 = \varepsilon_2 = 1.0$), a temperature ratio of the cold and hot plates equal to 0.5 and temperature independent thermal conductivity. In this figure, $\theta = \frac{T}{T_{\max}}$ is the non-dimensional temperature and $\xi = \frac{x}{L}$ the non-dimensional distance across the slab. For decreasing values of N, the temperature distributions are shown to depart progressively from the linear profile corresponding to pure conduction towards the pure radiation profile. Monte Carlo results are also shown in Fig. 2. They have been obtained by discretising the geometry into $21 \times 11 \times 11$ elements. The elements are equidistant in each direction and they have an aspect ratio of 10, in reference to their thickness, to represent the infinite width of the geometry. The direct exchange factors have been calculated after tracing 150,000 packets from each element. The data mentioned in this paragraph are common for all of the examined test cases. The agreement between the Monte Carlo results and the benchmark data shown in Fig. 2 is very satisfactory.

In order to evaluate the feasibility of the cutoff criteria, a new set of cases has been examined. In these cases, $N = 0.01$, while the cutoff criterion has been decreased from the value of 1.0, which corresponds to a complete database of the direct exchange factors, up to 0.85 by taking the intermediate values shown in Fig. 3. It is shown in this figure that the accuracy progressively diminishes with decreasing cutoff. The solutions are sufficiently accurate for a range between 0.99 and 0.95, while lower cutoff values produce less accurate results. The reduction in size of the database that is achieved for various values of the cutoff parameter is shown in Fig. 3. The results in this figure correspond to three values of the absorption coefficient and $N = 0.01$. The very rapid reduction of the database with decreasing cutoff can be easily observed in Fig. 3. For a moderately low criterion of 0.95 for example, the database is shrunk by a factor ranging from 3, for the optically thin medium, and up to 5 for the optically thick case.

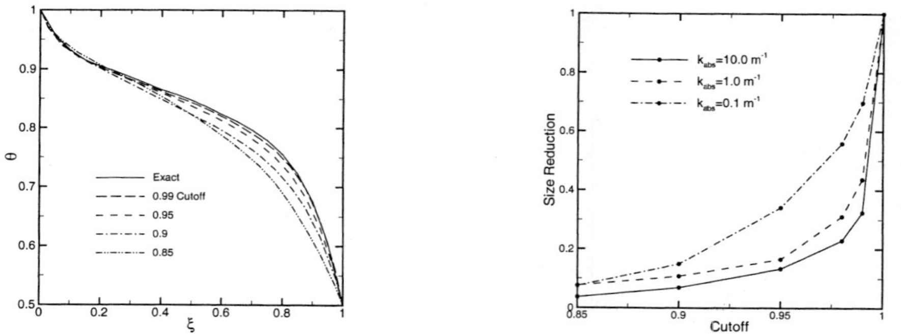

Fig. 3. The feasibility of the cutoff criterion; In the left: influence of the cutoff on the accuracy of the results; In the right: reduction of the size of the database achieved by decreasing the cutoff in reference with the optical thickness of the medium

The ray-tracing part of the algorithm has been parallelised following the Single-Program-Multiple-Data (SPMD) paradigm and using the Message Passing Interface (MPI) library. Since the packets are propagating independently of each other the parallelisation is made by tracing concurrently as many of them as possible in each of the available computing nodes. If p homogeneous processing nodes are available, then decomposing each sample of N energy packets into $N_p = \frac{N}{p}$ equal subsets accomplishes this task. This simple parallelisation strategy was tested in two very different homogeneous parallel computing architectures, namely a dedicated PC cluster made of 10 nodes and the Hitachi SR-8000 F1 supercomputer.

For the PC cluster, Fig. 4 shows the relation between the number of processors and the size of the sample subset. The reduction of this sample is in principle desirable because it helps to avoid unnecessary calculations after a certain level of confidence has been reached. However, it is evident from Fig. 4 that it is also associated with a degradation of the parallel efficiency. The latter is attributed to the increased communication overhead that results from the more frequent gathering of data from all the processors. The same tendency has been observed in the Hitachi SR8000 supercomputer. Measurements taken in this machine are shown in Fig. 5. An almost ideal speed-up was achieved for very large subset sample size (0.5 million packets) for up to 512 processors. However, such a large sample is rarely used in practical calculations. More interesting is the range of 50,000 packets, where the total sample size remains in reasonable size, while the parallel efficiency, as shown in Fig. 5, is still satisfactorily high.

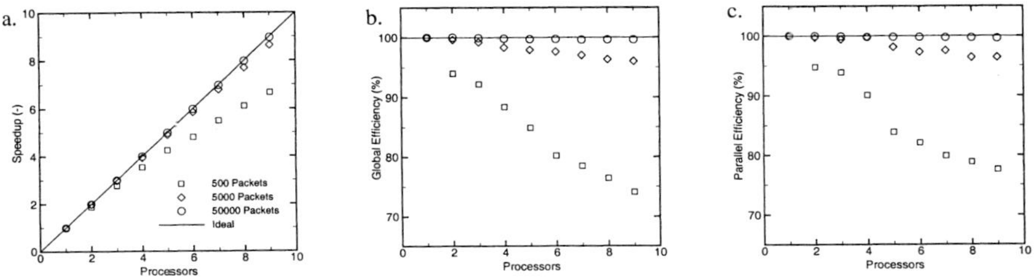

Fig. 4. Influence of the sample size on Speed Up (left), $S_p = \frac{t_1}{t_p}$, Global Efficiency (center), $E_g = \frac{t_1}{p t_p}$ and Parallel Efficiency (right), E_p, where t_1 and t_p are the execution times using 1 and p nodes, respectively, and $E_g = E_{load} E_{num} E_p$. The machine was a dedicated cluster of 10 PCs

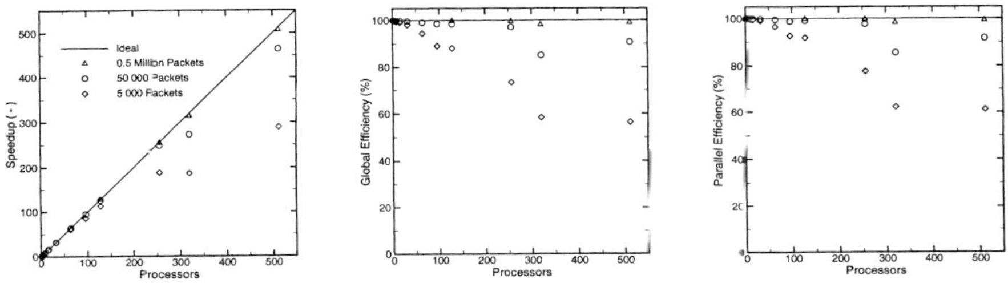

Fig. 5. Influence of the sample size on: Speedup, Global Efficiency and Parallel Efficiency for the Hitachi SR8000 -F1 with 96 nodes (512 processors)

4 Conclusions

A Monte Carlo method for combined radiative and conductive heat transfer has been applied in a simple configuration for which a semi-analytic solution is available. A very good agreement between the Monte Carlo results and the benchmark data was obtained. Then a series of test cases were examined aiming to determine whether the size of the direct exchange factor database can be reduced by introducing a cutoff parameter. It was shown that a very significant reduction is achievable for a usable range of the cutoff between 0.95 and 0.99. Further reduction of this parameter, though it continuously reduces the database size, is not recommended because it also reduces the accuracy of the solution.

The implementation of the ray-tracing algorithm in a dedicated network of workstations showed that very high efficiencies could be obtained from the computing resources readily available in most laboratories. This is because the histories of the energy packets are by definition independent of each other and therefore only the statistics of a sampling set need to be transferred among the processors. This minimal communication requirement between the nodes helps to avoid any potential network bottleneck caused by limited network bandwidth. The application of the ray-tracing algorithm in a supercomputer infrastructure showed that the very high parallel efficiency is maintained for as many as 512 processors. Therefore large and demanding in computational resources problems can now be solved using these high performance architectures.

Acknowledgements

This study has been financially supported by the Commission of the European Communities through the TMR-RADIARE network "Fundamental improvements in radiative heat transfer". The authors would also like to acknowledge the Leibniz Supercomputing Center in Munich for making available the Hitachi SR-8000 F1 computer as well as the team of the KONWIHR project BAUWIHR to support the optimization of the code.

References

1. Siegel, R., Howell, J.R. (1992), Thermal Radiation Heat Transfer. 3rd ed., Taylor and Francis, Washington DC
2. Modest, M.F. (1993), Radiative Heat Transfer. McGraw-Hill, New York
3. Haji-Sheikh, A. (1988), Monte Carlo Methods. in Minkowycz, W.J., Sparrow, E.M., Schneider, G. E. and Pletcher, R. H. (Eds.), Handbook of Numerical Heat Transfer, Wiley, New York
4. Yang, W.J., Tanigushi, H., Kudo, K. (1995), Radiative heat transfer by the Monte Carlo method. in Hartnett, J.P. and Irvine, T. (Eds.), Advances in Heat Transfer **27**, 1-215, Academic Press, San Diego
5. Howell, J.R. (1998), The Monte Carlo method in radiative heat transfer. Journal of Heat Transfer **120**, 547-560
6. Farmer, J.T. and Howell, J.R. (1998), Comparison of Monte Carlo strategies for radiative transfer in participating media, in Hartnett, J.P. and Irvine, T. (Eds.), Advances in Heat Transfer **31**, 1-97, Academic Press, San Diego
7. Marakis, J. G., Chamico, J., Brenner, H. and Durst, F. (2001), Parallel ray tracing for radiative heat transfer: Application in a distributed computing environment. Int. J. Numer. Methods Heat and Fluid Flow, accepted for publication
8. Kobiyama, M. (1989), Reduction of computing time and improvement of convergence stability of the Monte Carlo method applied to radiative heat transfer with variable properties. Journal of Heat Transfer **111**, 135-140
9. Schweiger, H., Oliva, A., Costa, M. and Pérez Segarra, C.D. (1999), A Monte Carlo method for the simulation of transient radiation heat transfer: Application to compound honeycomb transparent insulation. Numerical Heat Transfer, Part B **35**, 113-136
10. Hottel, H.C. and Sarofim, A.F. (1967), Radiative Transfer. McGraw-Hill, New York

Direct Numerical Simulation of Bubble Swarms with a Parallel Front-Tracking Method

M.F. Göz[1], B. Bunner[2], M. Sommerfeld[1], and G. Tryggvason[3]

[1] Institut für Verfahrenstechnik, Fachbereich Ingenieurwissenschaften, Martin-Luther-Universität Halle-Wittenberg, D-06099 Halle, Germany

[2] Microfluidics and Biotechnology Group, Coventor, 625 Mount Auburn Street, Cambridge MA 02138, USA

[3] Department of Mechanical Engineering, Worcester Polytechnic Institute, 100 Institute Road, Worcester, MA 01609, USA

Abstract. Direct numerical simulations are performed to study the behaviour of an important class of dispersed multiphase flows, namely gas bubbles rising in a liquid. The numerical method combines a finite difference scheme for solving the Navier-Stokes equations with a front tracking method for following the gas-liquid interfaces. The size of the problem as well as the simulation time requirements necessitate the use of large parallel computers. Sample simulation results are presented illustrating the evolution of such systems and the dependence of statistical quantities on the gas volume fraction. The goal of these numerical experiments is to gain insight into fundamental properties of bubbly flows and support the development of simplified models.

1 Background and Motivation

Bubbly gas-liquid flows play an important role in many natural and industrial processes. The presence of many small dispersed elements flowing through a continuous phase provides excellent heat and mass transfer characteristics due to the large interface-to-volume ratio. A widely used device in the chemical and biotechnological industries is the vertical bubble column, in which the hydrodynamics is determined by the bubble rise due to bouyancy. Rise, interaction, and resulting collective motion of bubbles depend strongly on gas volume fraction, bubble size and bubble size distribution [1,2]. In addition, the bubbles motion induces turbulence in the liquid through their wakes and oscillations, which in turn affects bubble dispersion and mass transfer. These effects increase with the bubble size as larger bubbles deform into ellipsoids and perform wobbling motions. Furthermore, a polydisperse bubble size distribution can be expected to have an additional effect on bubble-bubble interactions and the resulting collective behaviour due to different rise velocities and wakes. On the other hand, a higher gas volume fraction has a dampening influence on the large scale motion since the bubbles begin to hinder each other in their movement.

So far the interplay between these various effects has not been investigated in detail. Therefore, the mechanisms leading to experimentally observed behaviours in bubble columns are not well understood. This makes it difficult to develop practical models that can be used in engineering applications. Direct numerical simulations offer the opportunity to resolve all flow scales and provide a detailed picture of the flow field for a wide range of physical parameters. Specifically, this approach is used here to study the effect of bubble size, bubble deformation, and volume fraction on the flow of gas bubbles in a liquid. The full Navier-Stokes equations for a single flow field with variable density and viscosity are solved using a finite difference method combined with a front tracking algorithm allowing for a fully deformable interface between a bubble and the ambient liquid [3]. In order to account for the boundary conditions due to surface tension at the gas bubble-liquid interfaces, an appropriate source term of the form of an integral containing the local mean curvature over the bubble surfaces is included in the Navier-Stokes equations. Since it is computationally impossible to simulate the millions of bubbles that occupy an entire bubble column in this way, only a small section in the center of the column, where the flow is assumed to be homogeneous and not influenced by the walls, is studied. Typically between 10 and 100 bubbles are simulated in a periodic cell with periodic boundary conditions in all three spatial dimensions.

Simulating sufficiently many bubble systems allows to evaluate the influence of parameters like gas volume fraction, bubble size and deformability, and polydisperse bubble size distributions on bubble rise velocities, bubble interactions, resulting collective motions ('swarm effects'), and the induced liquid turbulence. For this purpose averaged and fluctuation quantities as well as the microstructure, e.g., the preferred alignment of bubble pairs, need to be analysed. As the bubble interactions taking place on the microscale influence the behaviour of the system on the mesoscale, the knowledge obtained from these simulations can be used to develop Euler/Euler and Euler/Lagrange models. These models are formulated on a mesoscale by averaging out the details of the microscopic flow and are critical to our ability to predict the macroscale behavior of a bubbly flow, for example in a full bubble column.

2 Numerical Solution Procedure

The principle of the numerical method consists of advecting the material boundary, determining surface tension, applying a Navier-Stokes solver, and repeating the whole cycle. The Navier-Stokes equations are solved by a second-order accurate projection method, using centered differences on a fixed, staggered, three-dimensional, Cartesian grid. The incompressibility condition leads to a non-separable (due to the variable density) elliptic equation for the pressure, whose solution is the most time-consuming operation and is usually done using a multigrid solver developed in-house.

The chief difficulty of multiphase flow simulations is to accurately track the interface between the different phases and impose the appropriate velocity and stress boundary conditions at this interface. This is achieved in this method by explicitly tracking the interface by a front, which is a deformable and moving two-dimensional mesh consisting of marker points connected by triangular elements. At the beginning of each time step, the marker points are advected with a velocity that is interpolated from the values of the velocity on the fixed, three-dimensional, Cartesian grid. The simplest interpolation method is volume-weighted averaging using the lever rule, which uses the velocities at the 8 grid nodes surrounding each marker point. A more accurate method uses the Peskin interpolation function [4] and the 64 grid points surrounding the marker point.

Next, the indicator function, which is a three-dimensional field whose values are defined at the nodes of the fixed grid, is determined from the new position of the front. The indicator function is equal to 1 in the liquid and 0 in the gas. Rather than jumping discontinuously across the interface, the indicator function transitions smoothly between 0 and 1 over approximately 3-4 grid points. This is achieved in the following way. The jump in the indicator function from 0 to 1 is distributed from the front onto the fixed grid using the same Peskin interpolation function mentioned above, in effect creating a discrete equivalent of the Dirac δ function. The distributed form of the jump, which represents the gradient of the indicator function, is then differentiated spatially to obtain what is formally the Laplacian of the indicator function. This Laplacian is then integrated twice, sometimes by a fast Fourier based method or a SOR method, but usually by the same multigrid solver used for the solution of the non-separable elliptic pressure equation. The result is a smooth indicator function field which is used to determine the density and viscosity by simple scaling.

The next step is the calculation of the surface tension force. The mean curvature on each front element is determined by using a local polynomial fit to the element. It is multiplied by the surface tension coefficient, the vector normal to the element, and the area of the element. The result is distributed from the front onto the grid, again using the Peskin interpolation function, and added to the right-hand side of the Navier-Stokes equation.

The last important operation associated with the front is the periodic restructuring of the front by addition of points and elements in areas where the elements are too large and deletion of points and elements from areas where the elements are too small. This crucial operation ensures that the front mesh remains approximately uniform throughout the simulation and necessitates the use of linked lists to store the data structures associated with the front.

Few simplications of the physical problem are made in the formulation and the numerical method beyond the usual assumptions present in the incompressible, isothermal Navier-Stokes equation. One simplification is that the gas inside the bubbles is assumed to be incompressible. Although this is obviously not valid if the motion of a bubble from the bottom to the top of an industrial bubble column is considered, where the volume of a bubble typically increases by a factor of two, it is a valid assumption when the motion of a bubble over a short distance only is considered, where the volume practically remains the same. Another, less valid but necessary simplification is that bubbles do not coalesce when they get close to each other. Although coalescence (and breakup, which is not an issue here because the bubbles are small and do not deform much) is known to play an important role in the dynamics of bubbly flows [5], two reasons prevent us from considering it. The first reason is that we are interested in obtaining time-averaged values of quantities like the bubble fluctuation velocities, for which the number of bubbles must remain constant. The second reason is that, as two bubbles get close to each other, the resolution in the thin film of liquid separating them, whose evolution determines whether the two bubbles coalesce or not, decreases because of the fixed Cartesian grid. Rather than allowing the two bubbles to coalesce when in fact they should not, we keep them separate.

Extensive resolution studies were performed to determine an adequate resolution so that the error on all important quantities (rise and fluctuation velocities, turbulent kinetic energy and Reynolds stresses...) is less than 5% [6]. Resolution requirements for bubbly flows result from the need to accurately simulate the flow in the boundary layer and are mainly a function of the Reynolds number. As the Reynolds number increases, the boundary layer becomes thinner and the number of grid points across the boundary layer and therefore across the diameter of the bubble must be increased. Additionally, resolution requirements are larger when the bubbles are deformable or the void fraction is large. For the parameters used in this work, a resolution of about 20-25 points across the diameter of the small bubbles proved to be sufficient.

The finite-difference/front-tracking method has been implemented for distributed-memory parallel computers using the standard Message Passing Interface protocol. The cubic domain is decomposed into equisized subdomains, each of which is assigned to a different processor. Boundary data are exchanged between neighbouring subdomains leading to small memory and communication time overheads. For the front treatment a dynamic master-slave approach has been chosen. This means that the processor corresponding to a subdomain, in which most of a front resides at a given timestep, collects all relevant data for this front, performs the necessary calculations, and redistributes the results to the other subdomains containing smaller parts of the front. This approach takes advantage of the fact that there are many small bubbles approximately homogeneously distributed in a large domain. The parallelization penalties are a slight communication time overhead due to front data exchange between various processors, and a memory overhead of about 10% for front data duplication on several processors.

3 Application to Bubble Swarm Simulations

This section contains the description of some general features of our bubble system calculations and previously obtained results, followed by sample simulation results from the most recent calculations.

3.1 General Features

Various calculations of mono- and bidisperse systems with nearly spherical and/or deformable (i.e. ellipsoidal) gas bubbles rising through an initially stagnant liquid have been carried out. The gas volume fraction was chosen to be 2, 6, or 12%, with the number of bubbles in the periodic cell ranging from 16 to 72 for the bidisperse simulations, and up to 216 for monodisperse simulations.

The number of bubbles in each periodic cell is an additional parameter of the simulation, which is obviously purely numerical and which the physical results, average rise and fluctuation velocities of the bubbles and average kinetic energy of the liquid, should be independent of. When this number is too small, the results depend strongly on it. However, simulating an arbitrarily large number of bubbles is computationally very expensive. A study of the dependence of the results on the number of bubbles was performed in [6]. It showed that accurate results are obtained for the rise velocity and kinetic energy with simulations using about 20 bubbles whereas larger numbers of bubbles are needed to obtain results for the fluctuation velocities that do not depend on the number of bubbles.

Between 8 and 18 processors were employed in the simulations. The grid size varied between 104^3 and 256^3 points depending on the number of bubbles in the computational domain and on the resolution requirements. The bubble rise Reynolds number, based on the effective diameter and the rise velocity of a bubble, lies between 10 and 120. Bubbles and liquid are released from rest. After an initial transient, the bubbles usually settle to a statistical steady state, in the sense that averaged quantities like the average bubble velocity approach a constant value, although the individual bubbles interact continually and move unsteadily. Illustrations of a sample bubble configuration, individual bubble rise velocities, and statistical quantities are provided in the figures below.

As reference results, to which all other cases may be compared, we use those obtained for monodisperse spherical bubble systems [6,7]. These systems settle fairly quickly towards a statistical steady state and show the following characteristic features. The average bubble rise velocity decreases as the gas volume fraction increases; this decrease can be described by a Richardson-Zaki relation (typical of the hindered settling of solid particles). The bubbles appear to be distributed uniformly throughout the computational domain with a slight tendency for pairs of bubbles to be aligned horizontally. At high gas volume fraction this leads to the formation of horizontal layers which are, however, not stable but break up and reform continuously. The Reynolds stresses and turbulent kinetic energy of the liquid increase with the gas volume fraction. The kinetic energy spectrum follows a -3.6 power law at large wavenumbers, which is not too far away from the value of $-8/3$ seen in two-phase flow experiments [8].

3.2 Effects of Bubble Deformability

The evolution of systems with deformable bubbles follows initially closely that of spherical bubble systems with the same gas volume fraction, but may experience a "streaming effect" where the bubbles suddenly accelerate sharply, turn and deform, and build streams of vertically aligned bubble clusters. This leads to a much larger swarm velocity compared to both the average rise velocity of the spherical bubble system and the terminal velocity of a corresponding single bubble; simultaneously the turbulence intensity of the liquid increases considerably. For more information about this phenomenon see [6,9]. The streaming was observed for a 6% system, and for a 2% system only when the bubbles were initially placed in a streaming configuration. There are indications, however, which suggest that streaming might be observed without special initial conditions for other gas volume fractions as well, if only the simulations would run long enough.

3.3 Effects of a Bidisperse Bubble Size Distribution

Another cause for larger than expected swarm velocities may lie in the interaction of bubbles of different sizes since a small bubble will likely be drawn into the wake of a larger bubble, such that the net effect is an increase of the average velocity. While a continuous bubble size distribution, often with two distinct peaks, is prevalent in experiments, our present focus is on bidisperse systems in order to assess the interaction of bubbles of different size and its effect on clustering, large-scale motion, and fluid turbulence. In a first series of simulations, the evolution of configurations with an equal number of large and small spherical bubbles with a volume ratio of 2 were calculated. The parameters were chosen such that gas volume fractions and mean bubble size are the same as for the monodisperse systems. The results of these bidisperse calculations show that the mean values of bubble rise velocity and turbulent kinetic energy of the liquid lie very close to those of the monodisperse systems, while the fluctuation velocities averaged over all bubbles are significantly larger in the bidisperse systems due to the differential rise velocity of the large and small bubbles [10].

3.4 Combination of Deformability and Bidispersity

In a second series of simulations, bidisperse systems are calculated with the same gas volume fractions but only a few large and many small bubbles corresponding more closely to experimental situations. Here, the volume ratio

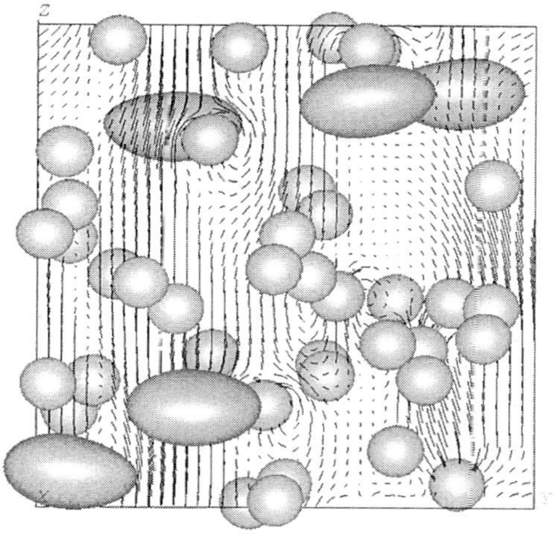

Fig. 1. Snapshot of a gas bubble-liquid system with 40 small spherical and 5 large ellipsoidal bubbles representing 6% volume fraction

has been chosen to be 8, such that the large bubbles deform into ellipsoids while the small ones remain nearly spherical. Figure 1 represents an instantanteous bubble configuration and a section of the velocity field of a system with 5 large and 40 small bubbles at a gas volume fraction of 6%.

The temporal evolution of all bubble rise velocities, scaled as Reynolds numbers, in a 2% system is shown in Fig. 2. Here, time was nondimensionalized using the constant of gravity, g, and a mean bubble diameter, d, according to $t^* = t \cdot \sqrt{g/d}$; the rise Reynolds number for each bubble is defined as $Re = \rho d_b u_b / \mu$, where ρ and μ are the density and viscosity of the liquid, and d_b and u_b are the individual bubble diameter and rise velocity. First the small and large bubbles rise mostly independently of each other, interacting mainly within their own size class. It is remarkable that the velocity fluctuations of the small bubbles are larger than those of the large bubbles. From time to time there are noticeable outbursts from the respective fluctuation ranges indicating strong interactions between small and large bubbles.

The 2% system displays some features that stand out from the other cases. In most simulations, the initial transient lasts until about $t = 10$ and corresponds to the time necessary for the formation of a boundary layer around the bubbles and to the beginning of interactions between bubbles. For the 2% case, the initial transient lasts much longer, until about $t = 100$. This is due to a mistake in the initialization of the position of the bubbles. The bubbles are usually distributed randomly throughout the computational domain at $t = 0$. In the 2% case, however, the bubbles were initially confined to a corner cube, i.e., 1/8 of the volume of the computational domain. The evolution of the bubbles between $t = 0$ and $t = 100$ reflects their dispersion from this corner cube to the entire domain. During this process, the average velocity of the small bubbles is larger than at statistical steady state, i.e.,

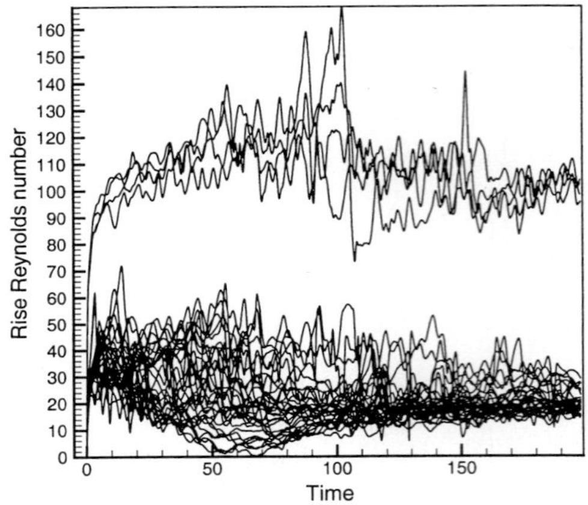

Fig. 2. Individual rise Reynolds numbers vs. time of 32 small spherical (lower band of curves) and 4 large ellipsoidal (upper band) bubbles in a system with 2% gas volume fraction

after $t = 100$, suggesting that the large bubbles pull the small bubbles along with them.

Given the relatively low Reynolds number, the flow field is laminar. However, the random motion of the bubbles induce in the liquid a fluctuating velocity field analogous to what is observed in turbulence. As in turbulence, these fluctuation velocities can be characterized by the turbulent kinetic energy, which is plotted versus nondimensional time in Fig. 3. In the monodisperse and bidisperse systems with volume ratio 2, it was found that the turbulent kinetic energy increases monotonically with gas volume fraction [10]. The results for the bidisperse systems with volume ratio 8 reported here

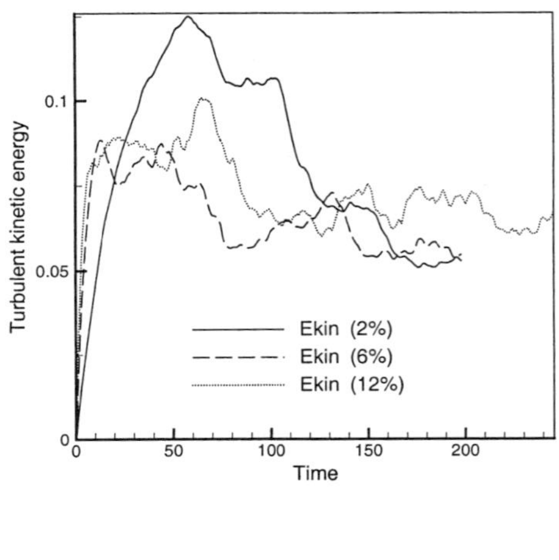

Fig. 3. Temporal evolution of the turbulent kinetic energy of the liquid in the three bidisperse systems with deformable large bubbles

seem quite different, although we note that they are not complete yet and that the 2% and 6% simulations, in particular, need to be pursued for a longer time. At the current stage, the average value of the turbulent kinetic energy for 12% is again the largest, but the values for 2% and 6% at $t = 200$ are approximately the same. The curve for the 2% system exhibits large values in the initial transient between $t = 0$ and $t = 100$. As the bubbles move into the entire computational domain, they create a strong stirring of the liquid. When they are distributed more evenly, the turbulent kinetic energy decreases strongly.

The kinetic energy spectra in the three bidisperse systems under consideration (Fig. 4) display very similar behaviour and are consistent with those found for the previous mono- and bidisperse simulations with spherical bubbles [6,10]. An exception is again the spectrum for the 2% system, which has a lot more energy in the low wavenumber range. This might indicate the formation of large-scale flow structures. Since it is not clear to which extent this behaviour is related to the special initial conditions, we are performing another calculation of the same system but with a more homogeneous initial bubble distribution.

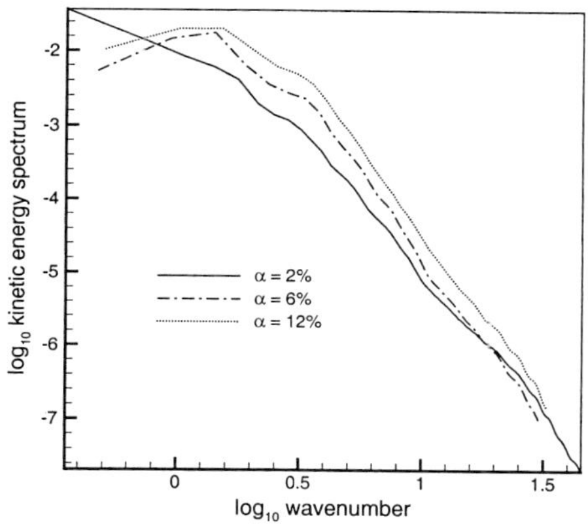

Fig. 4. Kinetic energy spectra of the three bidisperse systems with deformable large bubbles

4 Summary and Outlook

Using a parallel finite difference/front tracking method, direct numerical simulations of mono- and bidisperse bubble swarms rising in a liquid are carried out. From the temporal evolution of these systems integral and statistical quantities are derived providing information about how the microscopic flow behaviour determines that on the mesoscale. Important parameters are bubble size and deformability, bubble size distribution, and gas volume fraction.

These influence bubble interaction, swarm velocity, and bubble-induced liquid turbulence. It is hoped that these and future results, e.g. for higher rise Reynolds numbers, yield profound insight into the dynamics of such complex systems enabling us to improve their modelling on the mesoscale.

Acknowledgements

The authors gratefully acknowledge the generous grant of computer time on the IBM RS/6000 SP2 by the Computing Center of the University of Karlsruhe, Germany, and travel grants by the DAAD within the Foreign Exchange Visitor programme and the US National Science Foundation (grant INT-9726759). MFG and MS are supported by the Deutsche Forschungsgemeinschaft.

References

1. Schlüter M., Räbiger N. (1998) Bubble swarm velocity in two phase flows. In: Proc. of the ASME Heat Transfer Division, HTD-Vol. 361, Volume 5, Anaheim
2. Bröder D., Laín S., Sommerfeld M. (2000) Experimental studies of the hydrodynamics in a bubble column. In: Proc. 5th German-Japanese Symposium on Bubble Columns in Dresden, Germany, May 29–May 31, 2000, TU Bergakademie Freiberg, pp. 125–130
3. Tryggvason G., Bunner B., Ebrat O., Tauber W. (1998) Computations of Multiphase Flows by a Finite Difference/Front Tracking Method. I Multi-Fluid Flows. In: 29th Computational Fluid Dynamics. Lecture Series 1998-03. Von Karman Institute for Fluid Dynamics
4. Tryggvason, G., Bunner, B., Esmaeeli, A., Juric, D., Al-Rawahi, N., Tauber, W., Han, J., Nas, S., Jan, Y.-J.: A front tracking method for the computations of multiphase flows. Accepted for publication in J. Comp. Phys.
5. Lehr, F., Mewes, D. (2000) A transport equation for the interfacial area density applied to bubble columns, Proc. of ASME-FED June 2000, Boston
6. Bunner B. (2000) Numerical simulation of gas-liquid bubbly flows. Ph.D. thesis, The University of Michigan, Ann Arbor
7. Bunner B., Tryggvason G. (submitted) Direct numerical simulations of bubbly flows. Part I: Motion of the bubbles. J. Fluid. Mech.
8. Lance M., Bataille J. (1991) Turbulence in the liquid phase of a uniform bubbly air-water flow. J. Fluid Mech. **222**, 95–118
9. Göz M. F., Bunner B., Sommerfeld M., Tryggvason G. (2000) Direct numerical simulation of the interaction of gas bubbles in a liquid – the effects of deformability and bidispersity. In: Proc. 5th German-Japanese Symposium on Bubble Columns in Dresden, Germany, May 29–May 31, 2000, TU Bergakademie Freiberg, pp. 288–293
10. Göz M. F., Bunner B., Sommerfeld M., Tryggvason G. (2001) Simulation of bubbly gas-liquid flows by a parallel finite-difference/front-tracking method. In: E. Krause & W. Jäger (eds.), High Performance Computing in Science and Engineering 2000 (Springer), pp. 326–337

Symmetry-Preserving Discretization of Turbulent Channel Flow

R.W.C.P. Verstappen and A.E.P. Veldman

Research Institute for Mathematics and Computing Science
University of Groningen, P.O.Box 800, 9700 AV Groningen, The Netherlands.
Email: verstappen@math.rug.nl, veldman@math.rug.nl.

Abstract. We propose to perform turbulent flow simulations in such manner that the difference operators do have the same symmetry properties as the underlying differential operators, i.e. the convective operator is represented by a skew-symmetric matrix and the diffusive operator is approximated by a symmetric, positive-definite matrix. Such a symmetry-preserving discretization of the Navier-Stokes equations is stable on any grid, and conserves the total mass, momentum and kinetic energy (when the physical dissipation is turned off). Its accuracy is tested for a turbulent channel flow at Re=5,600 (based on the channel width and the mean bulk velocity) by comparing the results to those of physical experiments and previous numerical studies. This comparison shows that with a fourth-order, symmetry-preserving method a $64 \times 64 \times 32$ grid suffices to perform an accurate direct numerical simulation.

1 Introduction

Direct numerical simulation (DNS) is the most accurate way of computing turbulent flows. It does not invoke any turbulence model and thus requires a numerically reliable description of all dynamically significant scales of motion, making DNS rather expensive. In view of its computational complexity, our first concern is to reduce the number of grid points as far as we can get. But, how far can we go? What are the limits set by the discretization scheme? Can the computational mesh be adapted to the physical scales of the flow, or is its stretching limited, due to the stability of the spatial discretization, for instance?

There exist various ways to discretize convective and diffusive operators on nonuniform grids. Traditionally, the local truncation error is minimized. Other aspects of the discretization, like its conservation properties, caught barely attention. Recently, the design has changed. Conservation constraints are increasingly often enforced; see e.g. [1–3]. We propose to apply a high-order, finite-volume method that preserves the symmetries of the underlying differential operators, i.e. the convective operator is represented by a skew-symmetric coefficient matrix and the diffusive operator is approximated by a symmetric, positive-definite matrix (see Sect. 2).

Mimicing crucial properties of differential operators forms in itself a motivation for discretizing them in a certain manner. We give it a concrete form by noting that a symmetry-preserving discretization conserves the total mass, momentum and kinetic energy if the physical dissipation is turned off. When the physical dissipation is turned on, the energy of any solution of a symmetry-preserving discretization decreases in time unconditionally. This implies that a stable solution can be obtained on any grid. Hence, we need not add an artificial damping mechanism that interferes with the subtle balance between the production of turbulence and its dissipation at the smallest length scales. This has been our main motivation to investigate symmetry-preserving discretizations for DNS of turbulent flow.

Because stability is not an issue, the main question becomes how accurate is a symmetry-preserving discretization, or stated otherwise, how coarse may the grid be for a DNS? We will address this question in Sect. 3 by evaluating the results for a turbulent flow in a channel at Re=5,600.

2 Symmetry-Preserving Discretization

The temporal evolution of the discrete velocity vector $\boldsymbol{u}_h$ is governed by a finite-volume discretization of the incompressible Navier–Stokes equations:

$$\boldsymbol{\Omega}\frac{\mathrm{d}\boldsymbol{u}_h}{\mathrm{d}t} + \boldsymbol{C}\left(\boldsymbol{u}_h\right)\boldsymbol{u}_h + \boldsymbol{D}\boldsymbol{u}_h - \boldsymbol{M}^*\boldsymbol{p}_h = \boldsymbol{0} \qquad \boldsymbol{M}\boldsymbol{u}_h = \boldsymbol{0}, \tag{1}$$

where the vector $\boldsymbol{p}_h$ denotes the discrete pressure, $\boldsymbol{\Omega}$ is a (positive-definite) diagonal matrix representing the sizes of the control volumes, $\boldsymbol{C}\left(\boldsymbol{u}_h\right)$ is built from the convective flux contributions through the control faces, $\boldsymbol{D}$ contains the diffusive fluxes, and $\boldsymbol{M}$ is the coefficient matrix of the discretization of the integral form of the law of conservation of mass. We assume that the matrices $\boldsymbol{C}\left(\boldsymbol{u}_h\right)$, $\boldsymbol{D}$ and $\boldsymbol{M}$ are constructed such that the consistency conditions

$$\boldsymbol{C}^*\left(\boldsymbol{u}_h\right)\boldsymbol{1} = \boldsymbol{0}, \qquad \boldsymbol{D}^*\boldsymbol{1} = \boldsymbol{0} \quad \text{and} \quad \boldsymbol{M}\boldsymbol{1} = \boldsymbol{0} \tag{2}$$

are satisfied, and use the adjective 'symmetry-preserving' if

$$\boldsymbol{C}\left(\boldsymbol{u}_h\right) + \boldsymbol{C}^*\left(\boldsymbol{u}_h\right) = \boldsymbol{0}, \qquad \boldsymbol{D} + \boldsymbol{D}^* \text{ positive-definite.} \tag{3}$$

The two conditions on the coefficient matrix $\boldsymbol{C}\left(\boldsymbol{u}_h\right)$ reflect that $\boldsymbol{C}\left(u_h\right)$ represents a discrete gradient operator: its null space consists of the constant vectors and $\boldsymbol{C}\left(\boldsymbol{u}_h\right)$ is skew-symmetric like a first-order differential operator. The coefficient matrix $\boldsymbol{D}$ of the discrete diffusive operator inherits its positive-definiteness from the underlying Laplacian differential operator.

The semi-discretization (1) is fully-conservative and stable if the coefficient matrices satisfy (2)–(3). The total mass is trivially conserved. The total amount of momentum $\boldsymbol{1}^*\boldsymbol{\Omega}\boldsymbol{u}_h$ evolves in time according to

$$\frac{\mathrm{d}}{\mathrm{d}t}\left(\boldsymbol{1}^*\boldsymbol{\Omega}\boldsymbol{u}_h\right) \overset{(1)}{=} -\boldsymbol{1}^*\left(\boldsymbol{C}(\boldsymbol{u}_h) + \boldsymbol{D}\right)\boldsymbol{u}_h + \boldsymbol{1}^*\boldsymbol{M}^*\boldsymbol{p}_h \overset{(2)}{=} 0,$$

hence momentum is conserved if (2) holds. The evolution of the discrete energy $\boldsymbol{u}_h^* \boldsymbol{\Omega} \boldsymbol{u}_h$ of any solution $\boldsymbol{u}_h$ of (1) is governed by

$$\frac{\mathrm{d}}{\mathrm{d}t}(\boldsymbol{u}_h^* \boldsymbol{\Omega} \boldsymbol{u}_h) \stackrel{(1)}{=} -\boldsymbol{u}_h^*(\boldsymbol{C}+\boldsymbol{C}^*)\boldsymbol{u}_h - \boldsymbol{u}_h^*(\boldsymbol{D}+\boldsymbol{D}^*)\boldsymbol{u}_h + \underbrace{2\boldsymbol{p}_h^* \boldsymbol{M} \boldsymbol{u}_h}_{=0}$$

$$\stackrel{(3)}{=} -\boldsymbol{u}_h^*(\boldsymbol{D}+\boldsymbol{D}^*)\boldsymbol{u}_h \leq 0, \tag{4}$$

where we have used the skew-symmetry of $\boldsymbol{C}(\boldsymbol{u}_h)$. The right-hand side is zero if and only if $\boldsymbol{u}_h = \boldsymbol{0}$, or $\boldsymbol{D}+\boldsymbol{D}^* = \boldsymbol{0}$. Thus, the energy is conserved if the diffusion is turned off. Note that the pressure term $\boldsymbol{M}^* \boldsymbol{p}_h$ in (1) does not affect the evolution of the total kinetic energy (on condition that $\boldsymbol{M}\boldsymbol{u}_h = \boldsymbol{0}$), because the discrete pressure gradient is represented by the transpose of the coefficient matrix $\boldsymbol{M}$ of the incompressibility constraint.

With diffusion (that is for $\boldsymbol{D} \neq \boldsymbol{0}$) the right-hand side of (4) is negative for all $\boldsymbol{u}_h \neq \boldsymbol{0}$ provided that $\boldsymbol{D}+\boldsymbol{D}^*$ is positive-definite. So, the energy of the discrete system (1) decreases in time if (3) is satisfied. The semi-discrete system (1) is stable under this symmetry condition: a solution of (1) can be obtained on any grid, and there is no need to add an artificial damping mechanism to stabilize the spatial discretization.

Since these favorable conservation and stability properties are directly related to the symmetries of the coefficient matrices in (1), we want to construct these matrices such that they fulfil (2)–(3). To illustrate the way in which this may be achieved, we consider the approximation of the first-order derivative in one spatial dimension. The traditional method maximizes the order of the local truncation error. On three points, this leads to the second-order approximation

$$\partial_x u(x_i) \approx \begin{pmatrix} \dfrac{-r_i}{\delta x_{i+1}+\delta x_i} & \dfrac{r_i - r_i^{-1}}{\delta x_{i+1}+\delta x_i} & \dfrac{r_i^{-1}}{\delta x_{i+1}+\delta x_i} \end{pmatrix} \begin{pmatrix} u_{i-1} \\ u_i \\ u_{i+1} \end{pmatrix}, \tag{5}$$

where the parameter r_i is defined as the ratio between the grid spacings $\delta x_{i+1} = x_{i+1} - x_i$ and δx_i. The essence of our method is that the coefficient matrix $\boldsymbol{C}$ of the convective term is skew-symmetric whereas the coefficient matrix $\boldsymbol{D}$ of the diffusive term is symmetric and positive-definite. Since the diagonal-entry of operator in the right-hand side of (5) is non-zero for $r_1 \neq 1$, the standard discretization method breaks the skew-symmetry of the coefficient matrix of the convective term on a nonuniform grid. Therefore, the standard method does not conserve the energy and is not conditionally stable on nonuniform meshes. In our discretization, we take $r_1 = 1$ independent of the local grid spacing. Consequently, the resulting coefficient matrix $\boldsymbol{C}$ is skew-symmetric:

$$\left(\boldsymbol{\Omega}^{-1}\boldsymbol{C}\boldsymbol{u}_h\right)_i = \frac{2}{\delta x_{i+1}+\delta x_i} \begin{pmatrix} -\dfrac{1}{2} & 0 & \dfrac{1}{2} \end{pmatrix} \begin{pmatrix} u_{i-1} \\ u_i \\ u_{i+1} \end{pmatrix}, \tag{6}$$

and the local truncation error of this approximation

$$\tau_h(x_i) = \tfrac{1}{2}(\delta x_{i+1} - \delta x_i)\partial_{xx} u(x_i) + \mathcal{O}(\delta x_{\max}^2) \tag{7}$$

is first-order, unless the grid is (almost) uniform. Given stability, a sufficient condition for second-order accuracy of the discrete solution u_i is that the local truncation error $\boldsymbol{\tau}_h$ be second order. Then the error $\boldsymbol{\epsilon}_h$ in the solution $\boldsymbol{u}_h$, given by $\boldsymbol{\Omega}^{-1}\boldsymbol{C}\boldsymbol{\epsilon}_h = \boldsymbol{\tau}_h$, is second-order. Yet, this is not a necessary condition, as is emphasized by Manteufel and White [4]. They have proven that the approximation (6) yields second-order accurate solutions on uniform as well as on nonuniform meshes, even though its local truncation error $\boldsymbol{\tau}_h$ is formally only first-order on nonuniform meshes.

So, we do not try to minimize the local truncation error on nonuniform meshes. Instead we choose to preserve the (skew-)symmetry of the underlying differential operator. On a three-dimensional, uniform grid, the second-order scheme of Harlow and Welsh [5] preserves the symmetries of the convective and diffusive operator, and thus conserves the mass, momentum and energy of an inviscid flow. We generalize Harlow and Welsh's scheme to nonuniform meshes in such a manner that the symmetries are not broken. The accuracy of this basic scheme has been improved by means of a Richardson extrapolation, just like in [6]. The resulting discretization is fourth-order accurate. The discretization of a first-order derivative becomes

$$\partial_x u(x_i) \approx \tfrac{1}{2} H_i^{-1}\left(-u_{i+2} + 8u_{i+1} - 8u_{i-1} + u_{i-2}\right)$$

where

$$H_i = \tfrac{1}{2}\left(-x_{i+2} + 8x_{i+1} - 8x_{i-1} + x_{i-2}\right).$$

The diffusive operator undergoes a similar treatment, leading to

$$H_i \partial_{xx} u(x_i) = 8\left(\frac{u_{i+1} - u_i}{x_{i+1} - x_i} - \frac{u_i - u_{i-1}}{x_i - x_{i-1}}\right) - \left(\frac{u_{i+2} - u_i}{x_{i+2} - x_i} - \frac{u_i - u_{i-2}}{x_i - x_{i-2}}\right)$$

For more details, we refer to [7,8]. A variant of our approach for collocated grids has been developed at Delft University [9].

3 A Test-Case: Turbulent Channel Flow

The symmetry-preserving discretization is tested for turbulent channel flow. The Reynolds number is set equal to Re = 5,600 (based on the channel width and the bulk velocity), a Reynolds number at which direct numerical simulations have been performed by several research groups; see [10–12]. In addition we can compare the numerical results to experimental data [13].

As usual, the flow is assumed to be periodic in the stream- and spanwise direction. Consequently, the computational domain may be confined to a channel unit of dimension $2\pi \times 1 \times \pi$, where the width of the channel is normalized. All computations presented in this section have been performed with 64 (uniformly distributed) streamwise grid points and 32 (uniformly distributed) spanwise points. The channel walls are given by $y = 0$ and $y = 1$. For $0 \leq y \leq 1/2$ the wall-normal grid points are computed according to

$$y_j = \frac{\sinh(\gamma j/N_y)}{2\sinh(\gamma/2)} \qquad \text{with} \qquad j = 0, 1, ..., N_y/2,$$

where N_y denotes the number of grid points in the wall-normal direction. The stretching parameter γ is taken equal to 6.5. The grid points in the remaining half of the channel ($1/2 < y \leq 1$) are computed by means of symmetry: $y_j = 1 - y_{N_y - j}$, for $j = N_y/2 + 1, .., N_y$.

The temporal integration of (1) is performed with the help of a one-leg method that is tuned to improve its convective stability (see [8]). The non-dimensional time step is set equal to $\delta t = 1.25\ 10^{-3}$. Mean values of computational results are obtained by averaging the results over the directions of periodicity, the two symmetrical halves of the channel, and over time. The averaging over time starts after a start-up period. The start-up period as well as the time-span over which the results are averaged, 1500 non-dimensional time-units, are identical for all the results shown is this section.

Figure 1 shows a comparison of the mean velocity profile as obtained

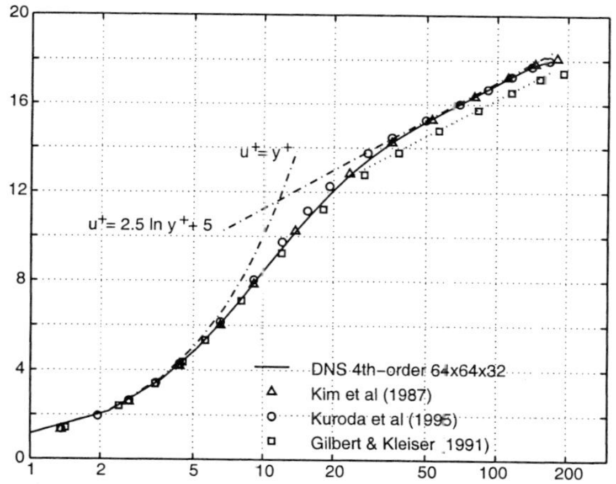

Fig. 1. The mean streamwise velocity u^+ versus y^+. The dashed lines represent the law of the wall and the log law. The markers represent DNS-results that are taken from the ERCOFTAC Database

from our fourth-order symmetry-preserving simulation ($N_y = 64$) with those of other direct numerical simulations. Here it may be stressed that the grids used by the DNS's that we compare with have typically about 128^3 grid points, that is 16 times more grid points than our grid has. Nevertheless, the agreement is excellent. Thus, we may conclude that a $64 \times 64 \times 32$ grid suffices for the fourth-order symmetry-preserving method.

To investigate the convergence of the fourth-order method upon grid refinement, we have monitored the skin friction coefficient C_f as obtained from simulations on four different grids. We will denote these grids by A, B, C and D. Their spacings differ only in the direction normal to the wall. They have $N_y = 96$ (grid A), $N_y = 64$ (B), $N_y = 56$ (C) and $N_y = 48$ (D) points in the wall-normal direction, respectively. The grids are stretched normal to the wall by means of the *sinh*-transformation given above. The first (counted from the wall) grid line used for the convergence study is located at $y_1^+ \approx 0.95$ (grid A), $y_1^+ \approx 1.4$ (B), $y_1^+ \approx 1.6$ (C), and $y_1^+ \approx 1.9$ (D), respectively. Figure 2 shows the skin friction coefficient C_f as function of the fourth power of

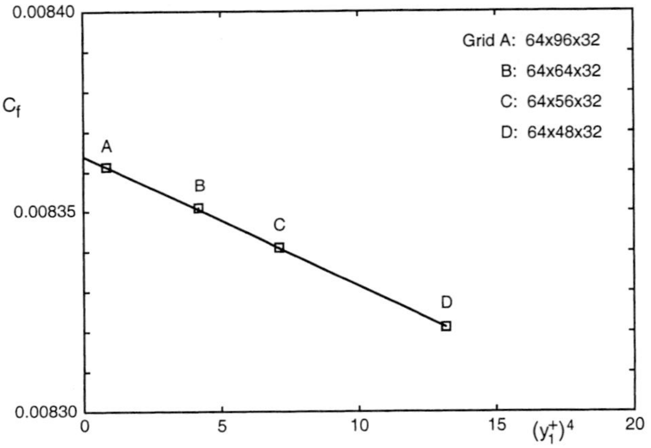

Fig. 2. Convergence of the skin friction coefficient C_f upon grid refinement. The figure displays C_f versus the fourth power of the first grid point y_1^+

y_1^+. As can be seen, C_f is (approximately) a linear function of $(y_1^+)^4$. Hence, the discretization scheme is indeed fourth-order accurate (on a nonuniform mesh). The straight line in Fig. 2 crosses $y_1^+ = 0$ at $C_f \approx 0.00836$. This extrapolated value lies in between the C_f reported in Kim et al. [10] (0.00818) and Dean's [14] correlation of $C_f = 0.073\, Re^{-1/4} \approx 0.00844$.

The results for the fluctuating streamwise velocity $u_{\rm rms}$ are compared to the experimental data of Kreplin and Eckelmann [13] in Fig. 3. This comparison affirms that the fourth-order, symmetry-preserving method is more accurate than the second-order method. The near-wall velocity fluctuation predicted by the fourth-order simulation fits the experiment data nicely, even up to very coarse grids with only 32 grid points in the wall-normal direction. Viewed through physical glasses the energy of the resolved scales of motion, the coarse grid ($N_y = 32$) solution, is convected in a stable manner, because

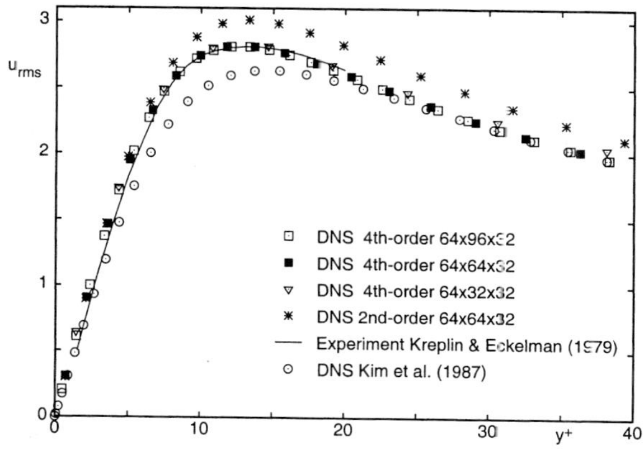

Fig. 3. Comparison of the mean-square of the streamwise fluctuating velocity as function of y^+ (linear scale)

it is conserved by the discrete convective operator. Therefore, we think that the symmetry-preserving discretization forms a solid basis for testing sub-grid scale models. The discrete convective operator transports energy from a resolved scale of motion to other resolved scales without dissipating energy, as it should do from a physical point of view. The test for a sub-grid scale model reads then: does the addition of the dissipative sub-grid model to the conservative convection of the resolved scales improve their accuracy. There is some room for improvement for larger values of y^+: the results with $N_y = 32$ become less accurate when compared to the DNS of Kim et al. [10] for $y^+ > 20$. Yet, with more grid points, $N_y = 64$, the agreement with Kim et al. is excellent in the region $y^+ > 20$. The turbulence intensity in the sublayer ($0 < y^+ < 5$) predicted by the simulations is higher than that in the experiment. According to the fourth-order simulation the root-mean-square approaches the wall like $u_{rms} \approx 0.38y^+$ ($N_y = 64$). The exact value of this slope is hard to pin-point experimentally. Hanratty et al. [15] have fitted experimental data of several investigators, and thus came to 0.3. Most direct numerical simulations yield higher values. Kim et al. [10] and Gilbert & Kleiser [11] have found slopes of 0.3637 and 0.3824 respectively, which is in close agreement with the present findings.

So, in conclusion, the results of the fourth-order symmetry-preserving discretization agree better with the available reference data than those of its second-order counterpart, and with the fourth-order method a 64×64×32 grid suffices to perform an accurate DNS of a turbulent channel flow at Re=5,600.

References

1. Morinishi, Y., Lund, T. S., Vasilyev, O. V. and Moin, P. (1998) Fully conservative higher order finite difference schemes for incompressible flow. J. Comp. Phys. **143**, 90–124
2. Vasilyev, O .V. (2000) High order finite difference schemes on non-uniform meshes with good conservation properties. J. Comp. Phys. **157**, 746–761
3. Ducros, F., Laporte, F., Soulères, T., Guinot, V. Moinat, P. and Caruelle, B. (2000) High-order fluxes for conservative skew-symmetric-like schemes in structured meshes: application to compressible flows. J. Comp. Phys. **161**, 114–139
4. Manteufel, T. A. and White, Jr., A.B. (1986) The numerical solution of second-order boundary value problems on nonuniform meshes. Math. of Comp. **47**, 511–535
5. Harlow, F. H. and Welsh, J. E. (1965) Numerical calculation of time-dependent viscous incompressible flow of fluid with free surface. Phys. Fluids **8**, 2182–2189
6. Antonopoulos-Domis, M. (1981) Large-eddy simulation of a passive scalar in isotropic turbulence. J. Fluid Mech. **104**, 55–79
7. Verstappen, R. W. C. P. and Veldman, A. E. P. Symmetry-preserving discretization of turbulent flow, submitted to J. Comp. Phys.
8. Verstappen, R. W. C. P. and Veldman, A. E. P. (1997) Direct numerical simulation of turbulence at lower costs. J. Engng. Math. **32**, 143–159
9. A. Twerda, Advanced computational methods for complex flow simulation. Delft University of Technology, PhD thesis (2000).
10. Kim, J., Moin, P. and Moser, R. (1987) Turbulence statistics in fully developed channel flow at low Reynolds number. J. Fluid Mech. **177**, 133–166
11. Gilbert, N. and Kleiser, L. (1991) Turbulence model testing with the aid of direct numerical simulation results. Proc. Turb. Shear Flows 8, Paper 26-1
12. Kuroda, A., Kasagi, N. and Hirata, M. (1995) Direct numerical simulation of turbulent plane Couette-Poisseuille flows: effect of mean shear rate on the near-wall turbulence structures. In: F. Durst et al. (Eds.) Proc Turb. Shear Flows, Springer-Verlag, Berlin, 241–257
13. Kreplin, H. P. and Eckelmann, H. (1979) Behavior of the three fluctuating velocity components in the wall region of a turbulent channel flow. Phys. Fluids **22**, 1233–1239
14. Dean, R. B. (1978) Reynolds number dependence of skin friction and other bulk flow variables in two-dimensional rectangular duct flow. J. Fluids Engng. **100**, 215–223
15. Hanratty, T. J., Chorn, L. G. and Hatziavramidis, D. T. (1977) Turbulent fluctuations in the viscous wall region for Newtonian and drag reducing fluids. Phys. Fluids **20**, S112

Parallelization Strategies and Efficiency of CFD Computations in Complex Geometries Using Lattice Boltzmann Methods on High-Performance Computers

M. Schulz, M. Krafczyk, J. Tölke, and E. Rank

Lehrstuhl für Bauinformatik, Technische Universität München, 80290 München, Germany

Abstract. A frequently stated property of the Lattice Boltzmann (LB) method is, that it is easy to implement and that the generation of computational grids is trivial even for three-dimensional problems. This is mainly due to the usually chosen approach of using full matrices to store the primary variables of the scheme. However this kind of implementation has severe disadvantages for simulations, where the volume of the bounding box of the flow domain is large compared to the actual volume of the flow domain. Thus the authors developed data structures which allow to discretize only the fluid volume including boundary conditions to minimize memory requirements, while retaining the excellent performance with respect to vectorization of standard LB-implementations on supercomputers. Due to extensive communication hiding using asynchronous non-blocking message transfer an almost linear parallel speedup is achieved.

1 Introduction

In the past years, the Lattice-Boltzmann method [1,2] was developed as a valuable complementary approach to classical CFD techniques for solving the Navier-Stokes equations. The primary variables of this scheme are normalized particle probability distributions $f_i(t, \boldsymbol{x}), i \in \{0, ..., N-1\}$ which represent the occupation number of particles on lattice nodes with a discrete velocity $\boldsymbol{e_i}$. The computations in this work are based on the three-dimensional D3Q15 model [1], where $N = 15$.

The evolution dynamics of the f_i is given by the Lattice-Boltzmann equation

$$f_i(\boldsymbol{x} + \boldsymbol{e_i}\triangle t, t + \triangle t) - f_i(\boldsymbol{x}, t) = \Omega_i(\boldsymbol{x}, t) \tag{1}$$

which implies that the computational grid is identical to the lattice defined by the set of vectors $\{\boldsymbol{e_i}\triangle t\}$.

The collision operator

$$\Omega_i = -\frac{1}{\tau}(f_i - f_i^{(eq)}) \tag{2}$$

is based on the Single Time Relaxation Aproximation (STRA) and represents the rate of changes due to particle collisions. The relaxation time

$$\tau = \frac{6\nu + 1}{2} \tag{3}$$

can be used to tune the kinematic viscosity and defines the rate of approach to the equilibrium state which is given by a polynomial of the macroscopic flow properties $\boldsymbol{u}$ and ρ. These can be defined as moments of the f_i:

$$\rho(t, \boldsymbol{x}) = \sum f_i(t, \boldsymbol{x}) \tag{4}$$

$$\boldsymbol{u}(t, \boldsymbol{x}) = \frac{1}{\rho} \sum f_i(t, \boldsymbol{x}) \boldsymbol{e}_i. \tag{5}$$

The dynamics of velocity and pressure (which is related to density by an equation of state, here $p = c_s^2 \rho$, c_s - speed of sound) for small Mach numbers satisfies the Navier-Stokes equations.

Obviously, (1) is an explicit Finite-Difference scheme based on a local stencil. For all grid nodes the collision term only depends on the local set of probability distributions of the actual timestep and the computation of the left-hand side of (1) requires information from the neighbouring nodes.

A straightforward approach to implement (1) is to map the discretized flow domain onto a uniform grid using full matrices to store f_i. The obvious advantages are:

- The generation of grids is trivial, even for complex geometries such as porous media. All nodes are marked with respect to being part of the flow or solid domain by using a flag matrix which can be directly obtained from e.g. computer based tomography.
- Neighbour relationships between nodes are known due to the topological simplicity of the full matrix approach and allow an easy and effective memory access. This in turn permits very effective vectorization and parallelization when using domain decomposition based on geometric slicing of the simulation domain.

Yet for many practical problems there is a serious drawback: If the volume of the flow domain is small compared to the volume of its bounding box, the use of full matrices results in a severe waste of memory.

Concerning parallelization, two additional points have to be taken into account:

- Load balancing:
  Especially when considering inhomogeneous simulation domains, for a huge number of parallel processes it is no longer possible to subdivide the computational domain into slices or cuboids without severe load imbalancing.

- Memory consumption:
  Even if for a moderate number of subdomains resonable load balancing with respect to an equal number of degrees of freedom (DOF) per subdomain can be achieved, the memory consumption per subdomain can vary significantly which may be critical for performance on distributed memory hardware.

To avoid these problems, we developed an efficient data structure which on the one hand maintains most of the advantages of a uniform grid, such as good performance in terms of Flop rate and which on the other hand in many cases drastically reduces memory consumption. In addition, the main problems of using full matrices for weakly connected flow domains in parallel simulations are completely cured.

2 A Data Structure Based on Indirect Addressing

The propagation step (evaluation of the left hand side of (1)) for an implementation using full matrices for e.g. f_1 assuming $\boldsymbol{e}_1 \equiv (1,0,0)$ typically could be implemented as (pseudo-code):

```
for(x=max_x-1 ; x>0 ; x--)   f1(x,y,z) = f1(x-1,y,z)
```

Using a topologically unstructured grid (i.e. a uniform cartesian grid with "holes"), neighbouring nodes are no longer immediately known. Having lost this natural ordering, one can take advantage to sort the distributions to optimize data flow for the computation of the collision term with respect to the boundary condition type as depicted in Fig. 1.

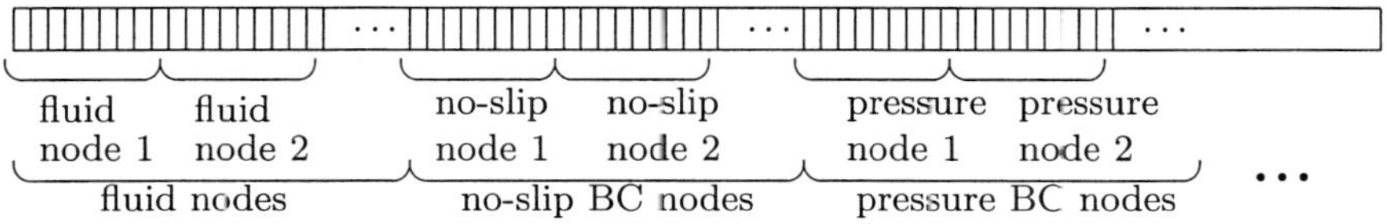

Fig. 1. Sorting within the one-dimensional array $f[]$ according to nodal boundary attributes

Propagation now implies copying distributions to arbitrary places within the one-dimensional array. The corresponding origins and destinations have to be stored in precomputed lists. Using only one array for the distributions this information has to be stored using two additional index arrays to store the target and source indices which minimizes memory consumption.

The propagation of e.g. f_1 assuming $\boldsymbol{e}_1 \equiv (1,0,0)$ results in

```
for(i=0 ; i<num_propagations ; i++)  f(target(i)) = f(source(i))
```

where *num_propagations* is the number of propagations required. This approach results in a moderate performance due to the two indirect addressing operations. Using an additional copy array for the f_i only one dereferencing is necessary:

```
for(i=0 ; i<num_propagations ; i++)  f_copy(target(i)) = f(i)
```

This roughly doubles the computational performance at the price of a slight increase in memory requirements.

The following example shows the advantage of using an unstructured grid with respect to memory requirement for a porous flow geometry with a porosity of 20% (Fig. 2, no-slip boundaries are shown in black and surround the flow channels) and the grid resolution of the x-ray tomography of $83 \times 83 \times 30$ was increased using three-dimensional interpolation [5] by a factor of five.

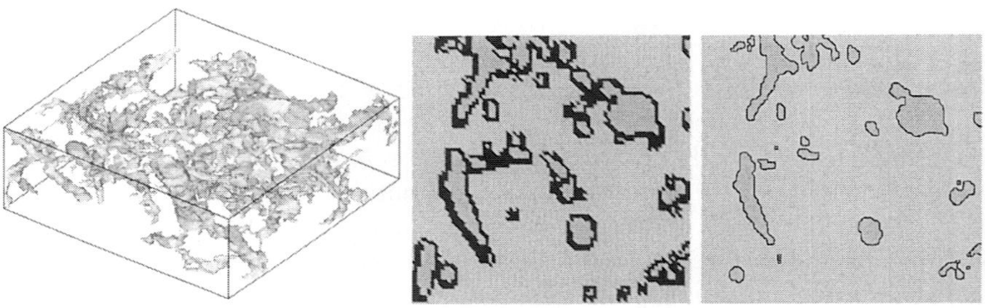

Fig. 2. Porous flow geometry: (**a**) 3D-view, (**b**) slicing for low resolution, (**c**) slicing for high resolution (factor 5)

Depending on the chosen scaling factor (i.e. grid resolution) the expected decrease in memory consumption as compared to the use of full matrices are shown in Fig. 3.

3 Domain Decomposition

Having abandoned the use of full matrices the geometric domain decomposition is no longer restricted to produce cuboidal subdomains. Thus we can utilize a state-of-the-art tool based on graph partitioning methods (METIS [3]) to obtain arbitrarily shaped subdomains which leads to very good load balancing even for a large number of subdomains for any flow geometry. An example is given in Fig. 4.

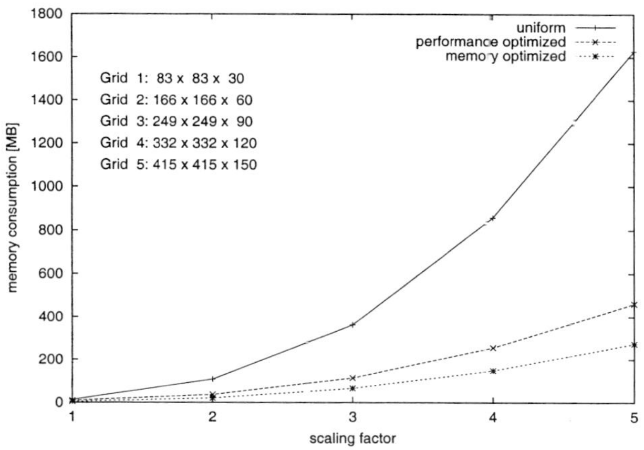

Fig. 3. Memory consumption for different grid resolutions

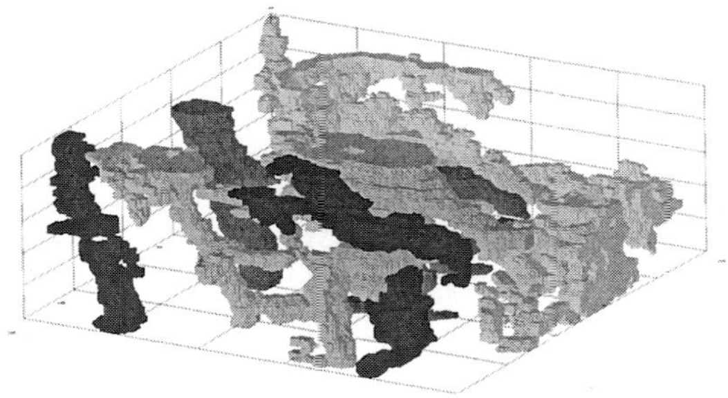

Fig. 4. METIS domain decomposition of a porous medium

4 Parallelization Strategy

As indicated above, all necessary information for a nodal update is basically obtained from next neighbouring nodes as is depicted in Fig. 5.

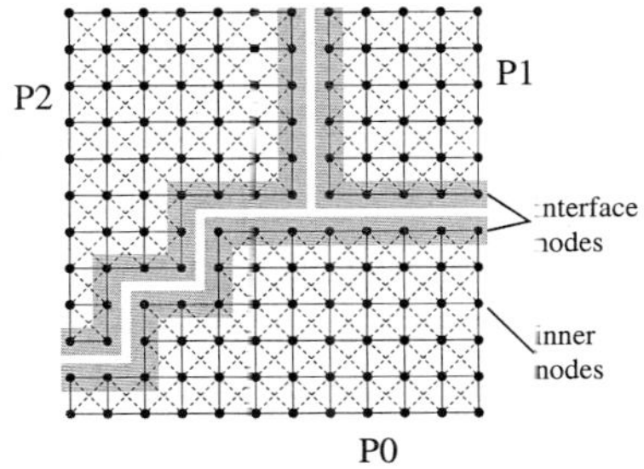

Fig. 5. Cut through the decomposed flow domain indicating the interface width

The information that has to be transferred between pairs of subdomains in each timestep depends on the length and width of the mutual interfaces. Asymptotically, the ratio of the computational effort for the inner and interface nodes is proportional to the inverse of the number of inner nodes, thus a reasonable parallel efficiency can be expected as the cost for the computational overhead due to data exchange becomes vanishingly small for subdomains of increasing size.

There is still room for fine tuning as it is well known that any parallel efficiency can be further improved by partially overlapping the process of data exchange between subdomains and the interface-independant computation (in case of appropriate network hardware). For the LBGK implementation proposed here this overlapping can be optimized by computing the collision term of the interface nodes and a subsequent initiation of their propagation to the corresponding subdomains by using asynchronous non-blocking message passing calls of the underlying MPI library [4]. After this initiation the collision and propagation of the inner nodal distributions (being by far the computationally dominant part) is done. Thus there is additional time left for the completion of the message transfer. Figure 6 shows the difference between the standard approach and the *optimal* one presented here.

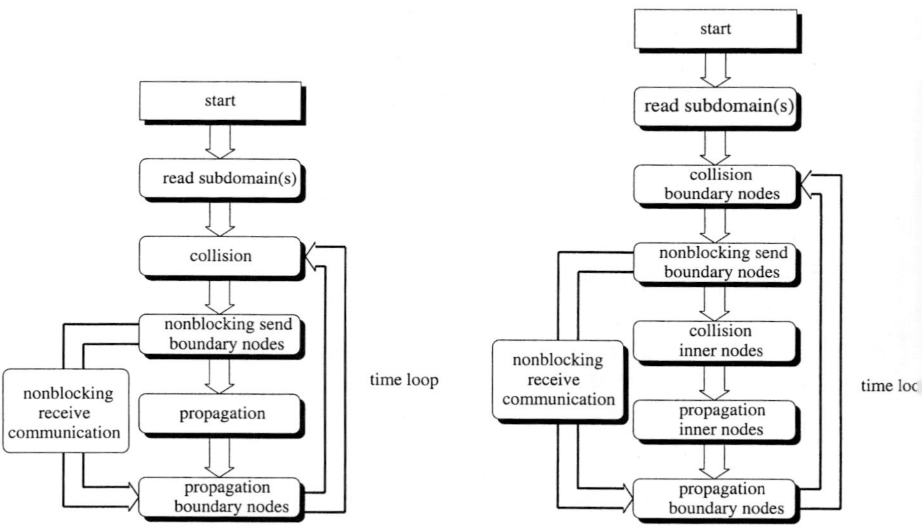

Fig. 6. Flow chart for the standard parallel-LB algorithm and the optimal implementation (right)

It should be noted, that the propagation of the interface distributions is naturally taken care of by the communication routine.

5 Results

The data structure explained above was implemented on an eight machine DEC EV 6 workstation cluster as well as on the Bundeshöchstleistungsrechner Hitachi SR8000-F1 being located at the Leibniz-Rechenzentrum Munich. On the Hitachi a two-level parallelization was implemented, as each node of this machine consists of eight processors sharing memory. Thus the auto-parallelizing compiler took care of the intra-node communication. On one computational node we obtain an update rate of up to 2×10^7 (in other words the computation of one timestep of (1) for a grid of 2×10^7 grid nodes takes about one second). This corresponds to about 3 GFlops. The inter-nodal communication was based on the message-passing scheme explained above.

A parallel efficiency of about 0.9 was achieved for various testruns (Fig. 7). Taking into account that the test cases investigated utilized at most 25 % of the nodal memory, the parallel efficiency for subdomains of maximum size can be estimated to be even closer to one.

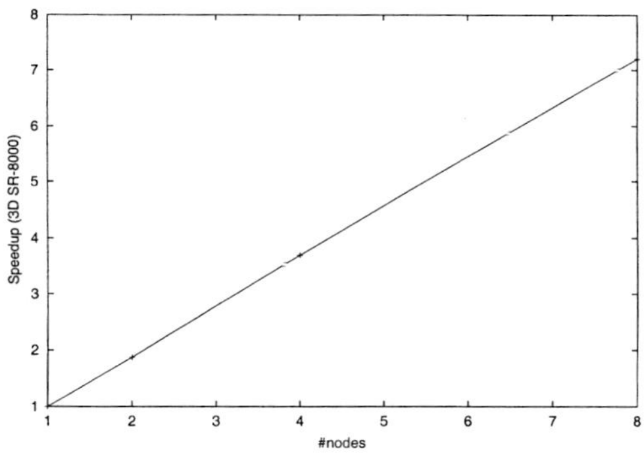

Fig. 7. Decomposed flow domain

As a numerical example we computed the friction factor (a non-dimensional measure for the permeability of a flow geometry) for a sphere packing generated by a molecular dynamics type simulation tool [6]. Figure 8 shows one half of the sphere packing inside a cylinder containing more than 1000 spheres. The diameter of one sphere corresponds to thirty grid nodes resulting in an overall system of about 1.7×10^7 nodes. The porosity with respect to the bounding box of the surrounding cylinder is 27 %, the bounding box volume would correspond to 4×10^7 nodes. The computation time (excluding data IO) using four nodes of the Hitachi SR8000-F1 was about 75 minutes.

The resulting friction factor coincides within 3 % with measurements obtained from experiment [7] and direct numerical simulation on a flow geometry with evenly distributed spheres [8].

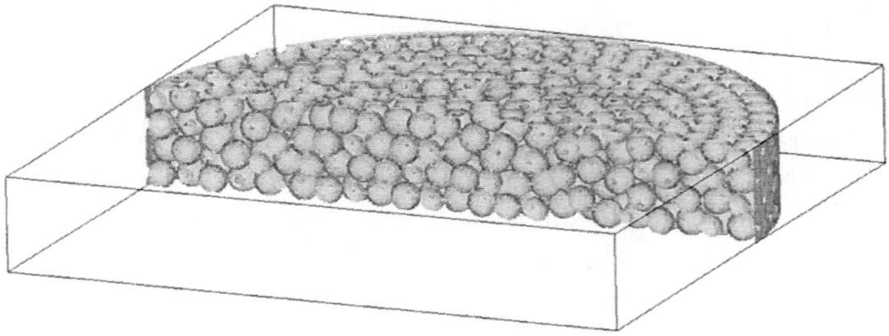

Fig. 8. Flow geometry

Acknowledgement

The authors would like to thank the HPC group at the Leibniz-Rechenzentrum for valuable technical support. The project is funded by the Deutsche Forschungsgemeinschaft (Ra 624/7-1).

References

1. Quian, Y.H., d'Humieres, D. and Lallemand, P. (1992) Lattice BGK models for N-S equation. Europhys.Lett. **17(6)**, 479–484
2. Krafczyk, M. (2001) Gitter-Boltzmann-Methoden - von der Theorie zur Anwendung. Professorial Thesis LS Bauinformatik TU München
3. Karypis, G., Kumar, V. (1998) Multilevel Algorithms for Multi-Constraint Graph Partitioning. http://www-users.cs.umn.edu/~karypis/publications/partitioning.html
4. http://www-unix.mcs.anl.gov/mpi/index.html
5. Engeln-Müllges, G., Reutter, F. (1996) Numerikalgorithmen. VDI Verlag 1996, ISBN 3-18-401539-4
6. Ristow, G. H. (1994) Granular Dynamics: a Review about recent Molecular Dynamics Simulations of Granular Materials. Annual Reviews of Computational Physics **I**, 275–308
7. Durst, F., Haas, R., Interthal, W. (1987) The Nature of Flows through Porous Media. J. Non Newtonian Fluid Mech **22** , pp. 169–189
8. Bernsdorf, J., Brenner, G., Durst, F. (2000) Numerical analysis of the pressure drop in porous media flow with lattice Boltzmann (BGK) automata. Comp. Phys. Com. **129**, 247–255

Applications of the Lattice Boltzmann Method to Complex and Turbulent Flows

L.-S. Luo[1], D. Qi[2], and L.-P. Wang[3]

[1] ICASE, MS 132C, NASA Langley Research Center
3 West Reid Street, Building 1152, Hampton, Virginia 23681-2199, USA
Email: luo@icase.edu; Homepapge: http://www.icase.edu/~luo

[2] Department of Paper and Printing Science and Engineering
Western Michigan University, Kalamazoo, Michigan 49008, USA

[3] Department of Mechanical Engineering
University of Delaware, Newark, Delaware 19716, USA

Abstract. We briefly review the method of the lattice Boltzmann equation (LBE). We show the three-dimensional LBE simulation results for a non-spherical particle in Couette flow and 16 particles in sedimentation in fluid. We compare the LBE simulation of the three-dimensional homogeneous isotropic turbulence flow in a periodic cubic box of the size 128^3 with the pseudo-spectral simulation, and find that the two results agree well with each other but the LBE method is more dissipative than the pseudo-spectral method in small scales, as expected.

1 Introduction

More than a decade ago, the lattice-gas automata (LGA) [5,22,6] and the lattice Boltzmann equation (LBE) [16,12,2,20] were proposed as alternatives for computational fluid dynamics (CFD). Since their inception, the lattice-gas and lattice Boltzmann methods have attracted much interest in the physics community. However, it was only very recently that the LGA and LBE methods started to gain the attention from CFD community. The lattice-gas and lattice Boltzmann methods have been particularly successful in simulations of fluid flow applications involving complicated boundaries or/and complex fluids, such as turbulent external flow over complicated structures, the Rayleigh-Taylor instability between two fluids, multi-component fluids through porous media, viscoelastic fluids, free boundaries in flow systems, particulate suspensions in fluid, chemical reactive flows and combustions, magnetohydrodynamics, crystallization, and other complex systems (see recent reviews [3,15] and references therein).

Historically, models of the lattice Boltzmann equation evolved from the lattice-gas automata [5,22,6]. Recently, it has been shown that the LBE is a special discretized form of the continuous Boltzmann equation [8,9]. For the sake of simplicity without loss of generality, we shall demonstrate an *a priori* derivation of the lattice Boltzmann equation from the continuous Boltzmann equation with the single relaxation time (Bhatnagar-Gross-Krook) approximation [1]. The Boltzmann BGK equation can be written in the form of an ordinary differential equation:

$$D_t f + \frac{1}{\lambda} f = \frac{1}{\lambda} f^{(0)}, \qquad f^{(0)} \equiv \frac{\rho}{(2\pi\theta)^{D/2}} \exp\left[-\frac{(\boldsymbol{\xi}-\boldsymbol{u})^2}{2\theta}\right], \tag{1}$$

where $D_t \equiv \partial_t + \boldsymbol{\xi}\cdot\boldsymbol{\nabla}$, $f \equiv f(\boldsymbol{x}, \boldsymbol{\xi}, t)$ is the single particle distribution function, λ is the relaxation time, and $f^{(0)}$ is the Boltzmann distribution function in D-dimensions, in which ρ, $\boldsymbol{u}$ and $\theta = k_B T/m$ are the macroscopic density of mass, the velocity, and the normalized temperature, respectively, T, k_B and m are temperature, the Boltzmann constant, and particle mass. The macroscopic variables are the moments of the distribution function f with respect to the molecular velocity $\boldsymbol{\xi}$:

$$\rho = \int f\, d\boldsymbol{\xi} = \int f^{(0)}\, d\boldsymbol{\xi}, \tag{2a}$$

$$\rho\boldsymbol{u} = \int \boldsymbol{\xi}\, f\, d\boldsymbol{\xi} = \int \boldsymbol{\xi}\, f^{(0)}\, d\boldsymbol{\xi}, \tag{2b}$$

$$\rho\theta = \frac{1}{2}\int (\boldsymbol{\xi}-\boldsymbol{u})^2 f\, d\boldsymbol{\xi} = \frac{1}{2}\int (\boldsymbol{\xi}-\boldsymbol{u})^2 f^{(0)}\, d\boldsymbol{\xi}. \tag{2c}$$

Equation (1) can be formally integrated over a time interval δ_t:

$$f(\boldsymbol{x}+\boldsymbol{\xi}\delta_t, \boldsymbol{\xi}, t+\delta_t) = e^{-\delta_t/\lambda} f(\boldsymbol{x}, \boldsymbol{\xi}, t) + \frac{1}{\lambda} e^{-\delta_t/\lambda} \int_0^{\delta_t} e^{t'/\lambda} f^{(0)}(\boldsymbol{x}+\boldsymbol{\xi}t', \boldsymbol{\xi}, t+t')\, dt'. \tag{3}$$

Assuming that δ_t is small enough and $f^{(0)}$ is smooth enough locally, and neglecting the terms of the order $O(\delta_t^2)$ or smaller in the Taylor expansion of the right hand side of (3), we obtain

$$f(\boldsymbol{x}+\boldsymbol{\xi}\delta_t, \boldsymbol{\xi}, t+\delta_t) - f(\boldsymbol{x}, \boldsymbol{\xi}, t) = -\frac{1}{\tau}[f(\boldsymbol{x}, \boldsymbol{\xi}, t) - f^{(0)}(\boldsymbol{x}, \boldsymbol{\xi}, t)], \tag{4}$$

where $\tau \equiv \lambda/\delta_t$ is the dimensionless relaxation time. The equilibrium $f^{(0)}$ can be expanded as a Taylor series in $\boldsymbol{u}$ up to $\boldsymbol{u}^2$

$$f^{(\mathrm{eq})} = \frac{\rho}{(2\pi\theta)^{D/2}} \exp\left(-\frac{\boldsymbol{\xi}^2}{2\theta}\right) \left[1 + \frac{(\boldsymbol{\xi}\cdot\boldsymbol{u})}{\theta} + \frac{(\boldsymbol{\xi}\cdot\boldsymbol{u})^2}{2\theta^2} - \frac{\boldsymbol{u}^2}{2\theta}\right]. \tag{5}$$

To obtain the Navier-Stokes equations, the hydrodynamic moments (ρ, $\rho\boldsymbol{u}$, and $\rho\theta$) and their fluxes must be preserved in finite discretized momentum space $\{\boldsymbol{\xi}_\alpha | \alpha = 1, 2, \ldots, b\}$, *i.e.*,

$$\rho = \sum_\alpha f_\alpha = \sum_\alpha f_\alpha^{(\mathrm{eq})}, \tag{6a}$$

$$\rho\boldsymbol{u} = \sum_\alpha \boldsymbol{\xi}_\alpha f_\alpha = \sum_\alpha \boldsymbol{\xi}_\alpha f_\alpha^{(\mathrm{eq})}, \tag{6b}$$

$$\rho\theta = \frac{1}{2}\sum_\alpha (\boldsymbol{\xi}_\alpha - \boldsymbol{u})^2 f_\alpha = \frac{1}{2}\sum_\alpha (\boldsymbol{\xi}_\alpha - \boldsymbol{u})^2 f_\alpha^{(\mathrm{eq})}, \tag{6c}$$

where $f_\alpha \equiv f_\alpha(\boldsymbol{x}, t) \equiv W_\alpha f(\boldsymbol{x}, \boldsymbol{\xi}_\alpha, t)$ [8,9]. It turns out that these moments can be evaluated *exactly* in discretized momentum space by using Gaussian-type quadrature [8,9,21].

We can derive the nine-velocity athermal LBE model on a square lattice in two-dimensions

$$f_\alpha(\boldsymbol{x}_i + \boldsymbol{e}_\alpha \delta_t, t + \delta_t) - f_\alpha(\boldsymbol{x}_i, t) = -\frac{1}{\tau}[f_\alpha(\boldsymbol{x}_i, t) - f_\alpha^{(\mathrm{eq})}(\boldsymbol{x}_i, t)], \quad (7)$$

where the equilibrium $f_\alpha^{(\mathrm{eq})}$, the discrete velocity set $\{\boldsymbol{e}_\alpha\}$, and the weight coefficients $\{w_\alpha\}$ are given by

$$f_\alpha^{(\mathrm{eq})} = w_\alpha\, \rho \left\{1 + \frac{3(\boldsymbol{e}_\alpha \cdot \boldsymbol{u})}{c^2} + \frac{9(\boldsymbol{e}_\alpha \cdot \boldsymbol{u})^2}{2c^4} - \frac{3\boldsymbol{u}^2}{2c^2}\right\}, \quad (8a)$$

$$\boldsymbol{e}_\alpha = \begin{cases} (0,\, 0), & \alpha = 0\,, \\ (\pm 1,\, 0)c,\ (0,\, \pm 1)c\,, & \alpha = 1,\, 2,\, 3,\, 4, \\ (\pm 1,\, \pm 1)\sqrt{2}c, & \alpha = 5,\, 6,\, 7,\, 8, \end{cases} \quad (8b)$$

$$w_\alpha = \begin{cases} 4/9, & \alpha = 0, \\ 1/9, & \alpha = 1,\, 2,\, 3,\, 4, \\ 1/36, & \alpha = 5,\, 6,\, 7,\, 8, \end{cases} \quad (8c)$$

and $c \equiv \delta_x/\delta_t$. Equation (7) involves only local calculations and uniform communications to the nearest neighbors. Therefore it is easy to implement and natural to massively parallel computers.

The (incompressible) Navier-Stokes equation derived from the above LBE model is:

$$\rho\partial_t \boldsymbol{u} + \rho \boldsymbol{u}\cdot\nabla \boldsymbol{u} = -\nabla P + \rho\nu\nabla^2 \boldsymbol{u}\,, \quad (9)$$

with the isothermal ideal gas equation of state, the viscosity, and the sound speed given by

$$P = c_s^2 \rho\,, \qquad \nu = \left(\tau - \frac{1}{2}\right) c_s^2 \delta_t\,, \qquad c_s = \frac{1}{\sqrt{3}}\, c\,. \quad (10)$$

It should be noted that the factor $-1/2$ in the above formula for ν accounts for the numerical viscosity due to the second order derivatives of f_α. This correction in ν formally makes the LGA and LBE methods second order accurate. Similarly, we can derive the six-velocity and seven-velocity models on a triangular lattice in two-dimensions, and the twenty-seven-velocity models on a cubic lattice in three-dimensions [9].

There have been some significant progress made recently to improve the lattice Boltzmann method: (i) the generalized lattice Boltzmann equation with multiple relaxation times which overcomes some shortcomings of the lattice BGK equation [14]; (ii) use of grid refinement [4] and body-fitted mesh [10,7] with interpolation/extrapolation techniques; (iii) adaptation of unstructured grid by using the finite element method or the characteristic

Galerkin method; (iv) application of implicit scheme for steady state calculation and multi-grid technique to accelerate convergence (see a recent review [15] for further references).

In what follows we shall demonstrate the applications of the LBE method to simulate the flow of non-spherical particulate suspensions in fluid and homogeneous isotropic turbulence in a periodic box.

2 LBE Simulation of Flows of Non-Spherical Particulate Suspensions

The flow of particulate suspensions in fluid is difficult to quantify experimentally and to simulate numerically in some cases. Yet the flow of particulate suspensions is important to industrial applications such as fluidized beds. There have been some successful simulations of the flow of spherical suspensions by using conventional CFD methods, such as the finite element method. However, the simulation of the flow of non-spherical suspensions still remains as a challenge to the conventional CFD methods. Recently the LBE method has been successfully applied to simulate the flow of non-spherical suspensions in three-dimensions [17,18]. The success of the LBE method to this problem relies on the fact that the LBE method can easily handle the particle-fluid interfaces.

We first simulate a single non-spherical particle in the Couette flow. The equilibrium states in a non-spherical particulate suspension in a 3D Couette flow are simulated for a particle Reynolds number up to 320. Particle geometries include prolate and oblate spheroids, cylinders and discs. We show that the inertial effect at any finite Reynolds number qualitatively changes the rotational motion of the suspension, contrary to Jeffery's theory at zero Reynolds number [13]. At a non-zero Reynolds number, a non-spherical particle reaches an equilibrium state in which its longest and shortest axes are aligned perpendicular and parallel to the vorticity vector of the flow, respectively. This equilibrium state is unique, dynamically stable, fully determined by the inertial effect, the maximum energy dissipation state. Systems of either fifty cylinders or fifty discs in Couette flow are also simulated. Multi-particle interactions significantly change the equilibrium orientation of solid particle. The effect is stronger for cylinders than for discs. The details of this work will be reported elsewhere [19].

Figure 1 shows a 3D LBE simulation of sixteen cylindrical particles falling under the influence of gravity. The left figure illustrates the time evolution of the entire system of sixteen particles, while the right figure demonstrates the formation of inverted T configurations in the sedimentation, which has been observed experimentally. To the best of our knowledge, this phenomenon was first reproduced numerically by the LBE direct numerical simulation [18].

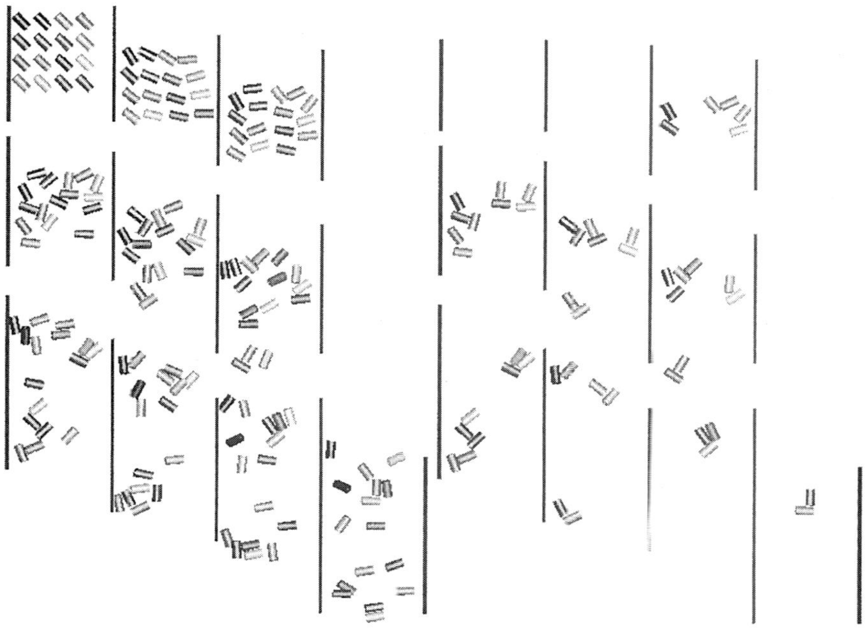

Fig. 1. 3D LBE simulation of particles sedimentation in fluid. Particle size is $D = 12$ and $L = 24$. System size is $N_x \times N_y \times N_z = 140 \times 150 \times 35$. The averaged single-particle Re ≈ 16.9. (left) Evolution of 16 particles (from left to right and top to bottom). (right) Formation of inverted T configurations which are also observed in experiment

3 LBE Simulation of 3D Homogeneous Isotropic Turbulence

Homogeneous isotropic turbulence in a three-dimensional periodic cubic box remains as a stand problem in the field of direct numerical simulation of turbulence. Due to the simplicity of the boundary conditions, the pseudo-spectral method can be easily used to simulate the flow. Because of its accuracy, the pseudo-spectral result is often used as a benchmark standard. Here the LBE simulation of the flow is compared with the pseudo-spectral simulation.

The parameters of the simulation are given in Table 1. The initial condition is a random velocity field with a Gaussian distribution and a compact energy spectrum. The boundary conditions are periodic in three dimensions. The Taylor microscale Reynolds number is defined as

$$\mathrm{Re}_\lambda = \sqrt{\frac{2K(t=0)}{3}}\frac{\lambda}{\nu} = \frac{u_{\mathrm{RMS}}\lambda}{\nu},$$

Table 1. Parameters in lattice Boltzmann and pseudo-spectral simulations: L is the length of box side; N^3 is the system size; ν is the viscosity; u' is the RMS fluctuation of the initial velocity field; dt is the time step size; T is total integration time, Re_λ is the Taylor microscale Reynolds number; and M is the Mach number

Method	L	N^3	ν	u'	dt	T	Re_λ	M
Spectral	2π	128^3	0.01189	0.993311	0.002	2	35.0	0
LBE	128	128^3	0.009869	0.040471	1	1000	35.0	0.0687

where $K(t=0) = \langle \boldsymbol{u}_0^2/2 \rangle_V = \langle 3u_{\mathrm{RMS}}^2/2 \rangle_V$ is the volume averaged kinetic energy (of the initial zero-mean Gaussian velocity field $\boldsymbol{u}_0$ with RMS component u_{RMS}), and λ is the transverse Taylor microscopic scale:

$$\lambda = \sqrt{15\nu u_{\mathrm{RMS}}^2/\epsilon},$$

where ϵ is the dissipation rate.

Figure 2 shows the energy spectrum $E(k)$ as function of time, and the time evolution of the mean kinetic energy K and dissipation rate ϵ. The lattice Boltzmann results (symbols) are compared with the pseudo-spectral results (lines). The LBE results agree well with the pseudo-spectral results. Obviously the LBE method is more dissipative, especially at high wave numbers $k > \frac{1}{3}k_{\mathrm{max}}$, where $k_{\mathrm{max}} = \frac{1}{3}N$, and N is the number of mesh nodes in each direction. This is because the LBE method is only second order accurate in space and time and thus more dissipative than the pseudo-spectral method.

4 Conclusions and Discussion

The above simulations were performed on a Beowulf cluster of Pentium CPUs. For the simulation of the particulate suspension, the code consists two part: the lattice Boltzmann method for the fluid and molecular dynamics (MD) for the solid particles [17]. Even though the MD part of the code is not yet parallelized, the speed of the code still scales well with the number of CPUs up to 32 CPUs when the system size is 64^3 and with fifty particles. Presently we can easily simulate a system of a few hundred particles on our Beowulf system.

As for the simulation of the 3D homogeneous isotropic turbulence, the LBE code without optimization has the same speed as the spectral code with a Beowulf cluster of eight CPUs (about 1s per time step). However, we do expect the LBE code will scale linearly with the number of CPUs, but not the spectral code.

Our current research includes particulate suspension in fluid with high volume fraction of particles, viscoelastic and non-Newtonian fluids, and forced or free-decay homogeneous isotropic turbulence in a periodic cube by using the lattice Boltzmann method on massively parallel computers.

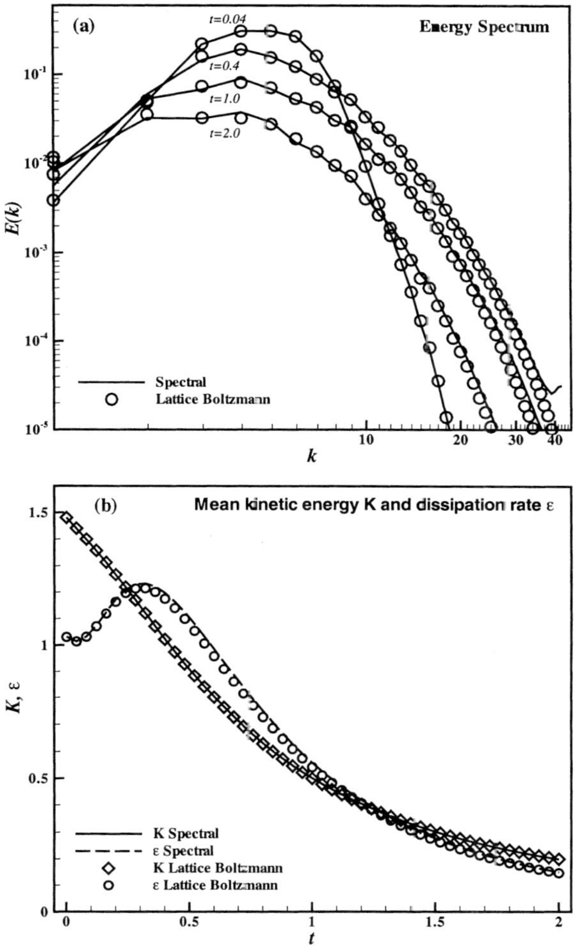

Fig. 2. LBE *vs.* Pseudo-spectral DNS of 3D homogeneous isotropic turbulence. System size is 128^3. $\mathrm{Re}_\lambda = 35$. (a) The energy spectrum $E(k)$ as a function of time. (b) The decay of the mean kinetic energy K and dissipation rate ϵ. The results from the LBE simulation are scaled according to the dimensions used in the spectral simulation

References

1. Bhatnagar P. L., Gross E. P., Krook M. (1954) A model for collision processes in gases. I. Small amplitude processes in charged and neutral one-component systems. Phys. Rev. **94**:511–525
2. Chen H., Chen S., Matthaeus W.H. (1992) Recovery of the Navier-Stokes equations using a lattice-gas Boltzmann method. Phys. Rev. A **45**:R5339–5342

3. Chen S., Doolen G.D. (1998) Lattice Boltzmann method for fluid flows. Ann. Rev. Fluid Mech. **30**:329–364
4. Filippova O., Hänel D. (1998) Grid refinement for lattice-BGK models. J. Comput. Phys. **147**:219–228
5. Frisch U., Hasslacher B., Pomeau Y. (1986) Lattice-gas automata for the Navier-Stokes equation. Phys. Rev. Lett. **56**:1505–1508
6. Frisch U., d'Humières D., Hasslacher B., Lallemand P., Pomeau Y., Rivet J.-P. (1987) Lattice gas hydrodynamics in two and three dimensions. Complex Systems **1**:649–707
7. He X., Doolen G. (1997) Lattice Boltzmann method on curvilinear coordinates system: Flow around a circular cylinder. J. Comput. Phys. **134**:306–315
8. He X., Luo L.-S. (1997) *A priori* derivation of the lattice Boltzmann equation. Phys. Rev. E **55**:R6333–R6336
9. He X., Luo L.-S. (1997) Theory of the lattice Boltzmann equation: From the Boltzmann equation to the lattice Boltzmann equation. Phys. Rev. E **56**:6811–6817
10. He X., Luo L.-S., Dembo M. (1996) Some progress in lattice Boltzmann method. Part I. Nonuniform mesh grids. J. Comput. Phys. **129**:357–363
11. He X., Luo L.-S., Dembo M. (1997) Some Progress in the lattice Boltzmann method. Reynolds number enhancement in simulations. Physica A **239**:276–285
12. Higuera F.J., Succi S., Benzi R. (1989) Lattice gas-dynamics with enhanced collisions. Europhys. Lett. **9**:345–349
13. Jeffery G.B. (1922) The motion of ellipsoidal particles immersed in a viscous fluid. Proc. R. London Ser. A **102**:161–179
14. Lallemand P., Luo L.-S. (2000) Theory of the lattice Boltzmann method: Dispersion, dissipation, isotropy, Galilean invariance, and stability. Phys. Rev. E **61**:6546–6562
15. Luo L.-S. (2000) The lattice-gas and lattice Boltzmann methods: Past, present, and future. In: Wu J.-H., Zhu Z.-J. (Eds) Proceedings of International Conference on Applied Computational Fluid Dynamics, Beijing, China, October 17–20, 2000. Beijing, 52–83
16. McNamara G. R., Zanetti G. (1988) Use of the Boltzmann equation to simulate lattice-gas automata. Phys. Rev. Lett. **61**:2332–2335
17. Qi D. (1999) Lattice-Boltzmann simulations of particles in non-zero-Reynolds-number flows. J. Fluid Mech. **385**:41–62
18. Qi D. (2001) Simulations of fluidization of cylindrical multiparticles in a three-dimensional space. Int. J. Multiphase Flow **27**:107–118
19. Qi D., Luo L.-S. (2001) Inertial effect of non-spherical suspension in 3D Couette flow. Submitted to Phys. Rev. Lett.
20. Qian Y.H., d'Humières D., Lallemand P. (1992) Lattice BGK models for Navier-Stokes equation. Europhys. Lett. **17**:479–484
21. Shan X., He X. (1998) Discretization of the velocity space in the solution of the Boltzmann equation. Phys. Rev. Lett. **80**:65–68
22. Wolfram S. (1986) Cellular automaton fluids 1: Basic theory. J. Stat. Phys. **45**:471–526

Computation of Flows Around Space Configurations

A. Henze, W. Schröder, and M. Meinke

Aerodynamisches Institut, RWTH-Aachen, Wüllnerstr. zw. 5 und 7, D-52062 Aachen

Abstract. Numerical investigations in the sub- and supersonic flowfield around space configurations are presented. The subsonic turbulent flow around the space plane configuration ELAC as well as the supersonic laminar flow around have been considered. The third configuration is a two-stage, i. e. EOS mounted on ELAC. system. The numerical results are obtained by solving the Navier-Stokes equations using a finite volume method based on node-centered blockstructured grids in curvilinear coordinates. The Baldwin-Lomax turbulence model was used for fully turbulent flow without transition. The results presented are compared with experimental findings and show satisfactory agreement.

1 Introduction

The development of aerospace vehicles requires accurate numerical and experimental investigations for the determination of aerodynamic forces and heat loads. The present paper deals with the computation of threedimensional flows around such configurations at low and high Mach numbers. Objects of the investigation are the hypersonic research configuration ELAC-1 (Fig. 1) as well as the prospective second stage EOS (Fig. 2). Both vehicles have been designed at Aachen in the Collaborative Research Center "Fundamentals of the Design of Aerospace Planes" (SFB 253) supported by the German Research Association (DFG). ELAC is a lifting body with a delta-wing shape consisting of two semi-elliptic cross-sections. This leads to rounded leading edges. The main cross-section is located at 2/3 chord length and its thickness is about 7%. The rear part of ELAC is an expansion ramp with two winglet-shaped control surfaces. The upper stage EOS is a body-wing configuration with a central fin. Most of the geometry can be described in terms of ellipses, parabolas, straight lines and circles. The wings and the control surface possess the contour of NACA-profiles. The configuration ELAC-1c is a piggyback system of both configurations (Fig. 3).

Usually, wind tunnel models are much smaller than the real configurations. Due to this fact the flows around the model and the full-scale configuration cannot be regarded as fully similar. The influence of the Reynolds number on transition and turbulence cannot be analyzed correctly. Therefore, to minimize the gap between real and model flow experimental results of subsonic flow around ELAC-1 have been obtained at high Reynolds numbers up to 40×10^6 with a large-scale model (1:12) in the German Dutch

Windtunnel (DNW) [2,3]. The numerical results of the present investigation will be compared with these experimental data.

First results for the laminar supersonic flow around EOS and ELAC-1c will be presented. At different angles of attack the separation of the flow on the leeside can be predicted. In former investigations of the ELAC-1 configuration the flow separation was already discussed in detail [4,5]. The new results for the supersonic flow around the piggyback system show the influence of the upper stage on the flow field.

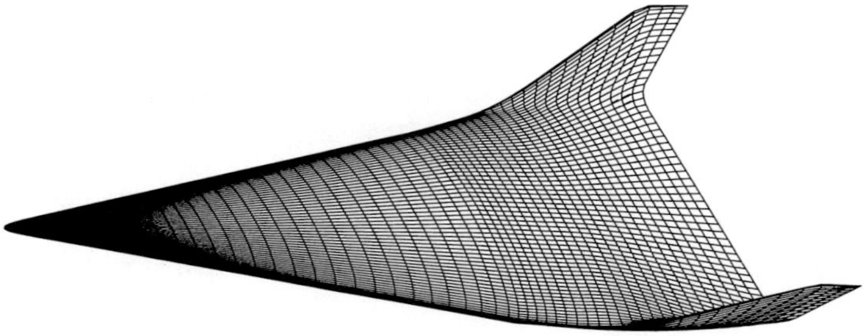

Fig. 1. Hypersonic research configuration ELAC-1

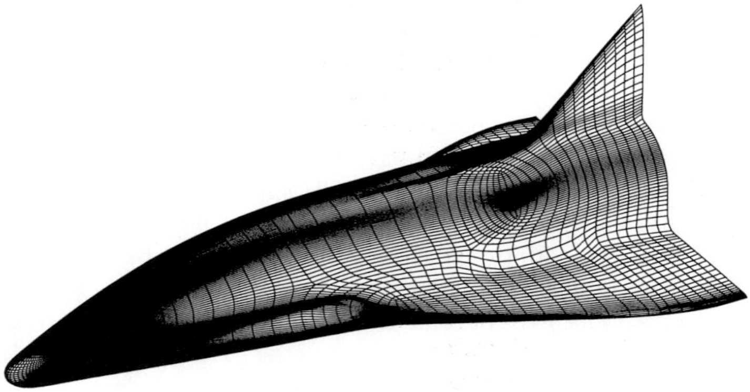

Fig. 2. **E**lac **O**rbital **S**tage

2 Numerical Scheme

2.1 Governing Equations

The Navier-Stokes read in tensor notation for a cartesian coordinate system

$$\frac{\partial \boldsymbol{Q}}{\partial t} + (\boldsymbol{F}^C_\beta - \boldsymbol{F}^D_\beta)_{,\beta} = 0 \quad , \qquad \boldsymbol{Q} = [\varrho, \varrho u_\alpha, \varrho E]^T \quad ,$$

with the vector of the conservative variables $\boldsymbol{Q}$ and the vector of convective ($\boldsymbol{F}^C_\beta$) and diffusive ($\boldsymbol{F}^D_\beta$) fluxes.

$$\boldsymbol{F}^C_\beta - \boldsymbol{F}^D_\beta = \begin{pmatrix} \varrho u_\beta \\ \varrho u_\alpha u_\beta + p\delta_{\alpha\beta} \\ u_\beta(\varrho E + p) \end{pmatrix} + \frac{1}{Re} \begin{pmatrix} 0 \\ \sigma_{\alpha\beta} \\ u_\alpha \sigma_{\alpha\beta} + q_\beta \end{pmatrix} \quad .$$

$$\sigma_{\alpha\beta} = -2\,\mu(S_{\alpha\beta} - \tfrac{1}{3} S_{\gamma\gamma}\delta_{\alpha\beta}) \qquad \text{with} \qquad S_{\alpha\beta} = \tfrac{1}{2}(u_{\alpha,\beta} + u_{\beta,\alpha}) \quad .$$

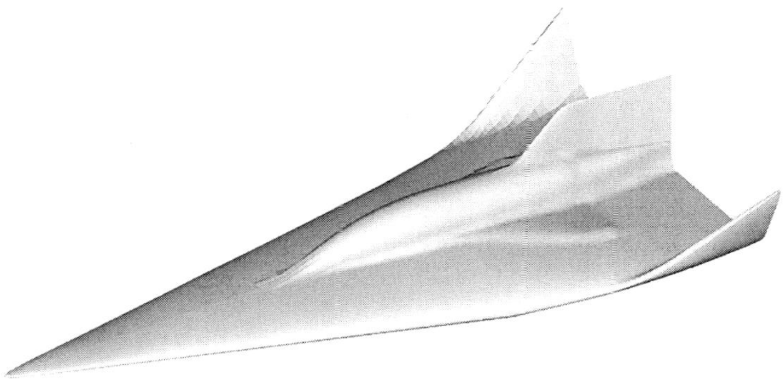

Fig. 3. Piggyback system of both configurations

The equations are closed using the thermal and calorical equations of state for an ideal gas as well as Fourier's law of heat conduction for the heat flux q_β. Assuming a constant Prandtl number the dimensionless heat conductivity equals the dynamic viscosity which is obtained using an exponential law

$$q_\beta = -\frac{k}{Pr(\gamma - 1)} T_\beta, \; p = \frac{1}{\gamma}\varrho T, \; p = (\gamma - 1)\left(\varrho e - \frac{\varrho}{2} u_\beta u_\beta\right), \; \mu = T^{0.72},$$

where γ is the ratio of specific heats, T is the temperature and Pr represents the Prandtl number. In turbulent flows the algebraic model of Baldwin and Lomax is used [1]. Since no analytical methods for the prediction of transition exist in threedimensional problems with strong crossflow, the flow is assumed to be fully turbulent.

2.2 Numerical Solution

Spatial Discretization The governing equations are discretized on a node-centered block-structured grid in general curvilinear coordinates with a finite-volume method. The convective fluxes are computed with a 2nd order upwind scheme (AUSM) [6]. To avoid oscillations near extrema the flux-limiter of van Albada has been implemented in the approach [7].

The viscous stresses are discretized using central differences of second-order accuracy.

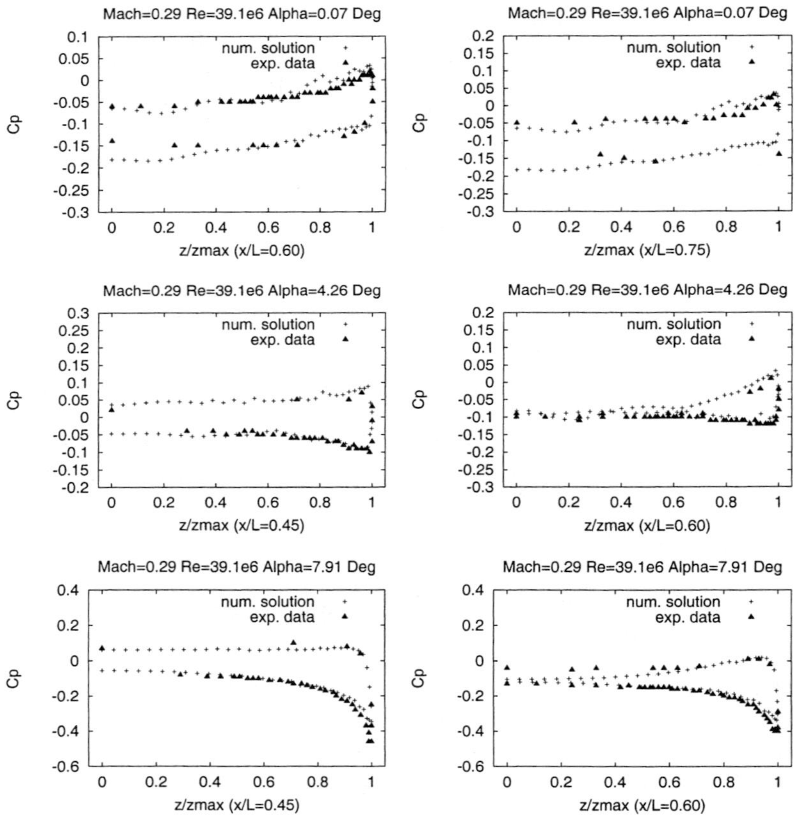

Fig. 4. Pressure distribution on the surface of Elac-1

Temporal Integration The integration in time of the differential equations is carried out using a 5-step low storage Runge-Kutta method. The scheme is adapted to a maximum Courant number of 4.0 for central and 3.5 for upwind schemes. The coefficients in the Runge-Kutta steps are chosen α_l=(0.059, 0.145, 0.273, 0.5, 1.) They are optimized for maximum stability for an upwind scheme. The accuracy in time is $O(\Delta t^2)$.

3 Results

3.1 ELAC-1

In the following table 1 the computed test cases are compiled. The grid constructed analytically has approx. 800,000 grid points and consists of 2 blocks. In Fig. 4 the pressure distribution in different cross-sections and at different

Table 1. Computed cases for ELAC

Ma_∞	Re_∞	angle of attack
0.29	39.1×10^6	0.07°
0.29	39.1×10^6	4.26°
0.29	39.1×10^6	7.91°

angles of attack is shown. The triangles denote the measured c_p-values and the +-signs represent the numerical solution. The agreement between experimental and computational results is entirely satisfactory. Only on the windward side at an angle of attack $\alpha = 7.91°$ the difference between the numerical and experimental data is slightly higher than for the other curves. In future investigations it has to be checked whether the use of the Baldwin and Lomax model causes this discrepancy. The agreement for the lift coefficient and the velocity profiles in Fig. 5 is definitely convincing.

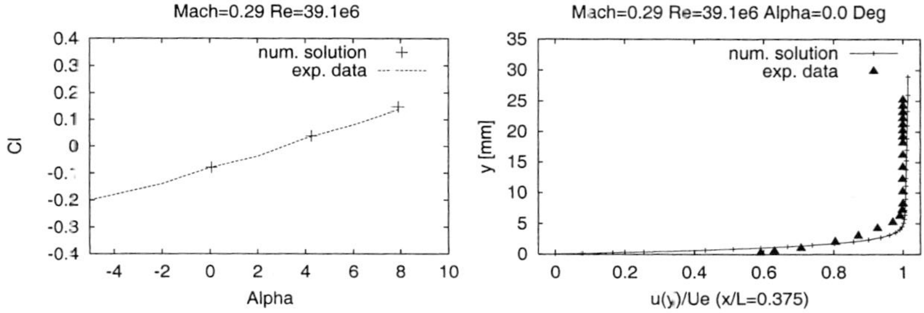

Fig. 5. Lift coefficient (left) and velocity profile in the plane of symmetry (right) of Elac-1

3.2 EOS

The computations of the supersonic flow around EOS again comprise the comparison between the computed and the measured c_p-values. The experiments have been performed in the trisonic windtunnel of the Aerodynamisches Institut of the RWTH Aachen. Details can be found in [8,9]. The grid has approx. 430,000 points on 6 blocks. It was generated using the commercial tool GridPro. The comparison of experimental and numerical data shows satisfactory agreement at $\alpha = 0°$ and $\alpha = 10°$ (Fig. 6). In Fig. 7 the computed wallstream patterns at angles of attack 0°, 5° and 10° are plotted. At 0° the attachment of streamlines at the leading edge of the wing is visible. This accumulation moves onto the lower side at 5° and at 10° a flow separation occurs on the upper side of the wing. This behaviour is confirmed by oil flow patterns.

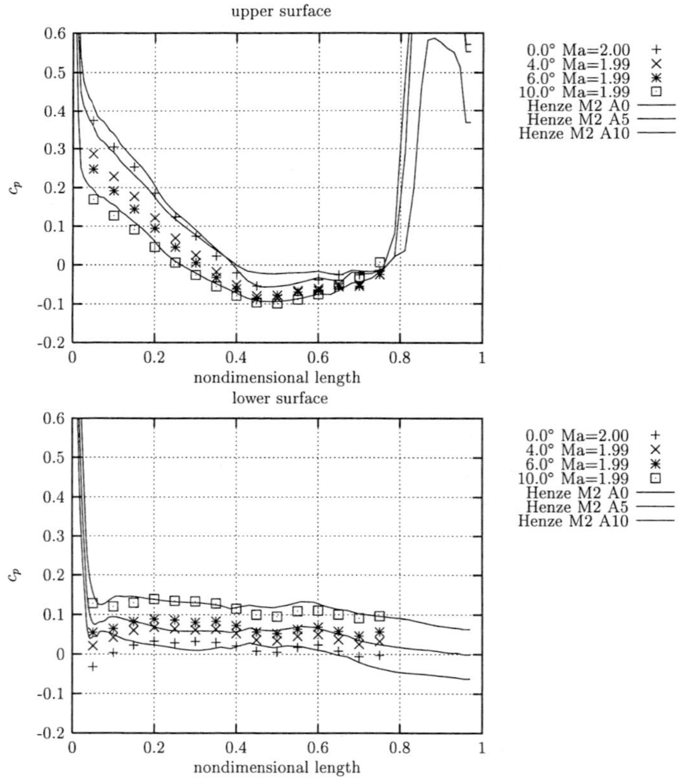

Fig. 6. Pressure distribution on EOS

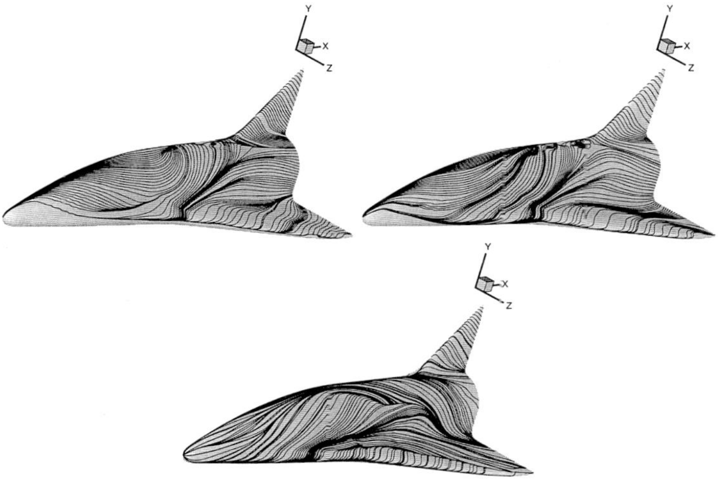

Fig. 7. Computed wallstream pattern on EOS at different angles of attack, $Ma_\infty = 2.0$, $Re_\infty = 2 \times 10^6$ (0° top left, 5° top right, 10° bottom center)

3.3 ELAC-1c

The investigation in the flow around the piggyback system Elac-1c has been performed to predict the influence of the upper stage on the flow. The grid has a total of 550,000 grid points in 2 blocks. In [4] the supersonic flow around ELAC was investigated experimentally and numerically. Oil flow patterns show that the lines of separation and reattachment are more or less straight lines if the flow on the upper side is undisturbed. In Fig. 8 the influence of the upper stage is visible. EOS causes a strong displacement of the flow at angles of attack up to $\alpha \approx 5°$. However, at higher angles of attack the displacement effect is reduced which is why at $\alpha = 10°$ the separation is approximately a straight line as it is for the single-stage configuration.

4 Conclusions

Surface pressure, lift coefficient and flow visualization results were obtained by means of numerical solutions of the Navier-Stokes equations for three-dimensional sub- and supersonic flows. The algebraic Baldwin-Lomax model was used for fully turbulent flow. A comparison of the data with experimental findings at the German Dutch Windtunnel and the results taken from experiments at the RWTH Aachen shows that the overall agreement is acceptable especially for the lift coefficient. The numerical results for EOS show good agreement with experimental data taken from the Aerodynamisches Institut of the RWTH Aachen. The separation on the leeside conformed the experimental findings. For the supersonic flow around ELAC-1c the influence of the upper stage on the leeside flow has been simulated numerically. Thus a large database ranging from incompressible to supersonic flow is available for the EOS-ELAC space transportation system.

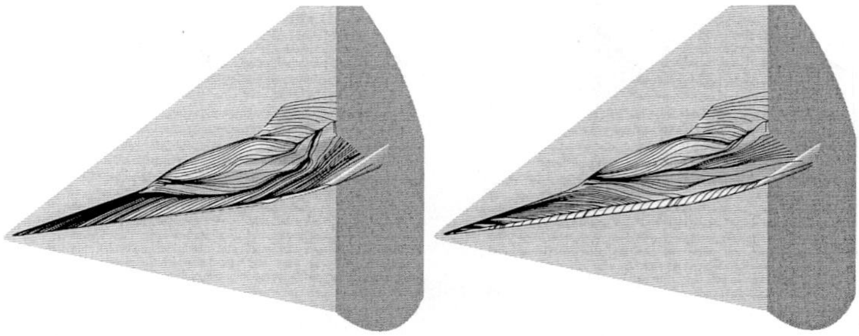

Fig. 8. Computed wallstream pattern on ELAC-1c at different angles of attack, $Ma_\infty = 2.0,\ Re_\infty = 3 \times 10^6$ (0° left, 10° right)

References

1. B. S. Baldwin and H. Lomax (1978) Thin Layer Approximation and Algebraic Model for Separated Turbulent Flows, AIAA 16th Aerospace Sciences Meeting, AIAA Pap. 78-257
2. G. Neuwerth, U. Peiter, F. Decker, and Jacob, D. (1998) Reynolds number effects on low speed aerodynamics of the hypersonic configuration ELAC, AIAA Proceedings, 8th International Space Planes and Hypersonic Systems and Technologies Conference, paper 98-1578
3. Krause, E., Abstiens, A., Fühling, S., and Vetlutsky, V.N. (2000) Boundary-layer Investigations on a Model of the ELAC 1 Configuration at High Reynolds Numbers in the DNW EJMB, vol 19:745–764
4. E. Krause, W. Limberg, A.M. Kharitonov, M.D. Brodetsky, and A. Henze (1999) An experimental investigation of the ELAC 1 configuration at supersonic speeds, Experiments in Fluids 26:423–436
5. D. Hänel, A. Henze, and E. Krause (1993) Supersonic and hypersonic flow computations for the research configuration ELAC-1 and comparison to experimental data, ZfW April 1993, Vol. 17,2:90–98
6. Wada, Y. and Liou, M.-S. (1994) A Flux Splitting Scheme with High-Resolution and Robustness for Discontinuities, NASA Technical Memorandum No. 106452, Also an AIAA Pap. 94-0083
7. G.D. van Albada, B. van Leer, and J.W.W. Roberts (1982) A comparative study of computational methods in cosmic gas dynamics, Astron. Astrophysics, vol. 108:76–84
8. W. Limberg and A. Stromberg (1993) Pressure measurements at supersonic speeds on the research confuguration ELAC 1, ZfW April 1993, Vol. 17,2:82–89
9. A. Stromberg (1995) Experimentelle Untersuchungen an der Hyperschallkonfiguration ELAC 1 bei Unter- und Überschallanströmung, Thesis RWTH Aachen

Flow Visualization on Hierarchical Cartesian Grids

St. Roettger[1], M. Schulz[2], W. Bartelheimer[3], and Th. Ertl[1]

[1] Institut für Informatik (VIS), Universität Stuttgart, Germany
[2] Science+Computing GmbH, Tübingen, Germany
[3] BMW AG, München, Germany

Abstract. Due to the limitations of the traditional finite volume CFD approach modern Lattice-Boltzmann methods are becoming more and more widespread. The results of developing an efficient visualization and exploration tool based on the Lattice-Boltzmann solver PowerFlow are summarized here to give the reader a basic insight into the pros and cons of such an approach. Also, the implementation of an automotive soiling simulation, which has been incorporated into the visualization tool, is presented here.

1 Introduction and Motivation

In the flow visualization community many solutions have been presented to embed flow visualization techniques into interactive and immersive environments, which allow an intuitive exploration and manipulation of a CFD data set [6,5,3,1,2,10]. In order to solve the Navier-Stokes equations most of todays available simulation applications are based on the finite volume approach, which is well known and established in the automotive industry. More recent approaches such as PowerFlow from Exa Corporation [4] use Lattice-Boltzmann simulation algorithms based on hierarchically refined cartesian grids (see Fig. 1), which allow the simulation mesh to be generated automatically. Also, these methods are well suited for massive parallelization. Wind tunnel validations at BMW AG, Munich have shown that the simulated flow coincides very well with real-world measurements, so with PowerFlow now being used as a standard flow solver at BMW, there was the demand for a visualization system that could take advantage of the special properties of the hierarchical data. The cartesian representation of the CFD data, however, implies several restrictions. In particular, the car geometry has to be stored and handled explicitly, since the simulation grid is derived by voxelization. For the same reason special care has to be taken when modeling near surface effects in the visualization environment.

2 Hierarchical Representation

In our visualization application the hierarchical cartesian grid is stored in a compact way so that each voxel is primarily consuming memory for the scalar

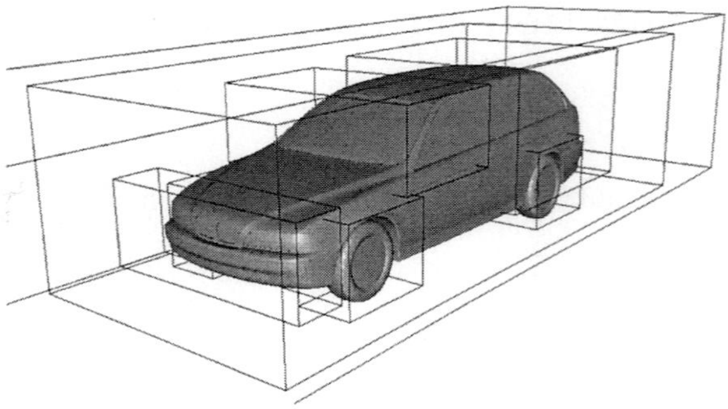

Fig. 1. Hierarchical cartesian grid, which is refined at the most interesting regions of a BMW 5 series for a Lattice-Boltzmann type CFD simulation

and vector components assigned with it. Each voxel can be accessed by a simple tree traversal of the hierarchy, which is illustrated in Fig. 2. A typical data set consists of about 4 nested hierarchies containing approximately 3 million voxels. With one vector and 5 scalars loaded these account for a total of 120 megabytes.

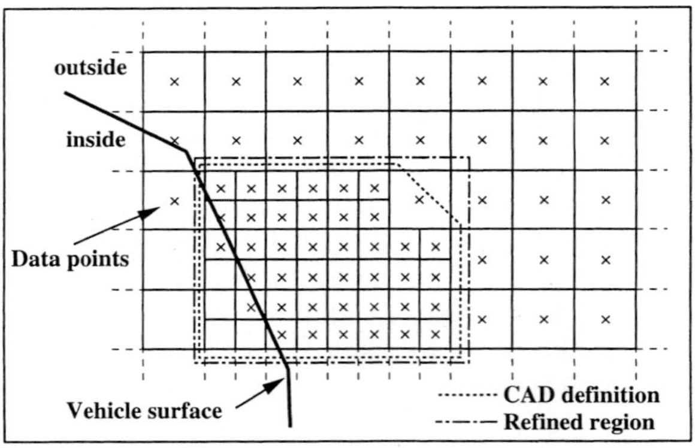

Fig. 2. Hierarchical grid representation with explicit vehicle surface

3 Visualization Methods

Because of the inherent hierarchical structure of the data set many of the established flow visualization techniques have to be adapted to the special properties of the hierarchical grids. For example, we are using an efficient Runge-Kutta particle tracer with adaptive step size control. The particle tracer utilizes an evaluation scheme of order 4(3), which has been shown to be optimally suited for streamline integration in cartesian grids [8,9]. For the purpose of particle tracing a method is also needed to detect the hit points of a particle trace on the surface of the car. To achieve this aim we are using an octree representation of the car surface as presented in [7], which reduces the total number of visited triangles to an average of 8 to 15 triangles per octree search for a model with approximately 70.000 triangles. We can speed up collision detection even further by storing an additional "geometry bit" per voxel, which denotes the presence of geometry inside each voxel. With these fundamental algorithms we can now compute stream lines, ribbons and glyphs interactively. An example of all three methods combined with a slice probe that shows the pressure inside the data set is given in Fig. 3. The stacked slice probe, as shown in Fig. 4, is an efficient extension to traditional isocontouring methods exploiting 2D texture mapping and transparency clipping. These stacked slice probes are well suited for exploring the whole extent of a vortex and not just a single plane that slices through it.

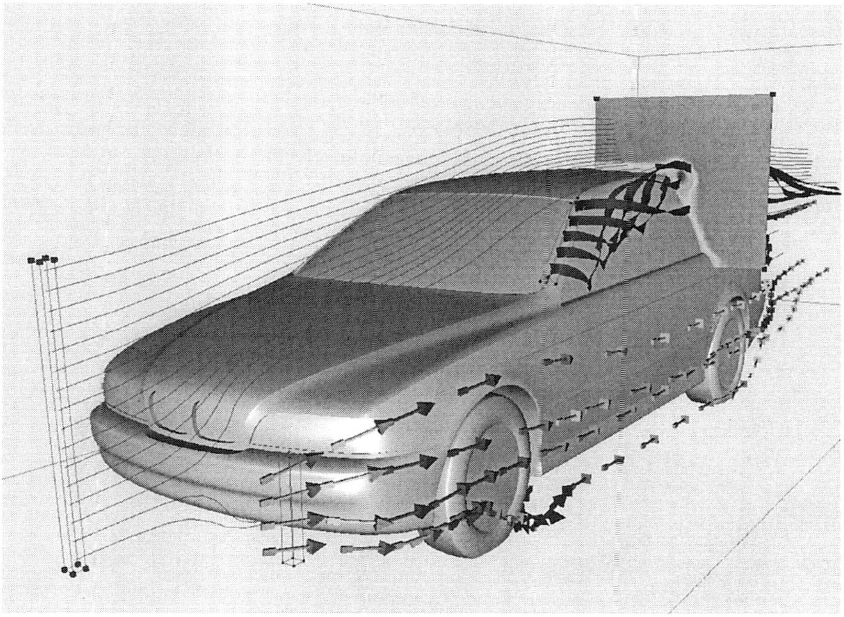

Fig. 3. Stream lines, ribbons, glyphs and a small pressure slice probe

Investigating scalar values on the surface is of great importance to the fluid flow engineer. In Fig. 5 for example the pressure on the car body is shown by means of coloring the so called surface elements (surfels). These

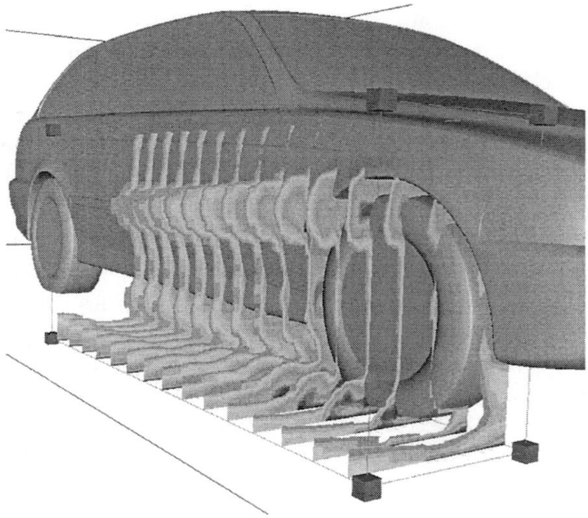

Fig. 4. Stacked slice probe

surfels are computed by PowerFlow along with the voxels to compensate for the limited accuracy of the voxelization of the CAD geometry.

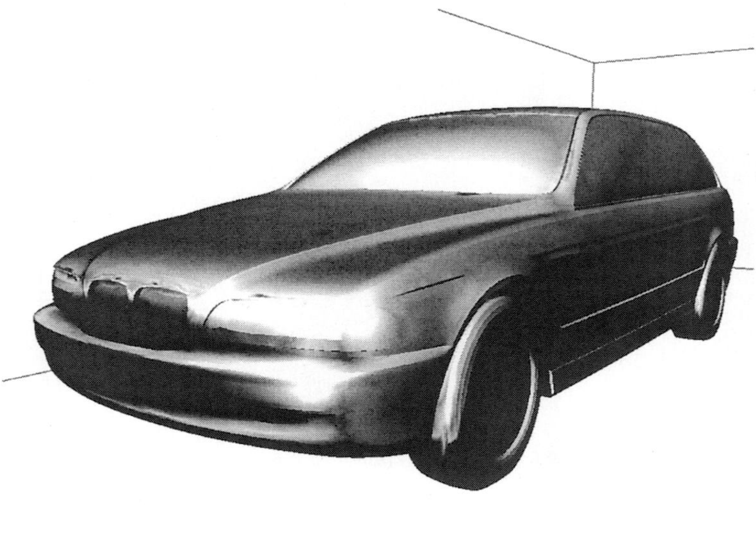

Fig. 5. Surface pressure (bright colors depict high pressure)

4 Interactive and Immersive Environment

All the aforementioned techniques can be used in an immersive environment like the PowerWall or the CAVE (see Fig. 6). Space mouse and a variety of tracking devices are supported to allow intuitive three-dimensional orientation and manipulation of objects. A CAD geometry that consists of about 70.000 triangles can be displayed at a frame rate of approximately 15 Hz on an SGI Octane MXI. The frame rate drops to still interactive 12 Hz when using a PowerWall and an SGI InfiniteReality2 with a single pipe for stereo rendering. The update times of the particle and slice probes depend on the specific settings but are interactive as well.

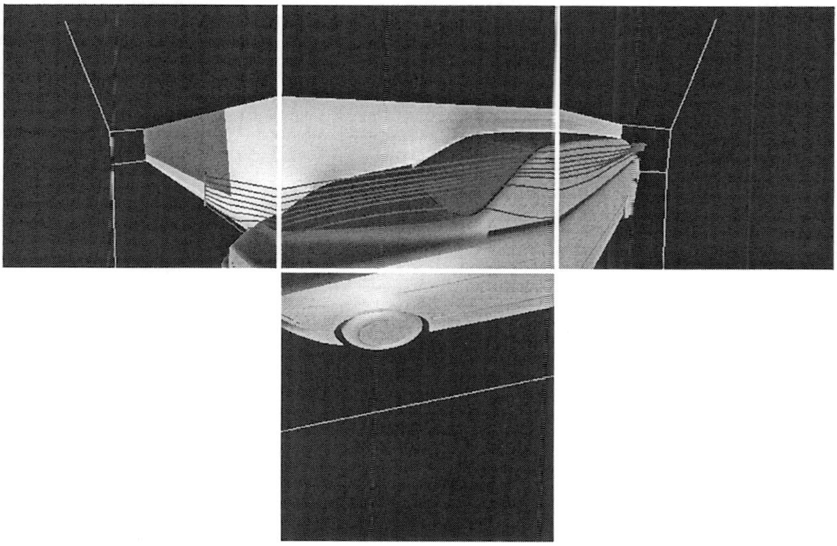

Fig. 6. The 4 side projections of a CAVE

5 Particle Animation

Reconsidering the basic visualization toolkit we thought that introducing interactively animated particle probes would give the viewer a better physical and three-dimensional understanding of the flow properties. Therefore an animated particle system with user manipulable particle emitters was developed. The emerging stream of particles is visualized by means of a virtual camera with constant exposure time, such that particles with higher velocities leave longer traces on each frame of the animation (see Fig. 7).

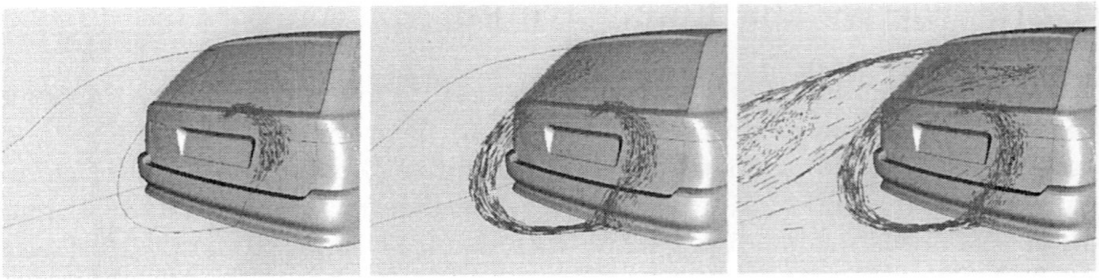

Fig. 7. Steps 1-3 of interactive particle animation

Furthermore, we extended the adaptive Runge-Kutta particle tracer for the integration of point masses, which lead to an embedded particle tracer of order 4(3). Here our main goal was to trace and simulate real-world dust particles in a soiling simulation. Each particle is emitted in the extent of a box probe with stochastically varying diameter, specific mass and initial velocity. The forces, which drive the point masses depend on the velocities relative to the flow stream and the c_w-value of each particle. Assuming that the particles can be idealized as spheres the c_w-value of a dust particle can be approximated by the formula of *O'Seen*. This formula is sufficiently exact for the low *Reynolds* numbers, which are encountered in automotive flow simulations.

With such a particle tracing system the accumulation of dirt on a car body can be visualized easily by coloring the hit points on the surface of the car. The adhesion probability of each particle depends mainly on the speed by which it is approaching the surface. Faster particles are more likely to hit the surface. Since flow velocity is always zero on the surface, the adhesion of each particle is driven by its momentum and by electrostatic effects. The latter effect is not yet completely understood and is subject to further research activities.

There are two other effects that introduce systematic errors when running a soiling simulation. First of all we are using a time-averaged flow field, which smoothes out the transient flow components and reduces the probability by which particles hit the surface. Secondly, the limited accuracy of the cartesian grid representation causes the flow velocity to be non-zero on the surface in contrast to the correct physical model. In order to compensate for that we define a small distance range around the car, in which the flow velocity is attenuated by a square root term depending on the distance to the surface of the car. Again, the distances can be calculated efficiently by using an octree representation of the surface triangles.

Our particle system is capable of tracing about 1000 point masses simultaneously on an SGI Octane MXI. With its dual 250 MHz R10K processors the refresh rate of the animation is approximately 20 Hz. The result of a dirt accumulation simulation employing the techniques mentioned above is shown in Fig. 8.

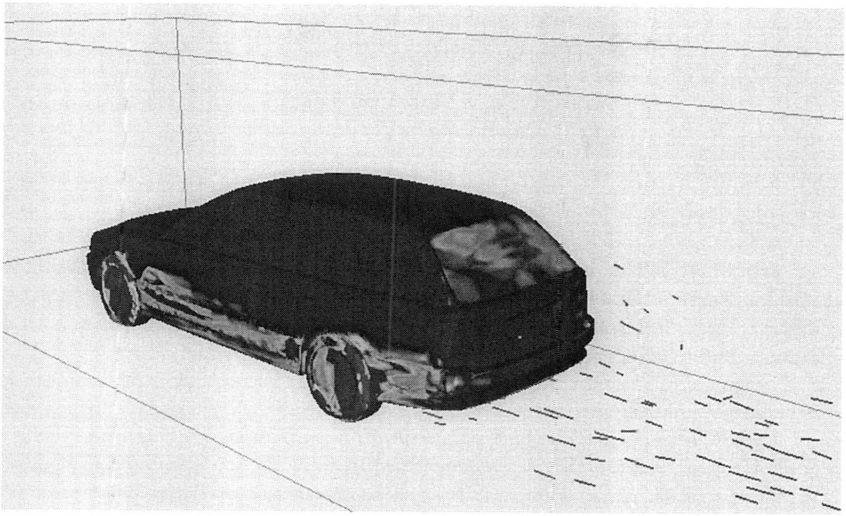

Fig. 8. Accumulated dirt after simulating dust particles

6 Conclusion

We have presented a flow visualization environment that is optimized for the hierarchical grid structure of the Lattice-Boltzmann CFD solver PowerFlow. The developement of a prototype of the visualization environment began in the second period of FORTWIHR, the *Bavarian Consortium for High Performance Scientific Computing*, and was sponsored subsequently by FORTWIHR III. By utilizing efficient memory structures, fast particle tracers and hardware texture mapping we are able to explore the hierarchical CFD data sets in real-time and in immersive environments. Furthermore, a particle tracing system has been developed that is utilized for soiling simulations.

References

1. S. Bryson and C. Levit. The Virtual Windtunnel. In *IEEE Computer Graphics and Applications, 12(4):25-34*, 1992.
2. Steve Bryson. Approaches to the Successful Design and Implementation of VR Applications. In *Proc. SIGGRAPH'94*, 1994.

3. Steve Bryson and Steven Feiner. Virtual Environments in Scientific Visualization. In *Virtual Reality for Visualization, Course Notes of Tut orial 5 at Visualization 95*, 1995.
4. Exa Corporation. *http://www.exa.com.*
5. D. A. Lane. Scientific Visualization of Large-Scale Unsteady Fluid Flows . In G. Nielson, H. Hagen, and H. Mueller, editors, *Scient ific Visualization*, pages 125–145. IEEE Computer Society, 1997.
6. F. Post and T. van Walsum. Fluid Flow Visualization. In H. Hagen, H. Mueller, and G. Nielson, editors, *Focus on Scientific Visualization*, pages 1–40. Springer Berlin, 1997.
7. M. Schulz, F. Reck, W. Bartelheimer, and Th. Ertl. Interactive Visualization of Fluid Dynamics Simulations in Lo cally Refined Cartesian Grids. In *Proc. Visualization '99*. IEEE, 1999.
8. C. Teitzel, R. Grosso, and T. Ertl. Efficient and Reliable Integration Methods for Particle Tracing in Unsteady Flows on Discrete Meshes. In W. Lefer and M. Grave, editors, *Visualization in Scientific Computing '97*, pages 31–41, Wien, 1997. Springer.
9. S.K. Ueng, C. Sikorski, and K.L. Ma. Efficient Construction of Streamlines, Streamribbons and Streamtubes on Unstructured Grids. *IEEE Transactions on Visualization and Graphics*, 2(2):100–109, June 1996.
10. S. P. Uselton. exVis: Developing A Wind Tunnel Data Visualization Tool. In R. Yagel and H. Hagen, editors, *Proc. Visualization '97*. IEEE, 1997.

Part II

Mathematical Methods

The Finite Mass Method – A New Approach to the Solution of Flow Problems

H. Yserentant

Mathematisches Institut, Universität Tübingen, 72076 Tübingen, Germany

Abstract. The finite mass method, a new Lagrangian method for the numerical simulation of gas flows, is presented. In contrast to the finite volume and the finite element method, the finite mass method is founded on a discretization of mass, not of space. Mass is subdivided into small mass packets of finite extension each of which is equipped with finitely many internal degrees of freedom. These mass packets move under the influence of internal and external forces and the laws of thermodynamics and can undergo arbitrary linear deformations. The basic reference is Gauger, Leinen, and Yserentant, SIAM J. Numer. Anal. 37 (2000), pp. 1768–1799. In the present note, a short survey is given.

1 Introduction

Fluid mechanics is usually stated in terms of conservation laws that link the change of a quantity like mass or momentum inside a given volume to a flux of this quantity across the boundary of the volume. Nevertheless, it is physically often much more natural to follow the motion of the mass itself and to start from the Lagrangian picture. This form of description can have many advantages. For example, there are no problems with free surfaces, and no convection terms arise. Such features make numerical methods based on the Lagrangian view especially attractive for free flows in vacuum.

The finite mass method presented here is such a Lagrangian approach. It has been developed at the Sonderforschungsbereich 382 of the Deutsche Forschungsgemeinschaft at the Universität Tübingen with the aim to substitute the Smoothed Particle Hydrodynamics, a Lagrangian method which is very popular in astrophysics, but which often is of poor accuracy. The finite mass method has been introduced in [1] and is based on earlier work [3,4] of the author. Convergence of the approximate solutions to limit functions satisfying the basic principles underlying fluid dynamics has been proven in [3,4]; in [5] the propagation of sound in such particle models is studied.

The main idea is to decompose the fluid into small mass packets with finitely many internal degrees of freedom that interact by continuum mechanical forces and the laws of thermodynamics. In the most simple case, the equations of motion for these mass packets, the particles, are derived from a Lagrange-function with the internal energy as potential energy. To incorporate entropy generation in shocks, additional frictional forces are added to these potential forces. The concept can be generalized to viscous fluids [1], which are not considered in this short survey.

2 The Finite Mass Method

The basic ingredient of the finite mass method is a normed shape function $\psi : \mathbb{R}^d \to \mathbb{R}_+$, $d = 1,2,3$ the space dimension, with compact support. This function describes the internal mass distribution inside the particles into which the fluid is subdivided and can be a B-spline, for example. The points $\boldsymbol{y}$ of the particle i move along the trajectories

$$t \to \boldsymbol{q}_i(t) + \boldsymbol{H}_i(t)\boldsymbol{y}\,, \quad \det \boldsymbol{H}_i(t) > 0\,. \tag{1}$$

The vector $\boldsymbol{q}_i(t)$ determines the actual position of the particle and matrix $\boldsymbol{H}_i(t)$ its size, shape, and orientation in space. The mass density of the particle at the point $\boldsymbol{x}$ in space and at time t then reads

$$\psi_i(\boldsymbol{x},t) = m_i\,[\det \boldsymbol{H}_i(t)]^{-1}\psi\big(\boldsymbol{H}_i(t)^{-1}(\boldsymbol{x}-\boldsymbol{q}_i(t))\big)\,. \tag{2}$$

The normalization factor $[\det \boldsymbol{H}_i(t)]^{-1}$ is chosen such that the mass

$$m_i = \int \psi_i(\boldsymbol{x},t)\,d\boldsymbol{x} \tag{3}$$

remains constant independent of $\boldsymbol{q}_i(t)$ and $\boldsymbol{H}_i(t)$. The total mass density

$$\rho(\boldsymbol{x},t) = \sum_{i=1}^{N} m_i\,[\det \boldsymbol{H}_i(t)]^{-1}\psi\big(\boldsymbol{H}_i(t)^{-1}(\boldsymbol{x}-\boldsymbol{q}_i(t))\big) \tag{4}$$

then results from the superposition of the mass densities of the single particles. The points $\boldsymbol{y}$ of the mass packet i have the velocity

$$t \to \boldsymbol{q}_i'(t) + \boldsymbol{H}_i'(t)\boldsymbol{y}\,. \tag{5}$$

Expressing the internal particle coordinates $\boldsymbol{y} = \boldsymbol{H}_i(t)^{-1}(\boldsymbol{x}-\boldsymbol{q}_i(t))$ by the space coordinates $\boldsymbol{x}$, one gets the velocity field

$$\boldsymbol{v}_i(\boldsymbol{x},t) = \boldsymbol{q}_i'(t) + \boldsymbol{H}_i'(t)\boldsymbol{H}_i(t)^{-1}(\boldsymbol{x}-\boldsymbol{q}_i(t)) \tag{6}$$

of the particle related to the space coordinates. Superimposing the mass flux densities $\psi_i\boldsymbol{v}_i$ of the single particles to the total mass flux density

$$\boldsymbol{j}(\boldsymbol{x},t) = \sum_{i=1}^{N} m_i\,[\det \boldsymbol{H}_i(t)]^{-1}\psi\big(\boldsymbol{H}_i(t)^{-1}(\boldsymbol{x}-\boldsymbol{q}_i(t))\big)\boldsymbol{v}_i(\boldsymbol{x},t)\,, \tag{7}$$

the continuity equation

$$\frac{\partial\rho}{\partial t} + \operatorname{div}\boldsymbol{j} = 0 \tag{8}$$

expressing the exact conservation of mass holds. The mass flux density $\boldsymbol{j}$ can be written in the form $\boldsymbol{j} = \rho\boldsymbol{v}$ where

$$\boldsymbol{v}(\boldsymbol{x},t) = \sum_{i=1}^{N} \chi_i(\boldsymbol{x},t)\boldsymbol{v}_i(\boldsymbol{x},t) \tag{9}$$

is the mean velocity resulting from the superposition of the particles and the

$$\chi_i(\boldsymbol{x},t) = \frac{\psi_i(\boldsymbol{x},t)}{\rho(\boldsymbol{x},t)} \tag{10}$$

are the local mass fractions of the single packets. Equation (8) then transfers into the usual form of the continuity equation.

To discuss the potential accuracy of the approach, we start from a given twice continuously differentiable velocity field $\boldsymbol{u}$. Then fixing the particle trajectories $\boldsymbol{q}_i(t)$ and the matrices $\boldsymbol{H}_i(t)$ for $t \geq t_0$ with given initial values, as solutions of the differential equations

$$\boldsymbol{q}_i'(t) = \boldsymbol{u}(\boldsymbol{q}_i(t),t)\,, \quad \boldsymbol{H}_i'(t) = (\nabla \boldsymbol{u})(\boldsymbol{q}_i(t),t)\,\boldsymbol{H}_i(t)\,, \tag{11}$$

the velocity field (6) of the particle i reads

$$\boldsymbol{v}_i(\boldsymbol{x},t) = \boldsymbol{u}(\boldsymbol{q}_i(t),t) + (\nabla \boldsymbol{u})(\boldsymbol{q}_i(t),t)(\boldsymbol{x}-\boldsymbol{q}_i(t)) \tag{12}$$

and is therefore a second order approximation of $\boldsymbol{u}$ in a neighborhood of $\boldsymbol{x} = \boldsymbol{q}_i(t)$. As the mass fractions (10) form a partition of unity, the resulting overall velocity field

$$\boldsymbol{v}(\boldsymbol{x},t) = \boldsymbol{u}(\boldsymbol{x},t) + \sum_{i=1}^{N} \chi_i(\boldsymbol{x},t)(\boldsymbol{v}_i - \boldsymbol{u})(\boldsymbol{x},t) \tag{13}$$

remains a second order approximation of $\boldsymbol{u}$ on the region occupied by mass, independent of where the particles are located. The mass density is an exact solution of the continuity equation with respect to this perturbed velocity field and therefore, with corresponding initial values, a good approximation of the true density in the sense of a backward error analysis.

For a complete description of the thermodynamic state of a compressible fluid, besides the mass density ρ a second thermodynamic quantity like temperature is needed. Most convenient in our context is the entropy s per unit volume that is given here the form

$$s(\boldsymbol{x},t) = \sum_{i=1}^{N} S_i(t) m_i \left[\det \boldsymbol{H}_i(t)\right]^{-1} \psi\big(\boldsymbol{H}_i(t)^{-1}(\boldsymbol{x}-\boldsymbol{q}_i(t))\big) \tag{14}$$

where the $S_i(t)$ denote the specific entropies of the single particles. The pressure, the absolute temperature and the internal energy per unit volume

$$\pi(\rho,s)\,, \quad \vartheta(\rho,s)\,, \quad \varepsilon(\rho,s) \tag{15}$$

are functions of the mass density and the entropy. These functions are not independent of each other but related by the basic law

$$\pi = \frac{\partial \varepsilon}{\partial \rho}\rho + \frac{\partial \varepsilon}{\partial s}s - \varepsilon\,, \quad \theta = \frac{\partial \varepsilon}{\partial s} \tag{16}$$

of thermodynamics, the Gibbs fundamental relation.

The particle i has the kinetic energy

$$E_i(t) = \frac{1}{2}\int \psi_i(\boldsymbol{x},t)|\boldsymbol{v}_i(\boldsymbol{x},t)|^2\,d\boldsymbol{x}\,. \tag{17}$$

Assuming without restriction that

$$\int \psi(\boldsymbol{y})\boldsymbol{y}\,d\boldsymbol{y} = \boldsymbol{0}\,, \quad \int \psi(\boldsymbol{y})\,y_k y_l\,d\boldsymbol{y} \;=\; \delta_{kl}\,, \tag{18}$$

it attains the closed representation

$$E_i = \frac{1}{2}m_i|\boldsymbol{q}_i'|^2 + \frac{1}{2}m_i|\boldsymbol{H}_i'|^2\,, \tag{19}$$

with $|\boldsymbol{H}_i'|$ the Frobenius norm of $\boldsymbol{H}_i'$. The total kinetic energy

$$E(t) = \sum_{i=1}^{N} E_i(t) \tag{20}$$

of the system is the sum of the kinetic energies of the single mass packets. Of this kinetic energy and the internal energy

$$V \;=\; \int \varepsilon(\rho(\boldsymbol{x},t), s(\boldsymbol{x},t))\,d\boldsymbol{x} \tag{21}$$

the Lagrange-function

$$\mathcal{L} = E - V \tag{22}$$

of the system is formed. For the completely adiabatic case as it is described by the Euler equations of gas dynamics, the equations of motion

$$\frac{d}{dt}\frac{\partial\mathcal{L}}{\partial\boldsymbol{q}_i'} - \frac{\partial\mathcal{L}}{\partial\boldsymbol{q}_i} = \boldsymbol{0}\,, \quad \frac{d}{dt}\frac{\partial\mathcal{L}}{\partial\boldsymbol{H}_i'} - \frac{\partial\mathcal{L}}{\partial\boldsymbol{H}_i} = \boldsymbol{0} \tag{23}$$

for the translational degrees of freedom and for the matrices describing the deformation and orientation of the particles are derived from this Lagrange-function. Written out, these equations read

$$\boldsymbol{q}_i'' = \boldsymbol{F}_i\,, \quad \boldsymbol{H}_i'' = \boldsymbol{M}_i\,, \tag{24}$$

where the normalized potential forces

$$\boldsymbol{F}_i = -\frac{1}{m_i}\frac{\partial V}{\partial\boldsymbol{q}_i}\,, \quad \boldsymbol{M}_i = -\frac{1}{m_i}\frac{\partial V}{\partial\boldsymbol{H}_i} \tag{25}$$

are derived from the internal energy. Should the occasion arise, the right hand sides of the equations of motion (24) and the potential energy (21), respectively, have to be complemented by further terms originating from gravitation, for example.

As long as the specific entropies S_i of the single particles are assumed to be constant, the time evolution of the system is completely determined by the equations of motion (23) and (24), respectively. This system is time-reversible, a fact that contradicts the behavior of actual fluids where heat is generated in shock fronts. The problem is that the kinetic energy (20) splits into the sum of the mean kinetic energy

$$\bar{E}(t) = \frac{1}{2} \int \rho(\boldsymbol{x},t) |\boldsymbol{v}(\boldsymbol{x},t)|^2 \, d\boldsymbol{x} \tag{26}$$

as known from continuum mechanics and the fluctuation energy

$$\bar{\bar{E}} = \frac{1}{4} \int \rho \sum_{i,j=1}^{N} \chi_i \chi_j \, |\boldsymbol{v}_i - \boldsymbol{v}_j|^2 \, d\boldsymbol{x} \tag{27}$$

into which the velocity differences of the overlapping particles enter. The surplus fluctuation energy (27) is a consequence of the kinetic approach and has no counterpart in continuum mechanics. Its influence is negligible in smooth flows but large in shock fronts where the particles collide.

Therefore a normalized frictional force

$$\boldsymbol{F}_i^{(r)} = -\frac{1}{2m_i} \int R\psi_i \, [\boldsymbol{v}_i - \boldsymbol{v}] \, d\boldsymbol{x} \tag{28}$$

is added to the right-hand side of the first equation in (24); the second equation is modified correspondingly. The forces (28) damp local fluctuations and softly couple the particles together. The difference between the velocity of the given particle and the surrounding velocity field of the flow determines the direction of the force (28) and the scalar function $R \geq 0$ its size. The frictional forces (28) transform the fluctuation energy into internal energy. Of course, the specific entropies of the particles then cannot longer remain constant; their time evolution is governed by the second law of thermodynamics [1]. With the frictional forces (28), shocks are surprisingly well reproduced as can be seen in Fig. 1.

The friction coefficient R determining the strength of the frictional forces (28) is of central importance. Up to now, we have mainly used constant R, which works well for many applications, but often requires a tedious adaption to the particle size. The essentially most general form of R invariant to a change in observer is

$$R = R\big(\rho, \theta, |\nabla\rho|, |\nabla\pi|, \imath_1(\boldsymbol{D}), \ldots, \imath_d(\boldsymbol{D})\big) \tag{29}$$

with $\imath_1(\boldsymbol{D}), \ldots, \imath_d(\boldsymbol{D})$ the principal invariants of the symmetric part $\boldsymbol{D}$ of the velocity gradient. To find out friction coefficients that properly scale in the particle size and work well both for shocks and contact discontinuities, is subject of current research.

The finite mass method as described here is restricted to free flows in vacuum, as arising in astrophysics. To incorporate boundaries, currently a new concept is developed [2] that is based on the use of mirror particles.

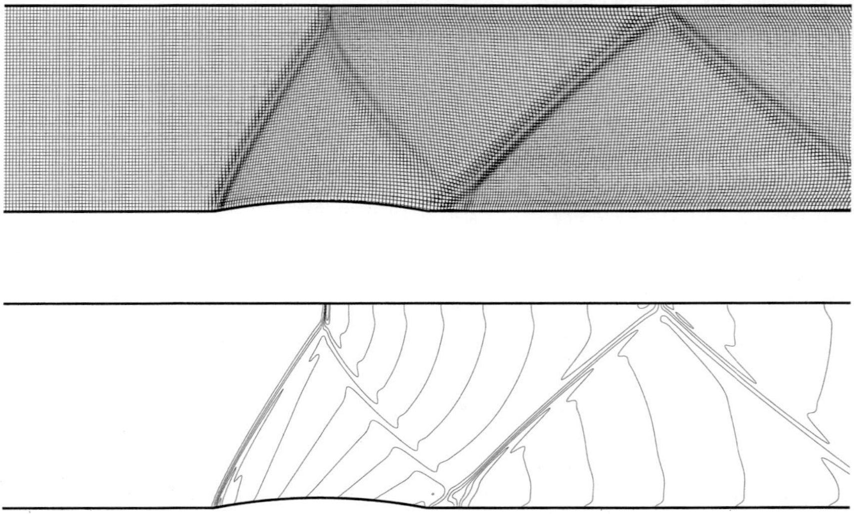

Fig. 1. Flow over a ramp, particle contours and isolines of density

3 Numerical Realization

To implement the method on a computer, the equations of motion for the particles have to be solved numerically. In [1], we developed a second order time integrator which is adapted to the special structure of the system, is linearly implicit in the frictional and viscous forces, and transfers into the Verlet method in case of pure pressure forces. Meanwhile we learnt that classical explicit Runge-Kutta methods work well, too, and are often superior when strong external forces are present.

A much more serious problem is the evaluation of the integrals forming the right-hand sides of the equations of motion. The usual cell-oriented quadrature formulas cannot be used here, because they destroy the invariance of the equations of motion to translations and rotations of space and therefore the inherent conservation properties of the method. Therefore we developed quadrature formulas which are based on the given decomposition

$$\int f\rho\,d\boldsymbol{x} = \sum_{i=1}^{N} m_i \int f(\boldsymbol{q}_i + \boldsymbol{H}_i\boldsymbol{y})\psi(\boldsymbol{y})\,d\boldsymbol{y} \tag{30}$$

of mass into the single mass packets. They read

$$\int f\rho\, d\boldsymbol{x} \;\rightarrow\; \sum_{i=1}^{N} m_i \sum_{\nu=1}^{n} \alpha_\nu f(\boldsymbol{q}_i + \boldsymbol{H}_i \boldsymbol{a}_\nu) \tag{31}$$

with fixed coefficients α_ν and nodes $\boldsymbol{a}_\nu$ and are used to discretize both the potential (21) from which the pressure forces are derived and the frictional and viscous forces.

The finite mass method can very efficiently be implemented both on serial and parallel computers. The essential observation is that the single mass packets do not directly interact with each other but only with global fields like the mass density or the velocity field. Therefore no interaction partners have to be determined.

The basic idea is to store the particle data and the quadrature points (together with the field information associated to them) separated from each other. The computation then splits into the following four phases.

- In the initial set up phase, the quadrature points $\boldsymbol{q}_i + \boldsymbol{H}_i \boldsymbol{a}_\nu$ and the assigned weights $m_i \alpha_\nu$ are generated.

- In the second phase, quantities like the mass density (4) or the mass flux density (7) are summed up at these quadrature points. In this phase, information is transferred from the particles to the quadrature points.

- In the third phase, only information attached to the single quadrature points is manipulated. For example, the values of the velocity field (9) at the quadrature points are calculated from the values of the mass flux density (7) and of the mass density (4). The operations in this phase neither require a data transfer between the single quadrature points nor between quadrature points and particles. Therefore they are very simple from an organizational point of view.

- In the concluding fourth phase, finally the forces acting upon the particles are computed. In this phase, information is transferred from the quadrature points contained in the support of the particles back to the particles themselves.

Phase 2 and phase 4 require an efficient access to the quadrature points contained in the support of a given particle. Search trees can be used for this purpose, which have to be set up in the initial phase.

References

1. Gauger, C., Leinen, P., Yserentant, H.: The finite mass method. SIAM J. Numer. Anal. **37** (2000), 1768–1799

2. Klingler, M., Leinen, P., Yserentant, H.: work in preparation
3. Yserentant, H.: A particle model of compressible fluids. Numer. Math. **76** (1997), 111–142
4. Yserentant, H.: Entropy generation and shock resolution in the particle model of compressible fluids. Numer. Math. **82** (1999), 161–177
5. Yserentant, H: The propagation of sound in particle models of compressible fluids. Numer. Math. (2000), DOI 10.1007/s002110000231

An Octree-Based Approach for Fast Elliptic Solvers

M. Bader[1], A.C. Frank[2], and Ch. Zenger[1]

[1] Institut für Informatik, TU München, 80290 München, Germany
[2] 4Soft GmbH, 80336 München, Germany

Abstract. We discuss an octree-based approach for the solution of elliptic partial differential equations, especially Poisson's equation and the convection diffusion equation. The discretization is derived from a starting discretization on a very fine octree grid. For the actual computation a discretization on a much coarser grid is generated by an accumulation process based on hierarchical transformation and partial elimination of unknowns. We also describe an efficient multigrid solver which takes advantage of the underlying octree structure. It is based on recursive substructuring of the domain and is very similar to the accumulation process. By adding additional unknowns to the coarse grids the resulting solver is robust even for the convection diffusion equation.

1 Introduction

Already in the seventies, simple methods for three-dimensional flow simulation on domains consisting of an assembly of regular cartesian grid cells were in use. Such an approach gives only a coarse, piecewise constant approximation of the boundary. Even for smooth boundaries, this results in an $\mathcal{O}(h)$ error of the solution (h being the meshsize of the grid). Interpolation can improve the solution, but the correct treatment of all occuring special cases is very difficult, especially for complicated geometries.

These problems can often be overcome by the use of unstructured grids. On the other hand, structured grids usually have big advantages regarding computational efficiency. This is mainly due to their simple regular data structures and therefore better performance, especially on vector computers. To achieve a better approximation of the boundary while retaining the structured grids, two possible approaches are for example an adaptive refinement of the grid or an homogenization approach to compute the discretization. Schroeder and Shepard [7] or Aftosmis et al. [6] use octree data structures as preliminary meshes in the grid generation process. These preliminary meshes are adapted afterwards to the actual geometry by using skewed cuboids or tetrahedrons to achieve a piecewise (bi-)linear interpolation of the boundaries. Hackbusch and Sauter [8] presented an approach *(Composite Finite Elements)* where a regular discretization mesh is generated by coarsening a discretization on a finer level. On this finer level, grid cells in the neighborhood of the boundary are mildly distorted such that a piecewise linear approximation of the boundary can be obtained from the corresponding grid points.

In other fields of computational engineering, like geometric modelling, digital mockup or visualization, octrees are becoming more and more common as a data structure both for computation and data exchange. To use octrees as tools for grid generation and discretization as well, it therefore seems worthwhile to consider approaches which are based on unmodified octree structures.

In this paper, we discuss an approach which is based on such an unmodified octree representation of the geometry, but on a very fine grid. In a second step, the discretization on this fine level approximation is reduced to a discretization on a coarser level in a systematic and robust way. This accumulation process can be restricted to the neighbourhood of the boundary which keeps the grid generation fast and efficient. By using an appropriate ratio between the coarse and fine level representation, we achieve an adequate compromise both with respect to accuracy and computational efficiency.

In the second part of this paper, we describe a solver which utilizes the octree structure for an efficient computation of the solution of the discretized problem. It is based on a recursive substructuring of the domain in combination with a change to hierarchical generating systems and the use of a partial elimination.

The octree representation of the geometry and the computation of the coarser level discretization is, in principle, independent of the dimension. However, our corresponding multigrid solver (see Sect. 4) is not yet available for the 3D case. We will therefore restrict most parts of this paper to a 2D discussion (which will also simplify the presentation). Nonetheless we will stick to a general 3D presentation of the grid generation where appropriate.

2 An Efficient Octree Structure

An unmodified octree structure can approximate a given geometry only with cubes of varying size. Thus, to obtain a sufficiently exact representation of the geometry, the octree has to be refined to a very fine level. Compared to a uniform fine grid, the storage complexity of such an octree structure is already reduced from $\mathcal{O}(n^3)$ to $\mathcal{O}(n^2)$, because the refinement is restricted to the boundary of the geometry. Nevertheless, this usually still consumes a great amount of memory.

In order to reduce the memory requirements, we use preorder traversal to get a linearized description of the octree structure which is then bit coded and stored on an external storage device. In a similar manner, the generated output—i.e. the computational grid and discretization—can be written directly to an external hard disk. In that case only a stack-like structure has to be held in main memory. The size of this stack structure is proportional to the maximum depth of the octree and therefore results in a memory requirement of only $\mathcal{O}(\log n)$. In practice, this has the advantage that the grid generation can be done on a comparably small (desktop) computer, while the actual solving of the problem can then be left to a high performance computer.

To discretize a partial differential equation on an octree grid, we have to describe not only the properties related to the volume but especially to the boundary of the geometry. That means, information about boundary conditions has to be provided at the faces of the corresponding octree cubes. Therefore, an extension of the octree structure is necessary which permits the correct assignment of the respective boundary conditions to the geometry. However, this can be done without a great loss in performance. Each of the octree's nodes is attributed with three quadtrees, one for each dimension, describing the faces of the octree's cubes. Since the tree structures are processed identically in every dimension, the dimension can be parametrized in the actual code. This approach allows a complete reuse of the program codes, so that the processing of the octree and quadtree structures is done by the same algorithms. As a consequence, the data structures and algorithms can even be applied to higher-dimensional problems [4].

3 Separation Approach and Geometry Accumulation

As mentioned above, the geometry mesh described by the octree structure consists of a huge amount of tiny elements *(micro cells)* in order to get a sufficiently exact representation of the geometry. Computing a solution on these elements would not be possible in a reasonable amount of time. Therefore, for the actual computation of the solution, we choose a much coarser grid composed of octree cells nearer to the root of the tree *(macro cells)*. The discretization on the computational grid defined by these macro cells is derived by a recursive accumulation process which is outlined in Fig. 1 and will be described in the following.

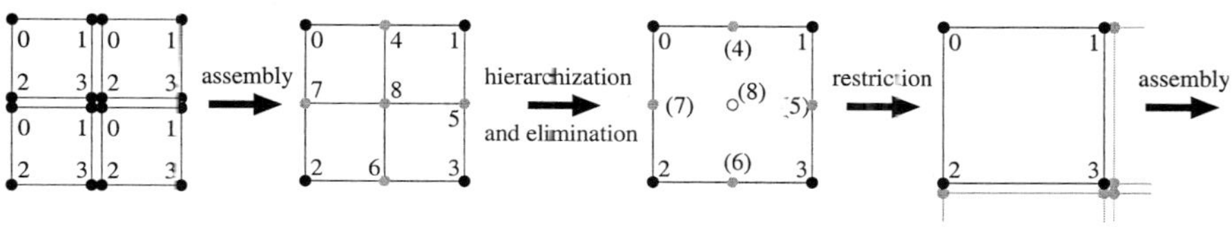

Fig. 1. The different steps of the geometry accumulation (2D example). For each step the numbering of the unknowns is given

We start with a simple finite difference or finite element discretization on the micro cells. In the 2D case for example, a finite element discretization using simple bilinear nodal basis functions leads to a local system of equations

$Ax = b$ of the form

$$\begin{pmatrix} 2 & -\frac{1}{2} & -\frac{1}{2} & -1 \\ -\frac{1}{2} & 2 & -1 & -\frac{1}{2} \\ -\frac{1}{2} & -1 & 2 & -\frac{1}{2} \\ -1 & -\frac{1}{2} & -\frac{1}{2} & 2 \end{pmatrix} \begin{pmatrix} x_0 \\ x_1 \\ x_2 \\ x_3 \end{pmatrix} = \begin{pmatrix} b_0 \\ b_1 \\ b_2 \\ b_3 \end{pmatrix} \tag{1}$$

for an interior micro cell. Note that this stencil differs from the standard finite element 9-point-stencil only as far as it just considers the interaction between the four unknowns of a single cell.

In the second step the linear systems of equations of four neighbouring child cells are assembled to a system for their parent cell. If the four child systems are given by $A_{(i)}x_{(i)} = b_{(i)}$, the resulting system $Ax = b$ for the parent cell is obtained by

$$A := \sum V_{(i)}^T A_{(i)} V_{(i)} \qquad b := \sum V_{(i)}^T b_{(i)}, \tag{2}$$

where each $V_{(i)}$ is a permutation matrix that performs the renumbering of the unknowns of child system i to the parent system.

After the assembly, the linear system of equations is transformed to an hierarchical basis:

$$H^T A H x_H = H^T b, \tag{3}$$

where H is the hierarchical transformation operator. If one or more of the four child cells contain the border of the domain, the hierarchization has to be modified accordingly (see [4] for detailed description). It is also necessary to include an additional elimination step

$$\underbrace{LH^T AHR}_{=:\,\widetilde{A}}\,\widetilde{x} = LH^T b =: \widetilde{b}, \tag{4}$$

which in the resulting stiffness matrix $\widetilde{A}$ explicitly eliminates the couplings between the central unknown and the four corner unknowns. Due to the hierarchization and the elimination, the essential information from the five interior unknowns is preserved in the four unknowns on the corners. Therefore, the five interior unknowns of the cell are no longer needed, and only the corner unknowns are kept for further processing. This is achieved by simply restricting the system of equations $\widetilde{A}\widetilde{x} = \widetilde{b}$ to those parts related to the four corner unknowns which leads to a discretization on the parent cell. This geometry accumulation process is then repeated until a macro element is reached in the tree structure.

In the absence of domain borders, the standard finite element stencil from (1) is reproduced (with mesh size doubled) on the parent cells if standard bilinear hierarchization is used. Thus, the accumulation process can indeed be restricted to the domain boundary.

The global linear system of equations has to be combined from all macro cells and can then be solved by any desired solver. In Sect. 4 we will present a multigrid solver that is particularly suitable, because it can work directly on the octree structure.

As only a piecewise constant approximation of the boundary is used, the approximation accuracy is only of $\mathcal{O}(h)$, where h is the size of the micro cells. However, if we choose h relative to the size H of the macro cells such that $h \propto H^2$, it can be shown that the $\mathcal{O}(H^2)$ accuracy of the starting discretization is still valid after the accumulation process [4]. In other words, to achieve a macro discretization which is as accurate at the boundary as inside the domain, we need to refine the grid to a resolution of $\mathcal{O}(H^2)$. That means that the octree has to be twice as deep at the boundary. The additional computational effort for the refinement and the accumulation process is surprisingly small, because the refinement is restricted to the boundary cells. It is therefore of order $\mathcal{O}(H^{-2})$ in the 2D case (which is no more expensive than the discretization inside the domain) and still only $\mathcal{O}(H^{-4})$ in the 3D case (instead of $\mathcal{O}(H^{-3})$ for the interior domain). For example, the discretization and accumulation for the domain pictured in Fig. 2 takes just a few seconds on a modern PC for a geometry grid with one million micro cells.

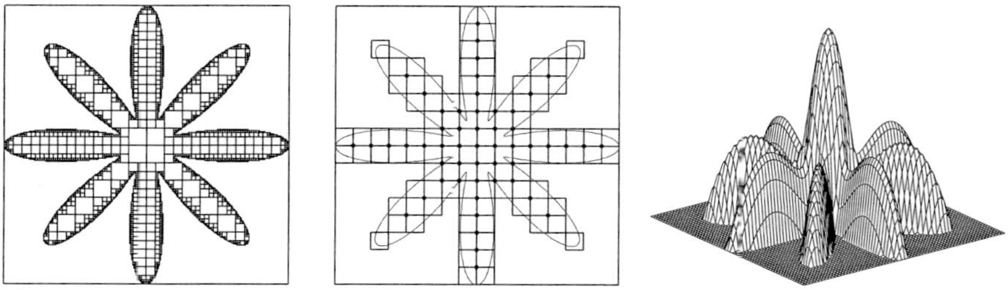

(**a**) octree representation of the geometry (**b**) computational grid of macro cells (**c**) exact solution

Fig. 2. Solving Poisson's equation on a star-shaped domain

As a numerical example, Fig. 2 shows a Poisson problem on a star-shaped domain with constant right hand side. Figure 2(a) shows the octree-coded approximation of the domain, Fig. 2(b) shows the actual computational grid of macro cells. In Fig. 2(c) the correct solution is depicted. This example is especially interesting, because its solution is very sensitive to a correct approximation of the domain, whereas the computational mesh is too coarse to resolve the correct curve of the boundary. Just the overall structure of the domain is preserved. In Fig. 3 the solution on the coarse grid with and without using the information gained from the accumulation process is presented. Underneath the errors to the analytical solution are shown. Since the computational costs are still dominated by the number of macro elements, we get a much better solution without a significant additional effort.

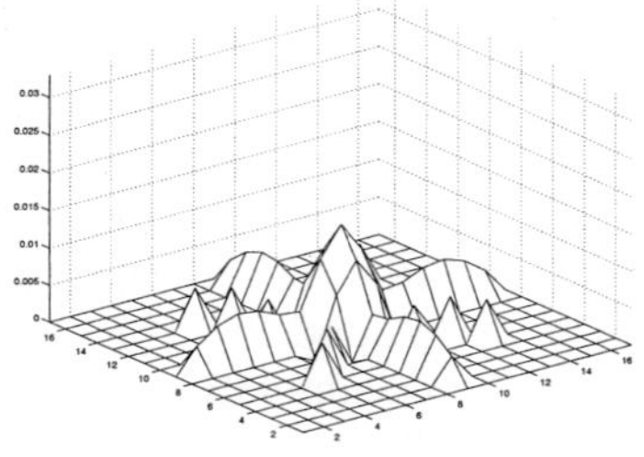

(**a**) solution on the macro cells as shown in Fig. 2(b)

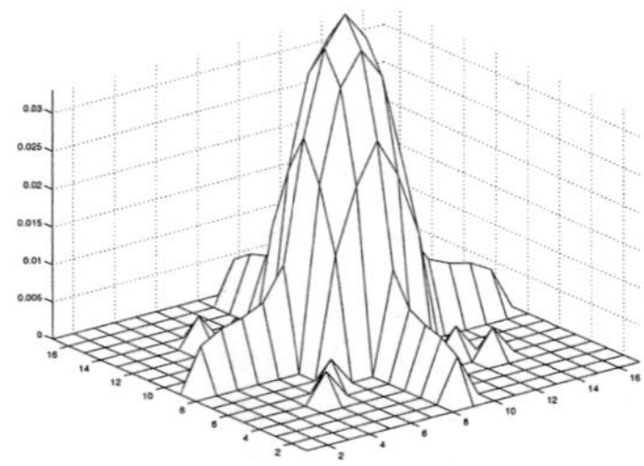

(**b**) solution with a standard discretization on a grid with the same mesh size

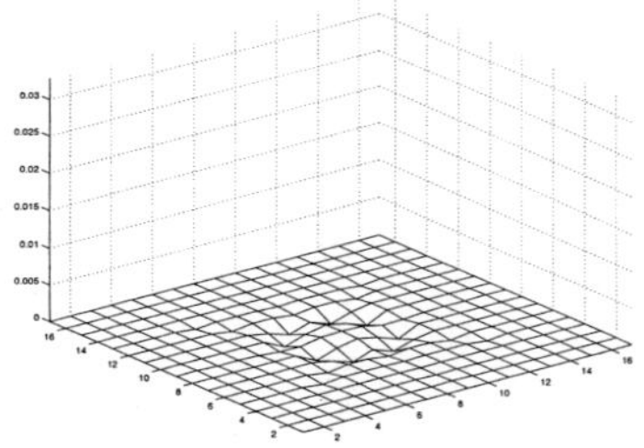

(**c**) error of Fig. 3(**a**)

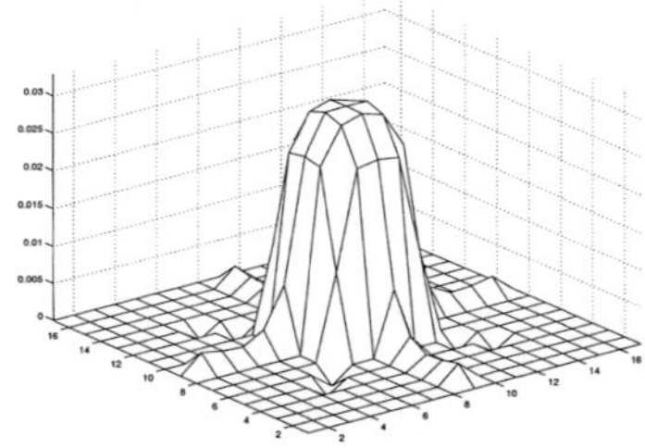

(**d**) error of Fig. 3(**b**)

Fig. 3. Solution with and without accumulated geometry information and its errors

4 Multigrid Solvers Based on Recursive Substructuring

The octree approach for the discretization suggests the use of a solution strategy that takes advantage of the given tree structure. The assembly of the local linear systems of equations in the geometry accumulation process is already similar to recursive substructuring methods, like for example the *nested dissection* method. The main difference is that in the latter all unknowns on the cell edges are kept, and only the interior unknowns are eliminated before moving to the parent cells. Compared to the accumulation process, we adopt a further slight modification regarding the substructuring of the subdomains. Instead of dividing each subdomain into four child domains (8 domains in

3D), we use an alternate bisection (*kd-tree*, see Fig. 5 on page 164). This is mainly done because it makes the implementation of the method easier, particularly with regard to the 3D case. Additionally, the amount of computational work per subdomain is less for the bisection approach, because the separators in the kd-tree contain less unknowns than in an octree substructuring. Therefore, the bisection approach also gives a slight advantage in the parallelization of the resulting algorithm. All the same, an octree substructuring is also appliable with only minor changes to the algorithm.

To obtain optimal performance, the recursive substructuring method is combined with the transformation to hierarchical bases and especially generating systems, which turns the algorithm into a true multigrid method [1]. Thus, instead of (4) in the geometry accumulation, we use the transformation to hierarchical generating systems

$$H_{GS}^T A H_{GS} = H_{GS}^T b. \tag{5}$$

This hierarchical multilevel-transformation is extended by a partial elimination between certain coarse grid unknowns. In contrast to the full elimination used by direct Schur complement methods, i.e. a transformation like

$$\underbrace{\begin{pmatrix} \mathrm{Id} & -A_{\mathcal{E}\mathcal{I}}A_{\mathcal{I}\mathcal{I}}^{-1} \\ 0 & \mathrm{Id} \end{pmatrix}}_{=:\,L^{-1}} \underbrace{\begin{pmatrix} A_{\mathcal{E}\mathcal{E}} & A_{\mathcal{E}\mathcal{I}} \\ A_{\mathcal{I}\mathcal{E}} & A_{\mathcal{I}\mathcal{I}} \end{pmatrix}}_{=\,A} \underbrace{\begin{pmatrix} \mathrm{Id} & 0 \\ -A_{\mathcal{I}\mathcal{I}}^{-1}A_{\mathcal{I}\mathcal{E}} & \mathrm{Id} \end{pmatrix}}_{=:\,R^{-1}} = \underbrace{\begin{pmatrix} \widehat{A}_{\mathcal{E}\mathcal{E}} & 0 \\ 0 & A_{\mathcal{I}\mathcal{I}} \end{pmatrix}}_{=:\,\widehat{A}}, \tag{6}$$

we use a transformation with modified elimination matrices $\widetilde{L}^{-1}$ and $\widetilde{R}^{-1}$ and obtain a transformed system $\widetilde{A}\widetilde{x} = \widetilde{b}$ with

$$\widetilde{A} := \widetilde{L}^{-1} H_{GS}^T \, A \, H_{GS} \widetilde{R}^{-1} \qquad \widetilde{b} := \widetilde{L}^{-1} H_{GS}^T \, b. \tag{7}$$

The elimination matrices $\widetilde{L}^{-1}$ and $\widetilde{R}^{-1}$ are chosen such that in the resulting matrix $\widetilde{A}$ the strongest couplings between unknowns on the separator ($\in \mathcal{I}$) and unknowns on the subdomain boundary ($\in \mathcal{E}$) are eliminated. The eliminated couplings are not chosen directly, but an a priori set of *coarse grid unknowns* is determined. Then all couplings between those unknowns are eliminated. The set of coarse grid unknowns is chosen with respect to the underlying physics of the differential equation. For both the Poisson equation and the convection diffusion equation, it is natural to choose the hierarchically coarse unknowns as coarse grid unknowns. However, the number of the unknowns that have to be included is different.

For the Poisson equation the hierarchical transformation makes it sufficient to eliminate the couplings between the central unknown on the separator and the four top level unknowns on the corners of the subdomain. This is identical to the geometry accumulation described in Sect. 3. It is also in accordance with the use of standard coarse grids in usual geometric multigrid methods.

In the convection-diffusion case however, the coarse grid mesh has to be refined on the edges of the coarse grid cells as shown in Fig. 4. The use of such coarse grids for the simulation of convection problems was for example suggested by Brandt (*total-semicoarsening*, [9]). It allows a correct description of the convective transport on the coarse grids without introducing additional numerical diffusion. The refinement is chosen such that the distance between the coarse grid unknowns grows with the square root of the size of the subdomain. This choice is due to the fact that for convection diffusion problems the isolines of the transported heat (or energy, substance, etc.) are typically parabolic. For a more detailed description of this approach see [3] or [2].

As a result of the additional partial elimination, the local systems of equations in the subdomains suffer from severe fill-in and can no longer be regarded as sparse matrices. Storing the entire system matrices would in fact destroy the optimal complexity of the algorithm. It is therefore necessary to use only one single iteration of the (weighted) Jacobi method as a smoother, because in that case only the main diagonal of the matrix and the current residual are required. The algorithm can then be implemented such that the computing time and the storage requirements grow only linearly with the number of unknowns. The convergence rates of the algorithm proved to be independent of the mesh size of the discretization but—as a result of the purely additive character of the multigrid scheme—are only in the range of about $\frac{2}{3}$ (compare Fig. 6).

Fig. 4. Different strategies for partial elimination: only the couplings between the black nodes are eliminated

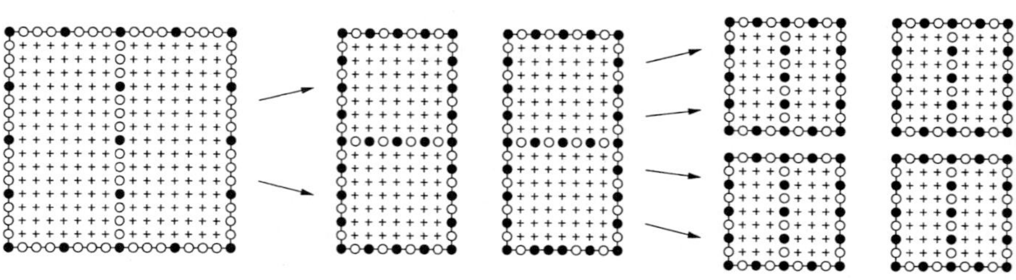

Fig. 5. Recursive substructuring of the domain. Only couplings between the coarse grid unknowns (represented as black nodes) are eliminated by $\widetilde{L}^{-1}$ and $\widetilde{R}^{-1}$

For the convection diffusion equation the number of required iterations does not depend on the geometry of the flow field. Moreover, it is also independent of the strength of convection as long as a certain amount of diffusion is present. More precisely, a certain mesh Péclet number must not be exceeded on the finest grid. Figure 6 shows the numerical results for the benchmark problem of a circular flow. The method was also applied as a preconditioner for the *Bi-CGSTAB*-method, which leads to faster convergence and also to an overall reduction of the computing time. Results for further numerical examples are given in [2] and [3].

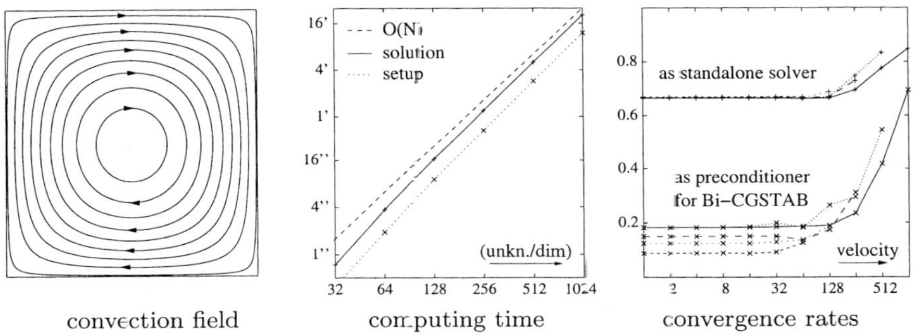

Fig. 6. Solving a convection diffusion equation with a circular convection field. The rightmost figure illustrates the convergence rates for increasing strength of convection on grids with 64×64 (- -) up to 1024×1024 (—) unknowns. The convergence slows down when the mesh Péclet number exceeds a certain value

For computational domains with complicated boundaries, the convergence rates can also degenerate, because the use of the standard hierarchization is no longer sufficient in that case. The partial elimination of couplings between the additional coarse grid unknowns can, at least to some extent, deal with this problem. Thus, for a broad range of test problems, a modified hierarchization is not necessary [3]. To achieve general robustness and optimal performance of the method, the hierarchization has to be modified like in the geometry accumulation.

5 Conclusion

The approach presented in this paper allows a straightforward procedure for the solution of elliptic partial differential equations using octree structures. It provides an automatic mesh generation for arbitrary complicated geometries which is both fast and simple. The discretization is not only of $\mathcal{O}(H^2)$ accuracy but can also—at least to some extent—be used to model geometry structures that are finer than the computational mesh. The hereby created octree structure of the generated computational mesh can be used directly by the multilevel solvers presented in Sect. 4. Due to the underlying octree structure, both the mesh generation and the multigrid solution are inherently adaptive and parallel (see [2] for parallelization). Furthermore, the application of such octree structures facilitates the connection of the numerical simulation with other fields of computational engineering, like geometric modelling, digital mockup and visualization [5].

References

1. Griebel, M., Multilevelmethoden als Iterationsverfahren auf Erzeugendensystemen, Teubner Skripten zur Numerik, Stuttgart, (1994)
2. Bader, M. and Zenger, C., A Fast Solver for Convection Diffusion Equations Based on Nested Dissection with Incomplete Elimination, Euro-Par 2000, Parallel Processing, Arndt Bode et. al., (2000), **795–805**
3. Bader, M., Robuste, parallele Mehrgitterverfahren für die Konvektions-Diffusions-Gleichung, TU München, (2001)
4. Frank, A. C., Organisationsprinzipien zur Integration von geometrischer Modellierung, numerischer Simulation und Visualisierung, TU München, (2000)
5. Breitling, P. and Bungartz, H.-J. and Frank, A. C., Hierarchical Concepts for Improved Interfaces between Modelling, Simulation, and Visualization, Vision Modeling and Visualization '99, B. Girod et. al., St. Augustin, (1999), **3-36**
6. M. J. Aftosmis and M. J. Berger and J. Melton, Robust and Efficient Cartesian Mesh Generation for Component-Based Geometry, AIAA-Paper 97-0196, (1997)
7. Schroeder, W.J. and Shepard, M.S., A Combined Octree/Delauney Method for Fully Automatic 3D Mesh Generation, Int. J. Numer. Met. Eng., **29**, (1990), **37-55**
8. Hackbusch, W. and Sauter, S.A., A New Finite Element Approach for Problems Containing Small Geometric Details, Arch. math., **34**, (1998), **105–117**
9. Achi Brandt, Multiscale Scientific Computation: Review 2000, Virtual proceedings of the 10th Copper Mountain Conference on Multigrid Methods, `http://www.mgnet.org/mgnet-cm2001.html`

A Variable Order Method of Lines: Accuracy, Conservation and Applications

H. Nishida and N. Satofuka

Kyoto Institute of Technology, Matsugasaki, Sakyo-ku, Kyoto 606-8585, Japan

Abstract. The variable order method of lines is presented for the DNS of incompressible flows. The present method is constructed by the spatial discretization, i.e., the variable order proper convective scheme and modified differential quadrature method, and time integration. The accuracy and conservation property are validated in the 2D Taylor-Green solutions and 3D homogeneous isotropic turbulence. As applications, the flows around a circular cylinder and a sphere are simulated by using Cartesian grid approach with virtual boundary method. Consequently, the present method is very promising for the DNS of the incompressible flows.

1 Introduction

In the practical direct numerical simulation (DNS) of incompressible flows, there are two requirements. One is the requirement of numerical scheme and another is the requirement of approach. In the numerical scheme, it is necessary that the scheme has the high accuracy in space and time, the high flexibility, the high efficiency, and the high conservation property, so that the variable order method of lines (VOMOL) [1] is intoroduced. On the other hand, the approach requires the easy generation of computational grid and the easy treatment of the complicated flows geometries. Usually the boundary fitted coordinate (BFC) system is adopted. However, in this BFC approach the computational grid generation is not easy for the complicated flow geometries. And it has larger computational effort than the Cartesian grid approach, because of transforming the physical plane to the computational plane. In this work, the Cartesian grid approach is used for the easy generation of the grid and the virtual boundary method [2–4] is introduced for the easy treatment of arbitrary flow geometries.

In this paper, the accuracy and conservation property of VOMOL are validated. The 2D Taylor-Green solutions for the accuracy and 3D homogeneous isotropic turbulence for the conservation property are considered. After validation of the method, the Cartesian grid approach with virtual boundary method based on the VOMOL is applied to the DNS of flows around a circular cylinder (2D) and a sphere (3D).

2 Variable Order Method of Lines

The incompressible Navier-Stokes equations expressed by the primitive variables in the Cartesian coordinates can be written by

$$\frac{\partial u_i}{\partial x_i} = 0, \tag{1}$$

$$\frac{\partial u_i}{\partial t} = F_i - \frac{\partial p}{\partial x_i}, \tag{2}$$

where F_i denotes the convection, diffusion and external forcing terms expressed by

$$F_i = -u_j \frac{\partial u_i}{\partial x_j} + \frac{1}{Re} \frac{\partial^2 u_i}{\partial {x_i}^2} + G_i. \tag{3}$$

Equation (1) denotes the continuity equation and (2) the momentum equations. The velocity and the pressure are expressed by u_i and p, respectively and Re is Reynolds number defined by $Re = UL/\nu$, where ν is the kinematic viscosity. The above equations are nondimensionalized by the reference length L, the reference velocity U and the reference pressure ρU^2.

The solution procedure of the incompressible Navier-Stokes equations is based on the fractional step approach. The flux F_i is discretized on the collocated grid points, in order to solve the momentum equation (2). The proper combination of discretized continuity and pressure terms is required for kinetic energy conservation. Then, the convection terms are approximated by the variable order proper convective scheme [1] for the consistency of the discrete continuity equation, conservation property, and variable order of spatial accuracy. This discretization technique is the extension of the proper convective scheme proposed by *Morinishi* [5] to the variable order. This scheme is constructed by the linear combination of the interpolation ($\overline{}^{\ell' x_j}$ and $\overline{u_j}^{x_j}$) and the discretization ($\delta_{\ell'} u_i / \delta_{\ell'} x_j$) as,

$$u_j \frac{\partial u_i}{\partial x_j}\Big|_{\mathbf{x}} = \sum_{\ell=1}^{M/2} c_{\ell'} \overline{\overline{u_j}^{x_j} \frac{\delta_{\ell'} u_i}{\delta_{\ell'} x_j}}^{\ell' x_j} \Big|_{\mathbf{x}}, \quad (\ell' = 2\ell - 1), \tag{4}$$

where M denotes the order of spatial accuracy, and the operators in (4) are defined by

$$\overline{f}^{\ell' x_j}\big|_{\mathbf{x}} = \frac{1}{2}(f_{\mathbf{x}_j + \ell'/2} + f_{\mathbf{x}_j - \ell'/2}), \tag{5}$$

$$\overline{u_j}^{x_j}\big|_{\mathbf{x}_j \pm \ell'/2} = \sum_{m=1}^{M/2} \frac{c_{m'}}{2} [u_j|_{\mathbf{x}_j \pm \{\ell' + m'\}/2} + u_j|_{\mathbf{x}_j \pm \{\ell' - m'\}/2}], \tag{6}$$

$$\frac{\delta_{\ell'} u_i}{\delta_{\ell'} x_j}\Big|_{\mathbf{x}_j \pm \ell'/2} = \frac{\pm 1}{\ell' \Delta x_j}(u_i|_{\mathbf{x}_j \pm \ell'} - u_i|_{\mathbf{x}_j}), \tag{7}$$

where $m' = 2m - 1$. In this technique, the arbitrary order of spatial accuracy can be obtained automatically by changing only one parameter M. The coefficients $c_{\ell'}$ and $c_{m'}$ are the weighting coefficients and Δx_j denotes the grid spacing in the x_j direction.

On the other hand, the diffusion terms are discretized by the modified differential quadrature (MDQ) method [1] as

$$\frac{\partial^2 u_i}{\partial x_i{}^2}|_{\mathbf{x}} = \sum_{m=-M/2}^{M/2} \Phi_m{}''(\mathbf{x}) u_i|_{\mathbf{x}_i+m}. \tag{8}$$

The coefficient $\Phi_m{}''(\mathbf{x})$ is the second derivative of the function $\Phi_m(\mathbf{x})$

$$\Phi_m(\mathbf{x}) = \frac{\Pi(\mathbf{x})}{(\mathbf{x} - \mathbf{x}_{i+m})\Pi'(\mathbf{x}_{i-m})}, \tag{9}$$

$$\Pi(\mathbf{x}) = (\mathbf{x} - \mathbf{x}_{i-M/2}) \cdots (\mathbf{x} - \mathbf{x}_i) \cdots (\mathbf{x} - \mathbf{x}_{i+M/2}). \tag{10}$$

The coefficients of the variable order proper convective scheme, $c_{\ell'}$, can be computed automatically by using the MDQ coefficients for the first derivatives. For example, the coefficients, $c_{\ell'}$, can be written as follows;

$$c_1 = 1, \quad (2nd\ order)$$
$$c_1 = 9/8, c_3 = -1/8, \quad (4th\ order)$$
$$c_1 = 150/128, c_3 = -25/128, c_5 = 3/128. \quad (6th\ order)$$

For the other cases, it is possible to determine the coefficients $c_{\ell'}$ by changing parameter M.

Then, the incompressible Navier-Stokes equations are reduced to the system of ordinary differential equations (ODEs) in time. This system of ODEs is integrated by the Runge-Kutta type scheme.

The pressure equation is solved by the variable order multigrid method [6] with the same order of spatial accuracy as the momentum equations. In the variable order multigrid method, the unsteady term in pseudo-time is added, so that the elliptic equation is replaced by the parabolic equation. Then, the equation can be solved by the VOMOL. As the pseudo-time integration, the rational Runge-Kutta scheme [7] is used because of the wider stability region. In order to accelerate the convergence, the multigrid technique is incorporated into the method. In this work, the full approximate storage (FAS) algorithm proposed by Brandt [8] is used as the multigrid technique. The variable order multigrid method gives the same high convergence efficiency as the original multigrid solver and the arbitrary order of spatial accuracy can be obtained by changing only the same one parameter M.

3 Cartesian Grid Approach with Virtual Boundary Method

In the virtual boundary method, the body surface is expressed by a set of plural points. For simplicity, we explain here in 2D, but the extension to 3D is straightforward.

First, the velocity components $\mathbf{U}(\mathbf{x}_s)$ on the virtual boundary point $\mathbf{x}_s = (x_s, y_s)$ are estimated by using the interpolation from four Cartesian grid points, i.e., (i, j), $(i+1, j)$, $(i, j+1)$, and $(i+1, j+1)$. Then, the velocity components $\mathbf{U}(\mathbf{x}_s)$ can be obtained by the relation,

$$\mathbf{U}(\mathbf{x}_s) = \sum_{i,j}^{i+1,j+1} W_{i,j}(\mathbf{x}_s)\mathbf{u}_{i,j}, \tag{11}$$

where $\mathbf{u}$ is the velocity on the Cartesian grid point and $W_{i,j}(\mathbf{x}_s)$ denotes the weighted coefficient defined by

$$W_{i,j}(\mathbf{x}_s) = w(x_s - x_i)w(y_s - y_j), \tag{12}$$

$$\begin{aligned} w(x_s - x_i) &= \frac{x_s - x_{i+1}}{x_i - x_{i+1}} \quad if \;\; x_i < x_s, \\ w(x_s - x_i) &= 1 \quad if \;\; x_i = x_s, \\ w(x_s - x_i) &= \frac{x_s - x_{i-1}}{x_i - x_{i-1}} \quad if \;\; x_i > x_s. \end{aligned} \tag{13}$$

Next, the additional forcing term $\mathbf{G}(\mathbf{x}_s, t)$ on the virtual boundary point $\mathbf{x}_s$ is determined by

$$\mathbf{G}(\mathbf{x}_s, t) = \alpha \int_0^t (\mathbf{U}(\mathbf{x}_s) - \mathbf{v}(\mathbf{x}_s))dt + \beta(\mathbf{U}(\mathbf{x}_s) - \mathbf{v}(\mathbf{x}_s)), \tag{14}$$

where $\mathbf{v}$ denotes the specified velocity vector on the virtual boundary point. The parameters α and β are specified by $\alpha = -400000$, $\beta = -600$ [4]. Then, the forcing terms on the Cartesian grid point $x_{i,j}$, $\mathbf{G}_{i,j}$, can be obtained by

$$\mathbf{G}_{i,j} = \frac{1}{N_s}\sum_{n=1}^{N_s} W_{i,j}(\mathbf{x}_s)\mathbf{G}(\mathbf{x}_s, t). \tag{15}$$

N_s is the total virtual boundary points in the cell.

4 Validation of Variable Order Method of Lines

4.1 Spatial Accuracy

In this paper, the spatial accuracy of the VOMOL is validated by considering the 2D Taylor-Green solutions [9]. In this problem, the analytic solutions are given by

$$\begin{aligned} u(x,y,t) &= -\cos(x)\sin(y)\exp(-2t/Re), \\ v(x,y,t) &= \sin(x)\cos(y)\exp(-2t/Re), \\ p(x,y,t) &= -1/4[\cos(2x)+\cos(2y)]\exp(-4t/Re). \end{aligned} \tag{16}$$

The computational domain is $(0,0) \leq (x,y) < (2\pi, 2\pi)$. The periodic boundary conditions are imposed in the x and y directions and the initial values can be given by the analytic solutions at $t = 0.0$. The convergence criterion of the pressure equation is set up $\varphi < 1.0 \times 10^{-10}$, where ϕ is the L2-norm of residual. The Reynolds number is $Re = 1000$ and $\Delta t = 1/1000$ is used here because of removal of the time integration error. The spatial accuracy is validated by L2-ratio. The L2-ratio is defined by the slope of the graph in which the L2-error versus mesh spacing is plotted on the log-log scale. Figures 1 and 2 show the L2-ratio of velocity u and pressure obtained by the present method at $t = 1.0$. The results show that each slope is equal to the specified spatial accuracy, that is, 2, 4, 6, 8, and 10, respectively. Therefore, the present VOMOL gives the arbitrary order of spatial accuracy.

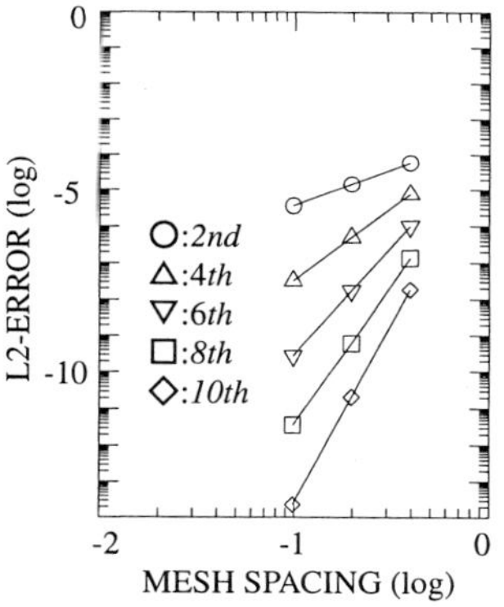

Fig. 1. L2-ratio of velocity u

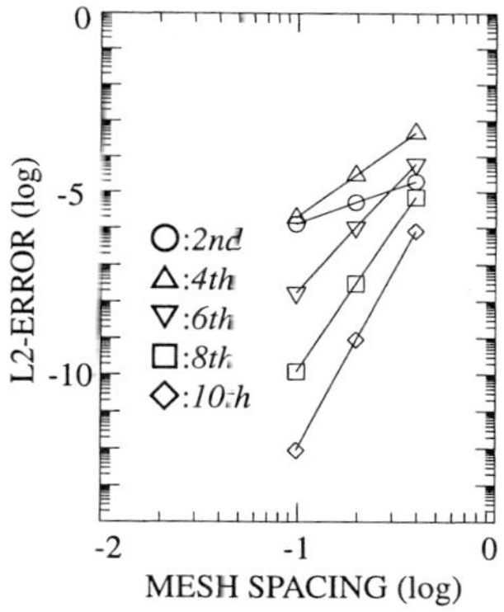

Fig. 2. L2-ratio of pressure

4.2 Conservation Property

In order to examine the conservation property, the DNS of 3D homogeneous isotropic turbulence is considered. The 3rd order Runge-Kutta time integration scheme is adopted with $\Delta t = 1/200$, and the computational domain is set to be $(0,0,0) \leq (x,y,z) < (2\pi, 2\pi, 2\pi)$. The energy and energy dissipation budgets are defined as

$$\frac{\partial k}{\partial t} = -\epsilon, \qquad \frac{\partial \epsilon}{\partial t} = P_\epsilon - D_\epsilon, \tag{17}$$

where

$$k = \frac{1}{2}\overline{u_i' u_i'}, \qquad \epsilon = \nu \overline{\frac{\partial u_i'}{\partial x_j}\frac{\partial u_i'}{\partial x_j}}$$

$$P_\epsilon = -2\nu \overline{\frac{\partial u_i'}{\partial x_j}\frac{\partial u_i'}{\partial x_k}\frac{\partial u_j'}{\partial x_k}}, \qquad D_\epsilon = 2\nu^2 \overline{\frac{\partial^2 u_i'}{\partial x_j \partial x_k}\frac{\partial^2 u_i'}{\partial x_j \partial x_k}}. \tag{18}$$

In (17), the prime denotes the fluctuation, and $^{-}$means the average. Figures 3 and 4 show the energy and energy dissipation budgets with $\nu = 1/41.1$ and $64 \times 64 \times 64$ uniform grid. The convergence criterion of pressure equation is specified by $\phi < 1.0 \times 10^{-8}$. The symbols of diamond, square, inverse triangle, triangle, and circle show the results of 2nd, 4th, 6th, 8th, and 10th order of accuracy, respectively. As the spatial accuracy becomes higher, the energy and energy dissipation budgets are satisfied very well. Therefore, the present higher order DNS give the solutions with conservation property of energy.

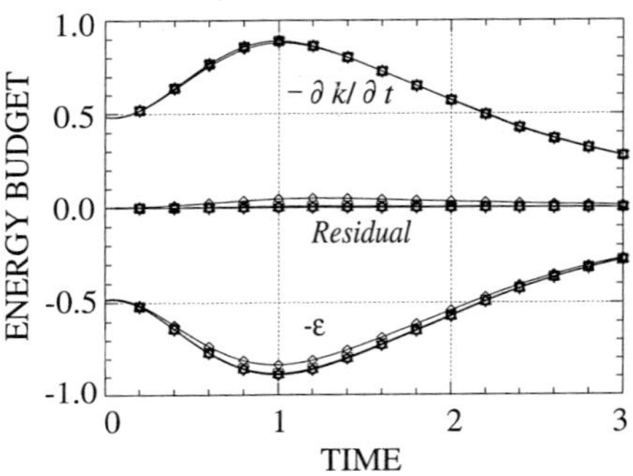

Fig. 3. Energy budget

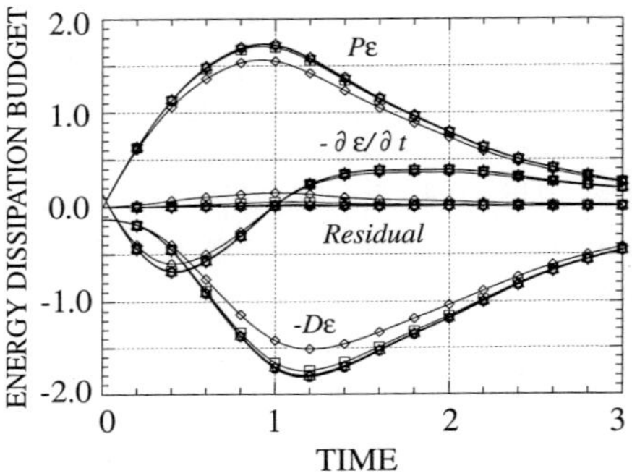

Fig. 4. Energy dissipation budget

5 Applications to External Flows

The Cartesian grid approach with virtual boundary method based on the VOMOL is applied to the DNS of flows around a circular cylinder and a sphere. In these DNS, the spatial accuracy is fixed by the 2nd order and the hierarchical Cartesian grid is introduced.

5.1 Flow Around a Circular Cylinder

The computational domain is shown in Fig. 5. The initial flow is fixed by the uniform flow, i.e., $u = 1$, $v = 0$. In the inflow boundary, the velocity components are fixed by the initial condition and the pressure is obtained by the Neumann condition. In the other boundaries, the pressure is estimeted by the Sommerfeld radiation condition and the velocity components are computed from the discrete continuity equation.

The first case is the steady flow, i.e., $Re = 40$. Figures 6 and 7 show the time history of virtual boundary velocity and the pressure contours. The non-slip condition is satisfied very well. The flow field (pressure) in the unsteady case is shown in Fig. 8. The number of Cartesian grid points are 16876 and

25424, respectively. And these results including the flow characteristics are in very good agreement with the other computational results.

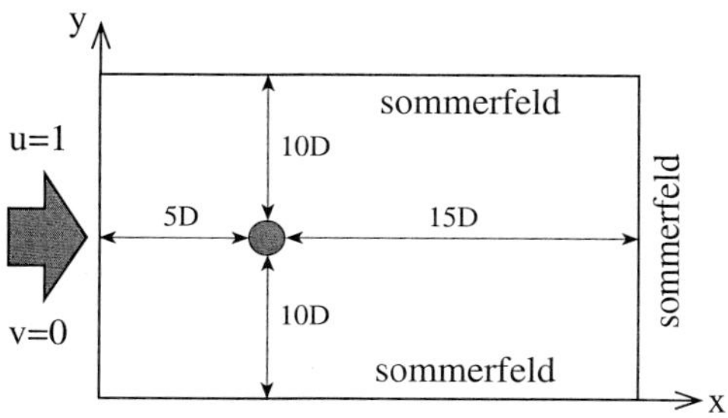

Fig. 5. Computational domain

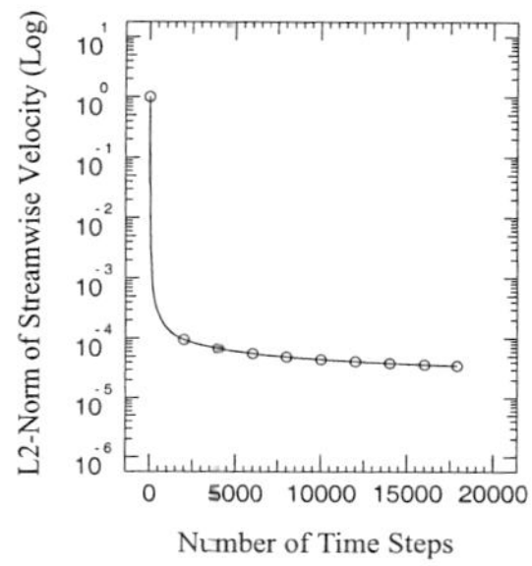

Fig. 6. Time history of virtual boundary velocity

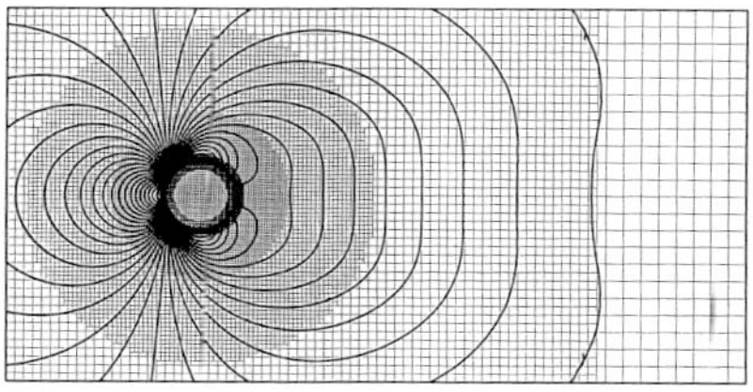

Fig. 7. Pressure contours ($Re = 40$)

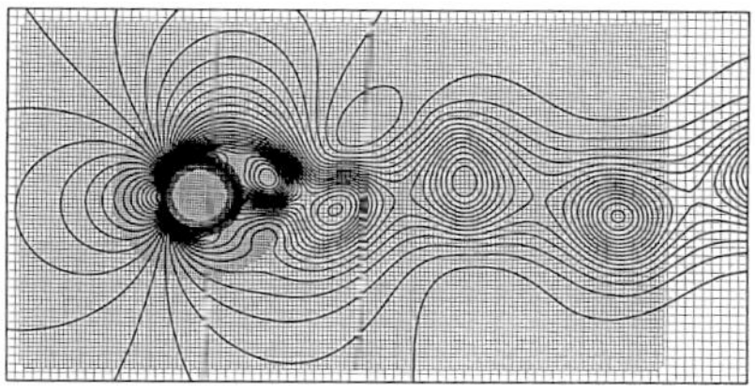

Fig. 8. Pressure contours ($Re = 200$)

5.2 Flow Around a Sphere

Figures 9 and 10 show the computational domain and the generated hierarchical Cartesian grid. In the laminar flow case with $Re = 300$ and 1855048 grid points, the iso-surface of $\nabla^2 p$ is shown in Fig. 11. It is very clearly observed that the hairpin vortex ring is released from the sphere. Next case is $Re = 1000$ with 3268428 grid points. In this case, the flow will transit from laminar to turbulent flow. The released vortex ring is brokendown in the downstream and many small eddies are generated in Fig. 12. The flow characteristics are in excellent agreement with the other results in both DNS.

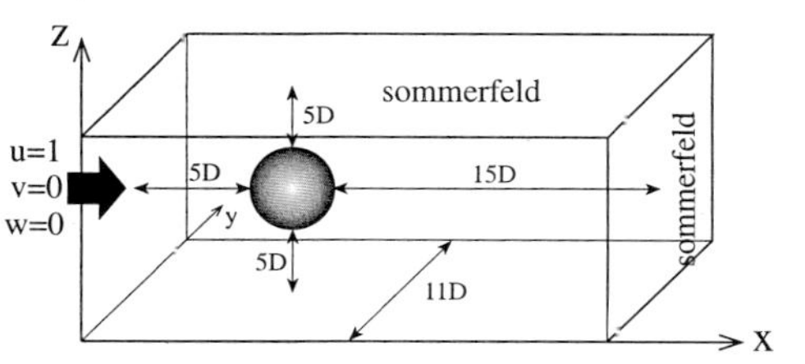

Fig. 9. Computational domain

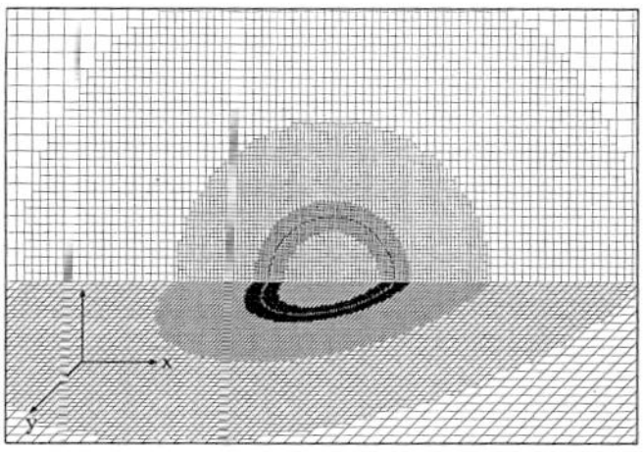

Fig. 10. Hierarchical Cartesian grid

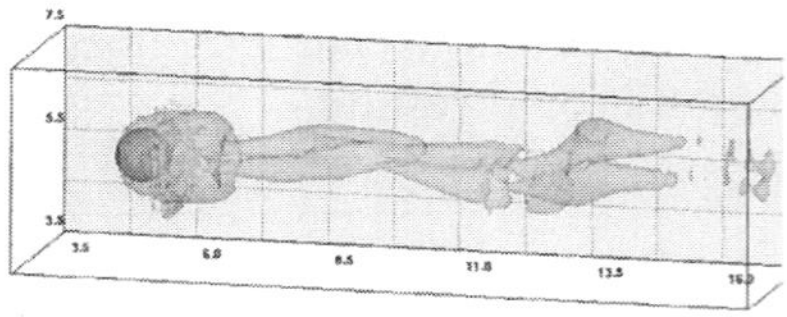

Fig. 11. Iso-surface of $\nabla^2 p$ ($Re = 300$)

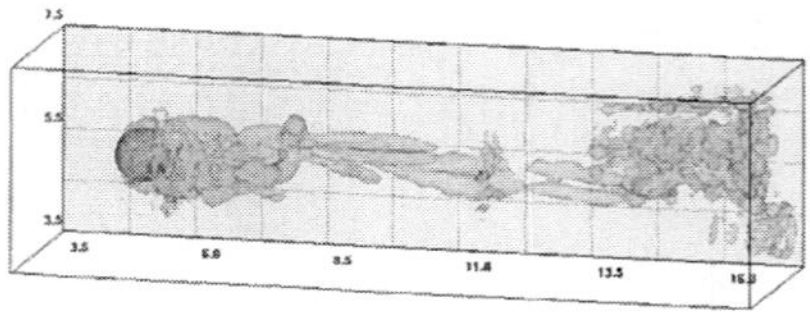

Fig. 12. Iso-surface of $\nabla^2 p$ ($Re = 1000$)

6 Conclusions

In this paper, the variable order method of lines (VOMOL) is presented for the DNS of incompressible flows. The present method gives the arbitrary order of spatial accuracy by changing only one parameter, and has the high conservation property. The Cartesian grid approach with virtual boundary method based on the VOMOL is applied to the DNS of incompressible external flows, i.e., flows around a circular cylinder and a sphere. The numerical results are in very good agreement with the other computational results and experiments. Therefore, it is concluded that the present method will be very promising for the DNS of incompressible complicated flows.

Acknowledgements

This work was supported in part by the Research for Future Program (97P01101) from Japan Society for the Promotion of Science and a Grant-in-Aid for Scientific Research (12650165) from the Ministry of Education, Culture, Sports, Science and Technology of the Japanease Government.

References

1. Nishida, H. (1999) Higher Order Direct Numerical Simulations of the Incompressible Navier-Stokes Equations Using a Variable Order Method of Lines. Proc. of the 8th ISCFD (CD-ROM)
2. Forrer, H., Jeltsch, R. (1998) A Higher-Order Boundary Treatment for Cartesian-Grid Method. J. Comput. Phys. **140**, 259–277
3. Goldstein, D. et al. (1993) Modeling a No-Slip Flow Boundary with an External Force Field. J. Comput. Phys. **105**, 354–366
4. Saiki, E.M. et al. (1996) Numerical Simulations of a Cylinder in Uniform Flow: Application of a Virtual Boundary Method. J. Comput. Phys. **123**, 450–465
5. Morinishi, Y. et al. (1998) Fully Conservative Higher Order Finite Differential Schemes for Incompressible Flow. J. Comput. Phys. **143**, 90–124
6. Nishida, H., Satofuka N. (1992) Higher Order Solutions of Square Driven Cavity Flow Using a Variable -Order Multi-Grid Mehtod. Int. J. for Num. Meth. in Eng. **34**, 637–653
7. Wambecq, A. (1978) Rational Runge-Kutta Methods for Solving Systems if Ordinary Differential Equations. Computing **20**, 333–342
8. Brandt, A. (1977) Multi-Level Adaptive Solutions to Boundary-Value Problems. Math. of Computation **31-138**, 333–390
9. Orszag, S.A. (1969) Numerical Methods for the Simulation of Turbulence. Phys. of Fluid Suppl. II **12**, 250–257

A Hybrid Direct/Iterative Algorithm for the Solution of Poisson's Equation Based on the Schur Complement Method

H.-J. Kaltenbach[1] and A. Jäkel[2*]

[1] Technische Universität Berlin, Sekr. HF 1, Hermann-Föttinger-Institut, Straße des 17. Juni 135, D-10623 Berlin
[2] Technische Universität Berlin, Sekr. MA 6-4, Fachbereich Mathematik

Abstract. An algorithm for efficient solution of large 3-D Poisson problems arising in the numerical simulation of the incompressible, unsteady Navier Stokes equations is presented. It is demonstrated that for a certain class of flows with one *homogeneous* coordinate direction an efficient load balancing on massively parallel machines can be achieved by simultaneous direct and iterative solution of 2-D Helmholtz problems. The use of the Schur complement method in conjunction with direct solution based on precomputed LU decomposition is capable to significantly reduce memory need and execution time.

1 Background and Objectives

In the simulation of unsteady, three-dimensional turbulent fluid motions based on 'Direct Simulation' (DNS) and 'Large-eddy simulation' (LES) easily $\mathcal{O}(10^7)$ unknowns arise in flows with moderate Reynolds numbers and the solution has to be advanced over $\mathcal{O}(10^4)$ timesteps. If the incompressible Navier-Stokes equations are solved based on the splitting approach (projection method) together with semi-implicit time advancement, solution of Poisson's equation carries the largest share of the overall cost of the algorithm. 'Fast' solvers for Poisson's equation have been devised in the past which utilize low operation counts of the fast Fourier transform (FFT) or cyclic reduction [1]. These methods are in general not applicable to problems with complex geometries.

Here we focus on a special class of '2D-complex' geometries – e.g. planar ducts, cylindrical obstacles, airfoils – for which the flow can be considered to be homogeneous in the 'spanwise' direction which allows use of a periodic boundary condition. Specifically adapted solvers for these type of geometries have been used in the past on vector computers [2–4] and parallel machines [5]. An efficient way of solving Poisson's equation on such a geometry consists in decoupling the 3D-Poisson problem via FFT along the

* This work was funded within "Universitärer Forschungsschwerpunkt UF 8" by Technische Universität Berlin. Computer time was provided by Konrad-Zuse-Zentrum (ZIB), Berlin.

'homogeneous' direction into a set of 2D-Helmholtz problems. The resulting Helmholtz problems are either solved iteratively or – less often – employing a direct solver for sparse matrices. Due to the availability of highly efficient FFT routines for problems where many transforms of the same size have to be carried out simultaneously, this hybrid approach is often more efficient than the use of a pure iterative solver for the 3D-Poisson problem.

The goal of the present study is to demonstrate how such an algorithm can be implemented on a massively parallel computer. The primary purpose of our investigation is to show that *direct solution* of large 2D-Helmholtz problems with $\mathcal{O}(10^5)$ unknowns is feasible and a competitive alternative to purely iterative methods. Novel aspects of our method include the simultaneous iterative and direct solution on a parallel machine and the use of the Schur complement method for solution of large 2D-Helmholtz problems.

2 Method

The incompressible Navier-Stokes equations in primitive variables (u, v, w, p) are solved on a staggered mesh with spatial discretization based on second order central differences. For the semi-implicit time-advancement we combine a low-storage third order Runge-Kutta with a Crank-Nicholson scheme [2]. The basic algorithm consists of the four steps

$$(1 - \beta_k \Delta t M)\, \tilde{u}_i^k = r_i^k \tag{1}$$

$$\frac{\partial^2 \Delta p}{\partial x_j \partial x_j} = \frac{1}{2\beta_k \Delta t} \frac{\partial \tilde{u}_i^k}{\partial x_i} \tag{2}$$

$$u_i^k = \tilde{u}_i^k - 2\beta_k \Delta t \frac{\partial \Delta p}{\partial x_i} \tag{3}$$

$$p^k = p^{k-1} + \Delta p \qquad \text{(k=1,2,3; i=1,2,3)} \quad . \tag{4}$$

Here, the index i refers to the coordinate direction whereas k denotes a substep of the Runge-Kutta integration cycle for each timestep Δt. The operator M involves differentiation with respect to the $y-$direction, resulting in tridiagonal systems of equations to be inverted in (1). Computation of right hand sides r_i is explicit with respect to variables u_i^{k-1} and p^{k-1} from the preceding substep, respectively timestep. Equation (4) is simply an update of the pressure since Poisson's equation is formulated in terms of the pressure difference Δp in order to retain second order accuracy in time. The algorithm applies to arbitrary non-equidistant Cartesian or curvilinear geometries in the x, y-plane. In the following we assume a single-connected region of width L_z consisting of N_z 'slabs' of size $N_x \times N_y$. Poisson's equation in discretized form reduces to a set of N_z Helmholtz problems

$$\left(\frac{\delta^2}{\delta x^2} + \frac{\delta^2}{\delta y^2} - k_m \right) \widehat{p}_{i,j,m} = \widehat{f}_{i,j,m} \tag{5}$$

after Fourier transformation of the discretized equivalent to (2) in the 'spanwise', respectively z-coordinate. Here, δ denotes the difference approximation and f is the right hand side of (2). A hat indicates Fourier transform, the index $\mathtt{m}$ denotes the Fourier mode and $k_m = 2/\Delta z^2\,[1 - \cos(2\pi m/N_z)]$ is the *modified wavenumber*, respectively the eigenvalue corresponding to second order central differentiation in the z-direction. Here, Δz denotes the grid spacing L_z/N_z. Since k_m is real, for each complex mode $\mathtt{m}$ with $1 \le \mathtt{m} \le \mathtt{N_z}/2 - 1$ the identical linear system of equations has to be solved with two different right hand sides.

3 Simultaneous Iterative and Direct Solution

The computational effort needed for solving 2-D Helmholtz problems changes with mode index $\mathtt{m}$ since k_m enters the diagonal of the matrix to be inverted in (5). Table 1 shows results from a model problem described in Sect. 5. With increasing index $\mathtt{m}$ the number of iterations decreases significantly which is mirrored by the corresponding decrease in the condition number C_m. Due to the increasing diagonal dominance the higher modes can be efficiently solved with SLOR based on Gauss-Seidel with red-black ordering and implicit treatment of the y-direction. Conversely, direct solution is found to be an efficient alternative to iterative solution for modes $m \le 3$. For our model problem, the cost of direct solution based on back-substitution with a precomputed factorization kept in core equals the cost of $\mathcal{O}(40)$ Gauss-Seidel iterations.

Table 1. Cost for iterative solution of 2D-Helmholtz problems

m	0	1	2	3	4	8	16	≥ 21
C_m^{a}	3.3×10^7	4.5×10^5	1.1×10^5	5013	2857	714	189	
I_m^{b}	1.3×10^5	750	260	114	58	24	10	≤ 6

[a] Condition number for a subdomain of size $N_x \times N_y = 280 \times 128$.
[b] Iteration count for iterative solution (SLOR) of (5) with relaxation parameter $\omega = 1.2$, time step $\Delta t = 0.015 H/U_\infty$, and residual $\varepsilon = 10^{-5}$.

The algorithm for solving (1)–(4) is based on unipartitioning of the domain in blocks of size $N_{x,b} \times N_y \times N_z$ as shown in Fig. 1. The pool of available processing elements (PEs) is split into two groups denoted as P_{RHS} and P_{LU}. The block size $N_{x,b} = N_x/P_{RHS}$ is determined by the number of PEs, P_{RHS}, which do the bulk of work in the algorithm. A second, much smaller group of PEs consists of P_{LU} members which – due to memory limitations – participate only in the direct solution of Helmholtz problems for modes $\mathtt{m} = 0 ... \mathtt{m_{max}}$. Usually, P_{LU} is much smaller than P_{RHS}, typical values being $\mathtt{m_{max}} = 1$, $P_{LU} = 4$, and $P_{RHS} = 60$ in our model problem. Direct solvers for banded non-symmetric matrices (SGBTRF and SGBTRS) from

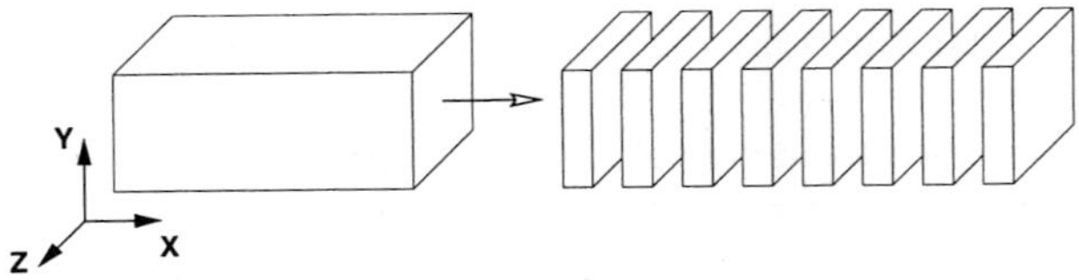

Fig. 1. Unipartitioning into blocks of size $N_{x,b} \times N_y \times N_z$

the LAPACK library are used. The LU factorization is computed once at the beginning of the temporal integration and kept in core memory to be used at each timestep with a new right hand side. Since the matrices are banded with a bandwidth $(2N_y + 1)$ for $N_y < N_x$, a compact storage format can be used. Still, the LU factorization that resides in core memory requires $M_{LU} = (3N_y + 1) \times (N_x N_y)$ words of memory.

Table 2 illustrates the different tasks and the communication patterns for both PE groups. After computing the FFT, global communication is required to transfer the right hand side for a given Helmholtz problem between processor groups P_{RHS} and P_{LU}. The solution of Helmholtz problems proceeds *simultaneously* on both processor groups. During the iterative solution, communication among members of group P_{RHS} involves the nearest neighbour only. Before backwards Fourier transformation of $\widehat{p}$, again global communication between both PE groups is required to put $\widehat{p}$ back into place on the corresponding subdomain locations assigned to members of group P_{RHS}.

Table 2. Individual tasks for both processor groups

eq.	tasks for PE group P_{RHS}		tasks for PE group P_{LU}
	idle		precompute LU factorization
	compute r.h.s. of momentum equation		idle
(1)	invert tridiag. system for $\tilde{u}_i$		idle
(2)	forward FFT of $\nabla \cdot \tilde{u}$		idle
	send FFT$(\nabla \cdot \tilde{u})$ for $\mathtt{m} = 0 \ldots \mathtt{m}_{\max}$	$\Longrightarrow$	receive r.h.s for $\mathtt{m} = 0 \ldots \mathtt{m}_{\max}$
(5)	iterative solution for mode $\mathtt{m} = \mathtt{m}_{\max} + 1 \ldots N_z/2$		direct solution for mode $\mathtt{m} = 0 \ldots \mathtt{m}_{\max}$
	receive $\widehat{p}$ for $\mathtt{m} = 0 \ldots \mathtt{m}_{\max}$	$\Longleftarrow$	send $\widehat{p}$ for $\mathtt{m} = 0 \ldots \mathtt{m}_{\max}$
(2)	backwards FFT of $\widehat{p}$		idle
(3)	obtain u_i^k from projection		idle
(4)	update pressure		idle

4 Schur Complement Method for 2D-Problems

The size of the available core memory limits the maximum size of the 2D-Helmholtz problem which can be solved on a single PE. We use an *additional* domain decomposition in conjunction with the Schur complement method to reduce both memory requirements and execution time by letting more than one PE participate in the direct solution of a single Helmholtz problem. The Schur complement method consists in decomposing the 2D-slab of size $N_x \times N_y$ corresponding to a Helmholtz problem in N_D subdomains, each being assigned to a single PE. This is achieved by formally reordering the system of equations in such a manner that the equations which describe the coupling between the subdomains appear in the lower right corner of the new system [6]. The resulting capacitance matrix or Schur complement is of size $M_{SCHUR} = ((N_D - 1) \times N_y)^2$. LU factorizations for the banded subdomain matrices as well as for the capacitance matrix are precomputed once and kept in core memory. A low storage version of this algorithm is commonly referred to as *implicit block factorization* [7]. It consists in solving twice the linear system on each subdomain.

The major part of the solution stage can be done in parallel for each subdomain, but the part of the backsubstitution process which involves solution of the capacitance system is essentially sequential. Because of this, the Schur complement method is limited to low granularities of $\mathcal{O}(10)$ PEs for a single Helmholtz problem for typical problem sizes with $300 < N_x < 1000$ and $N_y < N_x$. In order to use this method on massively parallel systems it is necessary to solve simultaneously for several Helmholtz problems.

5 Direct Solution of 2-D Helmholtz Problems

The hybrid iterative/direct algorithm has been used to simulate a flow problem discretized on $N_x \times N_y \times N_z = 560 \times 128 \times 128$ non-equidistant finite volumes on a rectangular domain of size $L_x = 21H, L_y = 11H, L_z = 6H$. The flow under consideration is a laminar boundary layer separating at the edge of a rearward facing step of height H for $Re_H = U_\infty H/\nu = 3000$. Details of the flow physics are described in [8]. The mesh is stretched in both the x and $y-$direction and the ratio of maximum to minimum spacing is $\Delta x/\Delta y = 75$. We report performance measurements obtained on a Cray T3E-900 LC 136-128 using DEC Alpha EV5 processors with a local memory of 128 MB. The implementation in FORTRAN follows the MPI standard.

We have evaluated the algorithm in terms of scalability and absolute performance, i.e. execution time, for a problem of constant size. This is commonly referred to as unscaled speed-up. For solution of 2D-Helmholtz problems we have also measured scaled speed-up. The minimum as well as maximum number of PEs that can be used for a problem of given size, is limited by constraints from hardware (memory) or coding (minimum block size $N_{x,b} = 4$). Therefore, the *relative* speed-up efficiency $S_\eta = (T_{min} \times P_{min})/(T_P \times P)$ is the appropriate performance measure. Here, P_{min} denotes the smallest number of PEs for which the problem was solved, T_{min} is the average CPU-time required on one of P_{min} PEs and T_P is the measured time for solution on P processors.

Table 3. Relative speed-up efficiency for direct solution

N_D	T_D^a (s)	M_D^b (words)	M_{SCHUR} (words)	C_D^c	C_{SCHUR}^c	S_η
2	0.570	13798400	16384	3333333.	33557.	1.00
3	0.383	9213696	65536	1639344.	69930.	0.99
4	0.293	6935040	147456	934579.	107527.	0.97
5	0.243	5498496	262144	609756.	142857.	0.94
6	0.212	4606848	409600	426221.	180148.	0.90
7	0.195	3913334	589824	313224.	216986.	0.84
8	0.187	3417984	802816	237665.	253749.	0.76
9	0.182	3071232	1048576	192626.	289981.	0.70
10	0.185	2724480	1327104	167670.	326552.	0.62
14	0.231	1931904	2768896	112285.	474856.	0.35

[a] CPU time needed for the direct solution of a 2D-Helmholtz problem of size $N_x \times N_y = 560 \times 128$ divided into N_D partitions
[b] Memory requirements for LU decomposes of subdomain (M_D) and capacitance matrix (M_{SCHUR})
[c] Condition numbers for subdomain and capacitance matrix corresponding to `m=0`

Table 3 shows that up to $N_D = 6$, the CPU time required for the direct solution decreases almost proportional to the number of subdomains, resulting in relative speed-up efficiencies above 90%. Because of the relative increase of the sequential part of the algorithm the speed-up efficiency drops significantly for higher values of N_D. In fact, we observe no further decrease in execution time for $N_D \geq 9$. The drop in efficiency might be tolerated in case the main motivation for the Schur method is reduction in memory requirements for an individual processor. A potential drawback of direct solution with the Schur complement method is the decrease of relative accuracy with increasing number of partitions. For our model problem the residuals remain below $\mathcal{O}(10^{-9})$.

The algorithm has been used to solve 2D-Helmholtz problems of various sizes as shown in Table 4. We find that the specific work W/M_H remains fairly constant over a one order of magnitude change in problem size. Also, the execution time can be kept constant if the number of PEs is increased accordingly to the problem size. The largest problem tested had $N_x \times N_y \times N_z = 845 \times 200 \times 256$ cells. Here, out of a total of 121 PEs a subset of $P_{LU} = 27$ was used to solve for modes $m = 0, 1, 2$ directly.

6 Performance of the Hybrid Iterative/Direct Navier-Stokes Solver in a 3-D Test Case

Table 5 contains a breakdown of absolute performance of different stages of the solution procedure. The times reported are those which occur at the PE

Table 4. Scaled speed-up of direct solution

N_x^a	N_y	N_D	M_H^b (Mwords)	T_D (s)	W^c (s)	W/M_H (s/Mwords)
257	90	2	6268230	0.135	0.27	4.31×10^{-8}
560	128	2	27596800	0.57	1.14	4.13×10^{-8}
845	200	8	101569000	0.54	4.31	4.25×10^{-8}

[a] A problem of size $N_x \times N_y$ is solved on N_D partitions
[b] The factorization of the matrix has a size of $M_H = (3N_y + 1) \times N_x N_y$
[c] A total work $W = T_D N_D$ is performed during direct solution

that finishes the algorithmic step last. The most economical use of computer resources is made for these configurations where the total work W is the smallest. Except for the case with $P_{RHS} = 15$ and m=2 we find in all configurations that the total work *decreases* if more modes are solved for directly, despite the fact that the total number of PEs involved, i.e. $P_{LU} + R_{RHS}$, *increases*. This clearly demonstrates that the hybrid approach is useful and economical. As expected, the 'saving', i.e. the reduction in W is the larger the more PEs are used in total. When m_{max} is increased from 0 to 1, the work

Table 5. Breakdown of the computational cost

P_{rhs}^a	P_{lu}	$m_{max}{}^b$	$N_{x,b}$	T_{tot}^c (s)	T_{rhs} (s)	T_{fft} (s)	T_{it} (s)	T_{lu} (s)	T_{com} (s)	W^d (s)	S_η
15	2	0	37	45.1	14.4	9.9	20.7	1.7	.011	767	0.96
15	4	1	37	37.5	14.4	9.9	13.1	3.7	.021	713	1.03
15	6	2	37	34.8	14.4	9.9	10.4	3.7	.030	731	1.01
30	2	0	19	24.1	7.4	5.1	11.5	1.7	.010	771	0.96
30	4	1	19	20.2	7.4	5.1	7.6	3.7	.020	685	1.08
30	6	2	19	18.8	7.4	5.1	6.2	3.7	.024	677	1.09
63	2	0	9	13.9	4.2	2.3	7.4	1.7	.014	905	0.81
63	4	1	9	11.7	4.2	2.3	5.2	3.7	.026	780	0.94
63	6	2	9	10.9	4.2	2.3	4.4	3.7	.029	749	0.98
112	2	0	5	10.7	3.3	1.3	6.0	1.7	.025	1219	0.60
112	4	1	5	9.2	3.3	1.3	4.5	3.7	.042	1061	0.69
112	6	2	5	8.7	3.3	1.3	3.7	3.7	.046	1024	0.72

[a] The table contains measured times for advancing the solution for one timestep – involving 3 Poisson solves – with a total of $P_{tot} = P_{rhs} + P_{lu}$ PEs
[b] Helmholtz problem for modes m=0 to m_{max} were solved on two subdomains ($N_D = 2$)
[c] CPU time with T_{rhs} denoting time to solve (1),(3), and (4), and T_{fft} summing up the total time for Fourier transformation. Iterative solution of all modes $m > m_{max}$ requires T_{it}, direct solution takes T_{lu} and distribution of the right hand side of (2) needs T_{com}.
[d] The 'work' W done by all PEs is the product of total time T_{tot} and P_{tot} and the relative speed-up efficiency S_η is computed with respect to $\overline{W} = 737s$.

is reduced by 7% for $P_{RHS} = 15$ and by 13% for $P_{RHS} = 112$. The reason for that is twofold: firstly the relative increase in total PE number is 13% respectively 2%; secondly, the iterative solver becomes less efficient for small block sizes $N_{x,b}$.

The last column of Table 5 gives insight into the scaling of the entire algorithm by comparing individual configurations with a 'reference' case which corresponds to the average work $\overline{W} = 737\,s$ of the three cases with $P_{RHS} = 15$. The most economical configuration is with $677\,s$ the one which utilizes 36 PEs. For $\mathtt{m_{max}} \geq 1$ and a block size of $N_{x,b} = 9$ corresponding to a total of $\mathcal{O}(70)$ PEs the loss in efficiency is 6% only. This shows that the overall algorithm exhibits excellent scaling behavior, allowing to economically compute large scale applications. If the block size is decreased to $N_{x,b} = 5$ the efficiency drops below 70%. The table also shows that further reduction in execution time is achieved by increasing $\mathtt{m_{max}}$ from 1 to 2, although at the price of reduced efficiency. Of course, comparison of total work performed is no longer a 'true' scaling in the sense of PEs. Rather, the algorithm, respectively the load balancing, is adjusted in such a way, that the overall time consumed is minimized.

For comparison with other solvers the CPU-time per node is reported: using a total of 69 PEs of Cray T3E, in order to advance the solution of the model problem with 9×10^6 cells for one timestep (involving three RK3 substeps with corresponding Poisson solves) approximately $1.2\,\mu s$ per grid point are needed. The corresponding value for a problem with 43×10^6 cells solved on 121 PEs is $0.8\,\mu s$.

References

1. Swarztrauber P., Sweet R. (1989) Vector and parallel methods for the direct solution of Poisson's equation. J Comp Appl Math 27:241–263
2. Akselvoll K., Moin P. (1996) An efficient method for temporal integration of the Navier-Stokes equations in confined axisymmetric geometries. J Comp Phys 125:454
3. Voke P., Yang Z. (1996) Computational methods for large-eddy simulation in 2D complex geometries. In: M. Deville I.R. S. Gavrilakis (Ed.) NNFM **53**: Computation of three-dimensional complex flows
4. Kaltenbach H.J., Fatica M., et al. (1999) Study of flow in a planar asymmetric diffuser using large eddy simulation. J Fluid Mech 390:151–185
5. Crawford C., Evangelinos C., et al. (1996) Parallel benchmarks of turbulence in complex geometries. Computers & Fluids 25(7):677–698
6. Golub G., Ortega J.M. (1993) Scientific computing - An introduction with parallel computing. Academic Press, San Diego
7. Duff I.S., Erisman A.M., et al. (1986) Direct methods for sparse matrices. Oxford University Press, Oxford
8. Kaltenbach H.J., Janke G. (2000) Direct numerical simulation of flow separation behind a swept, rearward-facing step at $Re_H = 3000$. Phys Fluids 12:2320–2337

Part III

Crystal Growth and Materials

High-Performance Computing, Multi-Scale Models for Crystal Growth Systems

J.J. Derby[1], P. Daoutidis[1], Y. Kwon[1], A. Pandy[1], P. Sonda[1], B. Vartak[1], A. Yeckel[1], M. Hainke[2], and G. Müller[2]

[1] Department of Chemical Engineering and Materials Science, Army HPC Research Center, and Minnesota Supercomputing Institute, University of Minnesota, Minneapolis, MN 55455, USA

[2] Department of Materials Science (WW6), University of Erlangen-Nürnberg, 91058 Erlangen, Germany

Abstract. Large-scale numerical simulation carried out via high performance computing is proving to be an increasingly useful approach to understand crystal growth systems. However, increasing realism demands new approaches for describing phenomena important at several disparate length scales. Of special importance is the ability to represent three-dimensional and transient continuum transport (flows, heat and mass transfer), phase-change phenomena (thermodynamics and kinetics), and system design (such as furnace heat transfer during melt growth). A brief overview is presented of mathematical models and numerical algorithms employed to include such multi-scale effects. Sample results are presented for Bridgman crystal growth and solution crystal growth systems.

1 Introduction

The growth of crystalline materials ranges from the one-time creation of milligrams of single-crystal protein pharmaceuticals to the annual production of metric tons of electronic-grade silicon. Due to these broad applications, the great variety of crystals needed, and the exacting quality typically required of single-crystal materials, their successful growth ranks among the most difficult challenges of modern materials processing.

Successful growth of a crystal is based upon whether certain special properties, e.g., those responsible for the performance of a crystal in a device, are obtained; these properties are determined by crystalline structure and composition. Often, these characteristics are affected during the growth of the crystal by processes which link molecular-scale growth events with macroscopic transport phenomena. Our long-term goal in crystal growth modeling is to understand the mechanisms which influence crystal quality through the hierarchy of length and time scales relevant to these molecular-scale and macroscale processes.

Mathematical modeling coupled with high-performance computing has emerged as a powerful tool to probe the complexities of crystal growth processes. The focus of this paper is on the challenges needed to represent realistic complexity in models for two different methods of crystal growth —

growth from the melt using the vertical Bridgman technique and growth from a liquid solution. We briefly describe our modeling approaches for each of these techniques in the following subsections, emphasizing the modeling issues appropriate for each system.

1.1 Bridgman Growth

Bridgman crystal growth is depicted schematically in Fig. 1(a). In this system, an ampoule filled with molten charge is slowly moved through a high-temperature furnace, causing solidification to proceed as the charge enters the cold zone. Below, we describe two modeling approaches for this system. The first is a detailed model for flow, heat transfer, segregation, and solidification within the ampoule which addresses time-dependent phenomena arising from nonlinear convection in the melt. The second is a new approach to multi-scale modeling whereby the crystal growth furnace is treated as axisymmetric, but transport phenomena within the ampoule is treated as three-dimensional.

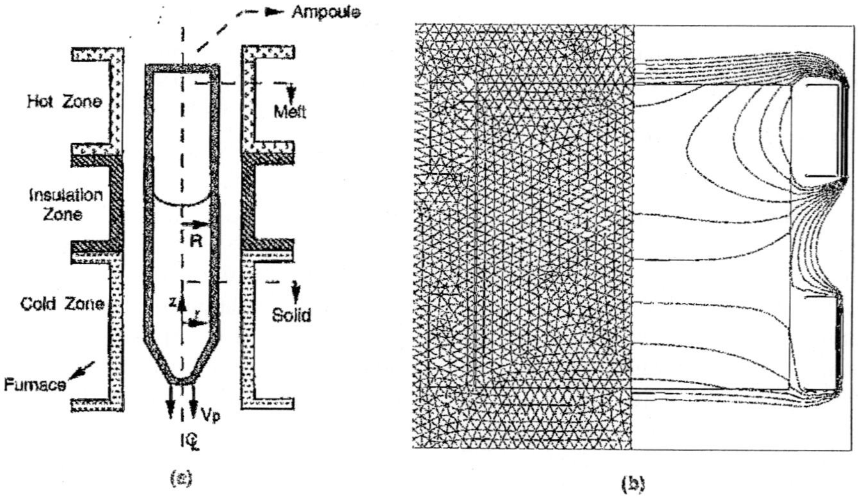

Fig. 1. (**a**) Schematic diagram of a general vertical Bridgman crystal growth system, with the crystal growing upward into the melt. (**b**) Mesh and temperature isotherms for idealized, two- zone furnace model meant to serve as a test configuration, solved using CrysVUN++

Detailed Local Model We consider time-dependent phenomena which arise during vertical Bridgman growth by modeling the ampoule and contents using mathematical equations of continuum transport, coupled with physically appropriate boundary and initial conditions and solved via numerical methods (see, e.g., [1,2]). Heat transfer in all domains is given by the transient dimensionless formulation of the energy balance, which accounts for conduction through all regions and convection in the melt phase. The temperature field is continuous and heat fluxes are balanced across all contiguous domains. At the solid-liquid interface, heat of fusion is released upon crystallization as a function of the local interface velocity. This interface is located by satisfying a time-dependent isotherm condition, which assumes that the equilibrium melting point temperature determines the interface position. At the exterior domain boundaries, we model the heat transferred from the charge and ampoule to the furnace via convection and radiation. The piecewise-linear furnace temperature profile corresponds to that assumed for the inner bore of the furnace; we address this idealization in the multi-scale model described in the next section.

The flow in the melt phase is described by the time-dependent Navier–Stokes equation with the Boussinesq approximation and the continuity equation for an incompressible fluid. This approximation well describes buoyant flow within the melt. No-slip conditions are imposed for melt flows along the crucible walls and the solidification interface. We assume the density of the liquid does not change appreciably upon solidification, so shrinkage and subsequent flow penetration into the solid/liquid interface are also assumed to be negligible.

We employ the Galerkin finite element method [3] to discretize the spatial dependence of the governing equations, yielding a coupled, nonlinear set of differential algebraic equations. Integration of these equations in time is implemented using a variable-time-step trapezoid method [4]. At each time step, the nonlinear set of equations obtained from the finite element discretization are solved with the Newton–Raphson method [4] employing an analytical Jacobian. The mesh deforms dynamically with each new solution of the governing equations to track the interface. We also employ computer-aided bifurcation theory employing the Newton–Raphson method to solve for a quasi-steady-state model based on the short-time-scale dynamics of the system; see [5,6] for more details.

Multi-Scale Global Model We desire to model realistic global heat transfer and three-dimensional phenomena which may arise during vertical Bridgman growth. We describe here an approach to multi-scale modeling whereby the furnace is treated as axisymmetric, but transport phenomena within the crucible or ampoule is treated as three-dimensional. This approach allows us to couple a sophisticated two-dimensional global furnace model, the CrysVUN++ code of Müller et al. [7,8], to our detailed finite element model of three-dimensional flows, segregation, and solidification during melt crystal growth, implemented in a parallel MPI-based code [9,10].

CrysVUN++ solves planar and axisymmetric equations for energy transport by conduction and radiation, using the finite volume method; a sample finite volume mesh and thermal computation are shown in Fig. 1(b). The program is specifically designed to analyze global heat transfer in crystal growth systems. A graphical user interface facilitates the preparation of input data. CrysVUN++ can import and edit data from several popular computer-aided drawing programs, simplifying data preparation for complicated furnace designs. CrysVUN++ is a powerful tool but has two key limitations: the model currently neglects convective heat transfer in the melt, and, by its axisymmetric nature, the model precludes study of three-dimensional effects.

The premise of our approach is that three-dimensional transport effects commonly arise due to flow phenomena occurring in the melt, and often are largely isolated to that region. This observation suggests dividing the system into two regions for the purpose of modeling: an inner region in which a *three-dimensional* model is used, and an outer region in which an *axisymmetric* model is used. The location of the boundary is chosen to be the outer surface of the ampoule or crucible. The key to the coupling is how best to match the inner and outer solutions at the boundary in order to preserve the three-dimensional character of the inner model.

Our first approach passes fluxes at the ampoule boundary computed by CrysVUN++ to be used as boundary conditions in our finite element codes. We believe that this approach has several merits. First, matching fluxes rather than temperatures assures that the global problem satisfies conservation of energy. Second, matching temperatures strongly enforces axisymmetry of the temperature field at the boundary, which is likely to delay or suppress altogether some three-dimensional flow phenomena, particularly axisymmetry-breaking instabilities; matching fluxes at the boundary does not impose any such axisymmetry *a priori*. However, a matching condition based purely on heat fluxes must overcome several complications. In the absence of any other model constraints, there is nothing to set the temperature level, i.e., the mathematical problem is under-specified and has no unique solution. Worse yet, if the flux data do not satisfy global energy conservation, the mathematical problem is inconsistent and has no solution at all. This latter concern is relevant because any numerical model can only supply approximate fluxes at the boundary, due to discretization error. We discuss some of the mathematical and implementation details of this matching in [11,12].

1.2 Solution Growth

Solution crystal growth relies on the controlled precipitation of a solute from a supersaturated liquid phase; a schematic diagram of the growth of potassium titanyl phosphate (KTP) for a system developed by Bordui et al. [13,14] is shown in Fig. 2(a). Growth is initiated from a seed crystal mounted on a stirring rod and is sustained by continually lowering the system temperature to keep a near-constant level of supersaturation. For large crystals, of the size of $\mathcal{O}$(cm), growth rates are typically limited by mass transfer through the solution phase [15], so some means of stirring is important for achieving high growth rates. In this system, the stirring rod and crystal are rotated using a predetermined cycle of forward and reverse directions. Flows past the facetted crystal can lead to significant inhomogeneity of mass transfer rates and surface supersaturation levels, which can lead to certain morphological instabilities and inclusion formation [15]. As for the prior models for Bridgman growth, we describe two modeling approaches. The first is a detailed model for flow and mass transfer at the continuum level; the second is a novel micro-scale model to describe the growth of a single vicinal facet from the solution phase.

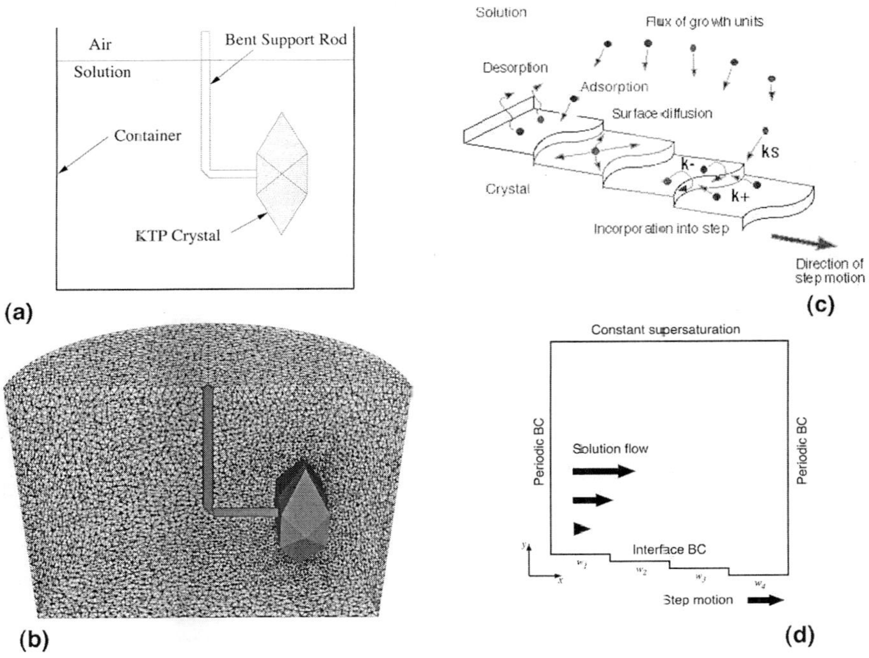

Fig. 2. (a) Schematic of the KTP growth system at the continuum level. (b) Schematic diagram of step growth model of a vicinal facet of a crystal which includes bulk transport, adsorption and desorption, surface diffusion, and incorporation into steps via the surface (k- and k+) and the overlying bulk fluid (ks)

Continuum Transport Model The model required here is somewhat different from that employed to model the Bridgman system discussed previously in Sect. 1.1. We are interested in the short-time-scale features of flow and mass transfer, so we assume that the geometry of the relatively slow-growing crystal is fixed. We solve the Navier-Stokes equations for an incompressible fluid and also assume the system to be isothermal. The supersaturation field is evaluated by solving the convection-diffusion equation with suitable flux conditions applied at the surface of the growing cystal; an extensive discussion of this model is provided in [16,17].

The most significant challenge is to accurately and affordably solve for the three-dimensional, time-dependent nature of the flow and supersaturation fields. To meet these challenges, we turn to parallel computing. We employ the finite element method; a typical mesh used for these computations is depicted by the cut-away view shown in Fig. 2(b). We employ an unstructured mesh of tetrahedral elements; particular care must be taken to construct meshes fine enough near solid surfaces to accurately capture boundary layers.

To compute the convection-dominated transport in this system, we employ stabilized, Galerkin Least-Squares (GLS) method [18,19]. The resulting discretizations are either solved in a steady-state formulation or integrated with an implicit Euler scheme in time. The nonlinear equation set resulting from the steady-state formulation or the time-marching scheme is solved using the Newton–Raphson method. We use the GMRES (generalized minimum residual) iterative technique [20] with diagonal preconditioning [21] to solve the linear equations arising at each Newton–Raphson iteration. Our formulation is implemented using data-parallel constructs on several parallel high performance computers; additional details of our approach are available in [9,21].

Facet Growth Model A crystal typically grows from solution by adding layer upon layer of molecules, which form steps and terraces on the facets of the growing crystal. The motion of these growth steps is governed by a complicated interplay among various transport and kinetic mechanisms acting near and on the surface of the crystal. The origin of these steps is typically a spiral or screw dislocation in the crystal. This scenario has subsequently been referred to as the Burton, Cabrera, Frank, or BCF mechanism [22]. To study such details, we have developed a comprehensive micro-scale model based on the BCF mechanism to describe the growth of a single vicinal facet from the solution phase, which includes bulk transport, adsorption and desorption of growth units to terraces, surface diffusion, and incorporation into steps from the surface and from bulk [23]; these processes are represented schematically in Fig. 2(c).

The coupled governing equations are written in a two-dimensional, Cartesian frame, where bulk fluid overlies terraces represented by flat line segments demarked on both sides by growth steps, which are represented as point discontinuities between the terraces; see Fig. 2(d). The compete set of equations is discretized by an efficient, moving-boundary, finite element numerical method, which solves for the bulk and surface supersaturation fields and simultaneously tracks the position of all growth steps. A detailed discussion of this model is provided in [23].

2 Sample Results

2.1 Bridgman Growth

Detailed Local Model We specifically desire to model the dynamical behavior observed in the classical growth experiments of Kim, Witt, and Gatos [24], who grew 1.3 cm diameter, tellurium-doped indium antimonide in a vertical Brdgman system. In this system, an ampoule filled with molten charge is slowly withdrawn from the top of the furnace, causing solidification to proceed downwards (opposite the direction depicted in Fig. 1a). Since the warm melt is located under the cooler crystal, this system has commonly been referred to as a *destabilized* system, analogous to the classical Rayleigh problem in fluid dynamics [25]; this system is prone to more intense convective flows than the *stabilized* system with the crystal placed below the melt.

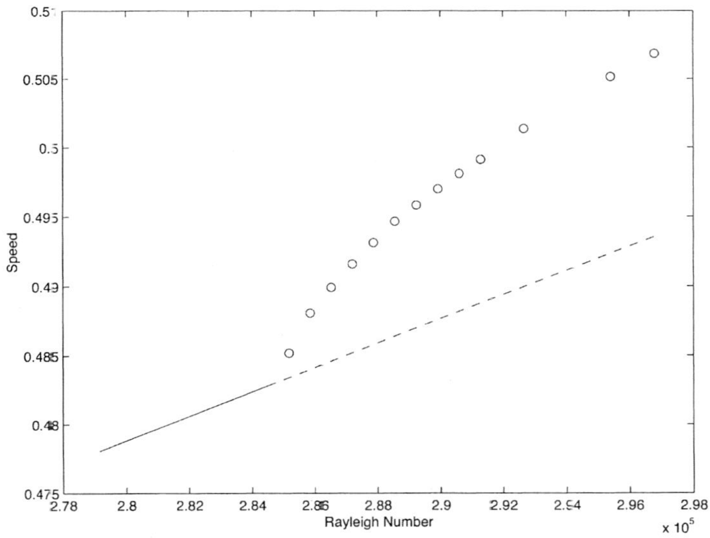

Fig. 3. Bifurcation diagram showing average speed of melt flow in Bridgman system plotted as a function of dimensionless Rayleigh number

Kim, Witt, and Gatos [24] document several flow transitions and their concomitant effect on solute segregation. Here, we compute one of these flow transitions, that of time-dependent to steady flows. We perform quasi-steady-state computations by holding the ampoule at a constant position in the furnace while imposing a heat flux along the melt-crystal interface consistent

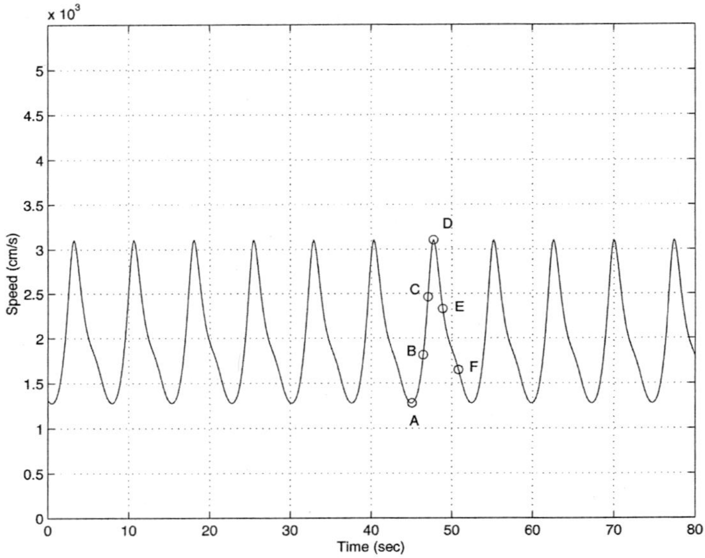

Fig. 4. Volume-averaged speed plotted as a function of time for the Bridgman system at a supercritical Rayleigh number

with the latent heat released at a constant growth rate. We perform several computations for various ampoule positions within the furnace; this strategy sets varying melt heights. Choosing the height of the melt as a characteristic distance, we define a dimensionless Rayleigh number, defined by $Ra_m = g\beta_T L_m^4 \gamma/\nu\alpha_m$, where g is the magnitude of gravitational acceleration, β_T is the thermal expansivity, L_m is the melt height, γ is the axial temperature gradient, ν is the kinematic viscosity, and α_m is the thermal diffusivity. With its fourth-order dependence on melt height, the Rayleigh number will decrease as the crystal is grown.

We represent these computations via a bifurcation diagram, depicted in Fig. 3, where a measure of the solution, here the flow speed averaged over the entire melt domain, is plotted as a function of the dimensionless Rayleigh number. Every point along the solid curve in Fig. 3 represents a steady state which is stable to temporal disturbances, as determined by an anlysis of the leading eigenvalues of the system. A critical Rayleigh number of $Ra_m = 2.843 \times 10^5$, marks the point where the steady flow becomes unstable as a complex-conjugate pair of eigenvalues cross into the right half plane. The family of unstable solutions is depicted by the dashed line in Fig. 3. At the critical point, known as a Hopf bifurcation, a family of time-periodic solutions arises; the maximum speed of this family is shown on the diagram by a series of open circles.

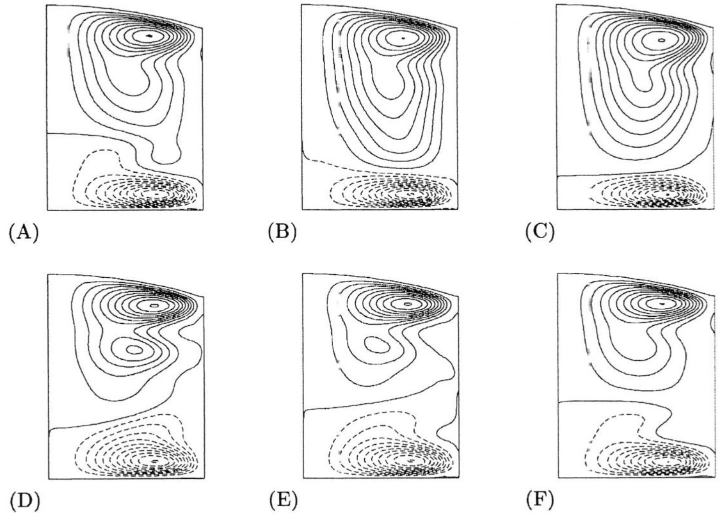

Fig. 5. Streamlines of flows for the Bridgman system at the points indicated in Fig. 4

In the context of a growth run, our results would indicate that a time-periodic flow exists at certain melt heights and that this flow decreases in strength as the crystal grows and the characteristic melt height (and subsequently Ra_m) decreases. As the system moves past the Hopf point, i.e., for $Ra_m < 2.843 \times 10^5$, the flow settles into a steady, time-independent mode. This transition corresponds well with the experimental findings of Kim, Witt, and Gatos [24].

Features of one of the time-periodic solutions is shown in Fig. 4 and 5. The average speed of the melt is plotted as a function of time in Fig. 4; the oscillatory nature of the flow is evident. The spatial structure of the flow is represented in Fig. 5, where instantaneous streamlines are shown for the corresponding points indicated in the prior figure. Solid streamlines indicate counterclockwise motion, and dotted streamlines show clockwise circulation. Beginning with (A), the upper vortex grows and strengthens to state (C), followed by a strengthening of the lower vortex from (D) to (F). This cycle repeats in an oscillatory manner. More extensive details are provided in [6].

Multi-Scale Global Model Preliminary results from the coupled, multi-scale code are shown are shown in Fig. 6. Here we have coupled the two-zone furnace computation shown in Fig. 1(b) with our 3D finite element code. In both codes, we assume simply that a cylindrical domain comprising one phase with only conduction acting is exchanging heat with the furnace via radiation. First, CrysVUN++ is run alone, providing the results shown in Fig. 1(b). Then, CrysVUN++ data are used to provide heat flux boundary conditions at the surface of the cylinder for our finite element, 3D code, which is set up to model conduction within an identical cylindrical domain. This idealized test problem is posed so that the 3D code should compute the same temperature profile in the cylinder as that predicted by CrysVUN++.

The temperature field computed by the 3D finite element code so obtained is shown in Fig. 6. The surface temperature profile obtained from the 3D code is compared to the CrysVUN++ results in Fig. 6(a), and temperature contours from the 3D code are shown on x-z and y-z planes in Fig. 6(b). The results predicted by both codes compare favorably; demonstrating that the first linkage between these codes is working well. The resulting temperature fields do not match exactly, since different discretizations and methods are used in the computations by CrysVUN++ (a finite volume code) and the 3D inner model (a finite element code). Our most recent work, solving for a 3D, crystal growth inner problem with an axisymmetric furnace heat transfer outer problem, is discussed in [11,12].

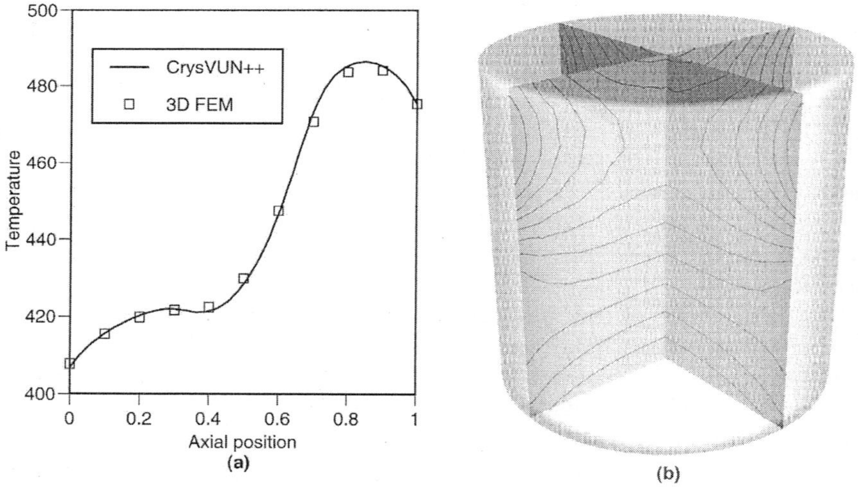

Fig. 6. (**a**) Comparison of surface temperature profiles computed by CrysVUN++ and 3D finite element code. (**b**) Temperature contours on x-z and y-z planes (compare to isotherms shown on the right half of Fig. 1(b)

2.2 Solution Growth

Continuum Transport Model Here, we desire to study the details of three-dimensional flow and mass transfer caused by steadily rotating the crystal in the system depicted in Fig. 2(a). Under the conditions employed by Bordui et al. [13,14], the system is characterized by a rotational Reynolds number of $Re = 252$ and a characteristic Peclet number for mass transfer of $Pe = 8.38 \times 10^6$. A more detailed description of these calculations is provided in [16].

Figure 7 shows pathlines computed for steady, three-dimensional flows driven by crystal rotation in two different growth system geometries — one with the crystal mounted directly onto a straight support rod (the 0° orientation) and one with the crystal mounted onto a rod bent at a right angle (the 90° case). Figure 7(a) shows two path lines that originate from symmetric points above and below the crystal (which is spinning steadily in a counter-clockwise direction when viewed from above). These path lines clearly reveal the large toroidal structures which are localized in space and do not connect, leading to relatively poor bulk mixing. Figure 7(b) shows results for a different crystal mounting geometry under the same operating conditions. A single path line is shown that originated from the same starting position as the lower path line in Fig. 7(a) and is integrated for an identical length of time. It is clear from the path line that, in this orientation, the rotating crystal breaks the non-mixing toroidal structures and promotes far more complete mixing in the bulk of the fluid. In a general sense, more complete mixing of the bulk solution phase is beneficial and leads to more homogeneous rates of solute transfer to the growing crystal, as well as higher overall rates of growth due to a reduction in the thickness of boundary layers adjacent to the crystal facets.

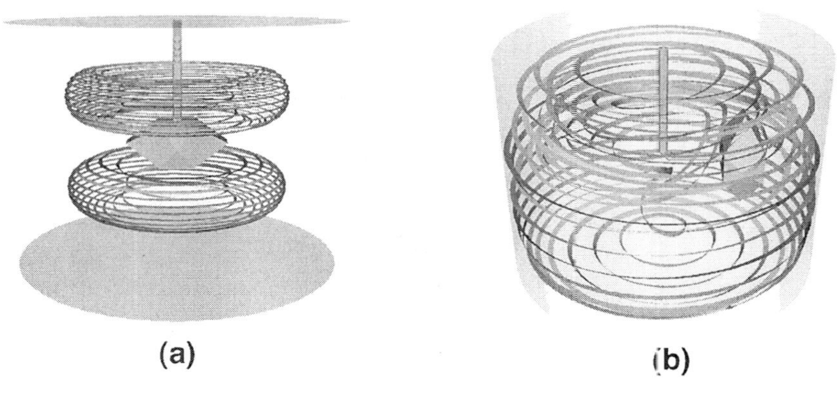

Fig. 7. Path lines show (**a**) large toroidal vortices above and below the crystal in the 0° orientation and (**b**) more complete mixing in the 90° case

Facet Growth Model A sample computation from our micro-scale facet growth model is shown in Fig. 8, where a snapshot is displayed of the transient response of a train of 21 growth steps moving from left to right with bulk and surface diffusion mechanisms acting and with surface and direct bulk incorporation of growth units to the steps. Initially, we impose a perturbation of wavelength equal to half the simulation box size to an equidistant configuration of steps.

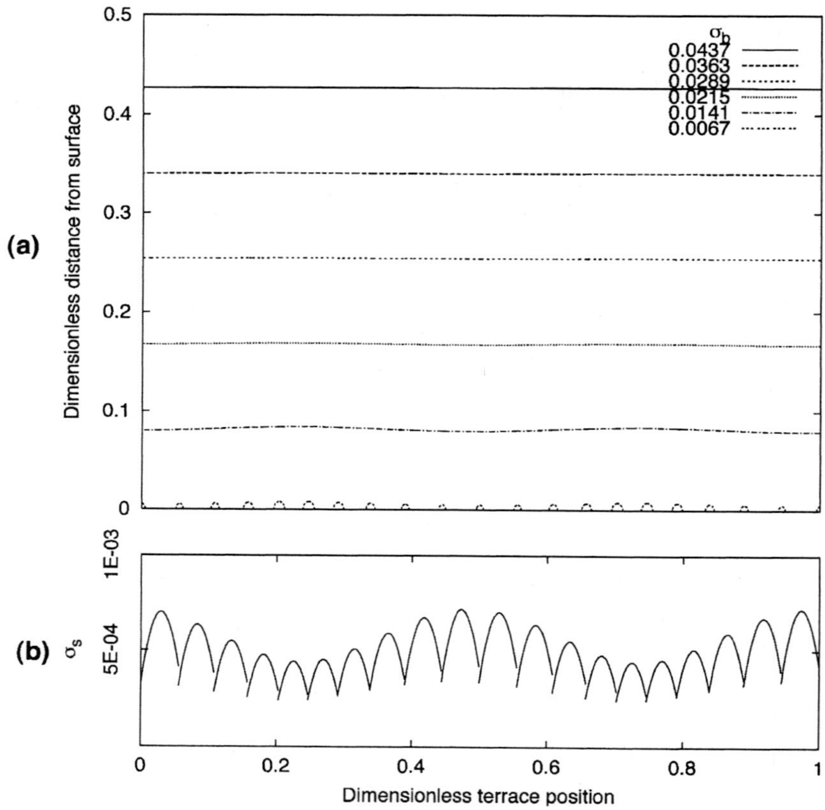

Fig. 8. Snapshot of (**a**) bulk (σ_b) and (**b**) surface (σ_s) supersaturation profiles for a train of 21 growth steps under conditions of bulk and surface transport and incorporation

Figure 8 shows instantaneous bulk and surface supersaturation profiles for this case after a short period of growth (σ_b and σ_s, respectively, plotted as functions of distance along the terrace). The supersaturation profiles across the individual terraces are caused by surface diffusion and incorporation into the steps, as shown in Fig. 8(b). These terrace profiles are also discontinuous across the steps.

The grouping of steps closer together or farther apart causes the overlying bulk fluid to become somewhat depleted or enriched of solute, respectively.

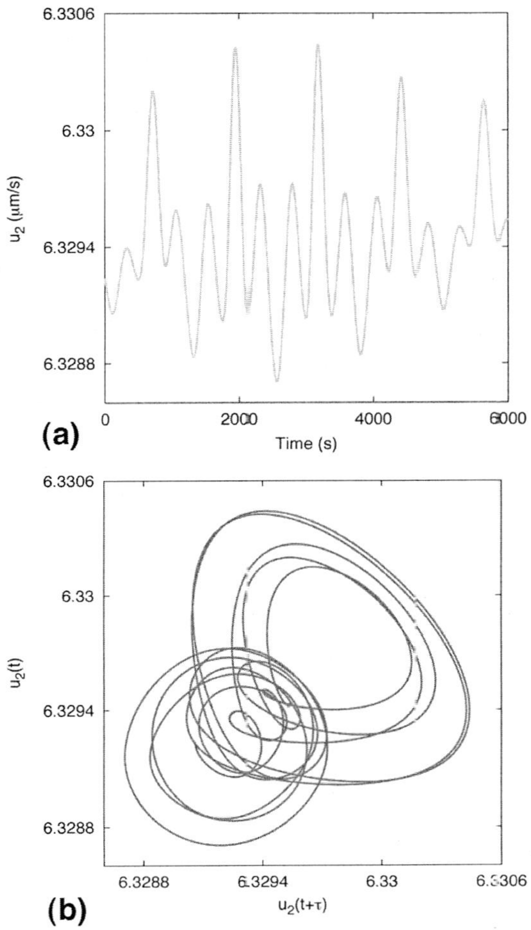

Fig. 9. (**a**) Step velocity plotted as a function of time for a system where bulk-diffusion effects dominate transport. (**b**) Time-delay plot of step velocity reveals the possible presence of a strange attractor

This is reflected by a longer wavelength variation of the levels of surface supersaturation maxima and minima for the different terraces in Fig. 8(b), as well as the slight variation in the bulk supersaturation field seen in Fig. 8(a). The slight waviness of the first iso-concetration contour away from the surface in Fig. 8(a) shows the greater depletion in the bulk phase caused by the greater uptake of solute by groups of steps which are more closely spaced.

The rate of motion of the growth steps is determined by solute transport to the steps and the kinetic processes for incorporation of growth units into the step, so the normal growth rate of the crystalline facet can be predicted from first-principles mechanisms. Of great interest is how transport and kinetic effects will cause the step density to change in time. Evolution to constant step spacing will stabilize the moving train of steps. However, nonlinear interactions may act to destabilize the train of steps by causing the spacing to change in time, which may lead to step-bunching instabilities or, potentially, more complicated dynamical interactions among the steps [26]. An example of the complicated dynamics which may exist in this system is shown in Fig. 9. Here, the step velocity is analyzed for a model system in which bulk-diffusion phenomena are controlling. When two steps are near each other, the overlap of the bulk supersaturation fields tends to slow the step in front, so that the pair travels together for some time. However, the pair travels more slowly than a single step, and when the pair is approached from behind by a single step, the middle step slows further and the front step moves away. The end result is a nearly periodic coupling and uncoupling of step pairs. The resulting motion of a single step appears to be chaotic (Fig. 9(a), leading to the possible presence of a strange attractor in the dynamical behavior of this system (Fig. 9(b)).

Future work will address such issues and the direct coupling of the above 3D flow fields with the kinetic model of step growth.

3 Final Comments

We have briefly discussed the modeling of several crystal growth systems with a focus on using high performance computing and the incorporation of multiple scales to increase the realism of the simulations. Predictive capability provided from such models will be of great utility for understanding crystal growth experiments and for the rational optimization of growth processes. More importantly, knowledge generated from more realistic, multiscale models for crystal growth will ultimately lead to the ability to link crystal properties with growth conditions and the macroscopic factors which influence them. In general terms, the understanding gained from successful modeling of crystal growth systems will lead to better process operation and design, ultimately yielding better quality crystals at higher production rates and lower costs. Indeed, the impact of numerical modeling to study crystal growth will become increasingly important as computers continue to grow in power and our models continue to become more sophisticated.

Acknowledgments

This work was supported in part by the United States National Aeronautics and Space Administration and National Science Foundation, the Minnesota Supercomputing Institute, and the Army High Performance Computing Research Center under the auspices of the Department of the Army, Army Research Laboratory cooperative agreement DAAH04-95-2-0003/ contract

DAAH04-95-C-0008, the content of which does not necessarily reflect the position or policy of the government, and no official endorsement should be inferred. JJD would also like to thank the Alexander von Humboldt Foundation for an award which initiated the collaboration between his research group in Minnesota and the research group of GM in Erlangen.

References

1. Kuppurao, S., Brandon, S., and Derby, J.J. (1995) Modeling the vertical Bridgman growth of cadmium zinc telluride I. Quasi-steady analysis of heat transfer and convection. *J. of Crystal Growth.* **155,** 93–102.
2. Kuppurao, S., Brandon, S., and Derby, J.J. (1995) Modeling the vertical Bridgman growth of cadmium zinc telluride II. Transient analysis of zinc segregation. *J. of Crystal Growth.* **155,** 103–111.
3. Hughes, T. J. R. (1987) *The Finite Element Method.* Prentice-Hall, Inc.
4. Dahlquist, G., and Björck, Å. (1974) *Numerical Methods* Prentice-Hall.
5. Sonda, P., Yeckel, A., Daoutidis, P., and Derby, J.J. (2001) Complex dynamics within the vertical Bridgman crystal growth process. In: *Proceedings of the IFAC Symposium: DYCOPS 6.*
6. Sonda, P., Yeckel, A., Daoutidis, P., and Derby, J.J. (2001) An analysis of time-dependent flows within a vertical Bridgman crystal growth system. *Chem. Eng. Sci.* Submitted.
7. Kurz, M., Pusztai, A., and Müller, G. (1998) Development of a new powerfull computer code CrysVUN++ especially designed for fast simulation of bulk crystal growth processes. Presentation at: ICCG-12, the Twelfth International Conference for Crystal Growth, Jerusalem.
8. Metzger, M., and Müller, G. (2001) Optimal Control of Crystal Growth Processes. *J. Crystal Growth.* Accepted for publication.
9. Salinger, A.G., Xiao, Q., Zhou, Y., and Derby, J.J. (1994) Massively parallel finite element computations of three-dimensional, time-dependent, incompressible flows in materials processing systems. *Comput. Methods Appl. Mech. Engrg.* **119,** 139–156.
10. Zhou, H., and Derby, J.J. (2000) Parallel Implementation of Finite Element Method with MPI: Application to Three-Dimensional Free Surface Stokes Flow. In: *Proceedings of the 2000 International Conference on Parallel and Distributed Processing Techniques and Applications,* Las Vegas, NV.
11. Yeckel, A., Pandy, A., and Derby, J.J. (2001) Representing Realistic Complexity in Numerical Models of Crystal Growth: Coupling of Global Furnace Modeling to Three-Dimensional Flows. In: *Proceedings of CHT'01 Advances in Computational Heat Transfer,* Cairns, Queensland, Australia.
12. Yeckel, A., Pandy, A., Hainke, M., Müller, G., and Derby, J.J. (2001) A coupled, multi-scale model for continuum transport in crystal growth systems. *J. Crystal Growth.* In preparation.
13. Bordui, P.F., and Jacco, J.G. (1987) Viscosity and density of solutions used in high-temperature solution growth of $KTiOPO_4$ (KTP). *J. Crystal Growth.* **82,** 351–355.

14. Bordui, P.F., Jacco, J.G., Loiacono, G.M., Stolzenberger, R.A., and Zola, J.J. (1987) Growth of large single crystals of $KTiOPO_4$ (KTP) from high-temperature solution using heat pipe based furnace system. *J. Crystal Growth.* **84,** 403–408.
15. Wilcox, W.R. (1983) Influence of convection on the growth of crystals from solution. *J. Crystal Growth.* **65,** 133–142.
16. Vartak, B., Kwon, Y.-I., Yeckel, A., and Derby, J.J. (2000) An analysis of flow and mass transfer during the solution growth of potassium titanyl phosphate. *J. Crystal Growth.* **210,** 704–718.
17. Vartak, B. (2001) *Three-Dimensional modeling of solution crystal growth via the finite element method.* PhD thesis, University of Minnesota.
18. Hughes, T.J.R., Franca, L.P., and Hulbert, G.M. (1989) A new finite element formulation for computational fluid dynamics: VIII. The Galerkin/least-squares method for advective-diffusive equations. *Comput. Methods Appl. Mech. Engrg.* **73,** 173–189.
19. Vartak, B., and Derby, J.J. (2001) On stable algorithms and accurate solutions for convection-dominated mass transfer in crystal growth modeling. *J. Crystal Growth.* **230,** 202–209.
20. Saad, Y., and Schultz, M. (1986) GMRES: A generalized minimum residual algorithm for solving nonsymmetric linear systems. *SIAM J. Sci. Stat. Comput.* **7,** 856–869.
21. Yeckel, A., and Derby, J.J. (1997) Parallel computation of incompressible flows in materials processing: Numerical experiments in diagonal preconditioning. *Parallel Computing.* **23,** 1379–1400.
22. Burton, W.K., Cabrera, N., and Frank, F.C. (1951) The growth of crystals and the equilibrium structure of their surfaces. *Phil. Trans. Roy. Soc. London.* **243**, 299–358.
23. Kwon, Y.-I., and Derby, J.J. (2001) Modeling the coupled effects of interfacial and bulk phenomena during solution crystal growth. *J. Crystal Growth.* **230,** 328–335.
24. Kim, K.M., Witt, A.F., and Gatos, H.C. (1972) Crystal growth from the melt under destabilizing thermal gradients. *J. Electrochem. Soc.* **119,** 1218–1226.
25. Turner, J.S. (1973) *Buoyancy effects in fluids.* Cambridge University Press.
26. Kwon, Y.-I., and Derby, J.J. (2001) Step growth dynamics during solution crystal growth. *J. Crystal Growth.* In preparation.

Semi-Direct Numerical Simulation of a Czochralski Melt Flow on High-Performance Computers

S. Enger, F. Schäfer, M. Breuer*, and F. Durst

Institute of Fluid Mechanics, University of Erlangen-Nürnberg, Cauerstr. 4, D-91058 Erlangen, Germany

Abstract. The three-dimensional and time-dependent turbulent flow field and heat transfer of the melt in a Czochralski crystal growth process were predicted using an efficient block-structured finite-volume Navier-Stokes solver. From semi-direct numerical simulations, detailed information of instantaneous and time-averaged quantities were obtained. Two different crucible rotation rates were considered, $\Omega_c = -2$ rpm and -5 rpm, whereas the crystal rotation was kept constant at $\Omega_s = 20$ rpm. Exact boundary conditions for the temperature were obtained from experiments. The time-averaged results are discussed and it is shown that, due to velocity and temperature fluctuations underneath the crystal, the growth conditions are superior at higher crucible rotation rates. The main reason for this is the stabilizing effect of the centrifugal forces. Furthermore, it is shown that the boundary layer below the crystal is very thin, so that the influence on the bulk flow is negligible. Rotation will mainly maintain the circular shape of the crystal and ensure the homogeneous distribution of dopants.

1 Introduction

Since the first silicon single crystal was pulled by Teal and Little [1] in 1950, the crystal diameter is growing constantly. From Fig. 1 it can be seen that the diameter is doubled every twelve years, with slightly faster development in the early years and, if the estimation by Zulehner [2] applies, somewhat slower for the years to come. It is believed that this growth rate cannot continue and that the 400 or 450 mm diameter wafer will be the last. The increasing size is driven by economic factors in device fabrication; e.g., growing chip sizes require larger wafers to fabricate the same number of chips per wafer. With increasing size of the silicon crystals grown by the Czochralski (CZ) method, the melt volume and the corresponding crucible diameter are growing accordingly. With increasing melt volume and melt height, the fluid flow in the crucible gets further into the turbulent regime with Rayleigh numbers beyond 10^8 [3].

Detailed literature surveys of experimental and numerical investigations of the CZ growth process were given in [4,5]. Therefore, here only a few issues

* e-mail: breuer@lstm.uni-erlangen.de

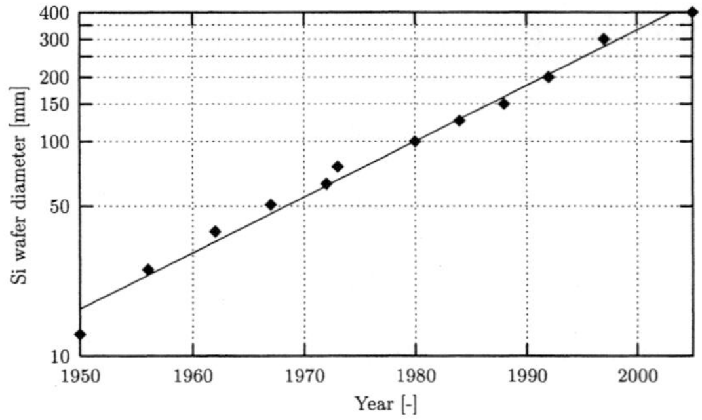

Fig. 1. Development of silicon wafer size from the beginning in 1950 [2]

mostly relevant to the present work are discussed briefly. There have been some attempts to simulate the turbulent flow in a CZ crucible by applying various two-equation turbulence models. This approach seems reasonable for a rough description of the two-dimensional time-averaged flow and heat transfer in the silicon melt [6], but fails to predict the three-dimensional, non-axisymmetric flow which is known to dominate in the crucible. For this kind of simulations, more accurate methods such as direct numerical simulation (DNS) or large-eddy simulation (LES) must be employed. Due to the high spatial and temporal resolution required, the simulations can only be carried out on modern vector-parallel high-performance computers, and even here, the simulation time will be in days or weeks, rather than hours.

In earlier work by the authors [7,8,5] it has been shown that a semi-direct numerical simulation (semi-DNS), i.e., simulations with very fine grids, but which do not entirely resolve the Kolmogorov scales, are in good agreement with experimental data. Breuer [9] gives a detailed survey of modern simulation strategies for turbulent flows including DNS, LES, and hybrid methods. It is shown that for moderate Reynolds or Rayleigh numbers the influence of an explicit subgrid scale model used in the LES approach is very small. However, an LES without any subgrid scale model is equivalent to the semi-DNS technique used in the present investigation. It has to be mentioned that this approach might not work properly for very large crucibles and crystals owing to much broader spectra of turbulent length scales at higher Rayleigh/Grashof numbers. For these cases an explicit subgrid scale model will be highly required.

Almost all simulations of the flow and heat transfer in a CZ melt are today carried out without any turbulence or subgrid scale model, but most of them with rather coarse grids. Only a few other investigations (e.g., [10]) have used high spatial and temporal resolution in order to be classified as semi-DNS or DNS, but these attempts were carried out with a simplified cylindrical crucible geometry.

The objective of the present work is to get a better description of the flow and thermal field inside an industrial CZ-crucible based on accurate numerical simulations and to enlighten the influence of velocity and temperature fluctuations, in terms of turbulent kinetic energy and temperature variance, underneath the growing crystal.

2 Governing Equations and Numerical Method

The flow and heat transfer in a crucible of a CZ system is governed by the three-dimensional Navier-Stokes equations for an incompressible fluid expressing the conservation of mass, momentum and energy. The coupling between momentum and energy is taken into account by the Boussinesq approximation, which is valid for moderate temperature differences, i.e., in this case less than about 50 K.

The solution domain was divided into a finite number of control volumes (CV) as indicated in Fig. 2. The conservation equations are approximated for each CV using a second-order accurate central difference scheme (CDS) for the convective and diffusive terms. For the time integration a fully implicit method of second-order accuracy was applied. The discretization results in a large system of equations which is solved iteratively with the strongly implicit procedure (SIP). In order to speed up the solution procedure, the multigrid method was applied. For transient simulations, the code employs a full approximation scheme (FAS) where the V-cycles in each time step start on the finest grid. A total of four grid levels were used in the present simulations. The finest and coarsest grids consist of 1 945 600 and 3800 CVs, respectively. The second coarsest grid (30 400 CVs) is shown in Fig. 2. More details about all features of the CFD code applied for the present simulations (FASTEST-3D) are given in [5,4,11–13].

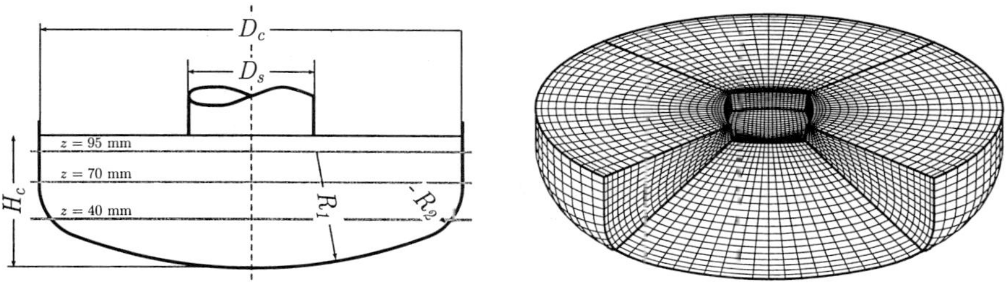

Fig. 2. Geometry and the corresponding grid of an industrial crucible. Only the second coarsest of a total of four grid levels is shown

3 Problem Description and Numerical Details

The flow and thermal field in an industrial crucible with melt height $H_c =$ 100 mm, diameter $D_c = 340$ mm and crystal diameter $D_s = 104$ mm according to Fig. 2 was simulated. The crystal and crucible counter rotate. The crystal rotation was kept constant at $\Omega_s = 20$ rpm in all simulations, whereas two different crucible rotation rates were considered; $\Omega_c = -2$ rpm and $\Omega_c = -5$ rpm. The crucible is filled with liquid silicon, which is characterized by a very low Prandtl number, see Table 1. The temperature at the wall was obtained from experiments (described by Enger et al. [7]), and due to the different flow and thermal fields in the two cases, different maximum vertical temperatures ΔT_{ver} were obtained, namely 27.0 K and 32.8 K for $\Omega_c = -2$ rpm and -5 rpm, respectively. Based on the physical properties of Si, the above temperature differences, the melt height H_c, and the rotation rate of the crucible ω_c (in rad/s), the most important dimensionless numbers can be determined according to Table 1.

Table 1. Dimensionless numbers for the considered simulations

Number	Definition	*Case L* $\Omega_c = -2$ rpm	*Case H* $\Omega_c = -5$ rpm
Grashof	$\mathrm{Gr} = g\,\beta\,H_c^3\,\Delta T_{\text{ver}}/\nu^2$	4.1×10^8	5.0×10^8
Taylor	$\mathrm{Ta} = 4\,\omega_c^2\,R_c^4/\nu^2$	1.6×10^9	1.0×10^{10}
Reynolds	$\mathrm{Re} = \omega_c\,R_c^2/\nu$	2.0×10^4	5.0×10^4
Rossby	$\mathrm{Ro} = g\,\beta\,H_c\,\Delta T_{\text{ver}}/(\omega_c^2\,R_c^2)$	2.95	0.57
Prandtl	$\mathrm{Pr} = \mu\,c_p/\lambda$	0.011	0.011

In the discussion of the results, the two cases with crucible rotation rate $\Omega_c = -2$ rpm and -5 rpm will be referred to as case L (low) and H (high), respectively.

A time step of $\Delta t = 0.01$ s was used, which results in a maximum Courant number of 2.5 in each control volume and time step. Since a typical pulling process lasts for several hours, only a part of this can be simulated within reasonable computational time. In order to compare the numerical results with experimental data, the simulations were continued up to ten minutes real time. In each time step, the iteration process was continued until the residual values of all variables had dropped at least four orders of magnitude.

The simulations were performed in a rotating frame of reference with the rotation rate given by the crucible rotation ω_c (in rad/s). Stokes' no-slip boundary condition was applied on the crucible walls and the crystal/melt interface which leads to $\mathbf{v} = 0$ on the wall of the crucible. The velocities at the plane crystal/melt interface were set according to $\mathbf{v} = (\boldsymbol{\omega}_s - \boldsymbol{\omega}_c) \times \mathbf{r}$. The free surface of the melt was assumed to be plane and free of shear stresses. Accordingly, a zero normal velocity was set at the surface and a symmetry condition was applied for the other velocity components. As mentioned above, the temperature on the crucible wall was taken from measurements [14] and at the crystal/melt interface the solidification temperature (T_s = 1687 K) for Si was chosen. A steady-state simulation of the pure conductive case was performed, which, together with a uniform solid body rotation, was assigned as initial conditions. An ambient temperature of 1600 K was used to calculate the radiative heat loss at the free melt surface.

4 Results and Discussion

From the instantaneous velocity and thermal field it is difficult to derive qualitative information about general transport phenomena in the crucible. Therefore, after an initial time-interval where the velocity and thermal fields develop (approximately 3 minutes), time-averaging of the velocities and the temperature was started. Furthermore, at this point, the calculation and averaging of the resolved Reynolds stresses $\overline{u_i'u_j'}$ and the temperature variance $\overline{T'T'}$ was started. This averaging procedure does not contradict the instantaneous simulation approach by semi-DNS and is an often applied modus operandi in turbulence research. Additionally, it has to be stated clearly, that averaging of the instantaneous three-dimensional flow/thermal field in time or space (azimuthal direction, see below) is not equivalent to solve the governing equations of the steady-state or axisymmetric problem.

The averaged velocity field in a vertical cut through the middle of the crucible after 7 minutes averaging time is shown in Fig. 3*(a)* and 3*(b)* for case *L* and *H*, respectively. The visualization method used for the velocity field (LIC[1]) only shows the structures of the flow and vortices. In order to quantify the velocity magnitude and the direction of the flow, it should be supported by, e.g., extracted values along radii at different heights, as shown in Figs. 4 and 5. Here, as well as for the temperature data in Fig. 7 and the turbulent kinetic energy in Fig. 8, the time-averaged values were also spatially averaged by applying data at every 5 degree in the azimuthal direction at z = 40, 70, and 95 mm melt height. Only cross-sectional plots are presented, since the azimuthal velocity field is dominated by the solid-body rotation. For a more detailed description, see Enger [5].

From the velocity field it can be seen that for both rotation rates, a strong downward flow is generated underneath the cold crystal. This downward plume induces a flow close to the free surface, directed towards the center of the crucible; in case *L* outside and case *H* just inside the edge of the crystal (indicated by "A" in Fig. 3).

In the region close to the hot crucible wall, a large dominant buoyancy-driven toroidal vortex can be recognized (indicated with "B" in Fig. 3). With increasing crucible rotation, this vortex is forced towards the wall by the

[1] "Line Integral Convolution" [15].

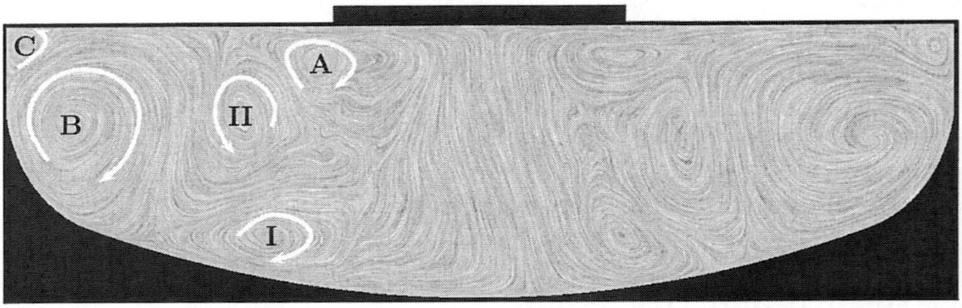

(a) $\Omega_c = -2$ rpm

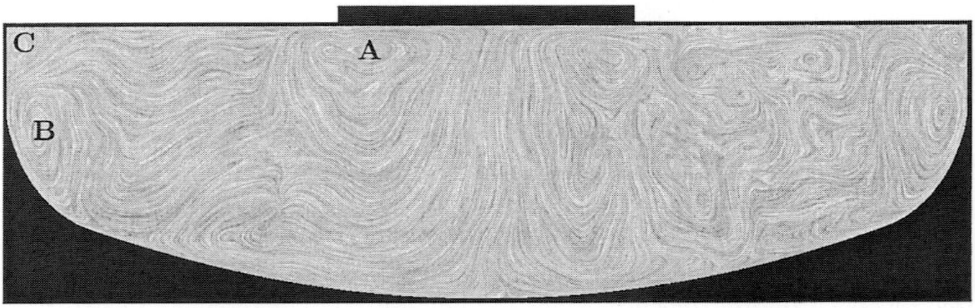

(b) $\Omega_c = -5$ rpm

Fig. 3. Time-averaged velocity field in coordinate system rotating with the crucible for $\Omega_c = -2$ rpm (**a**) and $\Omega_c = -5$ rpm (**b**)

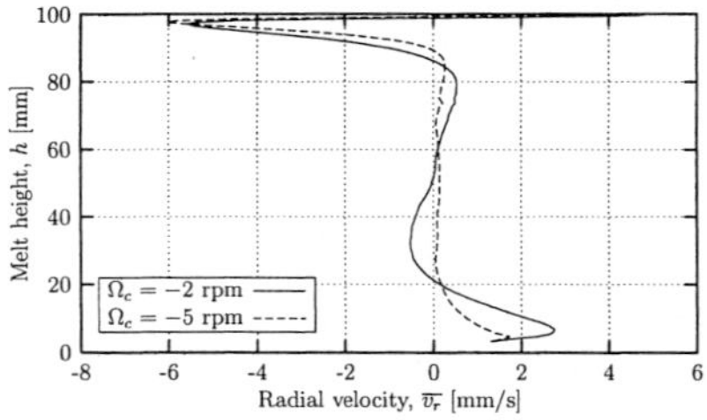

(a)

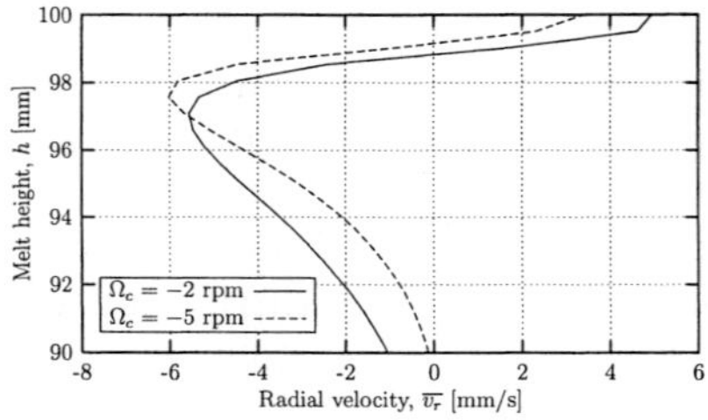

(b)

Fig. 4. Time- and azimuthal-averaged radial velocity v_r plotted versus the melt height h at crucible radius $r = 5$ mm for $\Omega_c = -2$ rpm and -5 rpm (**a**); with a close-up of the region in the vicinity of the rotating crystal in (**b**) centrifugal forces. Due to the negative temperature gradient at the crucible wall from about $z = 80$ mm towards the free melt surface ($z = 100$ mm), a small counter rotating vortex has developed at this position (region "C" in Fig. 3). In both cases the vortex is rotating downwards at the crucible wall, but it is more pronounced at the lower rotation rate, probably owing to damping effect of the crucible rotation.

In case L, two additional vortices can be observed in the vertical view in Fig. 3*(a)*, both at about 70 mm radius; the first ("I") close to the bottom of the crucible and the next ("II") at about $z = 60$ mm. The rotational direction of the two vortices is given from the other vortices with which they interact in a roller-like pattern. With increasing crucible rotation, the two latter vortices are damped, and the complete bulk is almost resting, as can be seen from the vertical velocity in Fig. 5*(b)*, where only the downward flow underneath the crystal can be recognized (Fig. 5*(d)*). In case L, the situation is similar in the bulk, i.e., until $r = 100$ mm, but closer to the wall, there are strong radial movements, at both heights (40 mm and 70 mm). The small magnitude of the velocity components in the bulk outside the radial edge of the crystal in case H explains why there is no clearly developed flow pattern in this region in the time-averaged flow field (Fig. 3*(b)*). It is questionable whether a longer averaging time would improve this, whereas an averaging time of 7 min seems to be sufficient for the low rotation case.

From the two-dimensional presentation of the radial velocity versus the melt height in Fig. 4 at crucible radius $r = 5$ mm — especially the close-up of the region directly below the rotating crystal in Fig. 4*(b)* — and the velocity plot in Fig. 5, it is evident that only a very thin boundary layer has developed under the rotating crystal. At a distance 5 mm below the crystal, the sucking effect of the rotating crystal is not even noticeable. The profile in Fig. 4 even indicates that the rotation of the crystal influences the flow in a boundary layer extending less than 2 mm from the crystal, where the flow is turned inwards (negative radial component) due to the downward flow below the cold crystal. The inaccuracy at the crystal edge appears since only cell-centered velocities were saved in the data file used for variable extraction. The thickness of the boundary layer originating from a rotating surface is known to depend on the Ekman number. It is therefore interesting to note that the thickness of the boundary layer is almost equal in case L and H, where the total rotation rate is very similar owing to an identical crystal rotation rate Ω_s.

As a result of the increasing crucible rotation rate, the buoyant plumes at the surface close to the crucible wall are suppressed, as can be seen in Fig 6. The higher rotation rate also leads to distinct horizontal temperature gradients at the crucible wall. This can be well recognized from the temperature plot in Fig. 7*(a)*, where the temperature rise in case L is equally distributed over the complete radius ($z = 70$ mm). The large buoyant vortex at the crucible wall in case L can be seen in the horizontal temperature

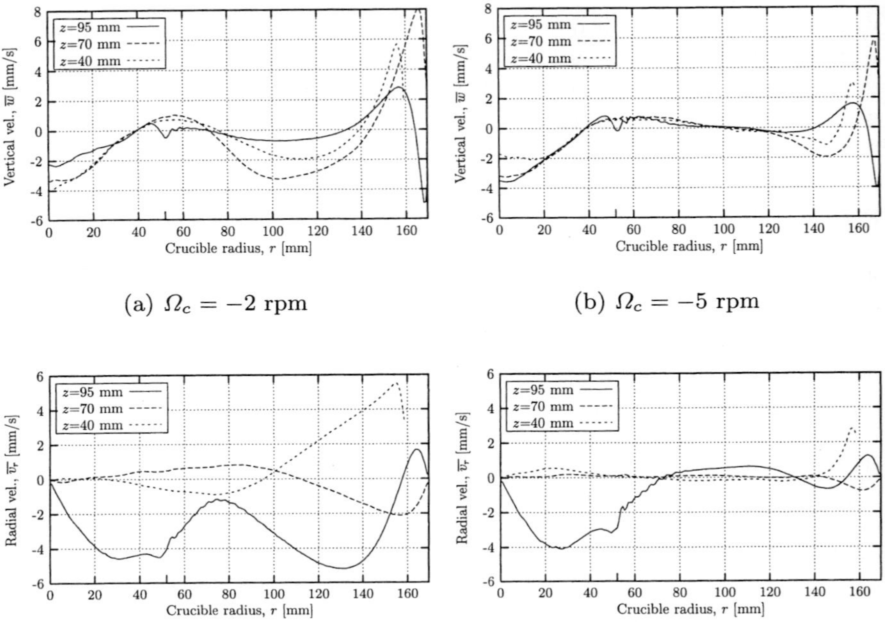

(a) $\Omega_c = -2$ rpm (b) $\Omega_c = -5$ rpm

(c) $\Omega_c = -2$ rpm (d) $\Omega_c = -5$ rpm

Fig. 5. Time- and azimuthal-averaged velocity (vertical and radial) plotted at $z =$ 40, 70 and 95 mm for $\Omega_c = -2$ rpm (left) and $\Omega_c = -5$ rpm (right)

plot at $z = 95$ mm, shown in Fig. 7*(a)*. From the small downward vortex at the top of the crucible wall, the main vortex is deflected inward, where the flow rises towards the free surface of the melt. At this height, the maximum temperature is reached at a radius of about 150 mm. In the other case, with higher rotation rate, the situation is analogous, but the temperature field is generally forced further outwards as was also the case with the main buoyant vortex at the crucible wall. This situation is even more lucid farther down in the crucible. In the bulk of the melt, between $z = 40$ mm and 70 mm, the temperature differs less than 4 K, indicating an almost radial temperature profile. In order to obtain a uniform crystal diameter, a large radial temperature gradient is desired, since this results in a small distinctive region where the phase change takes place. From Fig. 7 it can be seen that this is ensured in both cases, but with slightly better conditions in case *L*.

Investigations of grown crystals have shown, that the quality of the wafers with respect to dislocations is better when the temperature fluctuations underneath the crystal are small. Large temperature fluctuations generate ad-

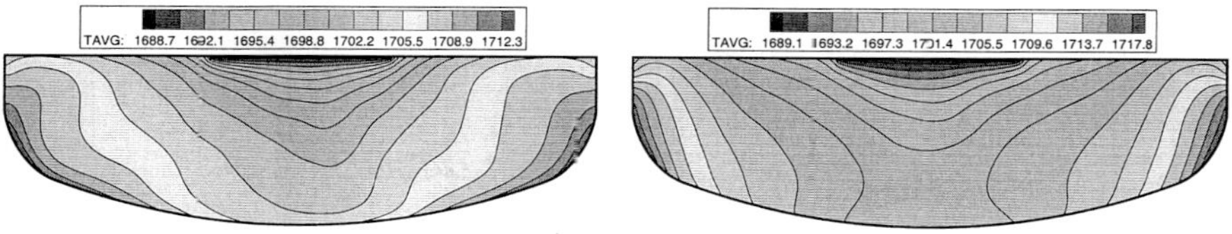

(a) $\Omega_c = -2$ rpm (b) $\Omega_c = -5$ rpm

Fig. 6. Time-averaged thermal field in a coordinate system rotating with the crucible for $\Omega_c = -2$ rpm (**a**) and $\Omega_c = -5$ rpm (**b**)

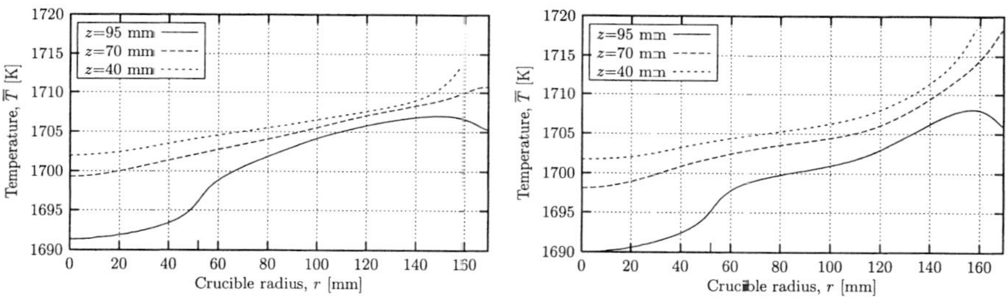

(a) $\Omega_c = -2$ rpm (b) $\Omega_c = -5$ rpm

Fig. 7. Time- and azimuthal-averaged temperature plotted at $z = 40$, 70 and 95 mm for $\Omega_c = -2$ rpm (**a**) and $\Omega_c = -5$ rpm (**b**)

ditional strain in the crystal, which can push the total strain over the yield stress, resulting in formation of dislocations [2]. Therefore, the velocity and temperature fluctuations under the crystal are of vital importance. Fig. 8 shows the turbulent kinetic energy, $k = \frac{1}{2}(\overline{u'^2} + \overline{v'^2} + \overline{w'^2})$, in the same horizontal planes as described above. Because the energy-carrying vortices are resolved in the present simulations, the predicted values of k (resolved part of the entire spectrum of length scales) should only slightly deviate from the total kinetic energy of all length scales in the flow.

From this diagram, it is evident that the velocity fluctuations are significantly damped with higher crucible rotation. Under the crystal the difference is up to 50%, but also in the bulk at $z = 70$ mm, the turbulent kinetic energy is much higher in case L than in case H. Closer to the bottom, at $z = 40$ mm, there is almost no difference between the two cases, most likely due to the well pronounced buoyancy-driven toroidal vortices at this height, resulting in a more distinct flow with smaller fluctuations. Normalized by the square

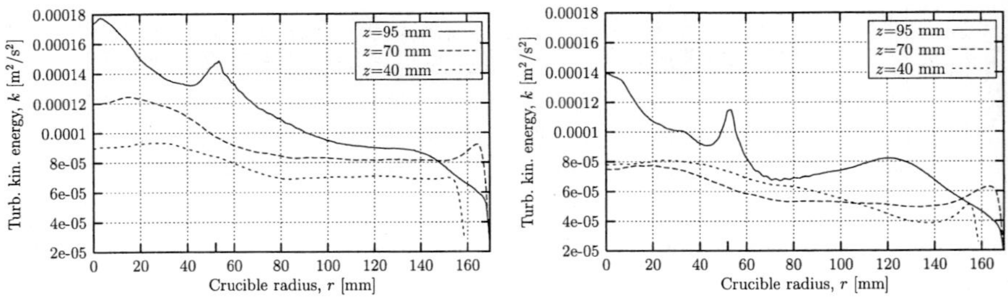

(a) $\Omega_c = -2$ rpm (b) $\Omega_c = -5$ rpm

Fig. 8. Time- and azimuthal-averaged turbulent kinetic energy k plotted at $z = 40$, 70 and 95 mm for $\Omega_c = -2$ rpm (left) and $\Omega_c = -5$ rpm (right)

of the mean azimuthal velocity of the crucible ($\frac{1}{2}(\omega_c D_c/2)^2$), the difference is even larger, with a turbulence intensity of 0.22 and 0.03 in case L and H, respectively. The temperature variance $\overline{T'T'}$ from Fig. 9 is slightly larger in the region under the crystal edge in case H, but clearly smaller directly underneath the crystal. Outside the crystal, the thermal stresses are also larger in the case with higher rotation rate. From this, it is tempting to conclude that the pulling conditions are superior in case H, but the frequency of the fluctuations also plays an important role. Enger et al. [7] showed that the frequency of the temperature fluctuations is lower and stronger in case L compared with case H. From experiments, it is a known phenomenon that the *low* frequencies are responsible for dislocations. From this point of view, the higher crucible rotation rate will provide superior growth conditions for Si crystals with a minimum number of dislocations, i.e., with a high quality.

5 Conclusions

The time-dependent three-dimensional flow and thermal field in a 14" industrial Czochralski crucible were calculated on a high-performance vector-parallel computer. The grid consisted of about 2×10^6 CVs, and the detailed results were obtained by applying the multigrid method. Two different crucible rotation rates were considered ($\Omega_c = -2$ rpm and -5 rpm), whereas the crystal rotation was kept constant at $\Omega_c = 20$ rpm. The simulations were performed with a plane interface between the crystal and melt. This might lead to minor differences, an effect which is currently investigated.

It was shown that an averaging time of about 7 minutes was sufficient to obtain both qualitative and quantitative results. The boundary layer under the rotating crystal was found to be very thin, about 2 mm. From the materials science point of view, a higher crucible rotation rate provides superior

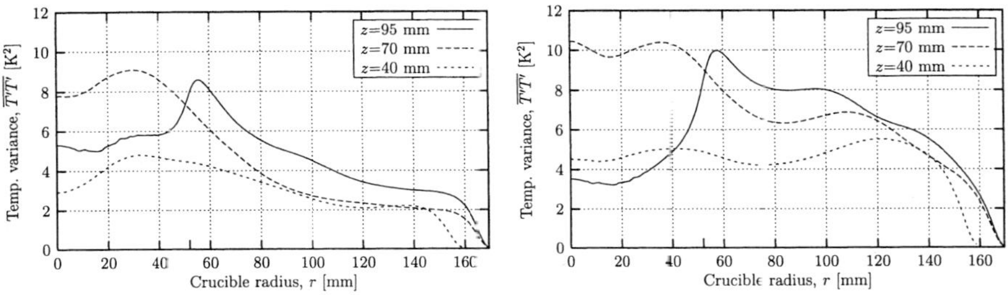

(a) $\Omega_c = -2$ rpm (b) $\Omega_c = -5$ rpm

Fig. 9. Time- and azimuthal-averaged temperature variance $\overline{T'T'}$ plotted at $z = 40$, 70 and 95 mm for $\Omega_c = -2$ rpm (left) and $\Omega_c = -5$ rpm (right)

growth conditions in terms of smaller fluctuations. The temperature variance was found to be larger in the high rotation rate case, but with a higher frequency, so that the probability of dislocation formation will still be smaller in this case.

Acknowledgements

Financial support by the Bayerische Forschungsstiftung in the Bavarian Consortium of High-Performance Scientific Computing (FORTWIHR) is gratefully acknowledged. The simulations were carried out on the NEC SX-4 and SX-5 at the High-Performance Computing Center, Stuttgart. This support is also gratefully acknowledged.

References

1. G.K. Teal and J.B. Little. Growth of Germanium Single Crystal. *Physical Review*, 78:657, 1950.
2. W. Zulehner. Historical Overview of Silicon Crystal Pulling Development. *Materials Science and Engineering*, B73:7–15, 2000.
3. O. Gräbner, A. Mühe, G. Müller, E. Tomzig, J. Virbulis, and W. von Ammon. Analysis of Turbulent Flow in Silicon Melts by Optical Temperature Measurement. *Materials Science and Engineering*, B73(1–3):130–133, 2000.
4. B. Basu, S. Enger, M. Breuer, and F. Durst. Three-Dimensional Simulation of Flow and Thermal Field in a Czochralski Melt Using a Block-Structured Finite-Volume Method. *Journal of Crystal Growth*, 219:123–143, 2000.
5. S. Enger. *Numerical Simulation of Flow and Heat Transfer in Czochralski Crucibles.* Ph.D. thesis, Technische Fakultät, Universität Erlangen-Nürnberg, 2001.

6. A. Lipchin and R.A. Brown. Comparison of Three Turbulence Models for Simulation of Melt Convection in Czochralski Crystal Growth of Silicon. *Journal of Crystal Growth*, 205:71–91, 1999.
7. S. Enger, O. Gräbner, G. Müller, M. Breuer, and F. Durst. Comparison of Measurements and Numerical Simulations of Melt Convection in Czochralski Crystal Growth of Silicon. *Journal of Crystal Growth*, 230:135–142, 2001.
8. S. Enger and M. Breuer. High Performance Computing: Numerical Simulation of the Melt Flow in an Industrial Czochralski Crucible. In E. Krause and W. Jäger, editors, *High Performance Computing in Science and Engineering 2000*, pages 290–304. Springer, 2001.
9. M. Breuer. *Direkte Numerische Simulation und Large–Eddy Simulation turbulenter Strömungen auf Hochleistungsrechnern.* Habilitationsschrift, Technische Fakultät, Universität Erlangen-Nürnberg, 2001.
10. C. Wagner and R. Friedrich. DNS of Turbulent Convective Flows in Idealized Czochralski Crystal Growth Configurations. In R. Friedrich and P. Bontoux, editors, *Notes on Numerical Fluid Mechanics*, volume 60, pages 367–380. Vieweg, 1997.
11. F. Durst, M. Perić, M. Schäfer, and E. Schreck. Parallelization of Efficient Numerical Methods for Flows in Complex Geometries. *Notes on Numerical Fluid Mechanics*, 38:79–92, 1993.
12. F. Durst and M. Schäfer. A Parallel Block–Structured Multigrid Method for the Prediction of Incompressible Flows. *Int. J. Num. Methods Fluids*, 22:549–565, 1996.
13. F. Durst, M. Schäfer, and K. Wechsler. Efficient Simulation of Incompressible Viscous Flows on Parallel Computers. *Notes on Numerical Fluid Mechanics*, 52:87–101, 1996.
14. O. Gräbner, G. Müller, J. Virbulis, E. Tomzig, and W. von Ammon. Effects of Various Magnetic Field Configuration on Temperature Distribution in Czochralski Silicon Melts. *Microelectronic Engineering*, 56:83–88, 2001.
15. B. Cabral and L. Leedom. Imaging Vector Fields Using Line Integral Convolution. In *Computer Graphics Proceedings*, volume 27,4 of *Annual Conference Series*, pages 263–270, Los Angeles, California, July 1993. ACM SIGGRAPH, Addison-Wesley Publishing Company, Inc.

High-Order Numerical Solutions for Rotating Flows with Walls

E. Serre[1], I. Raspo[1], O. Czarny[1], P. Bontoux[1], P. Droll[2], and M. Schäfer[2]

[1] L3M, FRE 2405 CNRS, Technopole de Chateau-Gombert, 38 rueJoliot-Curie, 13451 Marseille, France

[2] Dept. of Numerical Method inMechanical Engineering, Darmstadt University of Technology,Petersenstr. 30, D-64287 Darmstadt, Germany

Abstract. The study of rotating viscous flows with walls has significant importance for many industrial devices. In this approach subsystems of simple geometry coming from realistic geometries are studied using accurate methods (spectral). The three-dimensional incompressible Navier-Stokes equations are solved using a projection scheme. Depending on the aspect ratio of the cavity and on the Reynolds number, annular and spiral patterns of the generic types I and II boundary layer instabilities as well as vortex breakdown phenomena are investigated. Taylor-Couette flows in a finite-length cavity with counter-rotating walls, are also studied. Two complex regimes of wavy vortex and spirals are emphasized for the first time via direct numerical simulation in this configuration.

1 Introduction

The study of rotating viscous flows with walls has significant importance in particular in the cooling of gas turbines and turbomachinery [1, 2]. Typical configurations are cavities between compressors or turbine disks, between counter-rotating disks, or in a rotor-stator system, with or without throughflow. The motion is then driven by the differential rotation of the walls or by the throughflow. The major characteristics of these flows are known to be determined by two dimensionless parameters: the gap ratio L and the rotational Reynolds number ($Re = \Omega R_1^2/\nu$), where R_1 is the radius of the cylinder, Ω the speed velocity of the endwalls, and ν the kinematic viscosity of the fluid.

Complex fluid dynamics phenomena bringing about different physical mechanisms can occur for certain sets of (L, Re): Taylor-Couette instability, boundary layers type I and type II instabilities or vortex breakdown. These phenomena of industrial interest in the improvement of the turbine performance are also relevant of fundamental investigations in the study of the transition to turbulence in finite-dimensional systems.

The numerical study of these flows for industrial applications usually implies realistic but complex geometries. On the contrary, the approach of this study, combining fundamental and applicate investigations, consists in taking into account subsystems of simpler geometry in which these complex phenomena are studied using high-order numerical methods. Indeed, in these flows in which reside coexisting adjacent and coupled flow regions, greatly different in terms of the flow properties and of the length scales (as it is the case for the boundary layers and the geostrophic core region), high-order numerical methods are required to accurately describe them. The approximate solutions are computed with a spectral method based on a Chebyshev-Fourier approximation and the velocity-pressure coupling is performed with an efficient projection algorithm. Solutions have been computed on a NEC SX-5 super-computer. The estimates of the performance of the solver are for about 2.10^6 grid points are: 6 Gigaflops (2.5×10^{-6} s per time-step and per point), and 1.6 Gigabytes of memory.

2 Geometrical and Mathematical Models

The geometrical models correspond to two discs of radial extent $\Delta R = R_1 - R_0$, where R_0 and R_1 are the internal and external radii. The distance between these discs is H and they can be either open and submitted to a radial throughflow or enclosing by a cylindrical ($\mathrm{R_0} = 0$) or an annular domain. Three major types of configuration have been considered:

- *Rotor-stator*: the upper disc rotates (rotor) at uniform angular velocity $\Omega = \Omega e_z$, e_z being the unit vector on the axis and the other parts of the cavity are stationary (stator).
- *Opened rotating cavity*: the both discs rotate at uniform angular velocity $\Omega = \Omega e_z$.
- *Taylor-Couette cavity*: the cavity is annular delimited by two concentric cylinders that rotate independently from the other one at angular velocity Ω_1 and Ω_0. The cell is closed with two disks interdependent with the outer cylinder and then rotating at Ω_1.

The motion is governed by the three-dimensional Navier-Stokes equations written in primitive variables for an incompressible fluid. The two geometrical parameters refer to curvature and the aspect ratio, $R_m = (R_1 + R_0)/\Delta R$ and $L = H/\Delta R$. For the cylindrical cavity the only geometrical parameter is $L = H/\ R_1$, since $R_m = 1$. As it usual in the literature of Taylor-Couette flows, the curvature parameter is also defined by $\eta = R_0/R_1$ ($R_m = (1+\eta)/(1-\eta)$).

The scales for the dimensionless variables of space, time and velocity are $[H/2,\ \Omega^{-1},\ \Omega R_1]$ respectively. The relevant physical parameters are the rotational Reynolds number ($\mathrm{Re} = \Omega R_1^2/\nu$) and the mass flow rate, Q, made dimensionless as $C_w = Q/\nu R_1$, when the flow is submitted to a radial outflow (the flow is forced from $\mathrm{R_0}$ to $\mathrm{R_1}$ and parallel to the disks plane). In the Taylor-Couette configuration, there are two characteristics Reynolds numbers: $\mathrm{Re_0} = \Omega_0 R_1^2/\nu$ and $\mathrm{Re_1} = \Omega_1 R_1^2/\nu$.

In all cases, the boundary conditions correspond to no-slip conditions for u and w at the rigid walls. For the open cavity, an Ekman boundary layer flow is considered at $\mathrm{R_0}$ and $\mathrm{R_1}$ [3]. The azimuthal velocity profile is regularized by an exponantial function between the parts of the cavities which rotate at different speeds [4, 5].

3 Numerical Model

The solution method is based on a pseudo-spectral collocation Chebyshev-Galerkin Fourier method. This choice provides exponential convergence of the solution — referred to as spectral accuracy. The time scheme is based on the second-order backward differentiation formula for the diffusive term and on the Adams-Bashforth scheme for the nonlinear terms. The velocity-pressure coupling is performed with an improved projection algorithm which allows a possible temporal evolution of the normal pressure gradient at the boundaries. This original procedure is shown to reduce the slip-velocity on the boundaries by one order of magnitude (compared to that of the temporal scheme) and it improves the incompressibility condition without the need of a shifted grid for the velocity and the pressure. For the cylindrical cavity (involving the axis), the numerical approximation method was modified with a dependent variables transformation ($\widetilde{V} = \overline{r}V$ and $\widetilde{p} = \overline{r}p$). The details of this technique are developed elsewhere in [6].

Some time-dependent three-dimensional solutions have been recomputed with higher resolutions up to $121 \times 121 \times 96$. The space and time scales of theses solutions differ by less than 0.1% with the refinement in the grid spacing and in the time-step size. Additional results regarding the accuracy of the numerical methods are given by [6].

4 Rotor-Stator Cavity

The axisymmetric instability has been computed succeeding to the stationary basic flow. In the annular cavity ($R_m = 4$, $L = 0.2$), for Re = 62500 the solution is oscillatory ($\sigma = 2\pi\Omega^{-1}f = 4.7$). The axisymmetric vortices are visible along the two layers and travel following the flow as in recent experiments by [7]. In the cylindrical cavities closer to the axis, the axisymmetric instabilities also appears for $Re = 16000$, $L = 0.5$ and $Re = 35000$, $L = 0.2$. Differently to the annular cavity the Ekman layer over the rotating disk keeps stable. These vortices travel slowly inward in the Bödewadt layer down to near the axis which corresponds to a local Reynolds number $Re_\delta \simeq 27$ (with $\mathrm{Re}_\delta = \Omega\delta r/\nu$, where $\delta = (\nu/\Omega)^{0.5}$) . These results in cylindrical cavities are closely similar with experimental results of [8]. Then following the analysis of Savas [8] the instabilities are of type II in both the Bödewadt and the Ekman layers.

The rolls that progress under the form of rings in the axisymmetric solution, now constitute spirals arms. The transition to three-dimensional patterns was accelerated via an initial disturbance. Spiral patterns arise in both annular and cylindrical cavities. In the case of the annular cavity ($R_m = 5$), an oscillatory solution ($\sigma = 21.4$) is obtained with 22 spiral arms in both layers: inside the Ekman layer, the angle of the spiral wavefront ε (with respect to the azimuthal direction) decreases with $\overline{r}$ as $-15.3° \leq \varepsilon \leq -6.9°$ (referred as the type II [4]) and close to the stationary disk, $11° \leq \varepsilon \leq 23°$ (referred as the type I [4]). Mixed annular and spiral pattern is also observed in both cavities. For Re $= 62500$ in the annular cavity ($R_m = 4$, $L = 0.2$) [4], and in the cylindrical cavity ($L = 0.2$) for Re $= 35000$ (Fig. 1(a)). The temporal behavior exhibits an oscillatory regime ($\sigma \cong 1$). For $1.54 \leq \overline{r} \leq 5$, the vortices transform under the form of 5 annular structures while at larger distance from the axis rolls develop under the form of 6 spiral structures forming an angle $7° \leq \varepsilon \leq 28°$ and exhibit pairing phenomena.The spiral arms interact with the persisting axisymmetric structures and travel inward. The coexistence of these two types of waves was described by [8] who identifies these spirals to the type I instability of Bödewadt layer.

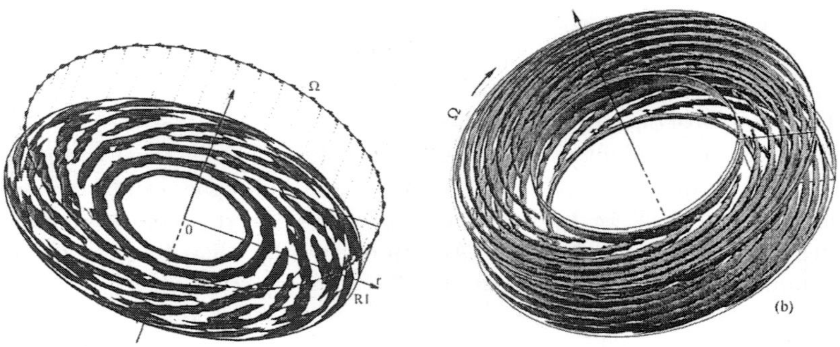

(a) (b)

Fig. 1. Three-dimensional mode of instabilities. (a) Display of iso-surfaces of the axial velocity component fluctuation showing spiral and annular patterns in the Bödewadt layer of the cylindrical cavity $L = 5$, at $Re = 35000$. (b) Displays of instantaneous iso-surfaces of the axial velocity component showing spiral patterns in the Ekman layers of the annular domain ($R_m = 5$, $L = 0.3$), for $C_w = 530$, Re $= 182500$

5 Rotating Cavity Submitted to a Radial Outflow

The base flow solution is steady and axisymmetric and organizes itself symmetrically and parallel Ekman layer flows form on the two disks with the same mass flow rate [3]. In all cases, the Reynolds number is Re $= 182500$, $R_m = 5$ and $L = 0.2$.

For $C_w = 530$ the flow is axisymmetric and oscillatory ($\sigma = 7.4$) [3]. We obtain good agreement between the computed value of σ and the frequency reported from experiments by [9]. In the numerical solution six pairs of counter-rotating axisymmetric rolls are exhibited in the Ekman layer. A good agreement with the theoretical results is obtained and the range of the characteristic parameters is relevant with the axisymmetric mode of the type II instability.

Multiple periodic solutions with different numbers of spiral arms following the periodicity of the disturbances are obtained (Fig. 1(b)). These 3D spiral patterns have already been observed in experiments of the Ekman layer (see a review in Ref. [3]) and the characteristic parameters are in good agreement with those obtained in the relevant experiments [10]. Thus, the spiral structure of the computed rotor layer shows the same characteristics as the type II standard instability of the Ekman layer.

6 Taylor-Couette Instability

Geometric parameters for our computations are $L = 6$ and $\eta = 0.75$.

Wavy Vortex Flow Regime

For $Re_0 = 750$, $Re_1 = -150$, 8 rolls are observed in the whole cavity and the regime is periodic, with a normalised dimensionless frequency $f^* = f/f_0 = 0.687$, where f_0 is the rotating frequency of the inner cylinder located at R_0. In order to favour the occurrence of three-dimensional structures, the rotation speed of the inner cylinder has been increased to $Re_0 = 940$. The main frequency of this solution is consistent with the phase frequency measured experimentally [11] or numerically [12]. About axial wavelength, λ is of the same order of Andereck's [11] and also of Neitzel's [13] in the case of Taylor vortices regime (for $Re_0 = 495$, $Re_1 = 0$, $\eta = 0.727$, $L = 23.25$).

In the counter-rotating WVF regime, the general structure remains quite similar to what observed experimentally by Andereck [11].

Spiral Flows

Three spiral flows have been obtained for different sets (Re_0, Re_1, L): Spiral flow, Interpenetrating Spiral regime (Fig. 2(a)) and Wavy Interpenetrating Spiral regime (Fig. 2(b)).

As in the previous case the initial condition is the axisymmetric solution at ($Re_0 = 750$, $Re_1 = -150$). Following the Andereck's results [11], the Reynolds numbers have been modified to ($Re_0 = 330$, $Re_1 = -500$) and the solution disturbed. The solution reaches an Interpenetrating Spiral regime (Fig. 3(a)) with complex structures similar to the ones observed in [11]. The spiral arms move in solid body rotation around the z-axis in the same direction than the inner cylinder. The axial confinement is increased by changing L down to 5, bringing about a new regime after a transition time. The solution is now periodic ($f^* = 0.96$) and the previous complex pattern switches to three regular spiral arms winding round the axis and forming an angle $\alpha = 18.5°$ with the horizontal plane. In order to obtain a Wavy Interpenetrating Spiral flow (Fig. 2(b)), the inner cylinder rotation is increased to $Re_0 = 394$ from the Interpenetrating Spiral regime at ($Re_0 = 330$, $Re_1 = -500$). A rotating wave of azimuthal wavelength equal to two, arises on the spiral arms. The spatial structure of this solution is qualitatively in good agreement with the Andereck's experimental shots of the same regime.

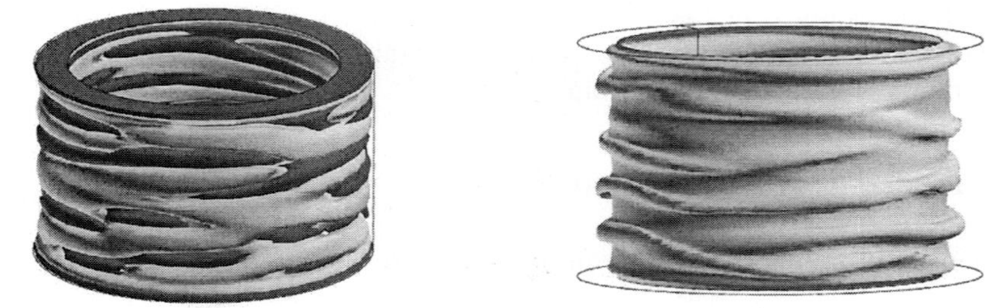

(a) (b)

Fig. 2. Three-dimensional time-dependent spiral regimes. Three-dimensional displays of instantaneous iso-surface of the radial component of velocity (**a**) Interpenetrating spiral regime ($Re_0 = 330$, $Re_1 = -500$, $\eta = 0.75$), (**b**) wavy spiral regime ($Re_0 = 394$, $Re_1 = -500$, $\eta = 0.75$)

7 Vortex Breakdown

The three-dimensional numerical simulations have been carried out in the same geometrical configuration than in the experiments of Escudier [14]. A cylinder of large aspect ratio, $L = 4$ was selected, providing the expectancy of a more unstable flow, the axial confinement being reduced.

As the flow spins up at $Re = 4500$, a small non-axisymmetric random perturbation is added at one given instant on the axial component w of the velocity. During the considered computation time $\Omega t = 1100$, no non-axisymmetric mode and the flow attains a purely periodic behaviour of low frequency $f\Omega^{-1} = 0.018$, favourably comparing with the recent computations of Blackburn and Lopez [15], $f\Omega^{-1} = 0.176$ for $H/R = 2.5$ and $Re = 3500$. The topology of the vortex breakdown bubbles is the same as that experimentally obtained [14], with two coalescing and separating axisymmetric bubbles.

In order to magnify the growth of three-dimensionnal modes, the rotation is instantaneously increased at $Re = 6500$. Helical vortices, associated with the centrifugal instability, emanating from the Bödewadt layer are convected in spirals by the main flow and disapear when they reach the geostrophic flow. After a long computational time, $\Omega t = 1300$, the entire flow region becomes completely three-dimensional and the structures of the centrifugal instability interact with the vortex breakdown. Then, the breakdown mutates from a bubble form into an elongated S-shape and girates about the central axis in the same sense than the main flow (Fig. 3). This topology of the breakdown is similar to the one found in a cavity with a rotating cone [16].

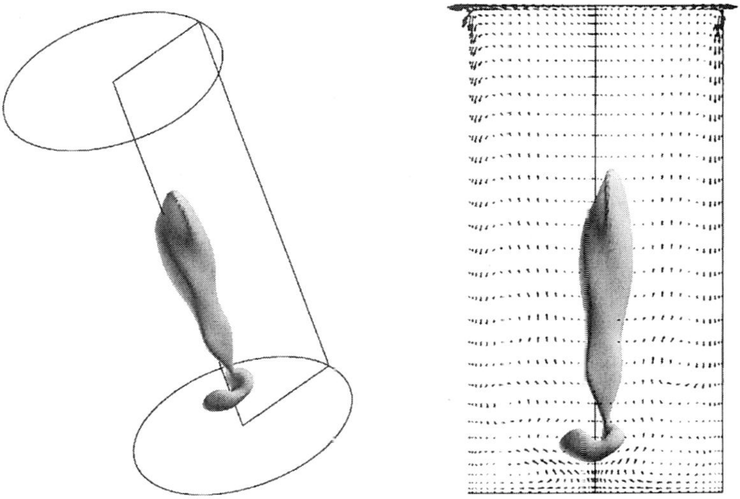

(a) (b)

Fig. 3. Visualization of a S-shaped breakdown, $Re = 6500$, $L = 4$. (**a**) iso-surface $w = 0$. (**b**) iso-surface $w = 0$ with the velocity field in the (r, z, $\pi/4$) plane

8 Conclusion

A Fourier-Chebyshev pseudospectral method has been proposed for solving the unsteady, 3D Navier Stokes equations in rotating cavities. The main computational difficulties come from the presence of the singularity at the axis and the incompressibility constraint. We obviated the first difficulty with a tranformation of the variables and the second one by means of an original projection scheme. Depending on the couple (Re, L), various types of problems have been investigated:

- For the smaller aspect ratii ($H/R_1 < 1$), Ekman and Bödewadt layers occur close to the disks. At high enough Reynolds number, instability arises under the form of annular and spiral vortices. The characteristic parameters such as wavelength, phase velocity and angle of spiral arms are relevant with available theoretical and experimental results about type I and type II instabilities.
- In the case of the counter-rotating Taylor-Couette cavity ($H/R_1 >> 1$), 3D-computations exhibit a wide range of steady or unsteady flow patterns

such as wavy vortices, interpenetrating or wavy spirals. Characteristic parameters of these regimes are shown in good agreement with those given in the literature.

- The confined vortex can also to undergo breakdown. Then, a vortex breakdown phenomenon occurs. This phenomenon is studied inside an enclosed cylindrical rotor-stator cavity with aspect ratio $H/R_1 = 4$, larger than the ones usually considered in the literature. The transition to a 3D flow associated with a precessing vortex breakdown is numerically investigated.

References

1. Owen, J. M., Rogers, R. H. (1989) Heat Transfer in Rotating Disk Systems, Vol. 1: Rotor-Stator Systems. Ed. W. D. Morris (Wiley, Taunton, Somerset, England).
2. Owen, J. M., Rogers, R. H. (1995) Heat Transfer in Rotating Disk Systems, Vol. 2: Rotating Cavities. Ed. W. D. Morris (Wiley, Taunton, Somerset, England).
3. Serre, E., Hugues, S. et al. (2001) Axisymmetric and three-dimensional instabilities in an Ekman boundary layer flow, Int. J. Heat Fluid Flows **22/1**, 82-93.
4. Serre, E., Crespo del Arco, E., Bontoux, P. (2001) Three-dimensional instabilities in an shrouded rotor-stator system, J. Fluid Mech. **434**, 65-100.
5. Czarny, O., Serre, E., Bontoux, P. (in print) Direct numerical simulation and identification of complex flows in Taylor-Couette counter-rotating cavities, C. R. Acad. Sci.
6. Serre, E., Pulicani, J. P. (2001) 3D pseudo-spectral method for convection in rotating cylinder, Int. J. of Computers and Fluids **30/4**, 491-519.
7. Gauthier, P., Gondret, P., Rabaud, M. (1999) Axisymmetric propagating vortices in the flow between a stationary and a rotating disk enclosed by a cylinder. J. Fluid Mech., **386**, 105-127.
8. Savas, O. (1987) Stability of Bödewadt Fow. J. Fluid Mech. **183**, 77-94.
9. Caldwell, D. R., Van Atta, C. W. (1970) Characteristics of Ekman boundary layer instabilities. J. Fluid Mech. **44**, 79-95.
10. Faller, A. J. (1991) Instability and transition of the disturbed flow over a rotating disc. J. Fluid Mech. **230**, 245-269.
11. Andereck, C.D., Liu, S.S., Swinney, H.L. (1986) Flow regimes in a circular Couette system with independently rotating cylinders. J. Fluid Mech. **164**, 155-183.
12. Schröder, W., Keller H.B. (1990) Wavy Taylor-Vortex flows via multigrid-continuation methods. J. Comput. Phys. **91**, 197-227.
13. Neitzel, G.P. (1984) Numerical computation of time-dependent Taylor-Vortex flows in finite-length geometries. J. Fluid Mech. **141**, 51-66.
14. Escudier, M.P. (1984) Observations of the flow produced in a cylindrical container by a rotating end wall, Exps. In Fluids **2**, 179-186.
15. Blackburn, H. M., Lopez, J. M. (2000) Symmetry breaking of the flow in a cylinder by a rotating end wall, Phys. of Fluids **12**, 2698-2701.
16. Pereira, J. C. F., Sousa, J. M. M. (1999) Confined vortex breakdown generated by a rotating cone, J. Fluid Mech. **385,** 287-323.

Parallel Coupled Simulation of Casting Processes on Cluster of PCs

P.A. Adamidis and M.M. Resch

High Performance Computing Center Stuttgart (HLRS), University of Stuttgart, Allmandring 30, 70550 Stuttgart, Germany

Abstract. This paper describes the parallelization of a thermo-mechanical software package that is used to simulate the solidification and cooling process in casting. Such problems require a substantial amount of computing power of high-end workstations which are typically not available in small and medium sized enterprises. In order to overcome this bottleneck this paper introduces clusters of PCs as a computing resource. To show the feasibility of the approach a casting code has been parallelized using a direct substructuring method first. Results show that this method is limited to small problems. Therefore, also an iterative method is presented. Results for realistic examples show a good speedup and give proof that clusters of PCs can make up for the lack of high-end computers in small and medium sized enterprises.

1 Introduction

The casting process in industrial production environments is characterized mainly by two physically relevant phenomena [7]. One is the flow of the cast into the mould. The important factor from the point of view of the simulation is to predict the accurate filling of the mould and to optimize the filling process. The second important question is the cooling of the cast in the mould. The key issue here is the change of shape that results from that cooling process which may yield to a shrinking of the part. This is caused by a thermo-mechanical process that has to be considered in the simulation. In order to predict the shrinking with sufficient accuracy substantial computer performance is required which is not available on a single PC or standard workstation today. High-end systems, however, are not affordable for the casting industry which typically are small and medium sized enterprises (SMEs).

Clusters of PCs have become both affordable and fast enough to be a reasonable alternative [2,3]. Today PCs allow to achieve a level of performance comparable to workstations. At the same time standard networks like Fast Ethernet allow for acceptable communication bandwidth. Given the increased performance of commodity parts like processors and networks the basic building blocks for clusters are very often already in place and are in any case a cheap alternative to high-end systems.

The aim of the project we present here is to solve the numerical problems of the cooling phase in casting processes on such cheap architectures. As a consequence it is necessary to parallelize the simulation code. A similar approach was chosen by [1]. Parallelization however, results in some challenges for the simulation that are not easy to overcome. On the one hand the still relatively slow communication of Ethernet is a big problem. It is mandatory to develop a parallel algorithm that avoids communication as much as possible but at the same time ensures convergence of the method. On the other hand the shrinking process during the cooling phase leads to a contact problem which is sequential by nature. This introduces a bottleneck which is not easy to overcome.

These problems were tackled in the frame of a European project called DECAST (Design Environment for industrial CASTing processes on parallel computing platform). Software developers, high performance computing experts and foundries worked together to develop a framework for casting simulations that should run on clusters of PCs.

2 Clusters of PCs

When doing cluster computing in industry it is necessary to employ standard methods and techniques that are already used in commercial environments. So far, even in small and medium sized enterprises high end PCs are used for development work. Furthermore, these are usually networked by Fast Ethernet. There is usually trained staff available familiar with handling such configurations. Typically, they run commercial packages for simulations either for Microsoft Windows or - in some cases - for Linux.

This is the standard basis for all efforts to introduce clusters in commercial environments as used in DECAST from which the results of this paper have been derived. The typical hardware configuration is:

- Pentium III processor
- 256 MB main memory
- 10 GB hard disk
- Ethernet or Fast Ethernet
- Windows98, WindowsNT or Windows2000

One way to improve performance in such environments is to replace existing mono-processor systems by dual-processor systems. These are relatively cheap and do not require any additional work for system administration since no networking issues have to be considered. In addition, OpenMP provides a programming model that allows to easily migrate from sequential to parallel code. However, the additional performance that can be achieved by a second processor in a box is reported to typically be in the range of 10% only [4].

Improving performance without buying new systems is possible if single processor systems are clustered using already existing Ethernet networks. Message passing is then the most portable approach. Although PVM is still used in some places, MPI has been accepted as the standard in this field. Its most widely used implementation [5] has been ported to almost all platforms. A version for Linux and Windows is available. The main disadvantage of MPI is the new complexity that is introduced to the code. It is mandatory to hide away these problems from the user. In the frame of the project a GUI was developed, which is not presented here. Based on this, the simulation package can still be presented to the user as a black-box solver. The key issue from the point of view of performance is the higher latency that is introduced by message-passing. In the following some techniques are described to cope with that.

3 Parallelization Strategy

3.1 Serial Coupled Algorithm

The solidification process is a coupled thermo-mechanical problem. The governing equations are the momentum and energy balance equations given in (1) and (2) where $\mathbf{B}$ are the prescribed forces, $\boldsymbol{\sigma}$ the Cauchy stress tensor, $\boldsymbol{\Theta}$ the absolute temperature, $\mathbf{S}$ the entropy, $\mathbf{q}$ the heat flux, R the heat source and D_{int} the internal dissipation per unit volume.

$$0 = \nabla \cdot \boldsymbol{\sigma} + \mathbf{B} \tag{1}$$

$$\boldsymbol{\Theta}\dot{\mathbf{S}} = -\nabla \cdot \mathbf{q} + R + D_{int} \tag{2}$$

The thermo-mechanical constitutive model is described in [8,9]. The coupled problem is solved using a staggered time-stepping algorithm. This approach is based on the use of an operator split, applied to the coupled system of non-linear equations, and a product formula algorithm [8]. According to this the original coupled problem is partitioned into two subproblems, the thermal and the mechanical, which are solved within the framework of classical fractional step methods. In Fig. 1 the flow chart of the algorithm is shown. Due to non-linearity for each subproblem an iterative Newton-Raphson procedure is performed. The simulation is done using the finite element method [10]. The contact problem is solved by penalty techniques. Nodes of the casting part, defined at the contact surfaces, are traced against penetration into the mould. In case of penetration correction terms of high order are introduced into the stiffness matrix. The underlying mesh is unstructured and the shape of the resulting system matrix is unstructured as well. In order to overcome this the substructering method has been used which leads to an arrow shaped matrix.

3.2 Direct Substructuring Method

Using domain decomposition techniques we can divide the original computational domain into a number of subdomains, equal to the number of available

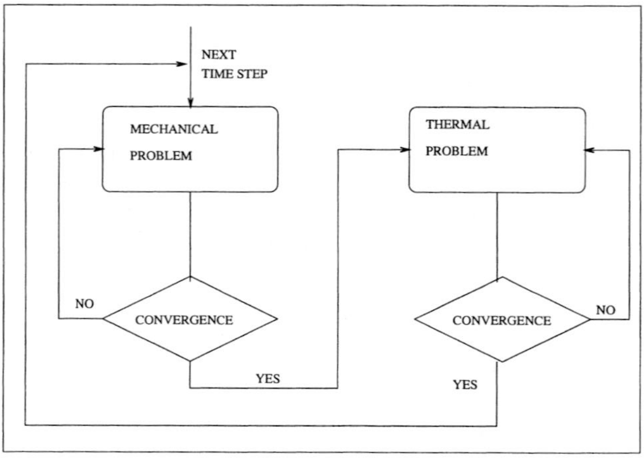

Fig. 1. Coupled algorithm

processors. The direct substructuring method is based on a non-overlapping domain decomposition and involves a renumbering of the grid points after the partitioning. Nodes residing in the interior of each subdomain are numbered locally starting from local number 1. Nodes residing on the artificial boundaries are numbered globally starting from the maximum of all inner nodes of all subdomains plus 1. This renumbering yields an arrow shaped system matrix as shown in (3).

$$A = \begin{pmatrix} A_{II}^1 & & & & A_{IB}^1 \\ & A_{II}^2 & & & A_{IB}^2 \\ & & \ddots & & \vdots \\ & & & A_{II}^n & A_{IB}^n \\ A_{BI}^1 & A_{BI}^2 & \cdots & A_{BI}^n & A_{BB} \end{pmatrix} \quad with \;\; A_{BB} = \sum_{i=1}^{n} A_{BB}^i \tag{3}$$

In (3) I stands for inner and B for boundary and n is the number of partitions. The equations that have to be solved after applying a direct LDU factorization are shown in (4) - (7).

$$A_{II}^i x_i = b_i - A_{IB}^i x_B \tag{4}$$

$$\hat{A} x_B = \hat{b} \tag{5}$$

with the Schur complement

$$\hat{A} = A_{BB} - \sum_{i=1}^{n} A_{BI}^i (A_{II}^i)^{-1} A_{IB}^i \tag{6}$$

and

$$\hat{b} = b_B - \sum_{i=1}^{n} A_{BI}^i (A_{II}^i)^{-1} b_i. \tag{7}$$

In the above equations x_i, b_i denote the inner unknowns and inner right hand side of each partition respectively and x_B and b_B are the vectors of the boundary unknowns and right hand side respectively. The Schur complement system (5) is the main bottleneck of this method. The matrix $\hat{A}$ is a dense matrix and its assembling requires a lot of communication. This method is described in more detail in [6]. In order to avoid the communication and mapping of the respective local A_{BB} parts(A_{BB}^i) in a slow network we define an extra subdomain consisting of all the elements which have one or more nodes on the boundaries. This subdomain is assigned to one processor which can perform the calculation of A_{BB} in (3) locally. At the end only two reduce operations are required to form the sums in (6) and (7). The algorithm de-

Table 1. Speed Up

number of processors	1	2	3	4
cpu time in seconds	70.68	46.50	28.27	18.21
theoretical speed up	1	2	3	4
actual speed up	1	1.52	2.5	3.88

scribed above was implemented and first tested with a 2D example consisting of 3339 dof's. The domain decomposition was done using the standard tool Jostle [11] giving different weights to each subdomain because of the serial contact part. The experiment was done on one dual processor and two single processor PCs. For the solution of the linear systems of each subproblem a direkt skyline LDU solver has been used. The cpu times and speed up for a run over 2 steps can be seen in Table 1. The average number of iterations per step is 4 for the mechanical and 2 for the thermal partition. However, this algorithm showed a low performance when tested with bigger problems due to the necessity of assembling and solving of the Schur complement system as described in equations (5) - (7). In order to avoid this an iterative algorithm was formulated which is described in the next section.

3.3 Parallel Iterative Algorithm

The convergence of an iterative method used to solve a linear system depends on the condition of the system matrix. In our case the matrix is ill conditioned due to the contact terms. The approach to overcome this problem in the serial algorithm is to separate the contact nodes from the remaining nodes.

Introducing this extra contact block in the decomposed arrow shaped system matrix in (3) yields a matrix as shown in (8).

$$A = \begin{pmatrix} A_{II}^1 & & & & A_{IB}^1 & A_{IC}^1 \\ & A_{II}^2 & & & A_{IB}^2 & A_{IC}^2 \\ & & \ddots & & \vdots & \vdots \\ & & & A_{II}^n & A_{IB}^n & A_{IC}^n \\ A_{BI}^1 & A_{BI}^2 & \cdots & A_{BI}^n & A_{BB} & A_{BC} \\ A_{CI}^1 & A_{CI}^2 & \cdots & A_{CI}^n & A_{CB} & A_{CC} \end{pmatrix} \tag{8}$$

In addition we now have the contact submatrix A_{CC} and the submatrices A_{CB},A_{CI}^i and their transposes A_{BC},A_{IC}^i which are the contributions of the inner and boundary nodes having a common edge with a contact node. Applying a Gauss-Seidel like splitting in this matrix we can formulate an iterative algorithm as shown in (9). Now we have to solve 3 equations. One for the interior unknowns (9a), one for the boundary (9b), and one for the contact (9c). In (9) m is the iteration index.

$$A_{II}^i x_i^{m+1} = b_i - A_{IB}^i x_b^m - A_{IC}^i x_c^m \tag{9a}$$

$$A_{BB} x_b^{m+1} = b_b - \sum_{i=1}^{n} A_{BI}^i x_i^{m+1} - A_{BC} x_c^m \tag{9b}$$

$$A_{CC} x_c^{m+1} = b_c - \sum_{i=1}^{n} A_{CI}^i x_i^{m+1} - A_{CB} x_b^{m+1} . \tag{9c}$$

This algorithm is tested on an inhomogeneous cluster of PCs consisting of one PC with an AMD Athlon processor operating at 1000 MHz one PC with an Athlon processor at 550 MHz and one dual processor PC with two Pentium II processors at 450 MHz. Due to that inhomogeneity we need to decompose the initial mesh into differently weighted subdomains. This is done with Jostle [11] by giving the required subdomains a fixed weight. For this we need to define the relative performance of each processor. The fastest processor in the cluster is the Athlon 1000 MHz. Compared to this the other 3 processors have a relative performance of approximately 50% each. Taking this and the number of elements of the problem into account we calculated the weights for each subdomain. The calculated 2D problem consist of 58959 dof's. For the solution of the linear systems in (9) a Conjugate Gradient iterative solver has been used. The convergence behaviour of the parallel iterative algorithm is compared to the serial one in Table 2 and Table 3 for the mechanical and thermal partitions respectively.

The serial run was made on the 1000 MHz Athlon and the parallel runs are compared to this. Since we add 3 slower processors to our fast one, we expect a speed up of 1.5 on two PCs, a speed up of 2 on three PCs and a speed up of 2.5 on four PCs. Considering the fact that the fourth processor

Table 2. Mechanical problem run over 10 steps

number of processors	1	2	3	4
min. number of iterations in a step	2	2	2	2
max. number of iterations in a step	4	5	5	5
average number of iterations/step	2.6	2.6	3	2.6

Table 3. Thermal problem run over 10 steps

number of processors	1	2	3	4
min. number of iterations in a step	2	2	2	2
max. number of iterations in a step	3	3	3	3
average number of iterations/step	2.1	2.1	2.1	2.1

Table 4. Speed Up

number of processors	1	2	3	4
cpu time in seconds	866.43	623.33	456.01	426.81
theoretical speed up	1	1.5	2	2.5
actual speed up	1	1.39	1.9	2.03

is the second of a dual processor machine it adds only 10% in sustained performance. Therefore, we expect a speed up of 2.1 on 4 processors. The cpu time and speed up is shown in Table 4.

4 Conclusion

Thermo-mechanical coupled simulation is necessary to accurately predict the outcome of casting processes in industry. Typically such simulations have to be done by small and medium sized enterprises which do not have access to high-end computing systems. However, clusters of PCs have become available for such companies in the last years. They offer the potential for parallel computing but communication may become a bottleneck. In our paper we present two different parallel numerical algorithms that allow to keep the coupled nature of the simulation but at the same time try to reduce the communication overhead for the simulation.

The parallel direct substructuring method shows low performance when applied to big problems. The parallel iterative algorithm shows rather good performance by reducing the communication overhead. However, it depends heavily on a good load distribution taking into account the inhomogeneity of the parallel computing environment as can be seen from our experiments in realistic environments.

Further work will therefore be done in developing an optimized load balancing algorithm for this type of problems according to the system parameters and the problem parameters. Summarizing we can say that the approach we follow has the potential to make coupled simulation a standard tool for real applications in industry.

Acknowledgment

The DECAST project is funded by the European Commission under ESPRIT contract Nr. 28144.

References

1. Franz-Josef Pfreundt, "HPCLine eine Bedrohung für die SP 2", Supercomputer'99, Mannheim/Germany, June 10-15, 1999.
2. Uwe Harms, "Clusters triumph over custom systems", Scientific Computing World, 49, October/November 1999.
3. Daniel Ridge, Donald Becker, Phillip Merkey, Thomas Sterling, "Beowulf: Harnessing the Power of Parallelism in a Pile-of-PCs", Proceedings IEEE Aerospace, 1997.
4. Robert Pennington, Patricia Kovatch, Barney Maccabe, David Bader, "Design and Analysis of High Performance Clusters", Supercomputing'2000, Dallas, USA, 2000.
5. W. Gropp, E. Lusk, N. Doss, A. Skjellum, "A high-performance, portable implementation of the MPI message passing interface standard", Parallel Computing, 22, 1996.
6. B. Smith, P. Bjorstad and W. Gropp, "Domain Decomposition : parallel multilevel methods for elliptic partial differential equations", Cambridge University Press,1996.
7. Panagiotis A. Adamidis and Michael M. Resch, "Low cost computing in casting industries", European Congress on Computational Methods in Applied Sciences and Engineering ECCOMAS 2000, Barcelona/Spain, 11-14 September, 2000.
8. M. Cervera, C. Agelet de Saracibar and M. Chiumenti, "Thermo-Mechanical Analysis of Industrial Solidification Processes", International Journal for Numerical Methods in Engineering, 46, 1575-1591, 1999.
9. M. Chiumenti, C. Agelet de Saracibar and M. Cervera, "Constitutive Modeling and Numerical Analysis of Thermo-Mechanical Phase-Change Systems", Monograph CIMNE, 48, March, 1999.
10. O.C. Zienkiewicz and R.L. Taylor, "The finite element method", McGraw Hill, Vol. I., 1989, Vol. II, 1991.
11. Chris Walshaw, "Serial Jostle Library Interface : Version 1.1.5", School of Computing & Mathematical Sciences, University of Greenwich, London, March 10,1990.

Controlling Point Defects in Single Silicon Crystals Grown by the Czochralski Method

A. Voigt, J. Nitschkowski, Ch. Weichmann, and K.-H. Hoffmann

research center caesar, Friedensplatz 16, 53111 Bonn, Germany

Abstract. Silicon is the basic material which makes the functions of semiconductors possible. Most of the conveniences we take for granted would simply not have existed without this essential resource. The need for more and more complex devices increases the size of the device and the requirements on the quality of the material. Crystal quality will therefore be of utmost importance in the coming years as these trends continue to decrease the tolerance on wafer specifications. We present a mathematical model and numerical simulations which help to design the growth process in order to improve the crystal quality. A macroscopic model for heat transfer is therefore combined with a microscopic model for intrinsic point defect diffusion.

1 Introduction

The growth of high quality silicon single crystals is one of the key processes that determine the yield and the profitability in semiconductor device manufacturing. Wafer suitability for micro-electronic devices is determined by several different measurements that include surface uniformity, lattice perfection, impurity concentration and crystal electrical properties such as resistivity. Increasing device size increases the requirements on the quality of the crystal. For a given concentration of defects, device failures are greater if fewer, more expensive devices are fabricated on the wafer. It is widely accepted that today without an effective accompanying process modeling no appropriate quality of the crystal can be achieved. Besides the help in understanding the complex processes in crystal growth the model can also be used in predicting optimal parameters to grow appropriate crystals. The model in this paper allows it to relate microscopic crystal quality to macroscopic growth parameters.

2 Modeling

2.1 Macroscopic Heat Transfer

The growth process of silicon in a Czochralski furnace includes a large number of physical transport mechanisms, such as convection, mass transport, interface dynamics and radiative heat transfer, which interact in complex ways, see Fig. 1. The governing equations can be derived from the basic principles

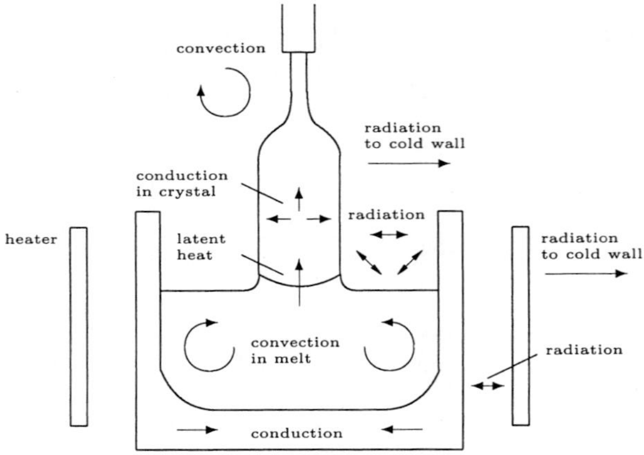

Fig. 1. Schematic visualization of heat exchange processes in Czochralski growth

of conservation of mass, momentum and energy. The melt flow is modeled by the Navier-Stokes equation

$$\frac{\partial \mathbf{u}_m}{\partial t} + (\mathbf{u}_m \cdot \nabla)\mathbf{u}_m - \nu \Delta \mathbf{u}_m + \nabla p = \beta(T_m - T_0)\mathbf{g},$$
$$\nabla \cdot \mathbf{u}_m = 0,$$

with $\mathbf{u}_m$ velocity, p pressure, T_m temperature, $\mathbf{g}$ gravitational force and β volume expansion coefficient, where we assume the melt being Newtonian and variations of melt density are negligible, except in the calculation of the body force induced by thermal buoyancy, which is the onset of Grashof convection. This simplification is known as Boussinesq approximation. Heat transfer in the apparatus occurs by conduction, convection and radiation. The equations for the temperature in all parts are

$$\rho_{i_0} c_{i_p} \left(\frac{\partial T_i}{\partial t} + \mathbf{u}_i \cdot \nabla T_i \right) - \nabla \cdot (k_i \nabla T_i) = \rho_{i_0} P_i, \tag{1}$$

where ρ_{i_0} is the density of the component, c_{i_p} is the heat capacity, u_i is the velocity, k_i is the thermal conductivity and P_i is a heat source term. i indicates the different components of the apparatus, like melt, crystal, crucible, heaters, heat shields and insulators. The heat sources P_i are nonzero only in the heaters. We neglect the influence of a gas flow in our model. Further we assume that almost no radiation is emitted or absorbed by the gas and that radiation is only diffuse. Moreover, we consider that emission, absorption and reflection of radiating waves occur only at the surfaces. The heat balance at the surface relating the heat flux caused by conduction to the surface q, the radiosity and the irradiation then becomes $q = \mathbf{G}(\sigma T^4)$, where $\mathbf{G} = (\mathbf{I} - \epsilon \mathbf{K}(\mathbf{I} - (1-\epsilon)\mathbf{K})^{-1})\epsilon$ is the infinite dimensional equivalent of the so called Gebhardt factor. ϵ is the emissivity, the operator $\mathbf{K}$ is defined as

$$\mathbf{K}\lambda(\mathbf{r}) = \int_{\Gamma} \omega(\mathbf{r},\mathbf{s})\Xi(\mathbf{r},\mathbf{s})\lambda(\mathbf{s})d\mathbf{s},$$

where $\omega(\mathbf{r},\mathbf{s})$ is the view factor, $\Xi(\mathbf{r},\mathbf{s})$ is the visibility factor and Γ defines the surface [8]. Along the free melt-crystal interface we have mass conservation and a no-slip boundary condition. Due to the rotation of the crystal this no-slip condition is the onset of forced convection by crystal rotation. Because the melt-crystal interface is a thermal free boundary the thermal conditions need to specify the temperature and the position and shape of the boundary, we set $T_m = T_c = T_{eq}$ at that boundary, where T_{eq} is the equilibrium melting temperature. The second thermal condition takes account of the heat flux balance between the two phases, the heat flux in the crystal is greater than in the melt by the amount of latent heat released during solidification, i.e. $k_m \nabla T_m \cdot \mathbf{n} - k_c \nabla T_c \cdot \mathbf{n} = \rho_{c_0} l(\mathbf{u}_c - \mathbf{u}_f) \cdot \mathbf{n}$, with l the latent heat, $\mathbf{n}$ the normal vector and $\mathbf{u}_f$ the speed of the phase-boundary. The meniscus separating the melt from the ambient gas is another free interface, whose shape is part of the problem unknowns. Here we again make a Boussinesq-type approximation, where we assume that the surface-tension coefficient γ equals a constant γ_0, except in the boundary condition for the tangential stress, where we assume a linear dependence $\gamma = \gamma_0(1 - \alpha(T_m - T_0))$ on temperature. The traction condition for tangential and normal stress reads $\boldsymbol{\sigma}\ \mathbf{n} \cdot \mathbf{t} = -\gamma_0(\alpha \nabla T_m \cdot \mathbf{t})$ and $\boldsymbol{\sigma}\ \mathbf{n} \cdot \mathbf{n} = \gamma_0 2H - p_g$, where p_g is the pressure in the gas phase and H is the mean curvature. The temperature differences at the interface have an influence on the transport of momentum and heat near the interface. The surface-tension gradient resulting from this differences acts like a shear stress on the melt-gas interface and thereby generates a surface flow. This phenomenon is known as Marangoni convection. We assume that the melt perfectly adheres to the crucible wall and specify a no-slip condition for the velocity which is the onset of forced convection due to rotation of the crucible. At the remaining boundaries we only need to specify a condition for the temperature. At the surfaces of the melt, the crystal, the crucible, the heaters, heat-shields, insulators and the inner surface of the enclosure we take $k_i \nabla T_i \cdot \mathbf{n} = -\mathbf{G}(\sigma T_i^4)$, where i stands for the different components and the approximation of a non-participating gas is used. We need to describe initial conditions for the temperature T_i in all components of the apparatus, conditions for the velocity $\mathbf{u}_m$ in the melt and an initial domain for melt and crystal.

2.2 Microscopic Defect Diffusion

Thermodynamic equilibrium conditions favor the formation of crystals with a certain concentration of lattice defects. For a detailed introduction in defect thermodynamics see [10] and the references therein. One of the most crucial measures of wafer suitability for device manufacture today is the defect distribution within the silicon material. We distinguish between point defects caused by misplaced silicon atoms and impurity defects formed by the presence of foreign atoms (Fig. 2). Point defects are incorporated into the crystal

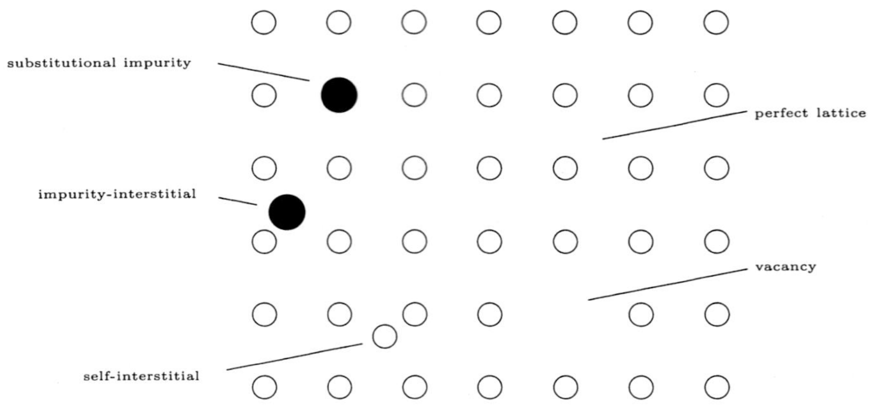

Fig. 2. Schematic representation of various point defects

in their equilibrium concentrations and then diffuse and recombine rapidly in the high temperature portion of the crystal. We consider convection, Fickian and thermodiffusion and reaction of point defects in the growing crystal. Fickian diffusion is driven by concentration gradients in the crystal, thermal diffusion is driven by temperature gradients in the direction of lower temperatures. Recombination is characterized by the annihilation of interstitials and vacancies and the creation of an undisturbed lattice site. It is the inverse process of the Frenkel pair generation. The equations can be derived from basic principles of thermodynamic, see [1]. The governing equations for the concentrations of interstitials and vacancies are

$$\frac{\partial C_i}{\partial t} + \mathbf{u}_c \cdot \nabla C_i - \nabla \cdot \left(D_i \nabla C_i + \frac{Q_i D_i C_i}{k T_c^2} \nabla T_c \right) = k_{rec} \left(C_i^{eq} C_v^{eq} - C_i C_v \right),$$

$$\frac{\partial C_v}{\partial t} + \mathbf{u}_c \cdot \nabla C_v - \nabla \cdot \left(D_v \nabla C_v + \frac{Q_v D_v C_v}{k T_c^2} \nabla T_c \right) = k_{rec} \left(C_i^{eq} C_v^{eq} - C_i C_v \right),$$

where C_i and C_v are the concentrations of interstitials and vacancies, D_i and D_v the temperature-dependent diffusion coefficients, Q_i and Q_v the activation enthalpy for thermal diffusion, C_i^{eq} and C_v^{eq} the equilibrium concentrations, k the Boltzmann constant, k_{rec} the reaction coefficient for recombination and $\mathbf{u}_c$ the velocity of the crystal. At the free phase boundary Dirichlet conditions with the equilibrium concentration are specified and at the remaining boundaries no-flux conditions are used. We need to describe initial conditions for the concentrations in the crystal. The thermophysical properties are experimentally because of the high temperatures not directly measurable. One way to circumvent this problem is to calculate the microscopic properties with the help of statistical mechanics and molecular-dynamic simulations [7].

3 Discretisation

3.1 Phase Transition

Instead of formulating the problem with the help of the heat equations in melt and crystal (1 we introduce an equation for enthalpy H in $\Omega_m \cup \Omega_c$ and treat enthalpy as the primary unknown. The energy density (or enthalpy) H is defined as:

$$H(T) = \int_{T_{ref}}^{T} \rho c(\tau) d\tau + l\eta \, (T - T_{eq}) \tag{2}$$

with the Heaviside function η. Instead of defining enthalpy as a function of temperature we can define temperature as the inverse of (2) and use a Kirchhoff transformation to define the Kirchhoff temperature $\beta(H)$. Now we can write the phase transition problem as

$$\frac{\partial H}{\partial t} + \mathbf{u}_i \cdot \nabla H - \Delta\beta(H) = 0 \tag{3}$$

The full discretisation of the phase transition problem (3) combines piecewise linear finite elements in space and a time discretisation including the convection that is based on the method of characteristics. Successive overrelaxation is used for solving the algebraic system.

3.2 Melt Flow

The behaviour of the free capillary boundary is central to the dynamics and the control of the diameter of a growing crystal. The shape of the growing crystal is determined by the angle at which the meniscus contacts the crystal at the three-phase boundary. Cylindrical growth will only be achieved when this angle of contact takes some unique value, see Fig. 3. In the absence of any volume change on solidification the system can be assumed to be in thermal equilibrium. The profile of the meniscus for a given contacting angle Θ_{Si} is then obtained by solving the Laplace-Young equation. An useful approximate solution to the Laplace-Young equation has been given in [3]. This approximation has been shown to compare well with the full solution of the free capillary surface problem. The capillary surface is therefore assumed to be fixed in each time step and the melt flow problem is restricted to solving the Navier-Stokes equation in a domain where only the phase-change boundary remains as a free boundary.

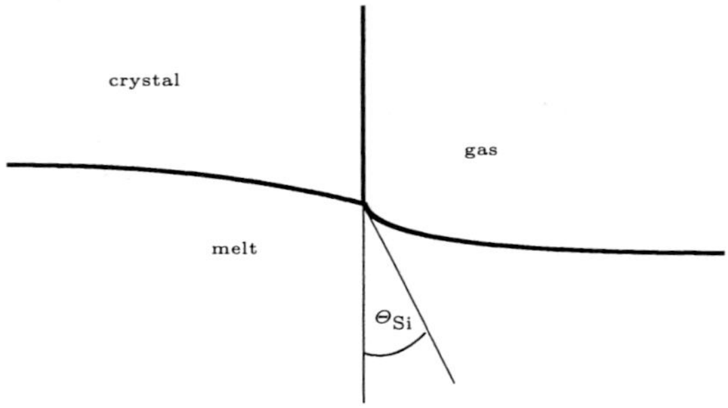

Fig. 3. Contacting angle at the three-phase boundary

One particular problem in computing the flow field is the unknown domain due to solidification. There are several ways to solve this problem in the discrete case, we use a fixed grid method with an explicit definition for the melt region and a no-slip condition. In [5] this approach was used for the simulation of dendritic crystal growth. In order to apply this method we use the following definition for melt and crystal: An element $S \in \mathcal{S}$ belongs to the melt, iff S lies completely in the melt region, i.e. for all $x \in S$ we have $T(x) > T_{eq}$. For a given temperature the Navier-Stokes equation can now be solved on a fixed grid at each time step.

We use a time discretisation based on the fractional θ-scheme. By this operator-splitting one reduces the Navier-Stokes equation to three considerably simpler subproblems. Numerical experiments show that this splitting scheme is second order accurate and nearly non-dissipative. For the space discretisation the Taylor-Hood element is used, i.e. piecewise quadratics for the velocity and piecewise linears for the pressure.

3.3 Radiation

The heat fluxes q in the radiation boundary conditions are related to the temperature, due to the Gebhardt factor $\mathbf{G}$. The method of view factors is used to compute the effective flux to the surface q which provides the boundary condition to the finite element model for the conduction process.

3.4 Defect Diffusion

The discretisation of the reaction-diffusion problem combines piecewise linear finite elements in space and a time discretisation including the convection that is based on the method of characteristics. Because of the nonlinearity in the recombination terms we use a Newton method to solve the reaction-diffusion system. The generalized minimum residual method (GMRES) is used for solving the linear system in each iteration.

3.5 Coupling of Equations

The overall numerical method is based on a semi implicit time discretisation scheme with a coupling of the subproblems of enthalpy, temperature, interface, geometry, velocity and defect evolution. Because of the different time and length scales in the subproblems the system can be decoupled at certain points. The velocity in the melt-flow is approximately 10^3 times higher than the growing velocity of the crystal. Therefore it is a good approximation to neglect geometry changes due to the growing of the crystal in the calculation of the velocity field in the melt. Over a long time period we need to solve in each time step:

(1) Using the old velocity $\mathbf{U}^n$, solve one timestep for the enthalpy equation, giving H^{n+1} and T^{n+1}.
(2) Using the new temperature T^{n+1} and the new interface (and thus the new liquid domain Ω_m^{n+1}), solve one timestep for the Navier-Stokes equation, giving $\mathbf{U}^{n+1}$ and P^{n+1}.
(3) Using the new temperature T^{n+1} and the new interface (and thus the new solid domain Ω_c^{n+1}), solve one timestep for the reaction-diffusion equation, giving C_i^{n+1} and C_v^{n+1}.
(4) Using the new temperature T^{n+1} at the surfaces to calculate the new effective fluxes to the surface q^{n+1}.

4 Results

The calculations are performed on a slightly modified industrial furnace for $100mm$ Si crystal growth. Due to the axisymmetry of the furnace the complexity of the problem can be reduced to 2D. The software CrysVUN [4] and ALBERT [6] are used for the computations. In order to verify the results an experimentally satisfied phenomenological law is used which qualitatively relates point defects to growth conditions [2]. Without being bound to any particular theory it is understood that the type and initial concentration of the point-defects in silicon are controlled by the ratio u_c/G, where u_c is the growth velocity and G is the instantaneous axial temperature gradient in the crystal at the time of solidification. As the value of u_c/G exceeds a critical value, the concentration of vacancies increases. Likewise, as the value of u_c/G falls below a critical value, the concentration of self-interstitials increases. At the critical value the crystal will be free of defects [9]. Because G varies considerably from the crystal center (low thermal gradient) to the surface (high thermal gradient), due to heat loss at the surface, the type of defects may varies as well and change from vacancy to interstitial. Figure 4 shows the type of defects in the growing crystal and the u_c/G curve over the crystal radius r for three different pulling velocities. The prediction of the numerical simulations for transition from vacancy to self-interstitial dominated regions are in good agreement with the u_c/G criteria even if material data are not

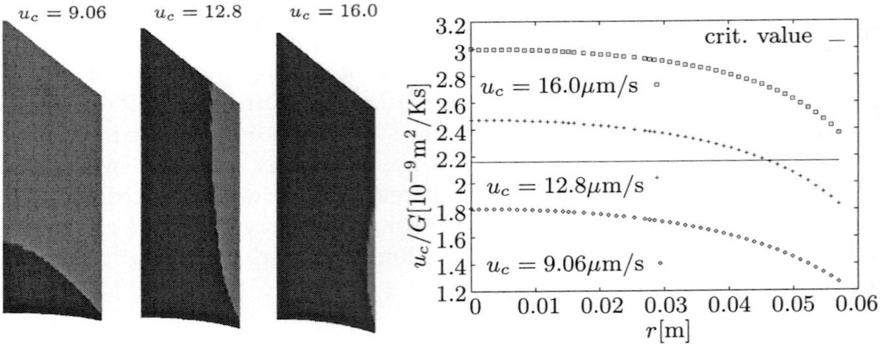

Fig. 4. Vacancies (blue) and self-interstitials (red), u_c/G curve, velocity in μm/s

known precisely and convection is not taken into account. The analysis of point defect dynamics is the starting point for predictions of density and size distributions of grown-in defects. For the first time a combined multi-scale model which combines heat transfer and defect dynamics in silicon crystal growth is presented. With this calculations macroscopic growth conditions can be related to microscopic quantities specifying the crystal quality and can therefore help to design growth processes.

References

1. de Groot S. R., Mazur P. (1984) Non-equilibrium Thermodynamics, North-Holland, Amsterdam
2. Dornberger E. (1997) Prediction of OSF Ring Dynamics and Grown-in Voids in Czochralski Silicon Crystals. PhD Thesis, Universite Catholique de Louvain
3. Hurle D. T. J. (1983) Analytic Representation of the Shape of the meniscus in Czochralski Growth. J. Cryst. Growth **63**, 13–17
4. Kurz M. R. H. (1998) Development of CrysVUN++, a Software System for Numerical Modelling and Control of Industrial Crystal Growth Processes. PhD Thesis, Universität Erlangen-Nürnberg
5. Schmidt A., Bänsch E. (1999) Simulation of dendritic crystal growth with thermal convection. Math. Tech. Report., Uni. Freiburg 4/1999
6. Schmidt A., Siebert K. G. (2000) ALBERT: An adaptive hierarchical finite element toolbox. Math. Tech. Report., Uni. Freiburg 6/2000
7. Sinno T. R. (1998) Defects in Crystalline Silicon: Intergated Atomistic and Continuum Modeling. PhD Thesis, MIT, Cambridge
8. Tiihonen T., Järvinen J., Nieminen R. (1997) Time-dependent simulation of Czochralski silicon crystal growth. J. Cryst. Growth **180**, 468–476
9. Voronkow V. (1982) The mechanism of swirl defects formation in silicon. J. Cryst. Growth **59**, 625–643
10. Wenzl H., Oates W. A., Mika K. (1994) Defect thermodynamics and phase diagrams in compound crystal. In: Hurle D. T. J. (Ed.) Handbook of Crystal Growth Vol. 1, Elesvier Science B.V., Amsterdam, 103–185

A Two–Scale Method for Liquid–Solid Phase Transitions with Dendritic Microstructure

Ch. Eck and P. Knabner

University of Erlangen–Nürnberg, Institute for Applied Mathematics,
Martensstr. 3, D-91058 Erlangen, Germany

Abstract. A two–scale model for liquid–solid phase transitions with equiaxed dendritic microstructure for binary material with slow solute diffusion is presented. The model consists of a macroscopic energy transport equation, coupled with local cell problems describing the evolution of the microstructure and the microsegregation. It is derived by an asymptotic expansion of a sharp interface model with Gibbs–Thomson effect. A discretization of the model leading to a two–scale method for such problems is presented, and a numerical example is given.

1 Introduction

In liquid–solid phase transitions as e.g. the casting of metals often a specific dendritic microstructure of the phase interface is observed. The two main morphologies of this microstructure are the columnar dendritic growth, where columnar dendrites are growing in the direction of the heat flow, and equiaxed growth, where small crystals nucleating in the melt are growing in all directions, c.f. Fig. 1. The microstructure has a profound influence on the properties and the quality of the material, in particular on the grain structure and the microsegregation (variations of the composition of the material on a small scale); and numerical methods for their computation are therefore highly desirable. However, the available numerical methods for the computation of dendritic growth are computationally expensive. Therefore its is not possible to resolve the microstructure of a given workpiece with a global grid. In this contribution, a new two–scale model for the evolution of equiaxed microstructure in binary material with "fast" heat and "slow" solute diffusion is presented. The model is derived by a two–scale expansion of a sharp interface model for binary material. It consists of a homogenized heat equation to be solved on the macroscopic domain, coupled with local cell problems to be solved on a up–scaled unit cell for each point of the global domain.

2 The Sharp Interface Model

In a sharp interface model the domain $\Omega \subset \mathbb{R}^N$ of space dimension $N \in \mathbb{N}$ is divided for every time t into the domains $\Omega_\ell(t)$ of liquid material and $\Omega_s(t)$ of solid material. Both domains are separated by the phase interface $\Gamma_I(t)$.

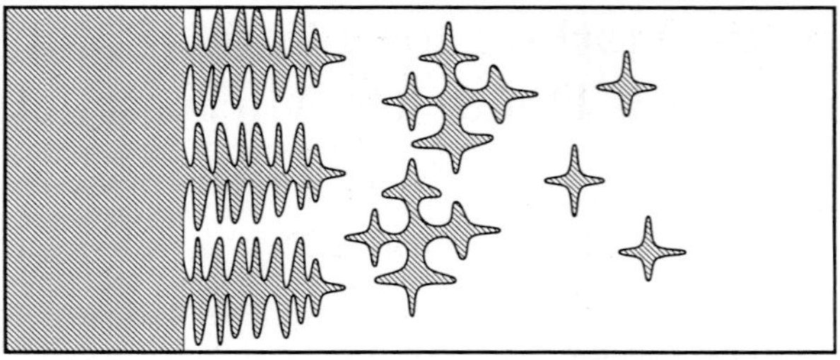

Fig. 1. Microstructure in liquid–solid phase transitions

The variables are temperature T, concentration c_k of one of the components of the material in liquid ($k = \ell$) and solid ($k = s$) phase and velocity v_s of the phase interface, measured in the direction of the outer normal n_s of the solid domain. The material properties are given by the density of internal energy $u_k = u_k(T, c_k)$, $k = \ell, s$, the chemical potential $\mu_k = \mu_k(T, c_k)$, the heat conductivity $K^{(k)}$, the solute diffusivity $D^{(k)}$, the curvature multiplier σ, the kinetic multiplier β, and a function $b(T, \mu)$ derived from the phase diagram. Then the model consists of the differential equations

$$\partial_t u_k - \nabla \cdot \left(K^{(k)} \nabla T\right) - \nabla \cdot \left(D^{(k)} \frac{\partial u_k}{\partial c_k} \nabla c_k\right) = f, \tag{1}$$

$$\partial_t c_k - \nabla \cdot \left(D^{(k)} \nabla c_k\right) = 0 \tag{2}$$

defined for $t \in I_T = [0; T]$ and $x \in \Omega_k(t)$, $k = \ell, s$; supplemented by boundary conditions on the phase interface $\Gamma_I(t)$,

$$\mu_\ell(T, c_\ell) = \mu_s(T, c_s), \tag{3}$$

$$\left[K \frac{\partial T}{\partial n} + D \frac{\partial u}{\partial c} \frac{\partial c}{\partial n}\right] = -[u] v_s, \tag{4}$$

$$\left[D \frac{\partial c}{\partial n}\right] = -[c] v_s, \tag{5}$$

$$b(T, \mu) = -\sigma \kappa - \beta v_s, \tag{6}$$

by boundary conditions on the boundary of the whole domain $\Gamma = \partial \Omega$

$$K \frac{\partial T}{\partial n} + \alpha T = f_\Gamma, \tag{7}$$

$$D \frac{\partial c}{\partial n} = 0, \tag{8}$$

and by initial conditions $T(t = 0, x) = T^{(0)}(x)$ and $c(t = 0, x) = c^{(0)}(x)$, $x \in \Omega$, for temperature and concentration and $\Omega_s(t = 0) = \Omega_s^{(0)}$ for the initial solid domain. Here, κ denotes the mean curvature, taken positive for Ω_s being convex, and the bracket $[\cdot]$ denotes the jump of the corresponding quantity at the phase interface, $\left[K\frac{\partial T}{\partial n}\right] := K^{(\ell)}\nabla T_\ell \cdot n_\ell + K^{(s)}\nabla T_s \cdot n_s$ and $[u] := u_s - u_\ell$. Equation (6) is a generalized Stefan condition with Gibbs–Thomson undercooling and kinetic undercooling adapted to the case of a binary material. The function b is specified from the phase diagram; in the simplest case of a linear model it is given by $b(T,\mu) = b_1 T + b_2\mu + b_3$ with coefficients b_1, b_2 and b_3. An existence result for this model is given in [4] for the case of vanishing kinetic coefficient $\beta = 0$.

3 Asymptotic Expansion and Two–Scale Model

Let us consider an idealized microstructure consisting of equiaxed crystals whose midpoints are located at the sites of a uniform grid. The scale of the microstructure is given by a small parameter $\varepsilon > 0$ signifying the spacing of this grid. This situation arises in the case of *instantaneous* nucleation, where small solid kernels of a given distribution and a given size nucleate at a given undercooling. The assumption of periodic distribution of the initial kernels is, of course, not very realistic, but it is frequently made in homogenization theory in order to derive macroscopic models which are then also used for more general situations. The initial conditions of the problem have to reflect this instantaneous periodic nucleation, they are given by

$$T_\varepsilon(t=0,x) = T^{(0)}(x,\tfrac{x}{\varepsilon}),\;\; c_\varepsilon(t=0,x) = c^{(0)}(x,\tfrac{x}{\varepsilon}) \tag{9}$$

with functions $T^{(0)}, c^{(0)} \in C^1\big(\Omega, L_{2\#}(Y)\big)$, where $Y \subset \mathbb{R}^N$ is a periodicity cell and $L_{2\#}(Y) := \{y \in L_2(\mathbb{R}^N) \,|\, y \text{ is } Y\text{–periodic}\}$; and an initial solid domain $\Omega_s^{(0)} := \Omega \cap \bigcup_{k\in\mathbb{Z}^N} \varepsilon\left(\{k\}\oplus Y_s^{(0)}\right)$ with unit solid domain $Y_s^{(0)} \subset Y$. Here, T_ε and c_ε denote solutions of the problem with scale parameter ε.

In the limit procedure $\varepsilon \to 0$ several physical parameters are scaled in dependence of ε. The density of surface energy is scaled proportional to ε in order to maintain an equilibrium between surface- and volume energy. The solute diffusivity is scaled proportional to ε^2, this is justified by the fact that it is usually much smaller than heat conductivity. The kinetic coefficient is scaled proportional to ε^{-1}, then the time scale in the evolution of the (up–scaled) single crystal does not change for varying ε. Let us remark that our approach also includes the case of vanishing kinetic coefficient $\beta = 0$. Summing up the scaling properties we have

$$\sigma = \varepsilon\sigma_0, \quad D^{(k)} = \varepsilon^2 D_0^{(k)} \text{ and } \beta = \varepsilon^{-1}\beta_0.$$

The limit problem is derived using the method of asymptotic expansions. It is assumed that the functions T_ε, c_ε and the velocity v_ε of the phase interface admit the asymptotic representations

$$\begin{aligned} T_\varepsilon(t,x) &= T_0\big(t,x,\tfrac{x}{\varepsilon}\big) + \varepsilon T_1\big(t,x,\tfrac{x}{\varepsilon}\big) + \varepsilon^2 T_2\big(t,x,\tfrac{x}{\varepsilon}\big) + \cdots, \\ c_\varepsilon(t,x) &= c_0\big(t,x,\tfrac{x}{\varepsilon}\big) + \varepsilon c_1\big(t,x,\tfrac{x}{\varepsilon}\big) + \varepsilon^2 c_2\big(t,x,\tfrac{x}{\varepsilon}\big) + \cdots, \\ v_\varepsilon(t,x) &= \varepsilon v_0\big(t,x,\tfrac{x}{\varepsilon}\big) + \varepsilon^2 v_1\big(t,x,\tfrac{x}{\varepsilon}\big) + \cdots \end{aligned} \tag{10}$$

valid for small ε. The gradient of a function $f(x, \frac{x}{\varepsilon})$ depending on a "slow" variable x and a "fast" variable $y = \frac{x}{\varepsilon}$ is given by $\nabla f = \nabla_x f + \frac{1}{\varepsilon}\nabla_y f_{|y=\frac{x}{\varepsilon}}$ or, shortly, $\nabla = \nabla_x + \frac{1}{\varepsilon}\nabla_y$. We insert these expansions into problem (1)–(8). This leads to a sequence of problems of different orders in ε. It turns out that the leading order term of the temperature is a purely macroscopic variable, $T_0 = T_0(t, x)$ independent of $y \in Y$. It is a solution of the homogenized heat equation

$$\partial_t u^*_{c_0,Y_s}(T_0) - \nabla \cdot \big(K^* \nabla T_0\big) = f \quad \text{in } \Omega, \tag{11}$$

with averaged density of internal energy $u^*_{c_0,Y_s}(T_0) = \sum\limits_{k=\ell,s} \int_{Y_k} u_k\big(T_0, c_0^{(k)}\big)\, dy$, effective heat conductivity $K^*_{ij} = \sum\limits_{k=\ell,s} \int_{Y_k} \left(K^{(k)}_{ij} + \sum\limits_{r=1}^{N} K^{(k)}_{ir}\partial_{y_r} H_j\right) dy$ and a possibly averaged heat source f. The functions H_j in the definition of K^* are solutions of the local cell problem

$$\begin{aligned} -\nabla_y \cdot \big(K^{(k)}\nabla_y H_j\big) &= \nabla_y \cdot \big(K^{(k)} e_j\big) && \text{in } Y_k(t,x),\ k = \ell, s, \\ [K\nabla_y H_j \cdot n_y] &= -[K e_j \cdot n_y] && \text{on } \Gamma_I(t,x) \end{aligned} \tag{12}$$

satisfying periodic boundary conditions on ∂Y. Here, e_j denotes the j–th unit vector in $\mathbb{R}^N$, Y_k, $k = \ell, s$, the liquid and solid part of the unit cell Y, Γ_I their common boundary and n_y the unit normal on Γ_I, pointing outward of Y_s. If the heat conductivity is constant and its value in liquid and solid is equal, $K^{(\ell)} = K^{(s)}$, then the effective heat conductivity is equal to the original one, $K^* = K$.

The leading order terms of the concentration field and the interface velocity are solutions of the local problems

$$\partial_t c_k(T_0; \mu_0) - \nabla_y \cdot \big(\widetilde{D}_0^{(k)} \nabla_y \mu_0\big) = 0 \text{ in } Y_k(t),\ k = \ell, s, \tag{13}$$

$$\Big[\widetilde{D}_0 \nabla_y \mu_0 \cdot n_y\Big] + \big[c(T_0; \mu_0)\big] v_0 = 0 \text{ on } \Gamma_I(t), \tag{14}$$

$$b(T_0, \mu_0) = -\sigma_0 \kappa_0 - \beta_0 v_0 \text{ on } \Gamma_I(t) \tag{15}$$

with $\widetilde{D}^{(k)} = D^{(k)} \frac{\partial c_k}{\partial \mu}$.

4 Numerical Method

The application of the two–scale model in a numerical algorithm is presented here for the simplest case with constant diffusivity tensors independent of the phases, with linear internal energy $u_k(T, c) = c_V T + \delta_{k\ell} L$ independent of the concentration, chemical potential $\mu = \mu_k(c_k) = c_k + \delta_{k\ell}\lambda$ with constant miscibility gap λ and linear function $b(T, \mu) = b_1 T + b_2 \mu$. The resulting problem consists of the global problem

$$\begin{aligned} \partial_t(c_V T - \varepsilon_s L) - \nabla \cdot (K \nabla T) &= f \quad \text{in } \Omega, \\ K\frac{\partial T}{\partial n} &= f_\Gamma \quad \text{on } \Gamma, \end{aligned} \tag{16}$$

with solid volume fraction $\varepsilon_s(t,x) = \frac{|Y_s(t,x)|}{|Y|}$, and the local problems

$$\begin{aligned} \partial_t \mu - \nabla_y \cdot (D \nabla_y \mu) &= 0 \ \text{ in } Y_k(t), \ k = \ell, s, \\ \left[D \frac{\partial \mu}{\partial n_y} \right] + \lambda v_s &= 0 \ \text{ on } \Gamma_I(t), \\ b_1 T + b_2 \mu &= -\sigma \kappa - \beta v_s \ \text{ on } \Gamma_I(t). \end{aligned} \tag{17}$$

Both problems are still coupled, the release of latent heat in the macroscopic problem depends on the microscopic solid volume fraction, and the equilibrium melting point in the microscopic problem depends on the macroscopic temperature. This model is not thermodynamically consistent, but it can be interpreted as a linearization of a thermodynamically consistent model.

Both the global problem (16) and the local problem (17) are discretized by an implicit Euler method with respect to the time and a finite element method using bilinear shape functions on a square grid with respect to the space variables. The local problems are given for every nodal point of the global grid. For their solution we need an algorithm computing the evolution of the solid phase from some discretized version of the Stefan condition in (17). For this task, we employ a *variational algorithm* due to Almgren [1].

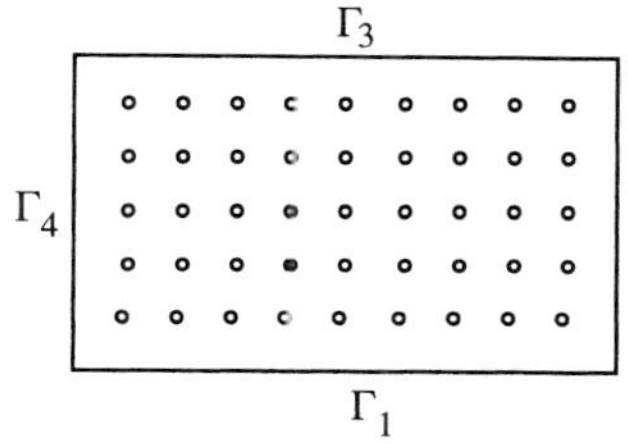

Fig. 2. Model problem for two–scale method.

The coupling of macroscopic and microscopic problem is done in a semi–implicit manner. In each time step, all the local cell problems are solved for the temperature of the preceding time step. Then, the macroscopic heat transport equation is solved with the release of latent heat computed from the new solid volume fraction. This approach is reasonable, because the time steps required for the solution of the crystal growth problem are already rather small.

5 Examples for the Two–Scale Method

For the numerical realization of the two–scale model we consider the problem (16), (17) defined on a rectangular domain Ω of size $[0;2] \times [0;0.5]$, c.f. Fig. 2, with material parameters $c_V = 1$, $K = 1$, $D = \varepsilon^2 \cdot D_0$ with $D_0 = 1$, $\lambda =$

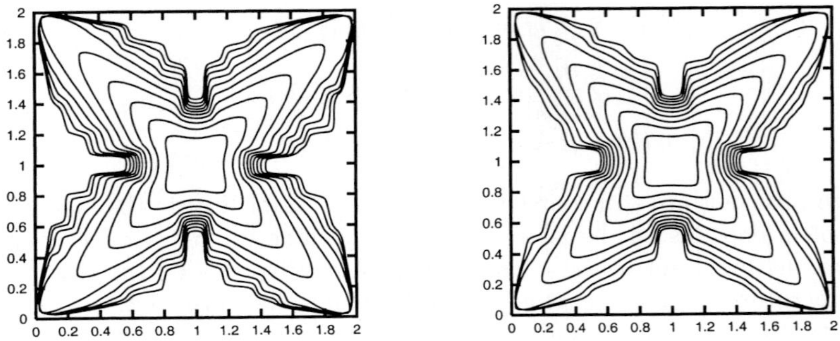

Fig. 3. Evolution of left and right crystal for latent heat $L = 10$

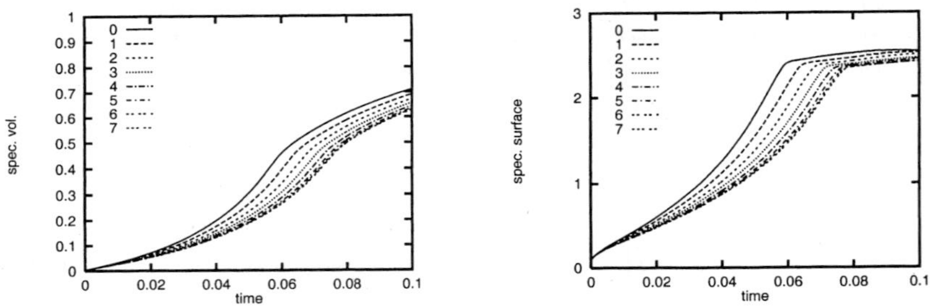

Fig. 4. Evolution of specific data for latent heat $L = 10$

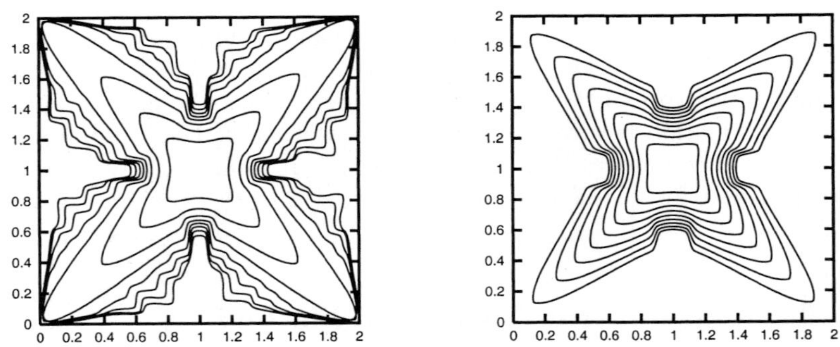

Fig. 5. Evolution of left and right crystal for latent heat $L = 5$

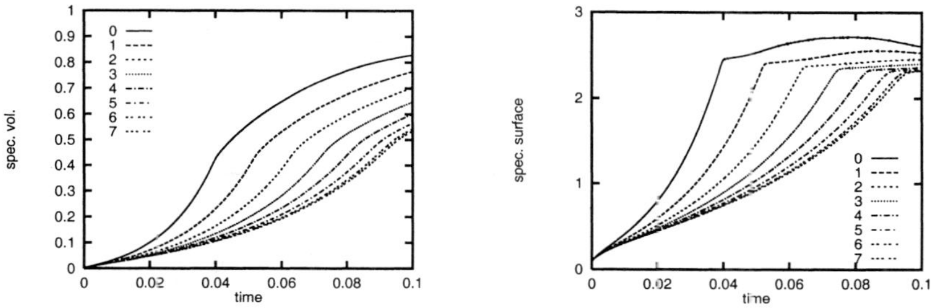

Fig. 6. Evolution of specific data for latent heat $L = 5$

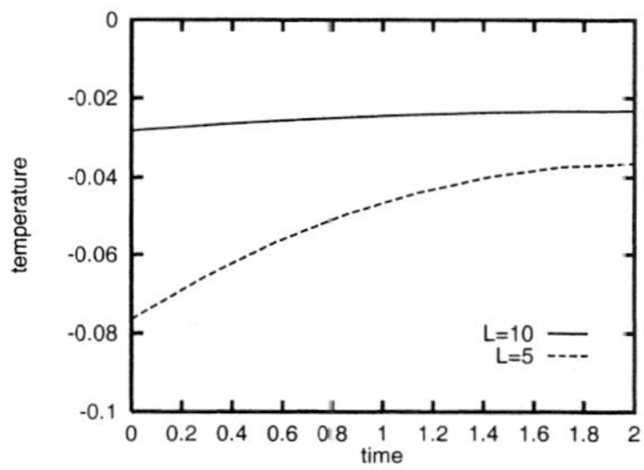

Fig. 7. Macroscopic temperature for latent heat $L = 10$ and $L = 5$ at time $t = 0.1$

0.5, anisotropic density of surface energy $\sigma(\varphi_n) = \varepsilon\sigma_0\big(1 - \frac{4}{45}\sin(4\varphi_n) - \frac{1}{189}\sin(8\varphi_n)\big)$ with $\sigma_0 = 0.002$ and φ_n denoting the angle between the normal vector of the interface and the x–axis, initial temperature $T^{(0)} = 0$ and initial chemical potential $\mu_s^{(0)} = 0.5$. The domain is cooled by the Neumann boundary condition $K\frac{\partial T}{\partial n} = -5$ on part of boundary Γ_4, while the remaining boundary is isolated. At initial time $t = 0$ small solid kernels having the shape of a sphere with radius 0.05ε are assumed to have just formed at the grid points of a uniform square grid with a density of $4/\varepsilon^2$. This gives a local cell problem defined on a square with edge length 2. The discretization of the macroscopic domain is done by a grid with 8×3 elements, that of the microscopic problem by a grid having 400×400 elements. The problem is essentially one–dimensional; in the graphs below we illustrate the properties of an array of 8 crystals in x–direction. In order to show the influence of macroscopic heat transport on the evolution of the solid crystals, we use two different values $L = 10$ and $L = 5$ for the latent heat. In both cases, the function $b(T, \mu)$ is defined as $b(T, \mu) = LT - \lambda\mu$.

Fig. 3 shows the evolution of the first crystal (that one at the cooled boundary Γ_4) and the last crystal (that one at boundary part Γ_2) for latent heat $L = 10$. In Fig. 4 the evolution of the macroscopic specific volume and specific surface is depicted. Fig. 5 and 6 show the corresponding graphs for latent heat $L = 5$. In this case, we observe a much bigger dependence of the microscopic data on the location of the crystal. The reason can be seen in the graph of the temperature distribution at a fixed time $t = 0.1$ in Fig. 7. For large value of the latent heat, the growth of the first crystal releases enough latent heat to almost compensate the cooling condition on the boundary, this leads to an almost uniform temperature field, and the main driving force of the crystal growth is the initial undercooling, which is essentially the same for all crystals. For small latent heat, the growth of the first crystals is limited by solute diffusion, hence they cannot release enough latent heat to consume up all the boundary cooling and the variation of the temperature field as well as that of the crystals is much bigger.

6 Conclusion

We have presented a two–scale model for liquid–solid phase transitions with equiaxed dendritic microstructure, consisting of a global homogenized heat equation and of local cell problems for the evolution of single crystals. The model is applicable to systems with crystal growth governed mainly by solute diffusion. Due to the necessity to solve the local cell problems, the model is still computationally expensive. Nevertheless the two–scale model is superior to the original model, if the mesh size for the global problem is not too small, in particular not much smaller than the size of the single crystals. A further increase of computational efficiency can be perhaps achieved by reducing the number of cell problems to be solved, e.g. by re–using data of already computed crystals with the help of an interpolatory data bank.

References

1. Almgren, R. (1993) Variational algorithms and pattern formation in dendritic solidification. J. Comput. Physics **106**, 337–354
2. Eck, Ch., Knabner, P., Korotov, S. (2001) A two–scale method for the computation of solid–liquid phase transitions with dendritic microstructure. Preprint, Institute for Applied Mathematics, University of Erlangen–Nürnberg
3. Escher, J., Prüss, J., Simonett, G. (1999) Analytical solutions for a Stefan problem with Gibbs–Thomson correction. Preprint No. 13/99, Fachbereich Mathematik/Informatik, Universität Gesamthochschule Kassel
4. Luckhaus, S. (1993) Solidification of alloys and the Gibbs–Thomson law. Preprint, University of Bonn
5. Visintin, A. (1996) Models of Phase Transitions. Birkhäuser, Boston

Application of Higher Order BDF Discretization of the Boussinesq Equation and the Heat Transport Equation

G. Bärwolff

TU Berlin, Inst. f. Mathematik, Sekr. MA 6-3, Str. d. 17. Juni 136, 10623 Berlin, Germany

Abstract. During the growth of crystals there were observed crystal defects under some conditions of the growth device. As a result of experiments a transition from the twodimensional flow regime of a crystal melt in axisymmetric zone melting devices to an unsteady threedimensional behavior of the velocity and temperature field was found. This behavior leads to striations as undesirable crystal defects. For the investigation of this symmetry break a mathematical model of the crystal melt was formulated for

i) the theoretical description of the experimentally observed behavior and
ii) the identification of critical parameters of the growth device. i.e. the evaluation of bifurcation points.

To describe and to avoid such a behavior it is necessary to solve the unsteady three-dimensional Boussinesq equation coupled with the heat transport equation efficiently.
To improve first and second Euler, leapfrog and Adams-Bashforth methods higher order explicit and BDF (Backward Differenciation Formulas) methods are applied und constructed for time dependend calculations and a Newton method is discussed for the resulting nonlinear equation systems for implicit integration methods and the steady state solution.

1 Mathematical Model

The crystal melt is described by the Navier-Stokes equation for an incompressible fluid using the Boussinesq approximation coupled with the heat transport equation. Heat conductivity and viscosity depend on the temperature slightly. Thus, we have the governing dimensionless equations in polar-cylindrical coordinates

$$
\begin{aligned}
u_t + (ruu)_r/r + (uv)_\varphi/r + (wu)_z - v^2/r &= -p_r + \nabla\cdot(2\mu \boldsymbol{S_r}) - 2\mu S_{\varphi\varphi}/r\,,\\
v_t + (ruv)_r/r + (vv)_\varphi/r + (wv)_z + uv/r &= -p_\varphi/r + \nabla\cdot(2\mu \boldsymbol{S_\varphi}) + 2\mu S_{r\varphi}/r\,,\\
w_t + (ruw)_r/r + (vw)_\varphi/r + (ww)_z &= -p_z + \nabla\cdot(2\mu \boldsymbol{S_z}) + Gr\theta\,,\\
(ru)_r/r + v_\varphi/r + w_z &= 0\,,\\
\theta_t + (ru\theta)_r/r + (v\theta)_\varphi + (w\theta)_z &= \frac{1}{Pr}(r\theta_r)_r/r + \frac{1}{Pr}(\theta_\varphi)_\varphi/r^2 + \frac{1}{Pr}(\theta_z)_z\,.
\end{aligned}
$$

in the cylindrical melt zone. $\boldsymbol{S}$ denotes the stress tensor, μ is a dimensionless viscosity and u, v, w and p are the primitive variables of the velocity vector and the pressure, θ denotes the dimensionless temperature. The occured material and geometrical parameters are noted in detail in [1].
For the velocity no slip boundary conditions are used. At the interfaces between the solid material and the melt we have for the temperature inhomogenous Dirichlet data, i.e. the melting point temperature. On the heated coat of the ampulla the heating procedure is described by Neumann conditions. Thus, the boundary conditions are of the form

$$\begin{aligned} u = v = w = 0 \quad &\text{on the whole boundary}, \\ -\frac{\partial \theta}{\partial r} = q \quad &\text{for } r = 1, 0 \leq z \leq 2\alpha, \varphi \in (0, 2\pi), \\ \theta = 0, \quad &\text{for } 0 \leq r \leq 1, z = 0,\ z = 2\alpha, \varphi \in (0, 2\pi). \end{aligned}$$

α is aspect ratio of the melting zone (height of the zone divided by the diameter of the ampulla). The initial state is assumed as the neutral position of the crystal melt ($\boldsymbol{u} = (u, v, w) = \boldsymbol{0}$) and a temperature field, which satisfies the non convective heat transport equation with the given temperature boundary conditions.

2 Numerical Solution Method and Mathematical Aspects

From mathematical point of view some bifurcation investigations are under consideration. But it's not possible to analyze the general nonlinear coupled problem described above. Simplifications with respect to the time dependency of the heat conductivity and the viscosity are necessary.
Quantitative bifurcation results of the general problem are only possible using numerical methods.
A three-dimensional finite volume method is used for the spatial discretization of the above described non linear initial boundary value problem. Four staggered grids are used for the three velocity components and the pressure. This guarantees the conservativity of the finite volume scheme and some useful a priori estimations for stability analysis are valid. A detailed overview to this topic is given in [2].
The aspect ratio α is found as the main critical growth parameter of the considered zone melting technique. To get larger sized crystals it is nessecary to have a high resultion of the unsteady velocity and temperature field in space and time in crystal melt zones considering technologies. Because of the axisymmetric geometry of the ampulla and the homogenous heating condition in the swirl direction we have the two major situations:

i) for subcritical geometrical growth parameters a steady state quasi two-dimensional regime of the flow field and the temperature field occurs, and

ii) for supercritical influence parameters the two-dimensional regime is not stable and an unsteady (oscillating or chaotic) three-dimensional behavior of the melt flow has to be considered.

Also in the i) case the above noted unsteady Boussinesq and heat transport equations can be used for the finding of steady state solutions. Without any respect to a time resolution, time marching methods with large timesteps tend to the steady state. Large timesteps without any strong restrictions mean implicit methods. Case ii) requires the time resolution which is necessary to describe the physics of the process, for example the frequency of occuring oscillations.

For the short explanation of the numerical methods we write the above noted momentum, mass and energy balance as

$$\boldsymbol{u}_t + K(\boldsymbol{u}, \boldsymbol{u}) = -\nabla p + A\boldsymbol{u} + \boldsymbol{f}(\theta) \,, \tag{1}$$

$$\nabla \cdot \boldsymbol{u} = 0 \,, \tag{2}$$

$$\vartheta_t + C(\boldsymbol{u}, \theta) = D\theta \tag{3}$$

closed by the discretized boundary conditions. K and A stand for the convective and viscous terms in the momentum balance in the result of the spatial finite volume (FV) discretization. C and D denote the results of the FV discretization of the convective and the heat conduction term in the energy balance. $\boldsymbol{u}$ is the velocity vector and $\boldsymbol{f}(\theta)$ is the body force vector depending on the temperaure.

Because of the small Prandtl number of the crystal material in the heat conduction equation the CFL-condition is not really a restriction for the timesteps. Thus, it is only necessary to handle the heat conduction term $div\, grad\, \theta$ implicitly to avoid remarkable timestep restrictions. The implicit handling of $div\, grad\, \theta$ is realized either by a leapfrog method [5] or fully implicit by an iterative method. These kinds of implicit handling are computationally cheap because of symmetry and positive definiteness of $div\, grad\, \theta$. Together with a semi-implicit time integration a solution method for case ii) problems is given with

$$\frac{\boldsymbol{u}^{n+1} - \boldsymbol{u}}{\tau} + \nabla p^{n+1} = A\boldsymbol{u} - K(\boldsymbol{u}, \boldsymbol{u}) + \boldsymbol{f}(\theta) \,, \tag{4}$$

$$\nabla \cdot \boldsymbol{u}^{n+1} = 0 \,, \tag{5}$$

$$\frac{\theta^{n+1} - \theta}{\tau} + C(\boldsymbol{u}^{n+1}, \theta) = D\theta^{n+1} \,. \tag{6}$$

$\boldsymbol{u}$ and θ are here the velocity and temperature values in the finite volumes at the time level n. The timestep τ must fulfill the CFL-condition and the restriction from the explicit handling of $A\boldsymbol{u}$ in the momentum equation.

The time integration scheme (4)-(6) is a method of first order in time and is straightforward with respect to a parallel implementation. This is a good approach if the numerical stability restrictions give timesteps close to those for the necessary physical time resolution. Besides the first order method for a more exact description of the unsteady behavior up to third order methods are tested. A theoretical analysis of higher order BDF methods was discussed in [6] and we use the ideas of [6] for the numerical experiments. The use of second and third order Adams-Bashforth methods allows us to work stable with timesteps which violate the CFL-condition slightly. Thus, the additional computational work of the third order method considering the timesteps n, n-1 and n-2 for the prediction of the new timestep n+1 can be paid by larger timesteps. At the end we have improved unsteady results without a substantial increasing of computational work. The third order Adams-Bashforth method (see for example [7]) for the momentum-mass equation system is of the form

$$\frac{\boldsymbol{u}^{n+1}-\boldsymbol{u}}{\tau}+\nabla p^{n+1}=\boldsymbol{f}(\theta)+\frac{23}{12}[A\boldsymbol{u}-K(\boldsymbol{u},\boldsymbol{u})]$$

$$-\frac{16}{12}[A\boldsymbol{u}^{n-1}-K(\boldsymbol{u}^{n-1},\boldsymbol{u}^{n-1})]+\frac{5}{12}[A\boldsymbol{u}^{n-2}-K(\boldsymbol{u}^{n-2},\boldsymbol{u}^{n-2})]\,, \qquad (7)$$

$$\nabla\cdot\boldsymbol{u}^{n+1}=0\,. \qquad (8)$$

The reason for the explicit approximation of the body force $\boldsymbol{f}(\theta)$ was the decoupling of the equations (7),(8) and the heat transport equation (6). This handling has no influence of the approximation order of the momentum transport equation for the velocity.
The solution of the equation system (4),(5) we get with Chorins projection method [3] which leads to a prediction and a correction step

$$\frac{\tilde{\boldsymbol{u}}-\boldsymbol{u}}{\tau}=A\boldsymbol{u}-K(\boldsymbol{u},\boldsymbol{u})+\boldsymbol{f}(\theta) \qquad (9)$$

$$\frac{\boldsymbol{u}^{n+1}-\tilde{\boldsymbol{u}}}{\tau}=-\nabla p^{n+1}\,. \qquad (10)$$

The application of the divergence operator $\nabla\cdot$ to (10) results in the pressure Poisson equation

$$-\Delta p^{n+1}=-\frac{1}{\tau}\nabla\cdot\tilde{\boldsymbol{u}}\,. \qquad (11)$$

With the solution p^{n+1} of (11) it is easy to compute $\boldsymbol{u}^{n+1}$ from (10). The solution procedure of the system (7),(8) is quite similar to the method (9),(10). Only the right hand side of (9) has to be reset by those of the equation (7). The implicit handling of the Boussinesq equation to have the possibility of larger timesteps is more complicated. The implicit handling of viscous terms leads to the prediction/correction method of the form

$$\frac{\tilde{\boldsymbol{u}}-\boldsymbol{u}}{\tau}=-K(\boldsymbol{u},\boldsymbol{u})+\boldsymbol{f}(\theta)\,, \qquad (12)$$

$$\frac{\boldsymbol{u}^{n+1}-\tilde{\boldsymbol{u}}}{\tau}-A\boldsymbol{u}^{n+1}=-\nabla p^{n+1}\,, \qquad (13)$$

which means the solution of three Helmholtz equations for the determination of the three velocity components using the calculated pressure p^{n+1} from (11). The pressure p^{n+1} is not unique solvable - only for the pressure gradient we have uniqueness. That means for the direct solution of the pressure Poisson equation an additional condition, for example the fixing of the pressure in a certain finite volume cell. Iterative methods find normally a solution without any additional condition depending on the start iteration.
In the case of higher order multi-step methods the projection methods must be modified only in the right hand side of the prediction equations (9) or (12) respectively. All the above discussed numerical solution methods are characterized by

I) explicit prediction steps and
II) linear equation systems with symmetric positive definite matrices for the numerical solution of Poisson or Helmholtz equations.

The solution methods for positive definite problems are well established and easy to realize. Iterative methods guarantee optimal load balance in parallel implementations and are often prefered to direct methods.
If the convective terms dominate the viscous terms in the momentum equation an implicit handling of $K(\boldsymbol{u}, \boldsymbol{u})$ is nessecary to overcome hard timestep restrictions. The consequences for the numerical work are huge. For a second order Crank-Nicolson method a nonlinear equation system of the form

$$\frac{\boldsymbol{u}^{n+1} - \boldsymbol{u}}{\tau} + \frac{1}{2}[K(\boldsymbol{u}^{n+1}, \boldsymbol{u}^{n+1}) - A\boldsymbol{u}^{n+1}] + \nabla p^{n+1} = \frac{1}{2}[A\boldsymbol{u} - K(\boldsymbol{u}, \boldsymbol{u})] + \boldsymbol{f}(\theta)\,, \tag{14}$$

$$\nabla \cdot \boldsymbol{u}^{n+1} = 0\,, \tag{15}$$

or

$$F(\boldsymbol{u}^{n+1}, p^{n+1}) = \boldsymbol{0} \tag{16}$$

with

$$F(\hat{\boldsymbol{u}}, \hat{p}) := \begin{bmatrix} \frac{1}{\tau}\hat{\boldsymbol{u}} + \frac{1}{2}[K(\hat{\boldsymbol{u}}, \hat{\boldsymbol{u}}) - A\hat{\boldsymbol{u}}] + \nabla\hat{p} - \boldsymbol{r} \\ \nabla \cdot \hat{\boldsymbol{u}} \end{bmatrix}$$

and

$$\boldsymbol{r} = \frac{1}{\tau}\boldsymbol{u} + \frac{1}{2}[A\boldsymbol{u} - K(\boldsymbol{u}, \boldsymbol{u})] + \boldsymbol{f}(\theta)$$

is to solve. With the derivative of F

$$DF(\boldsymbol{u}_o, p_o)\begin{pmatrix} \delta\boldsymbol{u} \\ \delta p \end{pmatrix} = \begin{bmatrix} \frac{1}{\tau}\delta\boldsymbol{u} + \frac{1}{2}[K(\boldsymbol{u}_o, \delta\boldsymbol{u}) + K(\delta\boldsymbol{u}, \boldsymbol{u}_o) - A\delta\boldsymbol{u}] + \nabla\delta p \\ \nabla \cdot \delta\boldsymbol{u} \end{bmatrix}$$

and

$$\begin{pmatrix} \delta\boldsymbol{u} \\ \delta p \end{pmatrix} = \begin{pmatrix} \boldsymbol{u}_n - \boldsymbol{u}_o \\ p_n - p_o \end{pmatrix}$$

we get the Newton linearization for (16)

$$DF(\boldsymbol{u}_o, p_o) \begin{pmatrix} \boldsymbol{u}_n - \boldsymbol{u}_o \\ p_n - p_o \end{pmatrix} = -F(\boldsymbol{u}_o, p_o)$$

or

$$\frac{\boldsymbol{u}_n}{\tau} + \frac{1}{2}[B(\boldsymbol{u}_n, \boldsymbol{u}_o) + K(\boldsymbol{u}_o, \boldsymbol{u}_n) - A\boldsymbol{u}_n] + \nabla p_n = \frac{1}{2}K(\boldsymbol{u}_o, \boldsymbol{u}_o) + \boldsymbol{r}\,, \quad (17)$$

$$\nabla \cdot \boldsymbol{u}_n = 0\,, \quad (18)$$

for the calculation of the new iteration

$$\begin{pmatrix} \boldsymbol{u}_n \\ p_n \end{pmatrix} \quad \text{starting from the old one} \quad \begin{pmatrix} \boldsymbol{u}_o \\ p_o \end{pmatrix}.$$

It is interesting to note that the second order Crank-Nicolson method and the resulting nonlinear system (14),(15) is of the same computational complexity as the first order implicit Euler method. The Newton linearization system (17),(18) has a nonsymmetric coefficient matrix and is hard to solve. For a fine spatial grid resolution the solution of (17),(18) requires a preconditioned GMRES iteration method which means a lot of computational work per one Newton iteration. Starting the Newton iteration with the information of time level n

$$\begin{pmatrix} \boldsymbol{u}_o \\ p_o \end{pmatrix} = \begin{pmatrix} \boldsymbol{u} \\ p \end{pmatrix}$$

one iteration gives with the solution $\begin{pmatrix} \boldsymbol{u}_n \\ p_n \end{pmatrix}$ of (16) an approximation of $\begin{pmatrix} \boldsymbol{u}^{n+1} \\ p^{n+1} \end{pmatrix}$. We make the same experiences as Jiang [8] that the result of one Newton iteration $\begin{pmatrix} \boldsymbol{u}_n \\ p_n \end{pmatrix}$ is really a good approximation of the velocity and pressure field $\begin{pmatrix} \boldsymbol{u}^{n+1} \\ p^{n+1} \end{pmatrix}$ of the next timestep. The good convergence property of Newton methods and the solution at the time level n as a good start iteration are reasons for that. And (17),(18) is in fact a consistent approximation of (14),(15), than we have really

$$\frac{\boldsymbol{u}_n - \boldsymbol{u}}{\tau} + \frac{1}{2}[K(\boldsymbol{u}_n, \boldsymbol{u}) + K(\boldsymbol{u}, \boldsymbol{u}_n) - A\boldsymbol{u}_n] + \nabla p_n = \frac{1}{2}A\boldsymbol{u} + \boldsymbol{f}(\theta)\,, \quad (19)$$

$$\nabla \cdot \boldsymbol{u}_n = 0\,. \quad (20)$$

Despite this sufficiency of one or only a few Newton iterations per timestep the computational amount for the implicit scheme (14),(15) is enormous. Thus, if there is any possibility of an explicit or semi-implicit method application one should do that. For the computation of the following results the semi-implicit (7), (8) and the onestep Newton method (19),(20) were used.

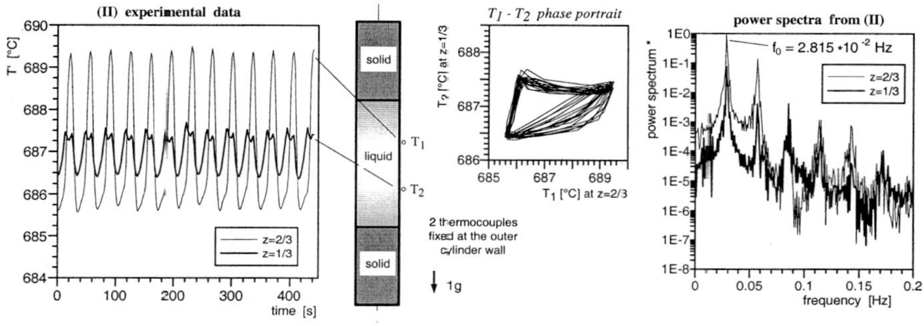

Fig. 1. Time history of the experiment

3 Results and Conclusion

A $(Bi_{1-x}Sb_x)_2Te_3$ crystal melt in an axisymmetric ampulla was investigated. We have the Prandtl number Pr, the Grashof number Gr and the aspect ratio α as parameters in the result of a dimension analysis.
With the numerical simulations using a threedimensional finite volume code critical parameters of the crystal growth configuration and bifurcation points of the time behavior of the velocity and temperatur could be identified and the experimental phenomena could be verified with finding a critical value of $\alpha = \alpha_c = 1.45$ for the problem parameters given in [1]. Thus, it was possible to define parameter intervals for a stable crystal growth configuration.
The above discussed higher order time integration methods were applied and for the given parameters a third order Adams-Bashforth method was to prefer. The numerical simulations were done on a massively parallel T3E computer. Equal parts of the FV grid were mapped to the available processors (max. 256) and the iterative methods were realized by a red-black ordering. The speedup was linear and the load balance was excellent. The finest spatial grid has had $128 \times 128 \times 64$ FV cells. On this grid only semi-implicit higher order methods were tested and a grid independece of the solution could be shown by a comparison to a solution for a $64 \times 64 \times 32$ spatial discretization.

The implicit scheme (14),(15) was tested using the Newton iteration method and the applicability for convective dominated problems was shown with respect to the time behavior for $32 \times 32 \times 16$ FV cells. This was also possible by the onestep Newton method. But compared to the semi-implicit time discretization (9),(10) and the possibility of using pressure velocity iteration methods according to [4] or methods according to [9] the amount for the solution of equations of the form (19)(20) was too large for FV grids of $64 \times 64 \times 32$, $128 \times 128 \times 64$ or more FV cells.
In conclusion there are necessary further investigations for improving methods for the solution of the linear equation systems (17),(18) or (19),(20) with non symmetric matrices which will be the base for an efficient use of the discussed BDF methods.

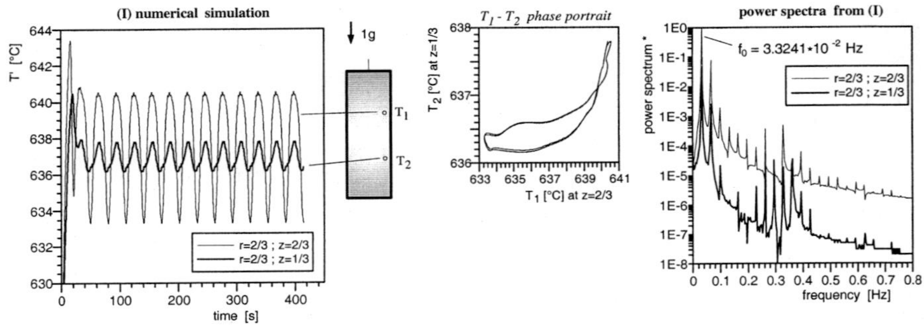

Fig. 2. Time history of the numerical simulation

Figures 1 and 2 show the time history evaluated with the third order Adams-Bashforth method at two monitoring points of the melt and the frequency spectra for an over critical value of α, $\alpha = 3$, both from the experiment and the numerical simulation. The results show a good qualitative agreement of the base frequency of the experiment and the simulation. The differences with respect to the amplitudes come from the fact, that the temperature can only be measured outside the ampulla, near the boundary.
The unsteady numerical results validate the theoretical analysis of [6] by the practical numerical simulation.

References

1. Bärwolff, G., König, F. and Seifert, G., (1997) Thermal buoyancy convection in vertical zone melting configurations, ZAMM **77** (1997) 10, 757-766
2. Bernert, K., (1990) Differenzenverfahren zur Lösung der Navier-Stokes-Gleichungen über orthogonalen Netzen, Wissenschaftliche Schriftenreihe der TU Chemnitz, Heft 10, Chemnitz
3. Chorin, A.J., (1968) Numerical Solution of Navier-Stokes Equation, Mathematics of computation, Vol. 22 p, 745-760
4. Butler, T.D., (1974) Recent Advances in Computational Fluid Dynamics, in: Computing Methods in Applied Sciences and Engineering, Part 2. Lecture Notes in Computer Science No. 11, Springer Berlin, Heidelberg, New York
5. Roache, P.D., (1972) Computational Fluid Dynamics, Hermosa publishers, Albuquerque
6. Emmrich, E. (2000) Error Analysis for Second Order BDF Discretization of the Incompressible Navier-Stokes Problem, Proc. of the 4th Summer Conference on Numerical Modelling in Continuum Mechanics, Prague, August 2000
7. Golub, G.H. and Ortega, J.M., (1992) Scientific Computing and Differential Equations, Academic Press, Inc.
8. Jiang, B., (1998) The Least-Squares Finite Element Method, Springer Berlin, Heidelberg, New York
9. Basu, B., Enger, S., Breuer, M. and Durst, F., (2000) Three-dimensional simulation of flow and thermal field in a Czochralski melt using block-structured finite-volume method, Journ. of Crystal Growth **219**(2000), 123-143

Spectral and Finite Volume Numerical Approximations for Solutal Convection in Melted Alloys

R. Guérin[1], M. El Ganaoui[2], and P. Haldenwang[3]

[1] L2MP, Faculté de St Jêrome, Marseille, France
[2] SPCTS, Université de Limoges, Faculté des Sciences, Limoges, France
[3] Université de Provence, Marseille, France

Abstract. This work concerns a numerical study of steady and time-dependent flow patterns induced by solutal free convection appearing during the directional solidification of binary alloys in $2D$ and cylindrical $3D$ domains. Numerical approximations are based on the spectral chebychev collocation and finite volume methods.

1 Introduction

Solutal convection can lead to segregation phenomena during solidification of alloy [1-4]. In this work, we are concerned with the directional solidification of the alloy $Pb - 30\%Tl$ characterized by the partition ratio at solid/liquid interface $k = 1.1$ [6]. Fundamental problem is governed by the solute equation coupled to Navier-Stokes equations. Solutions corresponding to uniform and refined finite volume grids are compared to spectral solution in terms of accuracy and ability to describe hydrodynamic transition in the melt. Results concerning the interaction of unsteady flow and segregation are illustrated in a $3D$ cylindrical configuration.

2 Model and Approximations

2.1 Conservation Equations

A cartesian domain is considered and the problem is scaled using the solutal length $l_s = D_l/V_0$ and the solutal time $\tau_s = D_l/V_0^2$ where V_0 is the pulling velocity and D_l the coefficient of the solute diffusion [6]. The mass fraction is reduced with $\Delta C = C_\infty(k-1)/k$, C_∞ is the mass fraction in liquid far from the interface. Using reduced variables the conservation equations of mass, momentum and solute mass fraction are :

$$\nabla.u = 0 \tag{1}$$

$$Sc^{-1}(\frac{\partial u}{\partial t} + u.\nabla u) - \Delta u + \nabla P + Ra_s C\widehat{g} = 0 \tag{2}$$

$$\frac{\partial C}{\partial t} + u.\nabla C - \triangle C = 0 \quad (3)$$

with

$$Ra_s = \frac{g\alpha_c C_\infty (k-1)}{k\nu \mathcal{D}_l} l_s^3 \quad (4)$$

α_c is the solutal expansion coefficient, ν is the kinematic viscosity and $\widehat{g}$ is the unit gravitational vector. The alloy $Pb - 30\%Tl$ corresponds to the Schmidt number $Sc = \nu/D_l = 121.5$. The solutal Rayleigh number Ra_s will be the control parameter for studying solutal transitions in the melt.

2.2 Boundary Conditions

For the velocity field, homogeneous Dirichlet conditions are imposed on the whole contour for the horizontal component ($2D$ case) and for the radial and the azimutal component ($3D$ case). The axial component is set opposite to the growth velocity.
For the mass fraction field, Dirichlet conditions are imposed at the top of the crucible ($C = C_\infty$), Neumann conditions on the lateral wall ($\partial_n C = 0$) (∂_n is the normal derivative operator) and the solute conservation condition at the bottom of the domain ($z = 0$):

$$\frac{\partial C}{\partial z} + [(1-k)V_0/D_l]C = 0 \quad (5)$$

this condition is equivalent to the solutal Stefan condition at the interface for solid/liquid phase change.

2.3 Striations Characteristics

Unsteady flow can have an important impact on the alloy composition illustrated by periodic longitudinal inhomogeneities named striations to be quantified using :

$$\widetilde{s_t} = Max_x \widetilde{X}(x) \quad (6)$$

$$\widetilde{X}(x) = \frac{Max_t X(x,t) - Min_t X(x,t)}{\langle X(x) \rangle_t} \quad (7)$$

This definition of the relative degree of longitudinal segregation expresses the fact that we first estimate (in the frame of moving front) $\widetilde{X}(x)$, the relative amplitude of the crystal composition time-fluctuation of the given abscissa x on the solid/liquid interface (x becoming (r, θ) in $3D$ case). Then we select the point where this time-fluctuation is maximum.
In other words, for a computed time-dependent growth situation, we estimate $Max_t X(x,t)$, $Min_t X(x,t)$ and $\langle X(x) \rangle_t$ which are respectively, the maximum, the minimum and the averaged thallium mass fraction in a given longitudinal

line in the crystal rod. We then take for S_t the maximum obtained when considering all longitudinal lines, defined by x in $2D$ and (r, θ) in $3D$ case [11].
For characterizing periodic striations occurring in the solidified crystal temporal and spatial periods are derived (see [11] for more details). The spatial T_s and temporal T_t periods are expressed using the solutal length l_s, the solutal time τ_s and the main frequency of the flow f:

$$T_s = l_s / f \tag{8}$$

$$T_t = \tau_s / f \tag{9}$$

The relation between T_s and T_t gives:

$$T_s / T_t = l_s / \tau_s = V_0 \tag{10}$$

2.4 Numerical Methods

Spatial discretization is carried out with two methods: in the first, discretization is performed in the framework of Chebychev collocation methods (on Gauss-Lobatto nodes in each direction). Reminding the reader that Chebychev collocation methods give an approximation of infinite order, we consider a 45×45 mesh to be greatly sufficient for insuring the accuracy of the results. A semi-implicit second-order time scheme is used.
For the finite volume method [7, 8], discretization is performed using a refined grid of 40×105 nodes. In order to describe better the boundary layer where are localized strong gradients of velocity and solute, the space step is refined progressively using ($p_{i+1} = p_i / q$, $q > 1$ and p_i is the ith space step). A second order Euler scheme is used for time discretisation.

3 Results and Discussion

3.1 $2D$ Case

First the dependency of the numerical solution on grid size and on time step for the finite volume discretization is compared to the solution obtained with the spectral discretization. Comparisons are achieved in terms of transition thresholds in the melt field and confirmed by comparing values of selected variables (the maximum intensity of stream-function and concentration at the solid/liquid interface). Table 1 illustrates this comparison for $Ra_s = 35$. It is observed that the solution obtained for the refined finite volume grid tend to the spectral one. Grid corresponding to $q = 1.07$ seems to ensure good compromise when considering time step simultaneously.

To evaluate the theoretical threshold of solutal convection, linear stability analysis has been performed by Guérin et al [5] for a large range of A_x ($A_x = l/l_s$, l is the domain length), considering a semi-infinite medium in the vertical direction and for alloy systems with $k = 0.3$ and $k = 1.1$. Two general features are exhibited due to lateral confinement A_x: a stabilizing effect on convection and a succession of odd and even convective critical modes. For the value $A_x = 9$ considered here, the analysis predicts a competition between the even and odd modes of convection. However, for $A_x < 10$, mode 1 corresponds to a mono-cellular asymmetrical and unstable mode. Mc Fadden et al [9] have shown the subcritical nature of the bifurcation node-coll for convective mono-cellular configurations. This bifurcation was then characterized by Impey et al using a continuation method [10]. The bifurcation has always a bounded amplitude, but the change of stability occurs in a point noted Ra_s^1 which is lower than the threshold of the first instability apparition Ra_s^2.
For the present simulation the control parameter is the solutal Rayleigh number, the linear analysis gives $Ra_s^1 = 20.38$ and $Ra_s^2 = 15.25$. We have considered a range of $Ra_s \in [0, 100]$. Starting with the diffusive solution $Ra_s = 0$, the other ones are obtained successively for higher value of Ra_s (see Table 2).

Table 1. Solution for four kinds of finite volume grids and a spectral grid for $Ra_s = 35$. Maximum of stream function ψ_{max} and minimum of interface mass fraction $C_{sl\,min}$ are compared

Method	Grid type	Time step	ψ_{max}	$C_{sl\,min}$
Finite volume	refined $q = 1.07$	5×10^{-2}	13.632	9.752
Finite volume	refined $q = 1.03$	5×10^{-3}	13.624	9.718
Finite volume	refined $q = 1.01$	4×10^{-4}	13.618	9.693
Finite volume	uniform $q = 1$ (40×105)	10^{-4}	14.112	9.906
Spectral	Collocation Chebychev (45×45)	10^{-4}	13.619	9.696

Table 2. Comparison of finite volume and spectral results : transitions for $Ra_s \in [10, 60]$; $m = 1$ for one cellular flow, and $m = 2$ for two cellular flow. For steady solution the maxima of the stream-function ψ_{max} are compared. For the periodic regime the frequencies f are compared

Ra_s	Finite volume	Spectral
10	Steady diffusive $\psi_{max} = 3.93 \times 10^{-3}$	Steady diffusive $\psi_{max} = 1 \times 10^{-7}$
20	Steady $m = 1$; $\psi_{max} = 8.58$	Steady $m = 1$; $\psi_{max} = 8.91$
35	Steady $m = 1$; $\psi_{max} = 13.63$	Steady $m = 1$; $\psi_{max} = 13.619$
41	Periodic $m = 2$; $f = 0.105$	Periodic $m = 2$; $f = 0.110$
60	Periodic $m = 2$; $f = 0.177$	Periodic $m = 2$; $f = 0.192$

3.2 Unsteadiness and Striations in 3*D* Cylindrical Case

Using the spectral method, we have studied the unsteadiness and striations occurring in 3*D* cylindrical configuration.
First we limit our considerations to the time-dependent 3*D* pattern of 'low symmetry'. The presented example corresponds to a periodic solution obtained for the following parameters: δ, the rod diameter reduced with the solutal length is 10 and $Ra_s = 100$. This unsteady solution is the asymptotic one resulting from time-integration starting with a steady pattern obtained for $Ra_s = 80$, as initial condition. The latter solution was characterized by one cell and can easily be related to the Fourier mode $m = 1$. Let us now to show that the mean pattern of the periodic solution - obtained after transition to unsteadiness - remains of the $m = 1$ type.

Consider first the mean fields (averaged over a period of the periodic regime). We illustrate the resulting pattern in Fig. 1 where the contour levels of $< uz >$, the mean axial velocity component, are drawn in two horizontal planes, $z = 2.43$ for Fig. 1 a and $z = 16$ for Fig. 1 (b). In Fig. 2, we report the contour levels of $< C >$, the averaged (reduced) concentration at the liquid-crystal interface ($z = 0$). From Figs. 1 and 2, we easily recognize the signature of pattern of $m = 1$ type corresponding to the mean flow obtained with $\delta = 10$ and $Ra_s = 100$ [11].

We now characterize the time behaviour of this 3*D* one-cell pattern. In Fig. 3 we have plotted the maximum of the vertical velocity uz. We can observe that, although oscillations are of weak amplitude (we are close to the threshold of unsteadiness), the temporal signal is far from the basic sine function. This is confirmed by Fig. 4 where the related frequency spectrum is given. It can be seen that the first and second harmonics are of the same order of magnitude. Next we consider the thickness of the striations induced in the crystal. By applying (4) for $Ra_s = 100$ and for the frequency 0.15 we get for the $Pb - 30\%Tl$ alloy, $l_s = 32[100]^{1/3}\ \mu m = 148\ \mu m$. The spatial striation period is then approximately $T_s = 990\ \mu m$. This corresponds to a length scale which seems to be observable form the experimental point of view. It still remains to estimate the order of magnitude of the striations amplitude. To characterize the influence of the time-dependent solutal convection on inhomogeneities along the vertical direction (i.e. transverse to the interface), we have computed several snapshots of the mass fraction profile in the solid at the interface, $C(r, \theta, z = 0, \tau_i)$. τ_i represents 10 equidistant time steps in the period. Fig. 5 shows the contour map of C and its maximum corresponds to S_t (6) that gives the order of magnitude of the striation amplitude. We obtain here a rather feeble quantity: $S_t = 7.5\ 10^{-4}$ which makes the striations difficult to detect.

4 Conclusion

We have presented a $2D$ numerical study of the free convection in the vertical upward Bridgman solidification of $Pb - 30\%Tl$ alloy by applying two numerical methods : spectral and finite volume approximations. Numerical solutions obtained by the finite volume methods is in good agreement with accurate spectral solution in the case of the time restricted model to fluid phase without phase change effects. The mean step of striations is estimated as a length scale of millimetric type for $3D$ unsteadiness case with low symmetry.

References

1. D. T. J. Hurle, 1977. *In crystal Growth and Materials (ed E. Kaldis & Scheel).* North-Holland 550.
2. M. E. Glicksman, S. R. Coriell and G. B. Mc Fadden 1986. *Ann. Rev. Fluid Mech.* **18**, 307.
3. S. R. Coriell, G. B. Mc Fadden and R. F. Sekerka 1985. *Ann. Rev. Mater. Sci.* **15**, 119.
4. Degenhart, A., Droll, P., El Ganaoui, M., Kadinski, L. et al 2001. In Numerical Simulation of Flows II, L.N. Physics, E.H. Hirschel, Springer Verlag, Berlin, 69-97.
5. R. Z. Guerin, B. Billia, P. Haldenwang 1991. *Phys. Fluids*A, **3**, p. 1873.
6. R. Guérin, C. Le Marec, P. Haldenwang. 1998 J. Crystal Growth **194**, 247.
7. R. Eymard, Th. Gallouet, and R. Herbin. 1999 *Finite Volume Methods*, in Handbook of Numerical Analysis, P. G. Ciarlet and J. L. Lions (Eds), North Holland. Preprint (to appear).
8. M. El Ganaoui, P. Bontoux and O. S. Mazhorova. 1999 Finite volume method for a fluid/solid phase change problem. In Finite Volume Method for Complex Applications II. Eds. R. Vilsmeir, F. Benkhaldoun and D. Hänel, Hermes, p. 559-566.
9. G. B. McFadden, S. R. Coriell and R. P. Boivert 1985. *Phys. Fluids*, **28**, p. 2716.
10. M. D. Impey, P., D. S. Riley, A. A. Wheeler and K. H. Winters 1991 Bifurcation analysis of solutal convection during directional solidification, *Phys. Fluids A*, **3**, p. 535.
11. C. Le Marec, R. Guérin and P. Haldenwang 1996. *Journal of Crystal Growth*, **169**, p. 147.

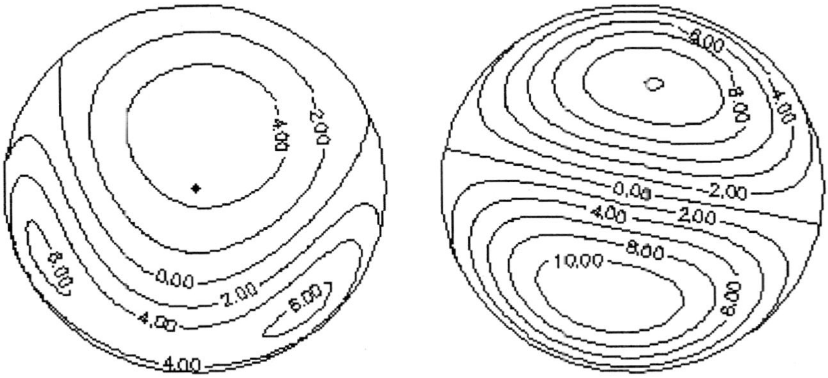

Fig. 1. Contour lines related to the mean axial velocity component in the plane z=2.43 **(a)**. z=16 **(b)**

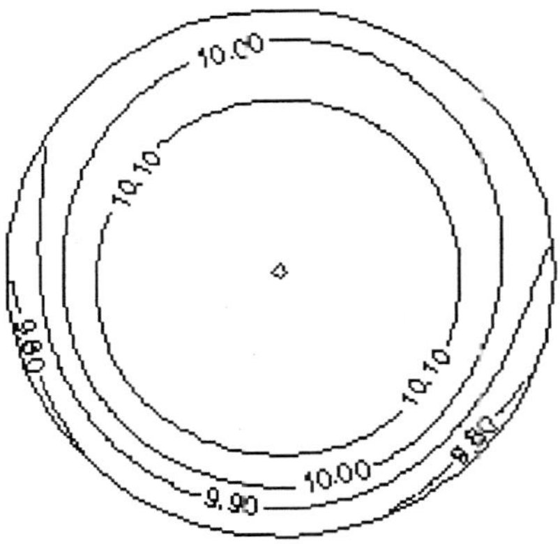

Fig. 2. Contour lines related to the mean thallium composition at the interface (z=0)

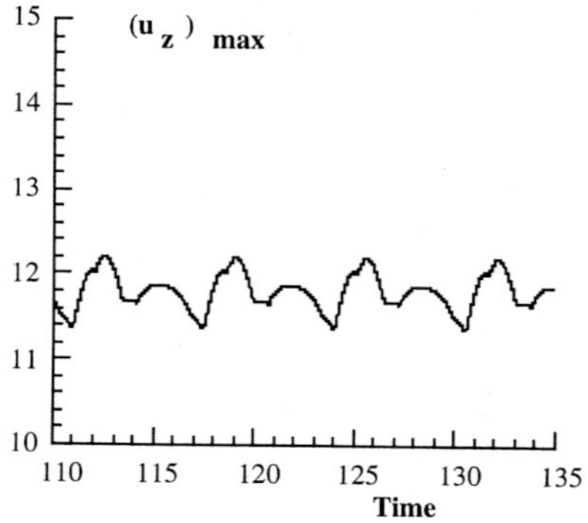

Fig. 3. Variation of the u_z, the maximum of the vertival velocity component versus time for $\delta = 10$ and $Ra_s = 100$

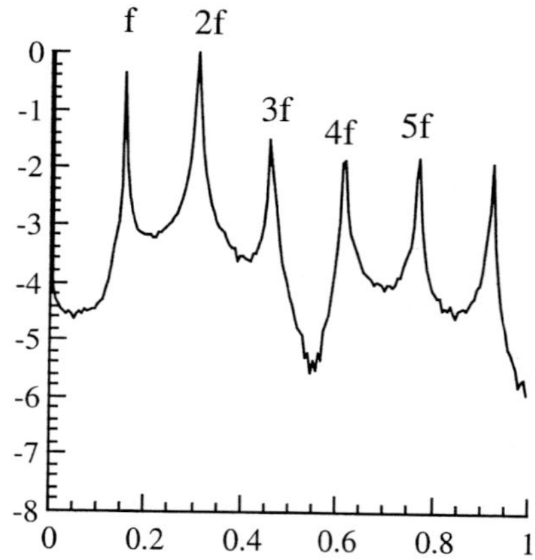

Fig. 4. Spectrum associated with signal of Fig. 3

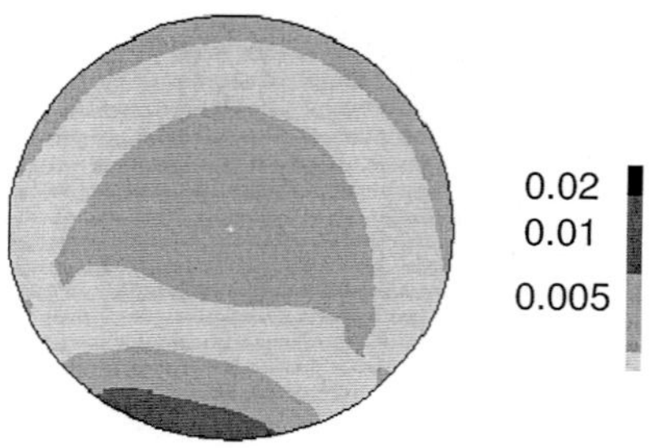

Fig. 5. Contour map of the relative amplitude of striation induced in solid ($\delta = 10$ and $Ra_s = 100$)

Numerical Simulation of Physical Vapour Transport Crystal Growth Processes by a Finite Volume Solution Algorithm

M. Selder, L. Kadinski, and F. Durst

Institute of Fluid Mechanics, University of Erlangen-Nürnberg
Cauerstr. 4, D–91058 Erlangen, Germany
Phone: +49-9131-761244, Fax: +49-9131-761275
E-mail: selder@lstm.uni-erlangen.de

Abstract. A mathematical model for the numerical simulation of physical vapour transport (PVT) crystal growth processes is presented in this paper. The model is based on the two-dimensional conservation equations for mass, momentum, energy and chemical species. Radiative heat transfer and species generation/consumption by heterogeneous chemical reactions are taken into account. The equations are solved by a finite volume algorithm on block-structured grids using the multi-grid technique to speed up convergence. The efficiency of the method is demonstrated. Results on the simulation of the SiC bulk growth process are given, and physical phenomena involved in the growth process are discussed.

1 Introduction

Single crystals with low defect densities are the base material for device production in many fields of semiconductor technology like optoelectronics, telecommunication, information technology, etc. For applications in high power/high temperature electronics and the production of LDs/LEDs emitting in the blue spectral range, silicon carbide (SiC) is a semiconductor possessing a promising combination of relevant material properties [1]. SiC bulk single crystals are usually produced by physical vapour transport (PVT) growth in inductively heated growth reactors [2]. The growth process involves inductive reactor heating, radiative and conductive heat transfer, diffusive and convective transport processes, advective flow, porous powder sublimation, crystallization at the seed crystal, and thermal stress formation [3,4].

A mathematical model of the growth process has to address, at least, all the mentioned phenomena. Due to the coupling of most of these processes, the global simulation of crystal growth is an enormous mathematical task making great demands on computational efficiency and computational power. The need for numerical modeling results from the fact that in the SiC growth reactor experimental devices have only limited applicability. Computer codes are therefore used to further improve reactor design and growth setup, particularly in respect of reducing defect densities and enlarging crystal sizes [5,6].

In this work, a mathematical model of the physical vapour transport SiC bulk crystal growth process is presented. The model is based on the two-dimensional conservation equations for mass, momentum, energy and chemical species and accounts for inductive reactor heating, radiative heat transfer and species generation/consumption by heterogeneous chemical reactions. The equations are solved by a finite volume algorithm on block-structured grids using the multi-grid technique to speed up convergence. The efficiency of the method is demonstrated. Results on the simulation of the SiC bulk growth process are given, and physical phenomena involved in the growth process are discussed.

2 Mathematical Model

A reactor configuration consisting of solid and gas parts is considered. In the solid parts, heat transfer by conduction and heat generation by external heat sources are taken into account. In the gas phase, heat transfer by convection, conduction and radiation and mass transfer by diffusion and convection is allowed for. The mass transfer problem is coupled with the generation/consumption of chemical species by heterogeneous chemical reactions. The gas flow is assumed to be sufficiently slow to use the model of low Mach number flows, i.e. the gas velocity is small compared to the speed of sound.

The mathematical model of the flow used in this work consists of the stationary conservation equations for mass, momentum, energy and chemical species [7]:

$$\nabla \cdot (\rho \mathrm{V}) = 0 \tag{1}$$

$$\nabla \cdot (\rho \mathrm{V}\,\mathrm{V}) = -\nabla \mathrm{p} + \rho\, \mathrm{g} + 2\,\nabla \cdot (\mu\, \dot{\mathrm{S}}) - \frac{2}{3}\nabla\,(\mu\, \nabla \cdot \mathrm{V}) \tag{2}$$

$$\mathrm{c_p}\, \nabla \cdot (\rho \mathrm{V}\,\mathrm{T}) = \nabla \cdot (\lambda\, \nabla \mathrm{T}) \tag{3}$$

$$\rho = \frac{\mathrm{P_0}}{\mathrm{R\,T}} \sum_{\mathrm{i}=0}^{\mathrm{N}} \mathrm{x_i}\, \mathrm{M_i}\ , \tag{4}$$

$$\nabla \cdot (\rho \mathrm{V}\,\omega_{\mathrm{i}}) = \nabla \cdot (\rho\, \mathrm{D_i}\, \nabla \omega_{\mathrm{i}}),\ \mathrm{i} = 1 \ldots \mathrm{N}\ , \tag{5}$$

where V: velocity vector, T: temperature, p: dynamic pressure, ρ: density, μ: dynamic viscosity, c_p: specific heat at constant pressure, λ: thermal conductivity, g: gravitation acceleration vector, $\dot{S}$: deformation rate tensor, P_0: hydrostatic pressure, x_i: molar fraction, ω_i: mass fraction, M_i: molar mass, D_i: diffusion coefficient of component i in the gas mixture, N: number of chemical species diluted in the carrier gas. It should be noted that the density is calculated by summing up the contributions of the diluted species and the carrier gas.

In the solid reactor parts, the basic equation is given by

$$\nabla \cdot (\lambda\, \nabla \mathrm{T}) = \mathrm{Q}\ , \tag{6}$$

where λ is the thermal conductivity of the solid material, and Q is the density of external heat sources.

Considering solid materials as opaque media and the gas mixture as transparent, radiative heat transfer is reduced to heat exchange between all internal surface elements of the reactor. The radiative heat flux between each two surface elements is calculated using a grey-diffuse, view-factor based radiation model. At internal solid/gas boundaries and at internal boundaries between different solid materials, the net heat fluxes at both sides of the interface are balanced [8].

The mathematical model of the mass transfer links the species conservation equations with the calculation of boundary mass fractions at side walls. The boundary mass fractions result from heterogeneous chemical equilibrium reactions between the gaseous species and solid materials [3]. For appropriate chosen chemical reactions, the mass action law equations are formulated,

$$\prod_{i=1}^{N} (x_i^{eq})^{\nu_{ij}} = K_j(T) , \tag{7}$$

where N is again the number of diluted gaseous chemical species, ν_{ij} is the stoichiometric coefficient of component i in reaction j, $K_j(T)$ is the reaction constant, and x_i^{eq} is the equilibrium molar fraction of component i. Using appropriate kinetic models, the molar fractions at the boundary are calculated from the equilibrium molar fractions [3].

Resulting from the chemical reactions, gaseous species are produced/consumed at reactive solid walls. Resulting from the volume change on the evaporation/crystallization processes, the normal component of the boundary velocity is different from zero at these walls [9]. Mass flow due to this velocity is called advective flow, the normal velocity component is given by

$$v_{adv} = -\sum_{i=1}^{N} \frac{D_i \nabla \omega_i}{1 - \sum_{j=1}^{N} \omega_j} , \tag{8}$$

where the carrier gas is excluded from the summations.

The mathematical formulation of thermoelastic stress formation is based on the governing equations of linear elasticity theory. The equilibrium conditions (Cauchy's law) are given in the axi-symmetric case in cylindrical coordinates by [10]

$$\frac{\partial \sigma_{rr}}{\partial r} + \frac{\partial \sigma_{rz}}{\partial z} + \frac{\sigma_{rr} - \sigma_{\varphi\varphi}}{r} = 0 \tag{9}$$

$$\frac{\partial \sigma_{rz}}{\partial r} + \frac{\partial \sigma_{zz}}{\partial z} + \frac{\sigma_{rz}}{r} = 0 , \tag{10}$$

where the elements of the stress matrix $\boldsymbol{\sigma}$ are determined by the strain tensor ϵ via the stress-strain relations. For thermoelastic anisotropic materials, the structure of the stress-strain relations depends on the crystal system [11].

The elements of the strain tensor are given by the displacement vector field u via the displacement-strain relations,

$$\epsilon_{rr} = \frac{\partial u_r}{\partial r}\,, \quad \epsilon_{\varphi\varphi} = \frac{u_r}{r}\,, \quad \epsilon_{zz} = \frac{\partial u_z}{\partial z}\,, \quad \epsilon_{rz} = \frac{1}{2}\left[\frac{\partial u_r}{\partial z} + \frac{\partial u_z}{\partial r}\right]\,. \tag{11}$$

3 Numerical Method

The coupled partial differential equations (1)–(3) and (5) are discretized on non-orthogonal block-structured grids. The block-structuring serves as base for the distinction between gas and solid reactor parts, and allows for the definition of material-dependent properties in the solid parts (e.g. thermal conductivity). The solution procedure is based on a finite volume approach using a colocated arrangement of the variables. Convective and diffusive fluxes at the control-volume faces are approximated by central differences using the deffered correction approach [12]. Sources are calculated using the variable values at each control-volume center as mean values of the whole cell.

Using an iterative solution procedure, non-linearities and the coupling of variables is accounted for through outer iterations. At the beginning of each outer iteration, the physical properties (viscosity, diffusion coefficients, density, etc.) are updated. Afterwards, the momentum equations are solved, and the SIMPLE-algorithm is used to simultaneously fulfill continuity and momentum equations via the coupling of pressure and mass fluxes [13]. Finally, the energy equation and the species equations are solved. In each of these steps, the actual equation is linearized using the respective variable values resulting from the preceding step/iteration. After convergence of the flow equations, the solution of the thermoelastic equations which are decoupled from the flow equations is determined (comp. Fig. 1). All equations are solved by the Strongly Implicit Procedure (SIP) of Stone [14]. Convergence of the procedure is ensured by employing under-relaxation of all variables. To speed up the calculations, a multi-grid algorithm based on the Full Approximation Scheme is used [7].

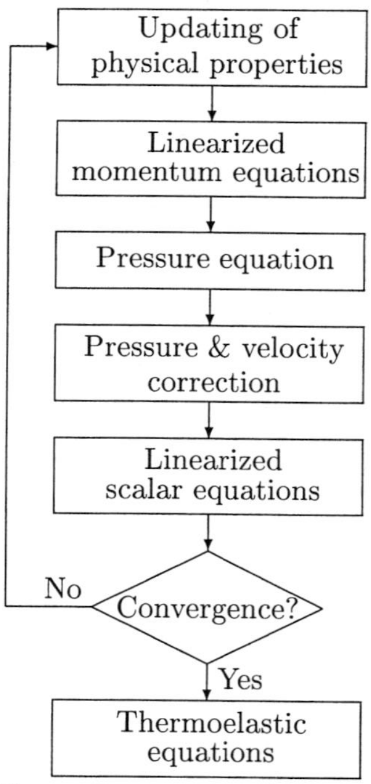

Fig. 1: Iterative solution procedure.

In each outer iteration, the values of temperature and species mass fractions at solid/gas interfaces are updated. The boundary values of species mass fractions are calculated by (7) using a quasi-Newton algorithm in each boundary point. The interface temperatures result from the balancing of radiative, convective and conductive heat flux contributions at both sides of each boundary segment. A Newton algorithm is used to linearize the radiation term in the balance equation. The view-factors used for the calculation of the radiative heat fluxes are calculated by numerical integration using a shadowing algorithm to take account of complex geometries [15].

4 Results

4.1 Numerical Efficiency

To analyze the efficiency of the multi-grid algorithm, calculations were performed for two different types of problem. In the first case, only the temperature equation was solved for a model reactor using four grid levels with the finest grid consisting of 114.688 control-volumes. In the second case, the full equation system (1)–(5) was solved on three grid levels with the finest grid consisting of 28.672 control-volumes. The equations for temperature, velocity and species mass fractions are strongly coupled due to the advective boundary velocity and the dependence of transport properties on temperature and gas phase composition.

The computing times on a SUN Ultra 10 workstation for the two cases using the single-grid (SG) and the multi-grid (MG) approach are given in Fig. 2 and 3, respectively. For each grid level, the calculations were stopped when the residual sums were reduced by eight orders of magnitude. The speedup by the multi-grid technique was 35 for the temperature calculations on four grid levels and 9 for the full equation system on three grid levels, respectively.

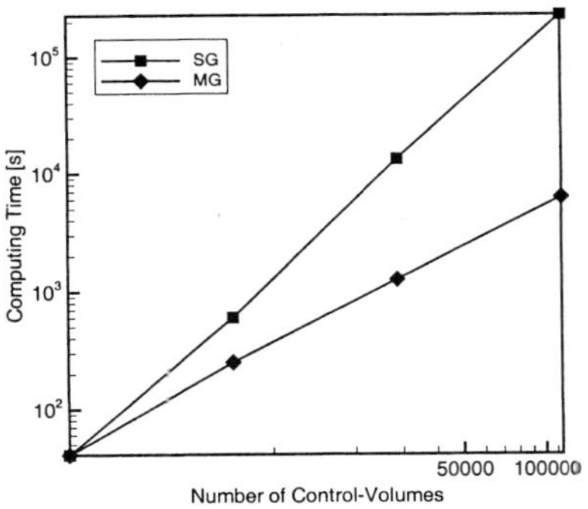

Fig. 2. Computing time for solving the temperature equation using the SG and the MG approach

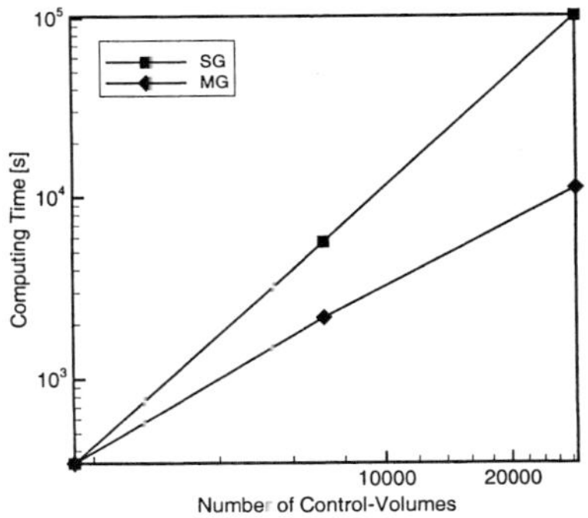

Fig. 3. Computing time for solving the full equation system using the SG and the MG approach

4.2 Simulation Results

The model was used to analyze the SiC bulk crystal growth process in inductively heated axi-symmetric reactors (Fig. 4). The main parts of the reactor are the induction coil, the insulation layer and the graphite growth cell which is partly filled with porous SiC powder. Chemical species (mainly Si, Si_2C and SiC_2) sublime in the powder source and crystallize at the seed crystal.

Using the solution of (1)–(5), the growth rate distribution at the surface of the growing crystal is calculated. Assuming that the macroscopic growth process proceeds significantly slower than the underlying physical and chemical processes, the changes in volume and shape of the growing crystal can be calculated with this growth rate distribution for a sufficiently small time step. The repeated application of this procedure allows the numerical prediction of the temporal crystal evolution (Fig. 5). The convex shape of the growing crystal mainly results from the radial non-uniformity of the growth rate at the beginning of the growth run. Due to the decreasing temperature variations at the crystal surface, nearly uniform growth is predicted for the end of the process. The modeling results have been shown to correspond well to experimental observations [3].

The growth process depends similarly on the growth kinetics at the seed, on material supply from the source, and on material transport through the gas enclosure. To analyze the nature of the transport mechanism, we examined the impact of the advective flow (comp. (8)) on the growth rate for different growth setups (Fig. 6). Neglection of the advective contribution obviously results in underestimation of the growth rates at low pressure levels. It should be mentioned that this effect is of minor importance if the molecular mass of the carrier gas is high, like in the case of Ar.

Temperature inhomogeneities inside the growing crystal induce thermal stress which is considered to be a main reason for a variety of crystal defects

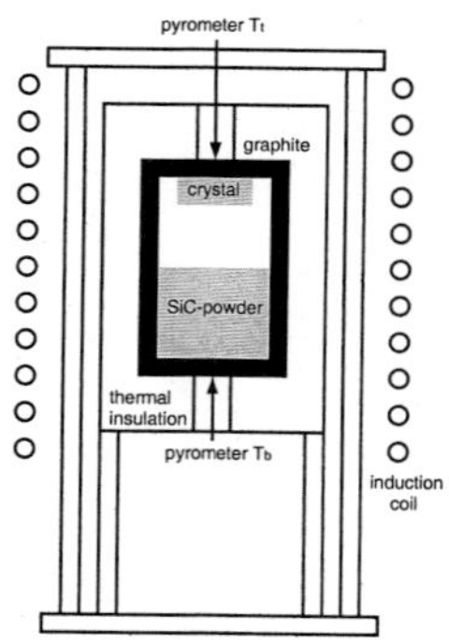

Fig. 4. Schematic model of the growth reactor

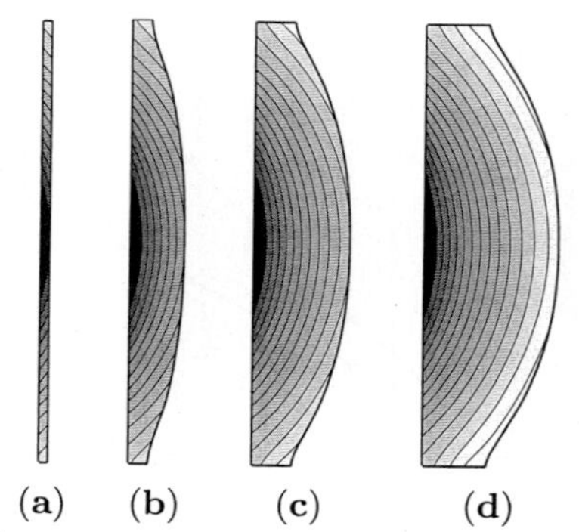

Fig. 5. Calculated shape of the crystal and temperature isolines after (**a**) 0 h, (**b**) 20 h, (**c**) 40 h, (**d**) 60 h

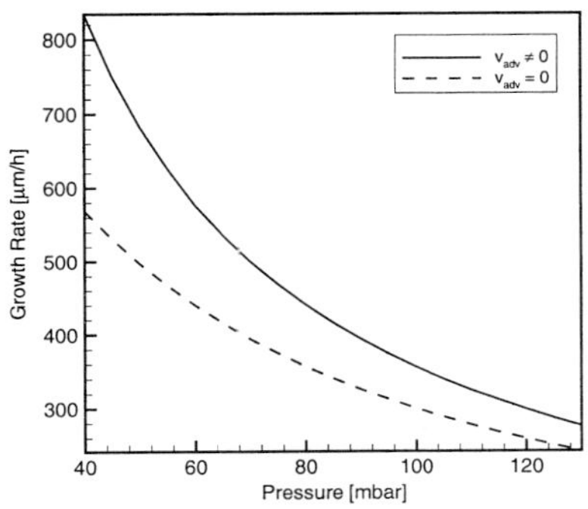

Fig. 6. Advective contribution to crystal growth in dependence on system pressure (carrier gas: helium)

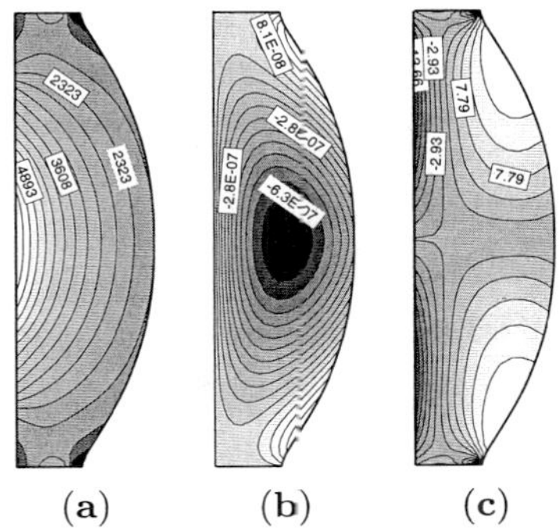

Fig. 7. Calculated distribution of (**a**) axial temperature gradient [K/m], (**b**) axial displacement [m], (**c**) shear stress [MPa]

resulting in serious material degradation. The distributions of axial temperature gradient, axial displacement and shear stress in the crystal at a fixed point in time are given in Fig. 7. Although the displacements are negligible from a macroscopic point of view, stress levels of several MPa are induced. To significantly lower the stress levels in the crystal, reduction of the temperature gradients by geometric reactor modifications is necessary. A detailed analysis of the stress distribution and of the impact of various parameters on the stress levels is given in [16].

5 Conclusion

A mathematical model of the PVT crystal growth process was presented. To solve the governing equations numerically, a finite volume algorithm on block-structured grids has been implemented and was discussed in this paper. The multi-grid technique was shown to be an efficient way to speed up the calculations. Using this technique, heat transfer problems as well as mass transfer problems including Stefan flow have been accelerated by one order of magnitude. The crystal growth process was discussed, and the thermal stress formation in the growing crystal was analyzed.

Acknowledgement

The authors want to thank Prof. M. Schäfer and Ms. I. Teschauer for their help in developing the routines for thermal stress calculations. Financial support by the Bayerische Forschungsstiftung and by the Deutsche Forschungsgemeinschaft is gratefully acknowledged.

References

1. P.A. Ivanov, V.E. Chelnokov: Recent developments in SiC single-crystal electronics. Semicond. Sci. Technol. 7 (1992) 863–880.
2. Y.M. Tairov, V.F. Tsvetkov: Investigation of growth processes of ingots of silicon carbide single crystals. J. Crystal Growth 43(2) (1978) 209.
3. M. Selder, L. Kadinski, Y. Makarov, F. Durst, P. Wellmann, T. Straubinger, D. Hofmann, S. Karpov, M. Ramm: Global numerical simulation of heat and mass transfer for SiC bulk crystal growth by PVT. J. Crystal Growth 211 (2000) 333–338.
4. M. Selder, L. Kadinski, F. Durst, T. Straubinger, P. Wellmann, D. Hofmann: Numerical simulation of thermal stress formation during PVT-growth of SiC bulk crystals. Mat. Sci. Forum 353–356 (2001) 65–68.
5. D. Hofmann, R. Eckstein, M. Kölbl, Y. Makarov, E. Schmitt, A. Winnacker, R. Rupp. R. Stein, J. Völkl: SiC bulk growth by physical vapour transport and its global modeling. J. Crystal Growth 174 (1997) 669–674.
6. E. Schmitt, M. Rasp, A.D. Weber, M. Kölbl, R. Eckstein, L. Kadinski, M. Selder: Defect reduction in sublimation grown silicon carbide crystals by adjustment of thermal boundary conditions. Mat. Sci. Forum 353–356 (2001) 15–20.
7. F. Durst, L. Kadinski, M. Perić, M. Schäfer: Numerical study of transport phenomena in MOCVD reactors using a finite volume multigrid solver. J. Crystal Growth 125 (1992) 612–626.
8. F. Durst, L. Kadinski, M. Schäfer: A multigrid solver for fluid flow and mass transfer coupled with grey-body surface radiation for the numerical simulation of chemical vapor deposition. J. Crystal Growth 146 (1995) 202–208.
9. D.W. Greenwell, B.L. Markham, F. Rosenberger: Numerical maodeling of diffusive physical vapor trnsport in cylindrical ampoules. J. Crystal Growth 51 (1981) 413–425.
10. J.P. Garandet: Thermal stress in vertical gradient freeze furnace. J. Crystal Growth 96 (1989) 680–684.
11. J. P. Hirth, J. Lothe: Theory of Dislocations. Wiley, New York (1982).
12. J.H. Ferziger, M. Perić: Computational Methods for Fluid Dynamics. Springer Verlag, Berlin (1999).
13. S.V. Patanker, D.B. Spalding: A calculation procedure for heat, mass and momentum transfer in three-dimensional parabolic flows. Int. J. Heat and Mass Transfer 15 (1972) 1787.
14. H.L. Stone: Iterative solution of implicit approximations of multidimensional partial differential equations. SIAM J. Numerical Analysis 5 (1968) 530–558.
15. F. Dupret, P. Nicodeme, Y. Ryckmans, P. Wouters: Global modelling of heat transfer in crystal growth furnaces. Int. J. Heat Mass Transfer 33 (1990) 1849–1871.
16. M. Selder, L. Kadinski, F. Durst, D. Hofmann: Global modeling of the SiC sublimation growth process: Prediction of thermoelastic stress and control of growth conditions. J. Crystal Growth 226 (2001) 501–510.

3D Block-Structured Grid Algorithms for the Numerical Simulation of Chemical Vapor Deposition in Horizontal Reactors

L. Kadinski, P. Kaufmann, C. Lindner, and F. Durst

Institute of Fluid Mechanics, University of Erlangen-Nürnberg
Cauerstr. 4, D–91058 Erlangen, Germany
Phone: +49-9131-761272, Fax: +49-9131-761275
E-mail: lev@lstm.uni-erlangen.de

Abstract. In this paper an advanced numerical approach for the simulation of epitaxial growth in metalorganic chemical vapor deposition (MOCVD) reactors is presented. The mathematical model is based on the conservation equations for momentum and heat transfer combined with mass transfer.
The heat conduction includes thermal solid/fluid interactions between the gas and solid parts of the reactor. The radiation heat transfer is coupled with convection and conduction. The radiation heat transfer modelling approach is based on the exchange of radiative heat fluxes between the internal solid surfaces in the computational domain using diffuse view–factors. The calculation of view factors is based on a double contour integral method. A shadowing algorithm is implemented for the calculation of view factors in complex geometries.
The model is implemented in a finite volume numerical solution procedure, incompressible or low Mach number flow on block-structured non-orthogonal grids for three-dimensional laminar flows.
In order to demonstrate the ability of the present method to analyse complex problems, investigations for horizontal MOCVD reactor configurations are presented. The simulated temperature distribution is compared with well-known temperature measurements and good agreement with them is achieved.
The growth of GaAs from trimethyl gallium (TMGa), arsine, and hydrogen was considered where the deposition process is assumed to be in the transport-limited regime. The predicted deposition rates in the reactor fairly well compare with the available experimental results from literature.

1 Mathematical Model

The chemically reacting gas flow with very low Mach numbers (the characteristic velocity of the gas is small compared to the speed of sound estimated using characteristic gas properties and temperature) and large temperature and density gradients is considered.
The mathematical model used in the present study is based on the solution of coupled flow, heat transfer and mass transport equations including multicomponent diffusion and chemical reactions [1].
In our model an enclosure with semi-transparent grey-diffuse radiating walls

and non-participating gas inside and ambient outside are assumed. In this case the energy flux to the inside wall is composed of two contributions: radiative transfer from surfaces facing the wall and conduction by the gas. Energy is lost from the wall to the ambient by radiation and cooling by the surrounding (ambient) gas. Thus, the wall boundary conditions take the form:
Flow–Wall:

$$\mathbf{n}\cdot\Big(-\lambda\nabla T\Big)_g = \mathbf{n}\cdot\Big(-\lambda\nabla T\Big)_w + q_i^{in} - q_o^{in}$$

Wall–Ambient:

$$\mathbf{n}\cdot\Big(-\lambda\nabla T\Big)_w = h(T_w - T_a) - q_i^{out} + q_o^{out} \tag{1}$$

where the indices "'w"' and "'g"' are related to the wall and to the gas, respectively, T_a is the temperature of the ambient, h is the overall heat transfer coefficient, q^{out} and q^{in} are the radiation fluxes to the wall (index i) and from the wall (index o) on the outer and inner side, respectively, and $\mathbf{n}$ is the unit vector orthogonal to the wall and directed outwards. In the case of thin walls, the boundary conditions at the wall take the form:

$$\mathbf{n}\cdot\Big(-\lambda\nabla T\Big) = h(T - T_a) - \underbrace{(q_i^{out} - q_o^{out} + q_i^{in} - q_o^{in})}_{q_{net}}, \tag{2}$$

where T is the temperature of the gas on the inner side of the wall. To close the model equations the radiation fluxes have to be specified. If radiative properties (E–emissivity, R–reflectivity, Tr–transmissivity) do not depend on wavelength, the net radiation flux q_{net} for the k–th surface element becomes:

$$q_{net,k} = E_k(q_{i,k}^{in} - 2\sigma T_k^4 + \sigma T_a^4) \tag{3}$$

where $q_{i,k}^{in}$ is defined by the system of equations providing the heat exchange relationship between the radiating boundaries:

$$q_{i,k}^{in} = \sum_{j=1}^{N} q_{o,j}^{in} F_{kj} \tag{4}$$

$$q_{o,k}^{in} = \sigma E_k T_k^4 + R_k q_{i,k}^{in} + \sigma \mathrm{Tr}_k T_a^4, \tag{5}$$

and σ is the Stefan-Boltzmann constant and F_{kj} are the view factors. For more details concerning the radiation model and equations, see [2,3]

2 Numerical Method

The governing equations are discretized on non-orthogonal grids using a finite volume approach with collocated arrangement of variables. Central differences are used to approximate both convection (in the form of the deffered

correction approach) and diffusion fluxes through control volume (CV) faces. The equations (including the non-linear system for the radiation fluxes) are solved sequentially, and the non-linearities and coupling of variables are accounted for through outer iterations (coefficients updating).
First, the mass transfer equations are solved sequentially by applying of the Strongly Implicit Procedure (SIP) of Stone [4].
After that thermal boundary conditions are determined. The discretized boundary conditions (2) link the energy equation and the non-linear system of heat fluxes (4), (5) and the temperature variables at all boundary nodes are coupled with each other resulting in a dense matrix. The system is not solved directly due to computational costs, but an iterative approach is employed, where heat fluxes are obtained according to (4) and (5) for given temperatures from the previous outer iteration cycle and specified as boundary conditions for the energy equation. This corresponds to a Gauss–Seidl method for solving the radiation exchange matrix.

The solution of the resulting coupled set of algebraic equations for momentum transport is achieved by a pressure correction scheme based on the SIMPLE algorithm in order to satisfy the continuity requirement. Based on these velocity, temperature and concentration fields, the coefficients of the difference equations are updated for the subsequent outer iteration cycle. The procedure is repeated until convergence of the whole system. Underrelaxation is employed to ensure convergence of the procedure. The blockstructuring is used for the treatment of flow problems in complex geometries including blocks with different physical properties as the considered solid/fluid or solid/solid interactions.

The calculation of view factors is based on the contour integral method (see e.g. [5] and the references therein). For the adjoining surfaces the formulation is singular, however, the method can be applied if one line integral is evaluated analytically. In order to reduce the view factor computation costs, the groups of surface elements (macrosurfaces, see e.g. [6]) consisting on 4, 16 etc. elementary surfaces are considered. First, the view factors are calculated between the macrosurfaces. Second, for the radiation exchange ((4),(5)), the radiation outcoming fluxes are summed up and incoming radiation fluxes are distributed between elementary surfaces proportional their areas. In that algorithm the additional assumption for uniform temperature for the macrosurface is not necessary. A shadowing algorithm is implemented for the calculation of view factors in complex geometries by determining a line connecting the centroids of two surfaces.

3 Results

3.1 Model Radiation Configuration

We consider a rectangular cube with the given temperature at the bottom $T = 1000$ K, and for the top side, the boundary conditions (2) are supposed.

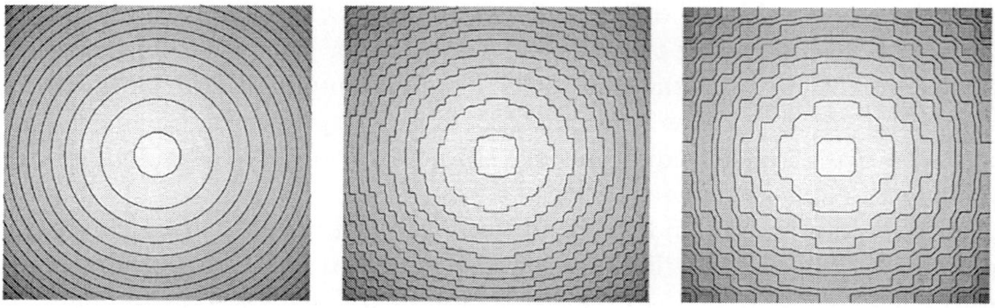

Fig. 1. Top temperature distributions for the cube 32^3 and view factors (VF) calculated using the same grid (left), cube 64^3 and VF 32^3 (middle) and cube 32^3 and VF 16^3 (right)

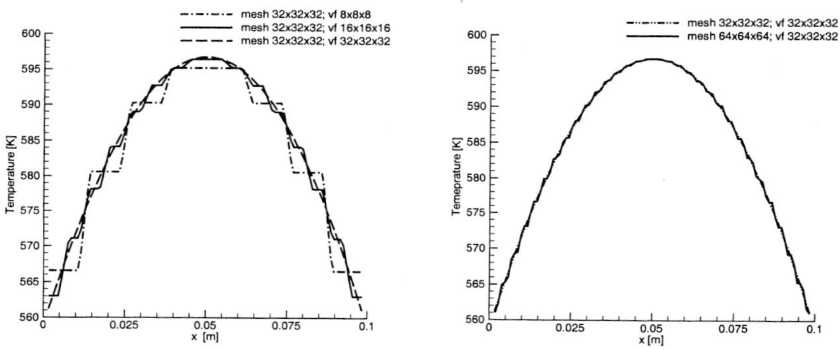

Fig. 2. Top temperature distibutions for the cube 32^3 and different VFs (left) and cubes 64^3 and 32^3 and VF 32^3 (right)

The computations have been done for different uniform grids using different grid subsets for the view factors (VF) computations. The finest grid was 64^3 and the finest grid for the VF was 32^3 because of the computer memory limitations. For the cube 64^3, one has only 262144 control volumes, but 24576 surface elements, and the number of VFs is $24576^2 = 603.979.776$. Therefore, it was not possible to use the local standard workstations with 1 GB RAM. The temperature distribution on the top wall is shown in the Fig. 1. The temperature maximum is precisely in the middle of the wall and the solution is nearly symmetrical. The calculated temperature maximum $T_{calc} = 596.7$ K matches the analytical temperature maximum $T_{analyt} = 596.801$ K very well. The temperature distribution is smooth in case when the same numerical grid is used for the VF and temperature computations as well. In case of using of macrosurfaces the solution is steplike because of the radiation heat flux approximation. Fig. 2 shows the comparison of the temparature with and without using of macrosurface algorithm. The computational results show that the macrosurfaces may be used to reduce VF computational costs and computer memory requirements without loosing the computational accuracy. The detailed analysis of the accuracy issues will be done in the forthcoming paper.

3.2 Simulation Results

In order to demonstrate the ability of the present method to analyse complex problems, investigations for horizontal MOCVD reactor configurations are presented. These problems include very complicated radiative heat transfer, buoyancy-driven flow, combined heat transfer in the reactor walls and the susceptor, as well as the transport of chemically reacting components.

Heat Transfer Computations To test the validity of the radiation heat transfer models employed, calculations are compared with earlier temperature measurements [7] in a horizontal MOCVD reactor. These measurements were used before for two-dimensional models testing purposes [8,2]. The reactor is shown in the Fig. 3 (with one traceline roughly demonstrating the flow in the reactor). Fig. 4 shows the isotherms in the middle cross section and on the side wall. The measured and predicted isotherms agree very well. Without going deeply into transport phenomena which are discussed in detail elsewhere [8], some remarks should be made. First, for this reactor the reasonable agreement has been early achieved by use of two-dimensional modelling [8,2]. On the other hand, two-dimensional modelling obviously is not able to predict the temperature distribution on the upper and side walls (Fig. 4) which may be crucial for so called "parasitic depositions" in the reactor[1]. Second, the radiation transport is predominant mode of heat transfer in the process temperature range and, therefore, has to be included into realistic reactor modelling.

Heat and Mass Transfer Computations To test the validity of the models employed, calculations were compared to results from MOCVD growth experiments described by van de Ven et al. [9]. The reactor is shown in the Fig. 5. For the all process conditions we refer to [10] where the reactor for the first time has been modelled by Kleijn and Hoogendoorn.

Following Kleijn and Hoogendoorn, we have done a study of the influence of the side wall temperatures on the flow and growth rate distribution in the reactor. In contrast to well defined temperature boundary conditions at the substrate (and, in this particular case, on the top wall), the temperature distribution on all other reactor walls is unknown in advance and can not be prescribed by the "standard" boundary conditions. The calculations have been made using both the prescribed boundary conditions (see Fig. 6) and the radiation boundary conditions ((4),(5)).
It can be seen that the flow distributions (Fig. 7) and, consequently, growth rates (Fig. 8) change drastically for altered temperature boundary conditions.

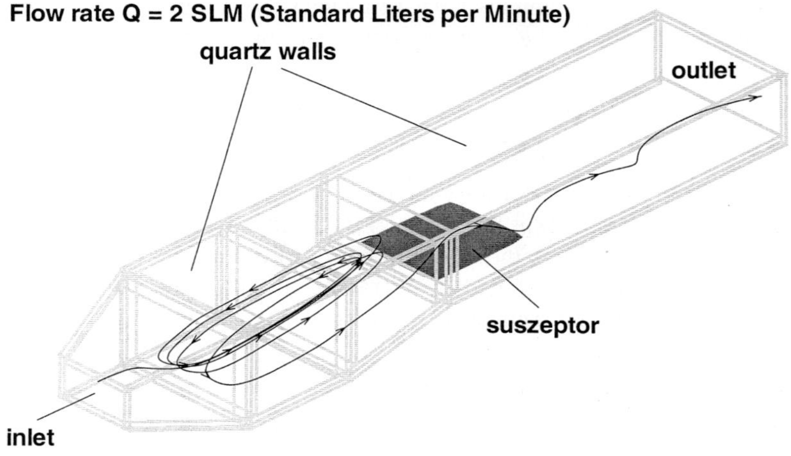

Fig. 3. Reactor geometry and process conditions

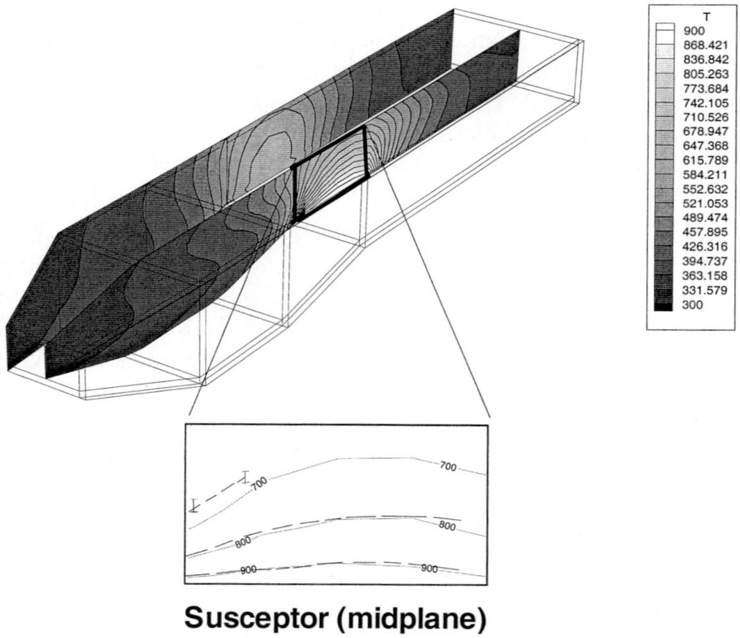

Fig. 4. Measured (dashed) and predicted (solid) isotherms above the susceptor

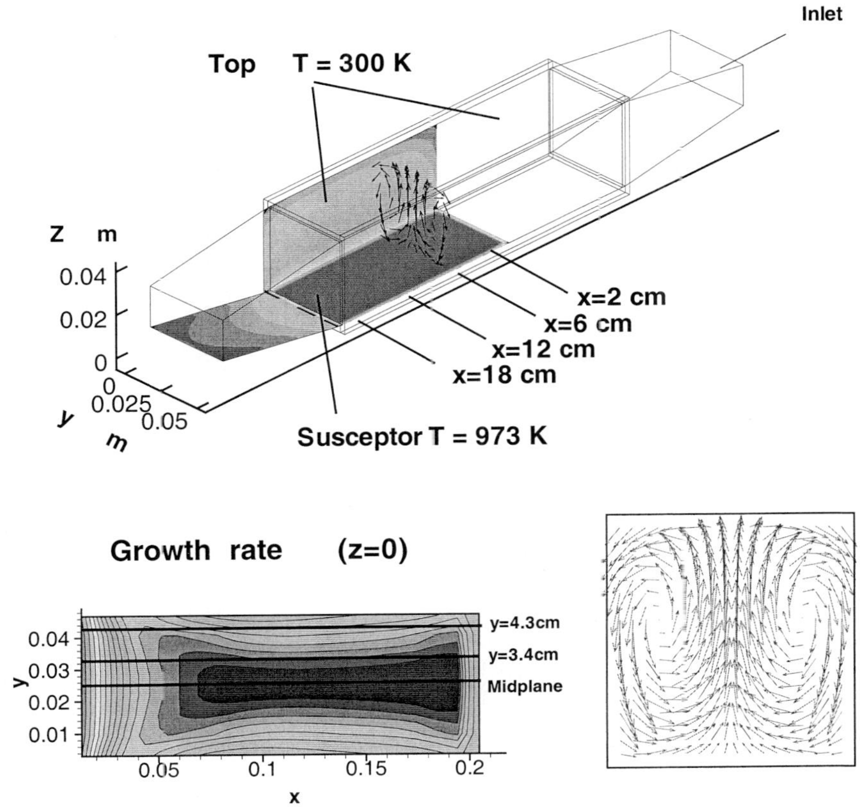

Fig. 5. Reactor geometry and boundary conditions (top), growth rate distribution (bottom left) and a flow field in a reactor cross section (bottom right)

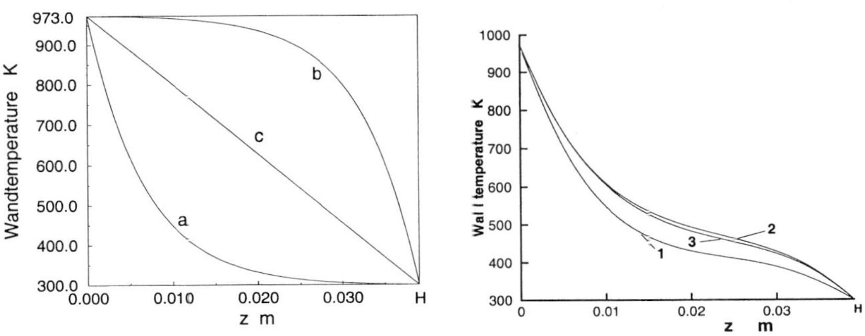

Fig. 6. Assumed (**a**) - cooled side wall, (**b**) - heated side wall (**c**) - linear; (left) and calculated (right) temperature profiles at the sidewalls: x=2 cm (1), x=10 cm (2), x=18 cm (3)

The "correct" calculated temperature distribution (Fig. 6) approximately corresponds to "cooled side wall". In that case the predicted growth rates are in a good agreement with experiments, but, the best agreement is achieved by use of the radiation model.
It should be noted that a simplified radiation model [10] which has been used for this particular geometry is obviously correct and the present results by the detailed radiation model are very similar to Kleijn and Hoogendoorn's ones.

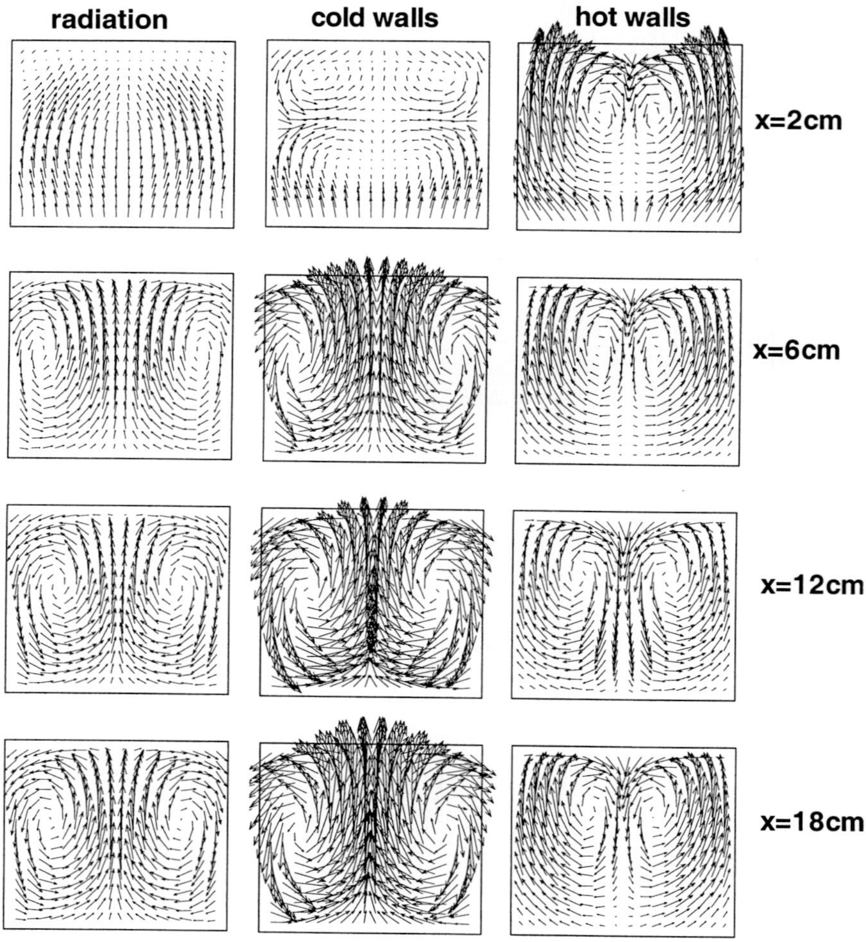

Fig. 7. Transverse velocities in four different y-z cross sections for different temperature boundary conditions

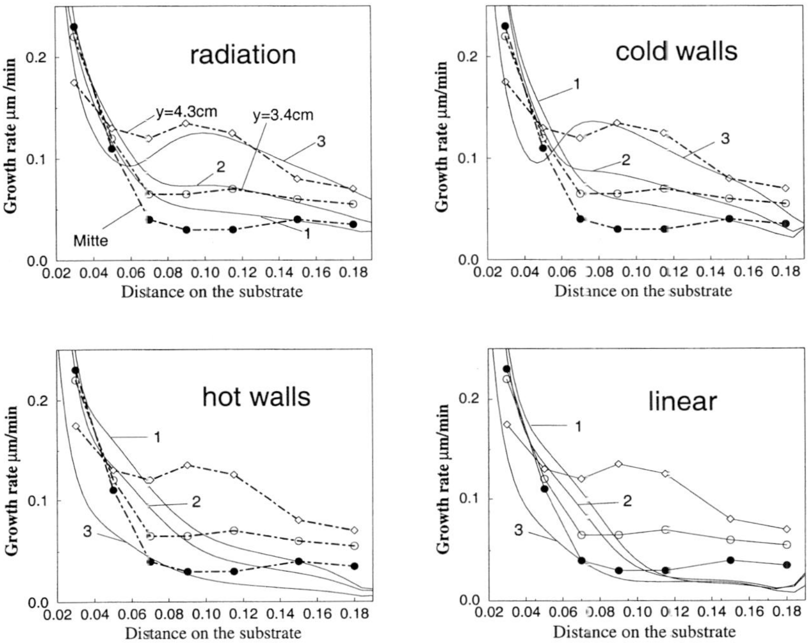

Fig. 8. Experimental data (- - -) midplane (•), y=3.4cm (○) and y=4.3cm (◇). Calculated growth rates (—) y=2.5cm (1), y=3.4cm (2), y=4.3cm (3)

The developed code has been also used for the process optimisation and reactor design in the industrial MOVPE reactors [11,12].

4 Conclusion

In this paper, a 3D mathematical model for grey-diffusive surface radiation combined with the transport equations for mass, momentum and heat was presented. The mathematical model was implemented in a solution procedure for 3D laminar flows. The algorithm for the calculation of view factors provided the needed accuracy for the complex geometries under the reduced memory requirements.
The three–dimensional modeling of coupled flow, heat and mass transfer of for the MOVPE of GaAs in the horizontal reactors has been performed. The predicted temperature distributions and growth rates in the reactors fairly well compare with the available experimental results from literature.

Acknowledgement

Financial support by the Bayerische Forschungsstiftung is gratefully acknowledged. The help of E. Mešić with the test calculations is also acknowledged.

References

1. Durst, F., Kadinski, L., Perić, M., and Schäfer, M. (1992) Numerical study of transport phenomena in MOCVD reactors using a finite volume multigrid solver. J. Crystal Growth, 125:612–626
2. Durst, F., Kadinski, L., and Schäfer, M. (1995) A multigrid solver for fluid flow and mass transfer coupled with grey-body surface radiation for the numerical simulation of chemical vapor deposition. J. Crystal Growth 146:202–208
3. Kadinski, L. and Perić, M. (1996) Numerical study of grey–body surface radiation coupled with fluid flow using a finite volume multigrid solver. Int. J. Num. Meth. Heat Fluid Flow, 6:3–11
4. Stone, H. L. (1968) Iterative solution of implicit approximations of multidimensional partial differential equations. SIAM J. Numerical Analysis, 5:530–558
5. Kuppurao, S., Tantra, I., and Derby, J. J. (1996) Parallel computation of radiation view factors between two arbitrarily oriented surfaces. Comm. Num. Meth. Eng., 12:43–50
6. Engelman, M. and Jamnia, M.-A. (1991) Grey–body surface radiation coupled with conduction and convection for general geometries. Int. J. Num. Meth in Fluids, 13:1029–1053
7. Stock, L. and Richter, W. (1986) Vertical versus horizontal reactor: an optical study of the gas phase in a MOCVD reactor. J. Crystal Growth, 77:144–150
8. Fotiadis, D. F., Boekholt, M., Jensen, K. F., and Richter, W. (1990) Flow and heat transfer in CVD reactors: comparison of Raman temperature measurements and finite element model predictions. J. Crystal Growth, 100:577–599
9. Ven, J. v., Rutten, G. M. J., Raaijmakers, M. J., and Giling, L. J. (1986) Gas phase depletion and flow dynamics in horizontal MOCVD reactors. J. Crystal Growth, 76:352–372
10. Kleijn, C. R. and Hoogendoorn, C. J. (1991) A study of 2– and 3–D transport phenomena in horizontal chemical vapor deposition reactors. Chemical Eng. Sci., 46:321–334
11. Hardtdegen, H., Kaluza, A., v.d.Ahe, M., Grimm, M., Kaufmann, P., and Kadinski, L. (2001) On the influence of gas inlet configuration with respect to homogeneity in a horizontal single wafer MOVPE reactor. J. Crystal Growth, 223:15–20
12. Dauelsberg, M., Hardtdegen, H., Kadinski, L., Kaluza, A., and Kaufmann, P. (2001) Modeling and experimental verification of deposition behavior during AlGaAs growth: a comparison for the carrier gases N_2 and H_2. J. Crystal Growth, 223:21–28

Electromagnetic Control of Electron Beam Evaporation: Numerical Simulation

U. Lüdtke[1] and Ch. Karcher[2]

[1] Department for Electroheat, Ilmenau University of Technology, P.O. Box 100565, D-98684 Ilmenau, Germany
[2] Department for Mechanical Engineering, Ilmenau University of Technology, P.O. Box 100565, D-98684 Ilmenau, Germany

Abstract. During electron beam evaporation of liquid metals, the strong energy input induces strong temperature gradients along the free surface and in the interior of the melt. Thus, the liquid metal is subject to both thermocapillary and natural convection. The vigorous convective motion within the melt leads to highly unwelcome heat losses through the walls of the crucible. The strong convective heat transfer limits the temperature rise in the hot spot and, therefore, the thermodynamic efficiency of the evaporation process. The present paper aims to demonstrate that melt-flow can be effectively controlled by using external magnetic fields in order to considerably reduce the convective heat losses. We employ numerical simulations based on the finite element method to study the effects of a rotating magnetic field on convective heat transfer in a liquid metal heated locally at its free surface.

1 Introduction

Electron beam evaporation of liquid metals (Schiller [1]) is an innovative PVD-technology increasingly used in industrial application to produce high-quality coatings. In this vacuum process a high-energy electron beam bombards the surface of a metal ingot. At the surface, the kinetic energy of the

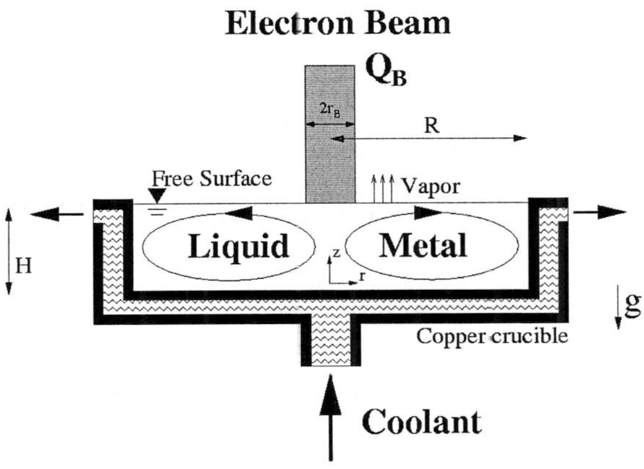

Fig. 1. Sketch of electron beam evaporation process

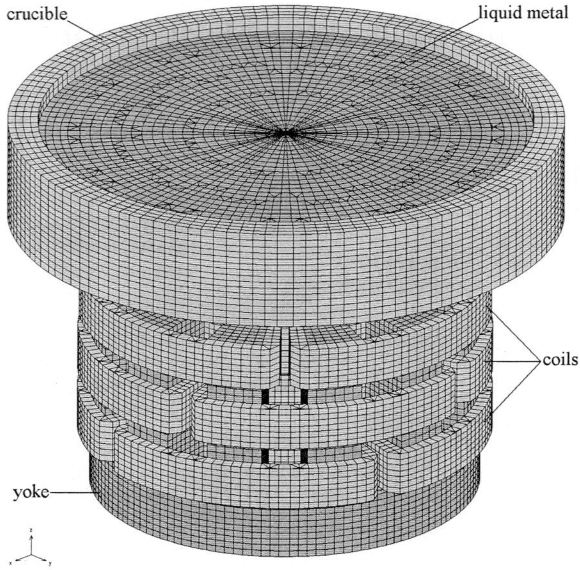

Fig. 2. Finite element mesh (without air) - PROMETHEUS

electrons is transformed into heating power. The ingot melts and forms a free surface. When the temperature of the free surface exceeds the actual (pressure-dependent) saturation temperature of the ambient gas, the liquid starts to vaporize. The rising vapor cloud condenses on a moving substrate located at some distance above. A sketch of the evaporation process is shown in Fig. 1. To guarantee a superior quality of the coating, the melt is typically confined in a water-cooled copper crucible. The intensive cooling prevents chemical reactions between the melt and the crucible walls. By that, crucible materials and their reaction products are practically excluded from evaporation.

A drawback of the process is that only a small portion ($< 3\%$) of the electron beam power is converted into vapor energy. This poor profit results from the fact that the strong energy flux from the electron beam induces strong temperature gradients along the free surface and in the interior of the melt. Hence, the liquid metal is subject to both surface-tension-driven (Davis [2]) and buoyancy-driven (Siggia [3]) convection. The strong convective heat transfer limits the temperature rise at the free surface and therefore the thermodynamic efficiency of the evaporation process. In typical applications these highly unwelcome heat losses amount for up to 70% of the power input.

Our aim is to reduce the conductive heat losses by controlling convection using an applied magnetic field. Here we exploit the physical mechanism that in electrically conducting fluids, interactions between the applied magnetic field and the fluid-flow, generate Lorentz forces (cf. Moreau [4]). This body force may serve to slow down the flow or to superimpose a secondary fluid motion that counteracts the convective transport. In this study we focus on the effect of a rotating magnetic field on surface-tension-driven convection in a liquid locally heated at its upper surface.

Our numerical investigations based on the finite element method consist of two steps. First we calculate the magnetic field lines in the melt by solving the Maxwell equations in quasi-static approximation. As a basic finding we obtain the resulting distribution of the Lorentz forces acting on the liquid metal. In a second step, we solve the Boussinesq equations describing the convective heat transfer within the melt. As an input we use the time-averaged distribution of the Lorentz forces obtained when a rotating magnetic field is applied.

2 Electromagnetic Simulations

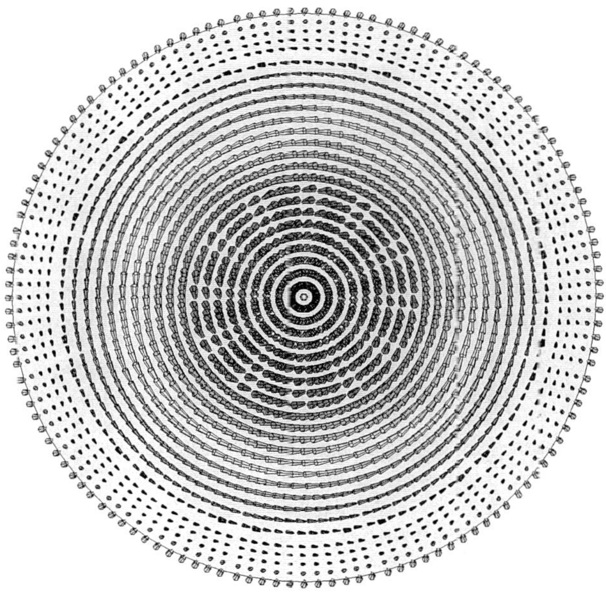

Fig. 3. Vector plot of the time-averaged Lorentz forces on the bottom of the liquid metal. The radial component of the Lorentz force vanishes at $r = 0$ and $r = R$. The maximum is at position $r/R = 0.6$

For the electromagnetic simulations we use the finite-element code PROMETHEUS [5]. This code allows solving the quasi-static three-dimensional Maxwell equations [6] for any geometric arrangement. We neglect the effects of the melt flow on the magnetic field. This is justified for small magnetic Reynolds numbers [4]. Once the magnetic field $\boldsymbol{B}$ and the electric current $\boldsymbol{j}$

are known, the induced time-averaged Lorentz force $\boldsymbol{F}$ can be calculated by evaluating the relation

$$\boldsymbol{F} = Re(\boldsymbol{j} \times \boldsymbol{B}), \tag{1}$$

where Re denotes the real part. Fig. 2 shows the arrangement for the generation of a rotating magnetic field with a part of the computational mesh. The inductor is placed under the crucible and is fed by a three-phase current of frequency $f = 50\text{Hz}$. The magnitude of the applied current has been chosen so that the magnitude of the magnetic field does not exceed $B = 5\text{mT}$ on the upper surface of the liquid metal. This limitation guarantees that the electron beam is compatible with the magnetic field. The inductor works like an asynchronous motor and induces a rotating magnetic field. Fig. 3 shows the Lorentz forces of the liquid metal at the bottom of the crucible. They mainly act in azimuthal direction and tend to stir the melt.

3 Fluid Dynamic Simulations

3.1 Governing Equations

For the numerical simulations of the fluid-flow problem we use a home-made finite-element code that solves the Boussinesq equations [7] in the streamfunction-vorticity formulation in an axisymmetric 2D coordinate system, cf. Fig. 1. Furthermore, we solve an additional transport equation for the angular momentum Γ of the melt-flow. This equation is equipped with an additional term that accounts for the Lorentz forces induced by the rotating magnetic field. Here, the vorticity ω, the streamfunction ψ, and the angular momentum Γ are defined as

$$\omega = \frac{\partial}{\partial z}u - \frac{\partial}{\partial r}w, \quad u = \frac{1}{r}\frac{\partial}{\partial z}\psi, \quad w = -\frac{1}{r}\frac{\partial}{\partial r}\psi, \quad \Gamma = v \cdot r, \tag{2}$$

where (u, v, w) denote the components of the velocity vector. We restrict the study to purely surface-tension-driven convection. We take into account the production of vorticity at the free surface due to the Marangoni effect. For simplicity we neglect the effects of surface deformation and the secondary contributions of thermal radiation and evaporation to the energy balance at the free surface. The crucible walls are assumed to be rigid ($u = v = w = 0$) and perfectly conducting ($T = 0$).

3.2 Parameters

The parameters of the present problem are the Prandtl number Pr, the Marangoni number Ma, and the Hartmann number Ha defined by

$$\text{Pr} = \frac{\nu}{\kappa}, \quad \text{Ma} = \frac{\gamma Q_B}{\rho \upsilon \kappa \lambda}, \quad \text{Ha} = B_0 H \left(\frac{\sigma}{\rho \upsilon}\right)^{1/2}. \tag{3}$$

Here, Q_B is the absorbed electron beam power and γ denotes the temperature coefficient of surface tension. Other fluid properties are the density ρ, the kinematic viscosity ν, the thermal diffusivity κ, the thermal conductivity λ, and the electrical conductivity σ. B_0 denotes the characteristic value of the magnetic induction. The specific geometry of the crucible and the electron beam are taken into account by the aspect ratios (cf. Fig. 1)

$$A_1 = R/H, \quad A_2 = r_B/H. \tag{4}$$

where r_H is the characteristic radius of the heated zone. Finally, the Nusselt number Nu characterizes the effects of the convective heat transfer:

$$\mathrm{Nu} = \Delta T/\Delta T_C. \tag{5}$$

where ΔT denotes the temperature difference in the hot spot and the index C refers to the state of pure heat conduction. When convection sets in we have $\Delta T < \Delta T_C$. Therefore, our goal shall be to minimize Nu.

3.3 Numerical Method

Our finite element method uses linear triangles for spatial representation and the classical backward Euler scheme for time discretization. The non-linear terms are treated explicitly. Starting from an appropriate initial condition, the governing equations are integrated in time until a (quasi) steady state emerges. During the integration the new time step is successively adjusted to satisfy the CFL stability criterion. At each time step the resulting sets of linear algebraic equations are solved iteratively using a Gauss-Seidel method with simultaneous over-relaxation (SOR), Chebyshev acceleration, and odd/even ordering [8]. The calculations were carried out on an equidistant grid of 161×161 elements. It turned out that this grid size is sufficient since the Nusselt number typically changes less than 1% upon doubling the number of elements in each direction.

4 Effect of a Rotating Magnetic Field

When a rotating magnetic field is present the induced time-averaged Lorentz forces mainly point in radial direction, see Fig. 3. They act as a source of angular momentum Γ, tending to stir the melt. This source term S_Γ scales as

$$S_\Gamma \propto \frac{\mathrm{Pr}}{\mathrm{Ma}} \mathrm{Ha}^2 \cdot F_0 \cdot f(r). \tag{6}$$

Here, $f(r)$ is a function of order unity that reflects the radial distribution of the Lorentz force with typical value F_0. As obvious, (6) represents the ratio

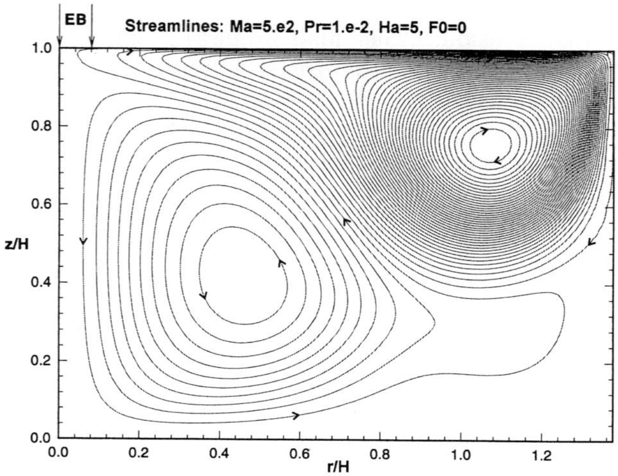

Fig. 4. Streamlines (without electromagnetic field)

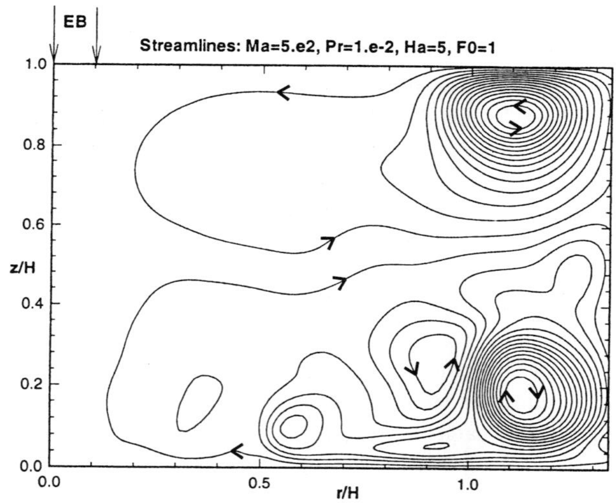

Fig. 5. Streamlines in the presence of a rotating magnetic field

of electromagentic to thermocapillary forces. In turn, the generated rotation acts as a source of vorticity, S_ω. We find

$$S_\omega \propto \frac{\Gamma}{r^3} \frac{\partial \Gamma}{\partial z}. \qquad (7)$$

Hence, vorticity is produced especially in regions where the magnitude of $\partial \Gamma / \partial z$ is large. As obvious, this is the case adjacent to the bottom of the crucible where a so-called Ekman boundary layer [9] is formed.

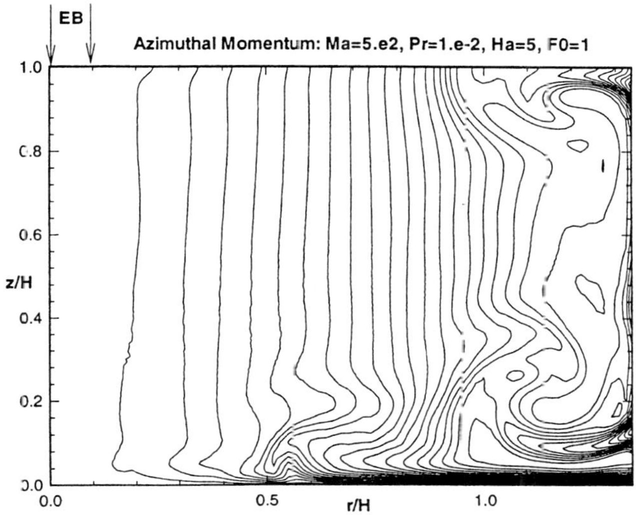

Fig. 6. Angular momentum Γ in the presence of a rotating magnetic field

As an example, we show results of a numerical simulation for the following set of parameters: Ma $= 0.5 \times 10^3$, Pr $= 0.01$, Ha $= 5$, $A_1 = 1.375$, and $A_2 = 0.25$. First, Fig. 4 shows the streamlines when the magnetic field is absent ($F_0 = 0$). The flow field consists of two global convection rolls of opposite circulation. Due to the Marangoni effect, a primary convection roll is induced at the free surface as fluid is pushed from the hot center towards the cold side wall. Here the liquid metal starts to sink, but separates from the wall at about half height. This separation effect creates the secondary convection roll in the lower part of the crucible. Fig. 5 clearly demonstrates that under the action of a rotating magnetic field ($F_0 = 1$) the flow field is completely rearranged. We observe two global convection rolls of opposite circulation which lie one on the top of the other. The centers of circulation are located at radial positions where the induced Lorentz forces are the greatest (cf. Fig. 3). In the Ekman layer adjacent to the bottom, fluid is pushed towards the center. Moreover, we conclude that the magnetic field effectively counteracts the Marangoni effect as the radial flow induced by the Lorentz forces and surface-tension-driven convection, respectively, point in opposite direction. Finally, Fig. 6 shows the distribution of the corresponding angular momentum Γ. We observe that in the bulk of the melt and at the free surface the liquid metal rotates like a rigid body. However, adjacent to the no-slip boundaries at the side wall and the bottom, the induced rotation dies out within thin boundary layers. The actual Nusselt number is Nu $= 1.0012$. This value indicates that convection is clearly suppressed by the rotating magnetic field. However, the actual reduction of the convective heat losses is just about 5%. This moderate reduction is because of the relatively low Marangoni number that is actually investigated in the numerical simulations.

5 Conclusions

We have investigated numerically the effects of a rotating magnetic field on surface-tension- driven convection in a liquid melt heated locally at its free surface. Due to the complexity of the problem we have split the numerical simulations into two steps. At the first, we perform an electromagnetic simulation to obtain the magnetic flux lines and resulting Lorentz forces acting on the melt. The second step is the simulation of the hydrodynamics of the melt affected by the induced Lorentz forces. The results show that the Lorentz forces generate a fluid motion that counteracts the Marangoni convection. However, the actual reduction of the convective heat losses was small. This is due to the relatively small Marangoni number of the flow. To study the effects at much higher Marangoni numbers typical in applications of electron beam technologies, would require incorporating into the simulations a fully three-dimensional model of the fluid dynamics, including turbulence modeling of free surface flows of liquid metals. This is a challenge for future research [10].

Acknowledgement

This work was sponsored by the Deutsche Forschungsgemeinschaft (DFG) within the Forschergruppe Magnetofluiddynamik under grant FOR 421/1-1.

References

1. S. Schiller, U. Heisig, S. Panzer: Electron beam technology, Technik Verlag Berlin (1982).
2. S.H. Davis: Ann. Rev. Fluid Mech. 19 (1987), 403.
3. E.D. Siggia: Ann. Rev. Fluid Mech. 26 (1994), 137.
4. R. Moreau: Magnetohydrodynamics, Kluwer, Dordrecht (1990).
5. U. L"udtke: Ilmenau University of Technology, PhD dissertation (1990).
6. K. J. Binns, P. J. Lawrensen, C. W. Trowbridge: The Analytical and Numerical Solution of Electric and Magnetic Fields, Wiley & Sons, New York (1992).
7. J. Platten, J. Legros: Convection in Liquids, Springer, New York (1984).
8. Press, W. H., Teukolsky, S. A., Vetterling, W. T., Flannery, B. B., 1992, Numerical recipes in Fortran 77, Cambridge Univ. Press, New York.
9. E. J. Hopfinger (Ed.): Rotating Fluids in Geophysical and Industrial Applications, CISM Courses and Lectures No. 329, Springer, New York (1992).
10. A. Thess, D. Schulze: Computational MHD Part I - Fundamentals, in: CISM Courses and Lectures, Springer, New York, (2001).

Part IV

Dynamic Systems and Optimal Control

Solution of a Hard Flight Path Optimization Problem by Different Optimization Codes

K. Chudej[1], Ch. Büskens[1], and T. Graf[2]

[1] Universität Bayreuth, Lehrstuhl für Ingenieurmathematik, 95440 Bayreuth
[2] Universität Bayreuth, Lehrstuhl für Angewandte Mathematik, 95440 Bayreuth

Abstract. Solar electric propulsion is the key technology to reduce propellant consumption for interplanetary missions. A number of studies of interplanetary and lunar missions are currently performed by the European Space Agency (ESA), which exploit the benefits of solar electric propulsion (e.g., [11,8]). Although solar electric propulsion has the disadvantage of low-thrust levels the high specific impulse leads to considerable reduction of propellant mass and therefore to an increase in payload mass. Trajectory optimization problems with solar electric propulsion are known to be extremly difficult (e.g., [3]). They have in the past been successfully solved by indirect methods while direct methods usually failed. Nevertheless the sophistication of direct solution methods has also permanently increased.

The interesting question is: Can low-thrust missions be solved today by direct methods? How precise are these solutions compared with an indirect solution? What time and requirements does it take for a successful solution?

A detailed numerical comparison of the direct solution code NUDOCCCS (Büskens [4]) and the indirect multiple shooting code MUMUS (Hiltmann [7]) is presented for a reference problem (a low thrust orbital transfer problem of a LISA spacecraft with constraints on the solar aspect angle) from [11].

1 Reference Problem – Low Thrust Orbital Transfer

The main objective of LISA (Laser Interferometer Space Antenna) is the detection and study of gravitational waves from sources involving massive black holes throughout the universe. LISA consists of three spacecraft flying 5 million kilometers apart in the shape of an equilateral triangle. The center of the triangle formation will be in the ecliptic plane 1 AU from the Sun and 20 degrees behind the Earth. A single launch vehicle will inject the three spacecraft into an Earth escape trajectory.

The reference problem consists of the optimization of the orbital transfer of a LISA spacecraft powered by a solar-electric propulsion module from a given Earth escape trajectory to its given final orbit [11]. The initial and final orbits will be slightly elliptical and slightly tilted with respect to each other and with respect to the plane of the Earth orbit. The aim of the optimization is to find the thrust control law that minimizes the propellant consumption subject to a constraint on the solar aspect angle.

For the thrust performance the following characteristics are assumed: power 470 [W], thrust level $u_{\max} = 18$ [mN], specific impulse $I_{\mathrm{sp}} = 3000$ [s].

We consider two mathematically equivalent problem formulations: At first we present the equations of motion in equinoctial elements [1,11,6]. These coordinates are adapted to elliptic orbits, which occur for coast arcs: Only the variable L changes on coast arcs. The state vector $\boldsymbol{x}$, which describes the position and orientation of the ellipse in three-dimensional space, is constant on a coast arc, i.e. $\dot{\boldsymbol{x}} = 0$.

The optimal control problem formulated in equinoctial elements uses as state variables the mass m, the vector $\boldsymbol{x} = [a, e_x, e_y, h_x, h_y]^\top$ and the argument of the true longitude L. Control variables are throttle setting δ and thrust angles ε, ζ. See [1,11,6] for a detailed description of the coordinate systems. Thus, the problem statement is

$$J[\delta, \varepsilon, \zeta] = m(t_\mathrm{f}) \to \max \tag{1}$$

subject to

$$\begin{aligned} \dot{\boldsymbol{x}} &= \boldsymbol{f}(\boldsymbol{x}, L)\, \boldsymbol{u}(\delta, \varepsilon, \zeta)/m, \\ \dot{L} &= g_0(\boldsymbol{x}, L) + \boldsymbol{g}_1(\boldsymbol{x}, L)\, \boldsymbol{u}(\delta, \varepsilon, \zeta)/m, \\ \dot{m} &= -u_{\max}\, \delta/(I_{\mathrm{sp}}\, g), \end{aligned} \tag{2}$$

$$0 \le \delta \le 1, \tag{3}$$

$$|\gamma(\boldsymbol{x}, L, \varepsilon, \zeta) - 90°| \le 25°, \tag{4}$$

and initial and final conditions

$$\begin{array}{lllllll} a(0) &=& 153\,233\,650.9\,[\mathrm{km}] &,& a(t_\mathrm{f}) &=& 149\,600\,000\,[\mathrm{km}] \;, \\ e_x(0) &=& -0.029157037 &,& e_x(t_\mathrm{f}) &=& 0.007056101 \;, \\ e_y(0) &=& -0.01386786 &,& e_y(t_\mathrm{f}) &=& 0.00657992 \;, \\ h_x(0) &=& 0 &,& h_x(t_\mathrm{f}) &=& -0.005695774 \;, \\ h_y(0) &=& 0 &,& h_y(t_\mathrm{f}) &=& 0.00610797 \;, \\ L(0) &=& -1.326712031 &,& L(t_\mathrm{f}) &=& 5.739777045 \;, \\ m(0) &=& 430\,[\mathrm{kg}] &,& t_\mathrm{f} &=& 430\,[\mathrm{d}] \;. \end{array} \tag{5}$$

Some abbreviations are used:

$$\mu = 1.327121 \cdot 10^{11}\,[\tfrac{\mathrm{km}^3}{\mathrm{s}^2}], \quad g = 9.81 \cdot 10^{-3}\,[\tfrac{\mathrm{km}}{\mathrm{s}^2}],$$

$$\boldsymbol{f}(\boldsymbol{x}, L) = \sqrt{\frac{a}{\mu}}\frac{A}{D} \begin{bmatrix} \frac{2aBD}{A^2} & 0 & 0 \\ \frac{2(e_x+c)D}{B} & -\frac{2e_xe_yc - s(e_x^2-e_y^2)+2e_y+s}{B} & -e_y(h_xs - h_yc) \\ \frac{2(e_y+s)D}{B} & \frac{2e_xe_ys + c(e_x^2-e_y^2)+2e_x+c}{B} & e_x(h_xs - h_yc) \\ 0 & 0 & \frac{1}{2}(1+h_x^2+h_y^2)c \\ 0 & 0 & \frac{1}{2}(1+h_x^2+h_y^2)s \end{bmatrix},$$

$$g_0(\boldsymbol{x}, L) = \sqrt{\frac{\mu}{a^3}}\frac{D^2}{A^3}, \quad \boldsymbol{g}_1(\boldsymbol{x}, L) = \sqrt{\frac{a}{\mu}}\frac{A}{D}\left[0, 0, (h_xs - h_yc)\right],$$

$$\boldsymbol{u} = \delta\, u_{\max}\, [\cos(\varepsilon)\cos(\zeta), \sin(\varepsilon), \cos(\varepsilon)\sin(\zeta)]^\top,$$

$$A = \sqrt{1 - e_x^2 - e_y^2}, \quad B = \sqrt{1 + 2e_xc + 2e_ys + e_x^2 + e_y^2},$$

$$D = 1 + e_xc + e_ys, \quad c = \cos(L), \quad s = \sin(L).$$

The constraint (4) on the solar aspect angle can be transformed with $\cos\gamma = -\cos(\varepsilon)\cos(\zeta)\sin(\xi) + \sin(\varepsilon)\cos(\xi)$ and $\xi = \xi(\boldsymbol{x}, L)$ into two control constraints

$$h(\boldsymbol{x}, L, \varepsilon, \zeta) = \begin{bmatrix} h_1 \\ h_2 \end{bmatrix} := \begin{bmatrix} \cos 115^\circ - \cos\gamma \\ \cos\gamma - \cos 65^\circ \end{bmatrix} \leq 0. \tag{6}$$

The optimal control problem can be formulated also in Cartesian coordinates: States are the mass m, the position vector $\boldsymbol{s} = [s_\mathrm{x}, s_\mathrm{y}, s_\mathrm{z}]^\top$ and the velocity vector $\boldsymbol{v} = [v_\mathrm{x}, v_\mathrm{y}, v_\mathrm{z}]^\top$. Controls are the throttle setting δ and the thrust angles α and β. A formula for the conversion between α, β and ε, ζ can be found in [6]. Thus, the problem statement is

$$J[\delta, \alpha, \beta] = m(t_\mathrm{f}) \rightarrow \max \tag{7}$$

subject to

$$\begin{aligned} &\dot{s}_\mathrm{x} = v_\mathrm{x}, \; \dot{v}_\mathrm{x} = -\mu\, s_\mathrm{x}/r^3 + \delta\, u_{\max} \cos(\beta)\cos(\alpha)/m, \\ &\dot{s}_\mathrm{y} = v_\mathrm{y}, \; \dot{v}_\mathrm{y} = -\mu\, s_\mathrm{y}/r^3 + \delta\, u_{\max} \cos(\beta)\sin(\alpha)/m, \\ &\dot{s}_\mathrm{z} = v_\mathrm{z}, \; \dot{v}_\mathrm{z} = -\mu\, s_\mathrm{z}/r^3 + \delta\, u_{\max} \sin(\beta)/m, \\ &\dot{m} = -u_{\max}\, \delta/(I_\mathrm{sp}\, g), \; r = \sqrt{s_\mathrm{x}^2 + s_\mathrm{y}^2 + s_\mathrm{z}^2}, \end{aligned} \tag{8}$$

$$0 \leq \delta \leq 1, \; |\gamma(\boldsymbol{s}, \boldsymbol{v}, \alpha, \beta) - 90^\circ| \leq 25^\circ, \tag{9}$$

and given initial conditions for $\boldsymbol{s}(0), \boldsymbol{v}(0), m(0)$ and given final conditions for $\boldsymbol{s}(t_\mathrm{f}), \boldsymbol{v}(t_\mathrm{f})$ and $t_\mathrm{f} = 430\,[\mathrm{d}]$. The boundary conditions are equivalent to (5).

2 Solution Methods

Optimal control problems, such as the above introduced trajectory optimization problems, can be solved by advanced direct or indirect solution methods (see e.g. Büskens [4], von Stryk [10], Chudej [5] and the survey articles of Betts [2] and Pesch [9]).

Direct methods transform the optimal control problem into a nonlinear programming problem (NLP) through discretisation [2,4]. E.g., the control functions can be approximated over a given time grid as a polygon line depending on a finite number of parameters. Then the ODE constraint (2) can be solved with an initial value problem solver or via collocation while the constraints (3)–(4) and the final conditions (5) are considered as additional finite dimensional constraints in the underlying NLP problem. The NLP problem is solved efficiently by SQP methods.

Indirect methods employ necessary optimality conditions (e.g. the maximum principle) in order to derive a (differential-algebraic) multipoint boundary value problem. This substitute problem is then solved, e.g., by sophisticated multiple shooting algorithms [7].

General characteristics of advanced direct methods are the large domain of convergence together with a good approximation of states and controls due to an automatic grid refinement. The advantage of the application of multiple shooting algorithms are the precise solutions, even for complicated optimal control problems.

Both methods allow the application of a number of a posteriori tests to eliminate nonoptimal candidate solutions.

Low thrust trajectory optimization problems are known to be very challenging, and have in the past been solved successfully by indirect methods, see e.g. [3]. The interesting question is now: Can these challenging problems be solved reliably by advanced direct methods, too?

3 Comparison of the Results

Results of a detailed comparison [6] for the reference problem with the direct method NUDOCCCS [4] and the indirect method MUMUS [7] are presented. Note that the controls ε, ζ and α, β are nonunique on coast arcs with $\delta = 0$.

Table 1. Switching times and final mass

		direct method NUDOCCCS		indirect method MUMUS	
		Cartesian coordinates	equinoctial elements	Cartesian coordinates	equinoctial elements
switching times of throttle setting δ					
t_1	[d]	2.50	2.24	2.208	2.208
t_2	[d]	31.35	31.40	31.476	31.476
t_3	[d]	99.40	99.40	99.506	99.506
t_4	[d]	197.10	196.00	195.902	195.902
t_5	[d]	407.60	407.66	408.052	408.052
switching times of the constraint on the solar aspect angle					
t_1^*	[d]	54.965	53.750	54.923	54.923
t_2^*	[d]	253.073	253.073	253.120	253.120
t_3^*	[d]	387.815	387.448	386.589	386.589
final mass					
$m(t_f)$	[kg]	415.077437	415.077429	415.077436	415.077436

The switching structure and the switching times of the bang-bang solution of the throttle setting δ and the activity of the control constraints are presented in Table 1 and Fig. 1. No differences between the two indirect solutions of the analytically equivalent problems appear. The switching times can

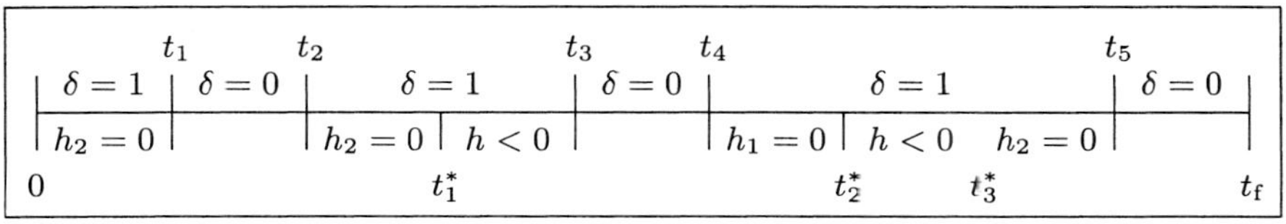

Fig. 1. Switching structure (not to scale)

be determined with a direct method in a first run of the method only up to a precision given by the time grid. Note that the code NUDOCCCS works with an automatic grid refinement algorithm which is linked to the discretisation error of the state ODE. Hence the individual grid refinement for the state ODE (2) and (8) explains the negligible deviations in the switching times of the direct method in Table 1.

These differences can be reduced by a second run of the direct method for a modified optimal control problem which is formulated as a multistage process with a given switching structure.

Figs. 2–3 present the optimal solution of the controls and states for the equinoctial elements and the Cartesian coordinates. The direct method NUDOCCCS and the indirect method MUMUS yield undistinguishable figures.

The Tables 2–3 describe the convergence of the performance index for finer non-equidistant time grids ($N \to \infty$) for the direct solution method NUDOCCCS. N denotes the number of grid points, M the number of optimization variables of the NLP problem, I the number of SQP iterations of the routine E04UCF in the NAG library.

Table 2. Convergence of NUDOCCCS with finer time grid (N) for equinoctial variables

N	M	I	CPU [s]	$m(t_f)$	Rel. Error
25	75	71	5.44	415.066184	$2.7 \cdot 10^{-5}$
46	138	43	14.35	415.075836	$3.9 \cdot 10^{-6}$
52	156	39	16.08	415.075959	$3.6 \cdot 10^{-6}$
65	195	60	38.71	415.076983	$1.1 \cdot 10^{-6}$
100	300	77	134.49	415.077255	$4.4 \cdot 10^{-7}$
122	366	88	255.14	415.077360	$1.8 \cdot 10^{-7}$
208	624	209	2586.00	415.077405	$7.5 \cdot 10^{-8}$
235	705	234	3953.62	415.077429	$1.7 \cdot 10^{-8}$
MUMUS				415.077436	

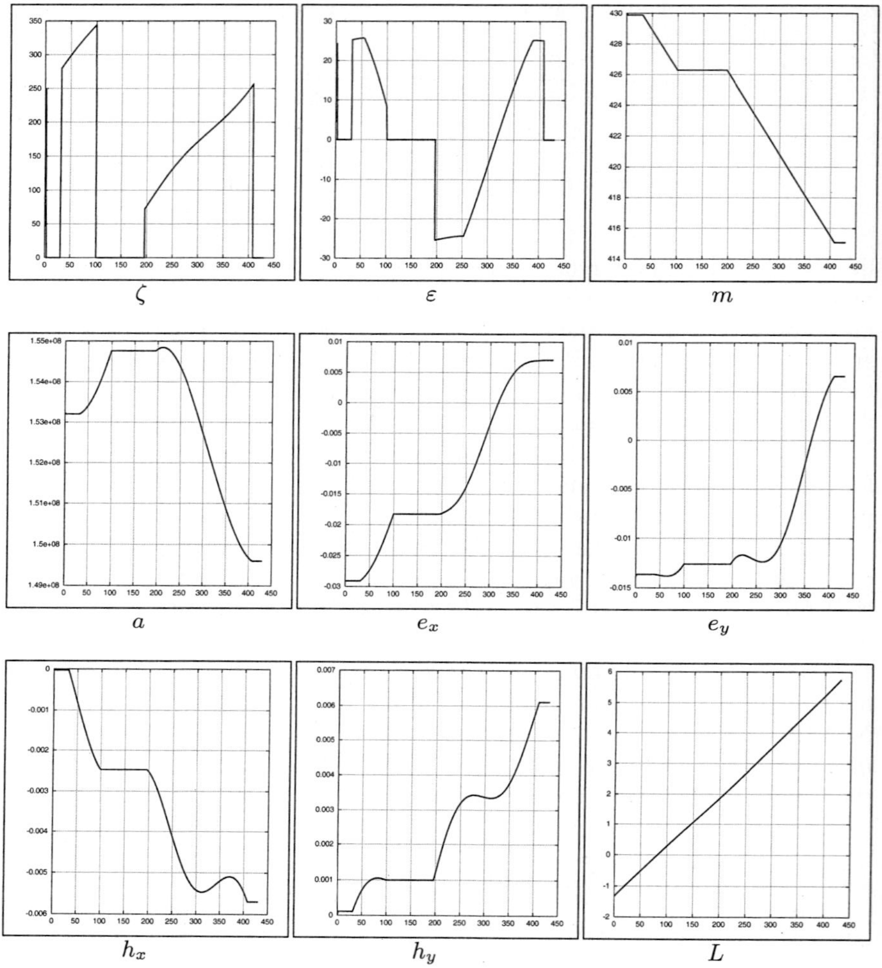

Fig. 2. Optimal solution (equinoctial elements)

Further results, including results for the trajectory optimization problem without constraint on the solar aspect angle, can be found in Vasile, Jehn [11] and Graf [6].

4 Conclusion

Advanced direct solution methods, such as NUDOCCCS, are able to solve reliably low-thrust missions today. Precise values of the performance index and precise solutions of states and controls are computed. This is due to the

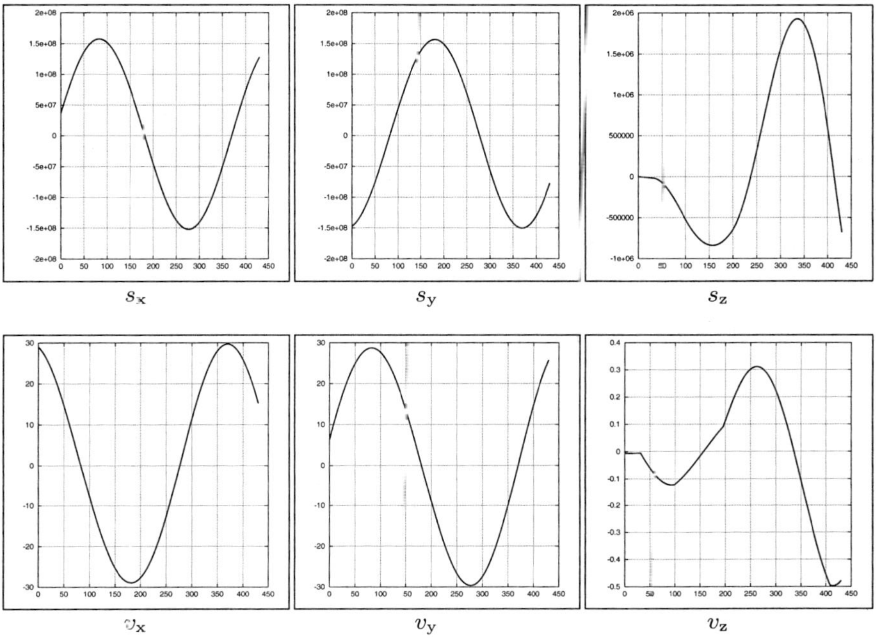

Fig. 3. Optimal solution (Cartesian coordinates)

Table 3. Convergence of NUDOCCCS with finer time grid for Cartesian coordinates

N	M	I	CPU [s]	$m(t_f)$	Rel. Error
25	75	87	4.65	415.050928	$6.4 \cdot 10^{-5}$
54	162	86	25.61	415.076139	$3.3 \cdot 10^{-6}$
108	324	115	186.07	415.077437	$2.2 \cdot 10^{-9}$
MUMUS				415.077436	

adaptive grid refinement. No detailed knowledge of the maximum principle for constrained optimal control problems is required for the user of the code.

A higher precision of the switching times is possible via a second run of the direct solution method. The switching structure is identified in the first run, and the optimal control problem is reformulated as a multistage process. This can be done without knowledge of optimal control theory.

Alternatively an indirect solution method, such as multiple shooting, can be employed in a second solution step. The solution of states and costates of the direct method is used as a good initial estimate for the indirect method.

This hybrid approach [10] ensures in general that the initial estimates needed for multiple shooting are already in its small region of convergence. Multiple shooting yields then a very precise solution of the states, controls and switching times in a negligible CPU time. The drawback of the indirect method is that the user needs a detailed knowledge of the maximum principle for constrained optimal control problems and a certain knowledge of numerical analysis [5]. Additional knowledge of a computer algebra system helps in formulating and coding the costate ODE and the optimal control laws.

References

1. Battin, R. H. (1987) An Introduction to the Mathematics and Methods of Astrodynamics, American Institute of Aeronautics and Astronautics, New York
2. Betts, J. T. (1998) Survey of Numerical Methods for Trajectory Optimization. Journal of Guidance, Control, and Dynamics, **21**, 193–207.
3. Bulirsch, R., Callies, R. (1992) Optimal Trajectories for a Multiple Rendezvous Mission to Asteroids. Acta Astronautica, **26**, 587–597.
4. Büskens, C. (1998) Optimierungmethoden und Sensitivitätsanalyse für optimale Steuerprozesse mit Steuer- und Zustands-Beschränkungen. Dissertation, Universität Münster
5. Chudej, K. (2000) Effiziente Lösung zustandsbeschränkter Optimalsteuerungsaufgaben. Habilitationsschrift, Universität Bayreuth
6. Graf, T. (2001) Flugbahnoptimierung eines Niedrig-Schub Raumfahrzeugs. Diplomarbeit, Lehrstuhl für Ingenieurmathematik, Universität Bayreuth
7. Hiltmann, P., Chudej, K., Breitner, M. (1993) Eine modifizierte Mehrzielmethode zur Lösung von Mehrpunkt-Randwertproblemen, Benutzeranleitung. Report No. 14, Sonderforschungsbereich 255 Transatmosphärische Flugsysteme, Lehrstuhl für Höhere Mathematik und Numerische Mathematik, Technische Universität München
8. Jehn, R., Hechler, M., Rodriguez-Canabal, J., Schoenmaekers, J., Cano, J. L. (2000) Trajectory Optimisation for ESA Low-Thrust Interplanetary and Lunar Missions. CNES Workshop on Low-Thrust Trajectory Optimisation, Toulouse
9. Pesch, H. J. (1994) A Practical Guide to the Solution of Real-Life Optimal Control Problems. Control and Cybernetics, **23**, 7–60.
10. von Stryk, O. (1993) Numerical solution of optimal control problems by direct collocation. In: R. Bulirsch, et al (Eds.) Optimal Control – Calculus of Variations, Optimal Control Theory and Numerical Methods, ISNM 111, Birkhäuser, Basel, 129–143.
11. Vasile, M., Jehn, R. (1999) Low Thrust Orbital Transfer for a LISA Spacecraft With Constraints on the Solar Aspect Angle. MAS Working Paper No. 424, ESOC Darmstadt

Adaptive Data Structures and Algorithms for Efficient Visualization and Data Management at Runtime of Terrain and Feature Data

K. Nothnagel[1], A. Paul[2], and G. Sachs[1]

[1] Institute of Flight Mechanics and Flight Control, Technische Universität München, Boltzmannstr. 15, 85748 Garching, Germany
[2] Institute of Informatics for Engineering and Science and Numerical Programming, Technische Universität München, Arcisstr. 21, 80333 München, Germany

Abstract. An approach is presented for rendering terrain and feature data in real time, applicable to flight simulators and to cockpit displays for generating synthetic vision. Basically, a hierarchical adaptive triangulation technique is used for dealing with the terrain elevation data in the memory. The interpolation hierarchical surplus is used as a criterion to detect redundant data and serves to control the error of the terrain approximation. With this level-of-detail mechanism and an easy visibility check due to the hierachical adaptive strategy, a high frame rate can be reached. A special organisation of the data is developed to exchange parts of the data at runtime in order to limit the required memory. On multiprocessor computer architectures, the reloading data process can be performed in the background without interrupting the rendering by starting parallel running tasks (threads).

1 Synthetic Vision Concept

Synthetic vision as considered in [1-3] is a promising means for improving guidance and safety of aircraft. It basically comprises a 3-dimensional image of the outside world and integrated guidance elements. This novel display type is considered to contribute to improve flight in poor visibility or flight close to the ground, particularly when approaching a landing site.

A synthetic vision system is shown in Fig. 1, providing full visual information to the pilot. It presents a 3-dimensional computer generated out-the-window scene and guidance elements including a tunnel-in-the-sky image.

2 DFAD and DTED Databases

The 3-dimensional image of the terrain is generated using a digital terrain database which basically consists of Digital Terrain Elevation Data (DTED) and Digital Feature Analysis Data (DFAD), Fig. 2. The database may be complemented for describing airports and their surroundings.

The elevation data is organized in a regular grid with an interval spacing of 1"x 1" which corresponds to 90 m x 60 m for locations in Central Europe.

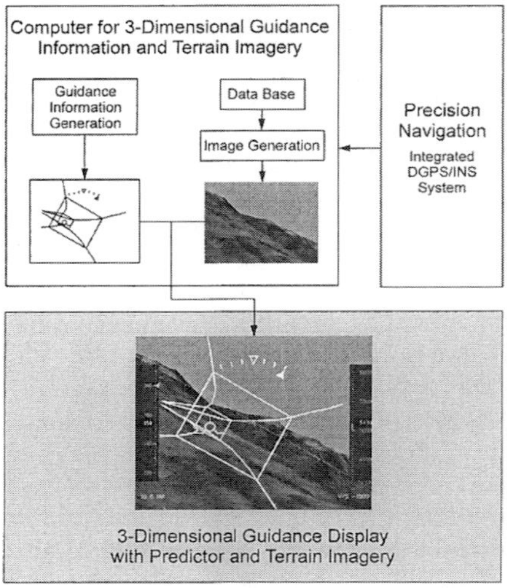

Fig. 1. Synthetic vision concept

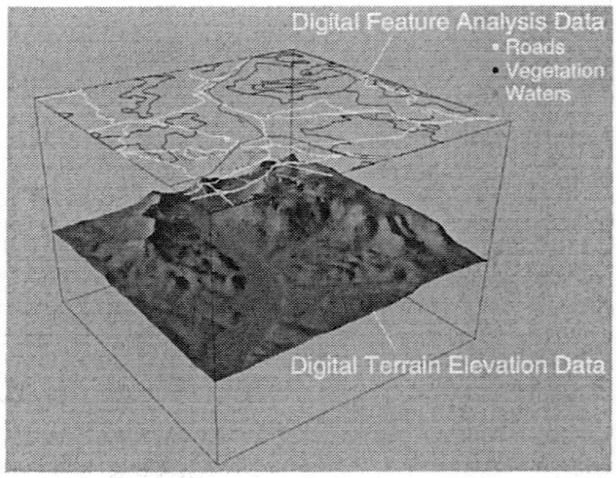

Fig. 2. DFAD and DTED databases

The feature analysis data contains information about areal, linear and point features. Areal features are given as polygons marking areas of certain surface properties to indicate fields, woods, seas, etc. For this kind of features, a texture map is generated that is drawn on the surface, thus coloring the elevation data from the DTED database. Linear features given as linestrips indicate objects like streets, rivers or railroad tracks. Point features mark the position of buildings relevant to air traffic such as towers, power lines or bridges. Linear and point feature types are treated as 3D-objects. A shading process is applied to the texture maps to make them more realistic. Therefore, an extra CPU-time consuming lighting of the scene is not needed (Fig. 3).

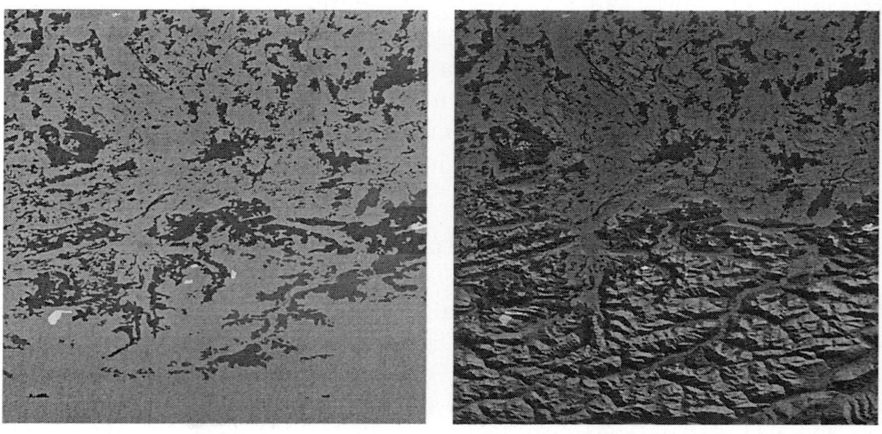

Fig. 3. Shading of the texture maps generated from areal features

3 Design Goals

There are several design goals for the synthetic vision concept in mind:

- Since the system will be used for visual information purposes, a frame rate of at least 25 frames per second is required.
- The rendered terrain should be close to reality in order to improve guidance and situational awareness.
- Since the databases are too large to be held in memory, there has to be a possibility to exchange data at runtime without reducing the frame rate. Therefore, on multiprocessor architectures independent threads shall be started.
- The system should include the possibility of compressing data as well as working directly with the compressed data in order to reduce disk space and memory requirements.

4 Reloading Event

With regard to a reloading event, a procedure checks the position of the aircraft against the borders of the data in memory loaded at startup. If the aircraft is out-of-bounds, a reloading is needed and a parallel running process (thread) is started for an update of the data in memory. There are independent events for reloading linear and point feature data, the textures made from the areal features and the elevation data. To prevent an overlapping of these threads, semaphores are used to assure that sufficient CPU-time is available in addition to the time needed for the continuous graphical redisplay.

For the future, a scheduler process is worked out which enables one processor architectures to simulate multiple threads by sharing CPU-time with the use of interrupts. This technique is considered feasible for one processor architectures because the bottleneck of the visualization pipeline lies in drawing the graphics and not in calculating the coordinates of the triangles. There is unused CPU-time remaining because the system has to wait for the graphics hardware to finish drawing.

5 Management of DFAD Data

While the data of the areal features is converted into bitmaps to cover the elevation data as textures, the linear and point feature data is changed into a new format stripped from unimportant feature types and attributes. The cells have the form of a square of 1/3 deg size. Nine cells are held in memory at runtime (Fig. 4), as described in [4]. A reloading event will occur when the aircraft gets out of the bounds of the safety cell. Then a thread takes care of the update of the data in memory, making the cell into which the aircraft has moved the new innermost cell X. The other cells are changed by swapping the pointers to their structures in memory or are loaded from harddisc if needed. Finally, unused cells are deleted from the memory.

For both the areal features and the linear/point features separate threads are used to reload data at runtime.

With a cell size of about 25 km x 25 km for Central Europe, a reloading event will occur about every five minutes for a straight flight at a speed of 300 km/h.

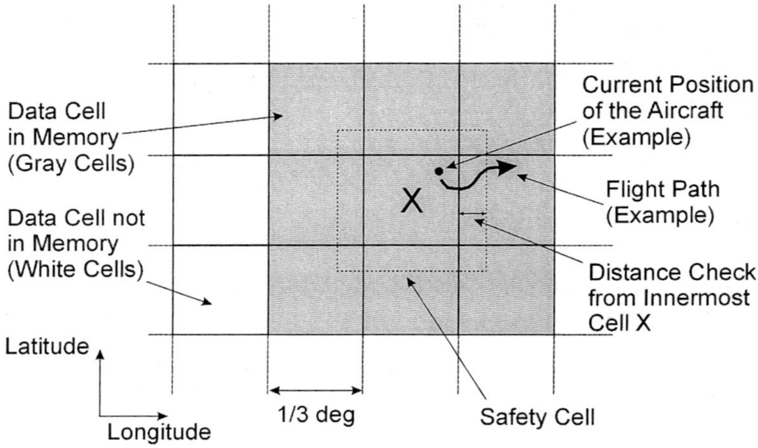

Fig. 4. Cell structure for DFAD data

6 Management of DTED Data

To achieve the goals concerning data compression, level-of-detail control and reloading of memory segments at runtime, a hierarchical interpolation approach, which was originally evaluated in [5] for the synthetic vision guidance application and which is similar to the technique proposed in [6], was used for terrain triangulation. Related work dealing with hierarchical interpolation techniques for datacompression is described in [7] and [8].

For the hierarchical interpolation of the data, the regular grid of the DTED database is split into rectangular areas of $(2^n + 1)$x$(2^n + 1)$ grid elements. The next step is to implicitly subdivide these grid elements into hierarchical levels, building a tree data structure (Fig. 5). In order to gain an indicator for its importance, each elevation value is linearly interpolated with regard to its hierarchical predecessors. The difference between interpolated and actual values, the hierarchical surplus of the underlying branch, is stored in the corresponding node. Every node is allocated separately in memory and connected with pointers in the tree. This has a high cost in memory usage, but the advantage is that only those parts of the tree are held

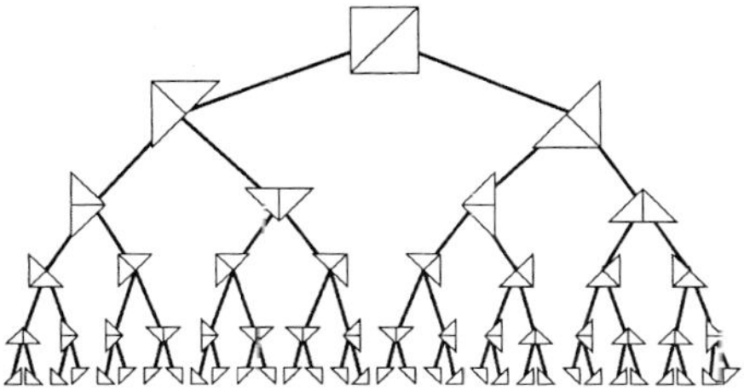

Fig. 5. Hierarchical tree structure

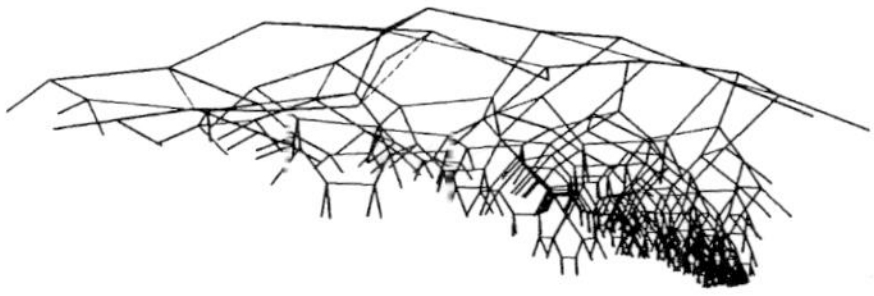

Fig. 6. Tree structure at runtime

in memory which are necessary. As another advantage, it allows a simple, straight forward and adaptive triangulation based on a divide-and-conquer scheme.

To make the algorithm more efficient, it is useful to make the error margin depending on the distance between the examined triangle and the observer. Objects farther away from the observer are drawn with a lower degree of detail so that less triangles are needed to generate the mesh. A level-of-detail function called `eps()`-function is used for this process which compares an error margin with the hierarchical surplus stored in every node. Thus, the level-of-detail can be toggled at which the mesh parts have to be drawn. Fig. 6 shows a display tree at runtime, with the observer on the right side.

Examples for the `eps()`-function are presented in Fig. 7. The left function is the one currently applied. The right function could be used to simulate three separate level-of-detail areas. There are various other level-of-detail possibilities which are easily controllable.

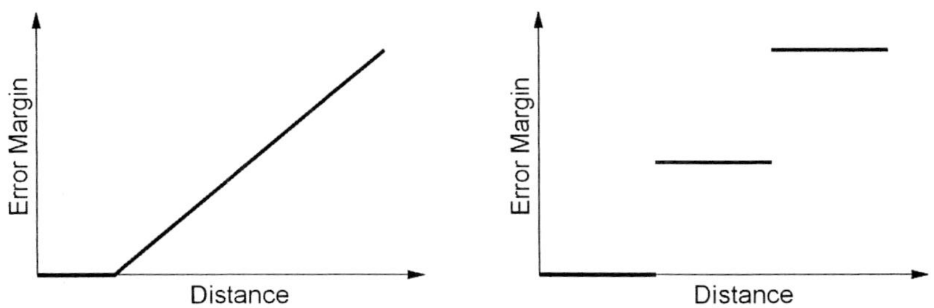

Fig. 7. Examples for `eps()`-function

An effect called *vertical holes* (Fig. 8), which appears when using a level-of-detail function, can be prevented by applying pointers between two corresponding triangles at the same hierarchical level. These pointers mark two branches which have to be triangulated in the same manner to avoid holes in the mesh.

For an easy access of the DTED data at runtime a new approach is used. The adaptive triangulation structure is stored on harddisc and can be ex-

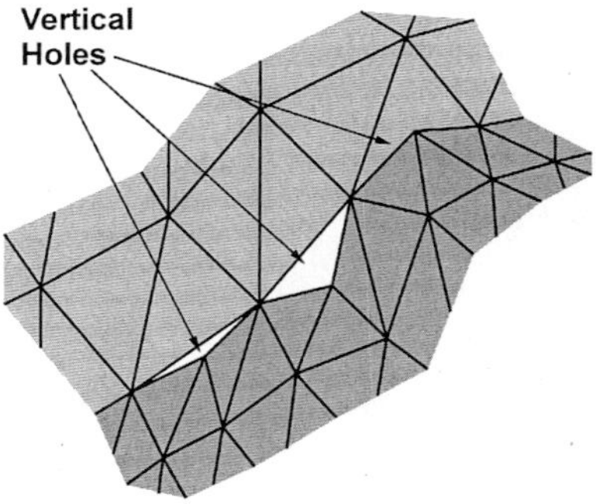

Fig. 8. Vertical holes

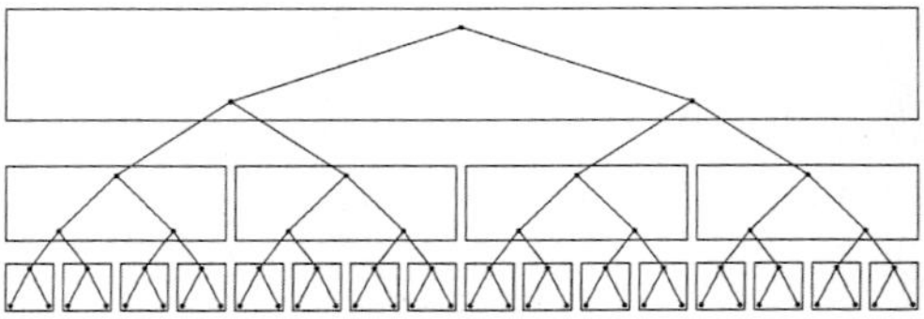

Fig. 9. Combining strategy

changed with the data in memory at runtime using functions which are similar to the swap-functions of most of the operating systems. To reduce the I/O-operations, branches of the tree are combined into cells as shown in Fig. 9. The result of this approach is that a fast update of the data in memory is possible without a retriangulation.

A detailed description of the algorithms and data structures for the DTED database can be found in [9-11].

7 Results

A mountain scene as it would appear in the synthetic vision environment with textured polygons is presented in Fig. 10. The left scene is rendered with 6500 triangles. For the right scene, only 600 triangles are used. The scenes look similar because a suitable `eps()`-function is applied. The refresh rate of the right scene is higher, corresponding to the reduction in the number of the triangles.

Fig. 10. Comparison of mountain scenes

The scene presented in Fig. 11 is used to show the capability of the system to hold large databases in the main memory, with Germany used as an example. On both snapshots the observer is considered to be high above Hamburg and to look towards the Alps, with a far clipping plane of 1000 km and an error level of 11 m on each of the six levels of the tree structure. The filled polygon picture is taken from a Silicon Graphics O2 workstation with a 300 MHz R12000 CPU and 384 MB main memory. The refresh rate of about 16000 triangles is 3 to 4 Hz. The left picture shows the triangulation mesh in detail.

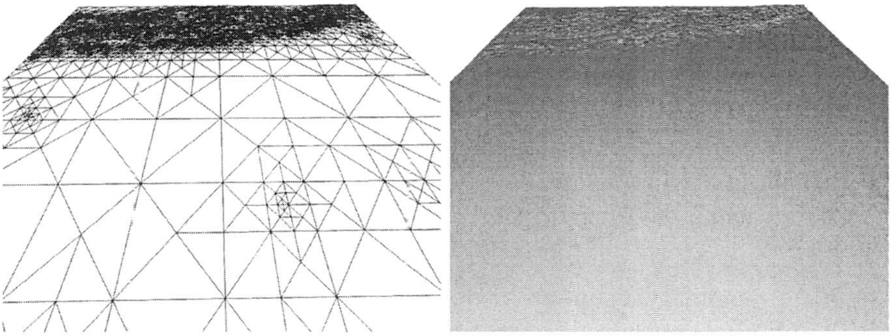

Fig. 11. View from the North Sea to the Alps

8 Conclusions

Computer generated synthetic vision, comprising a 3-dimensional image of the outside world and integrated guidance elements including a tunnel, is considered a means for enhancing flight in poor visibility and difficult terrain environment. For a synthetic vision system, algorithms and data structures are developed, showing the following properties:

- Possiblility of exchanging parts of the database in the main memory at runtime.
- Easy access at reloading events, using an appropriate storage format of DFAD and DTED data.
- User control of level-of-detail functions to speed up the frame rate.
- Optional use of threads on multiprocessor architectures.

The program is written in C++ using the OpenGL and Motif libraries, thus providing independence from hardware.

Acknowledgment

This work has been supported by the FORTWIHR Bavarian Research Center for Technical-Scientific High Performance Computing at Technische Universität München and Universität Erlangen-Nürnberg.

References

1. Möller H., Sachs G. (1994) Synthetic Vision for Enhancing Poor Visibility Operation. IEEE Aerospace and Electronics Magazine 9, 27–33.
2. Sachs G., Dobler K., Schänzer G., Dieroff M. (1997) Precision Navigation and Synthetic Vision for Poor Visibility Guidance. AGARD CP 593 "Advances in Flight Testing", 15-1–15-12.
3. Möller H. (1997) Computergenerierte synthetische Sicht zur Verbesserung der Flugführung bei schlechter Außensicht. PhD Thesis, Institut für Maschinenwesen, Technische Universität München.
4. Datenmanagement und Datenfusion für computergesteuerte visuelle Flugführungshilfen (2001) FORTWIHR (III) Transferprojekte, Abschlußbericht Sep. 1998–Feb. 2001. Herausgeber: FORTWIHR - Bayerischer Forschungsverbund für Technisch-Wissenschaftliches Hochleistungsrechnen, Erlangen, 102–121.
5. Gerstner T. (1995) Ein adaptives hierarchisches Verfahren zur Approximation und effizienten Visualisierung von Funktionen und seine Anwendung auf digitale 3-D Höhenmodelle. Master's Thesis, Institut für Informatik, Technische Universität München.
6. Lindstrom P., Koller D., Ribarsky W., Hodges L.F., Faust N., Turner G.A. (1996) Real-Time, Continuous Level of Detail Rendering of Height Fields. Computer Graphics (SIGGRAPH '96 Proceedings), 109–118.
7. Hiller K. (1993) Datenkompression mit dem Dünn-Gitter-Verfahren. Master's Thesis, Institut für Informatik, Technische Universität München.
8. Paul A. (1995) Kompression von Bildfolgen mit hierarchischen Basen. Master's Thesis, Institut für Informatik, Technische Universität München.
9. Paul A., Dobler K. (1997) Adaptive Realtime Terrain Triangulation. WSCG'97, The Fifth International Conference in Central Europe on Computer Graphics and Visualization '97, Conference Proceedings ISBN 80-7082-306-2, Volume II, University of West Bohemia, Plzen, Chech Republic, 404–413.
10. Paul A., Dobler K., Sachs G., Zenger Chr. (1997) Echtzeit-Gelände-Triangulierung mittels eines adaptiv-hierarchischen Interpolationsverfahrens. 5. Workshop Sichtsysteme - Visualisierung in der Simulatortechnik, Wuppertal, Reinhard Möller (Hrsg.), Shaker Verlag Aachen, 129–139.
11. Paul A. (2000) Entwicklung und Implementierung adaptiver Datenstrukturen zur effizienten Visualisierung und Speicherung digitaler Höhenmodelle. PhD Thesis, Institut für Informatik, Technische Universität München.

Recent Improvements in the Trajectory Optimization Software ASTOS

P. Gath[1], A. Wiegand[2], A. Markl[1], and K.H. Well[1]

[1] University of Stuttgart, Institute of Flight Mechanics and Control, Pfaffenwaldring 7a, 70550 Stuttgart, Germany
[2] TTI GmbH, Dept. of Optimization Guidance and Control, Pfaffenwaldring 7a, 70550 Stuttgart, Germany

Abstract. The Aerospace Trajectory Optimization Software ASTOS is using direct optimization methods such as multiple shooting and collocation to calculate the optimal control time history for an optimal control problem. The mathematical model is completely data driven, which means that the end-user need not write any code to specify the problem to solve.

This paper will describe some of the major technical improvements recently made to ASTOS, as there are: initial guess generation, branched trajectories and new optimization methods. In addition, some new applications are discussed which demonstrate the flexibility of the ASTOS Model Libary.

1 Introduction

ASTOS [1–3] has been developed since 1989. Since 1992, the Institute of Flight Mechanics and Control at the University of Stuttgart is in charge of the software development. In April 2000, commercial marketing of the software was started via the Technology Transfer Initiative GmbH at the University of Stuttgart, Department of Optimization, Guidance and Control.

Several improvements have recently been made in order to meet current and future customer requirements. Among the most important issues from a customer point of view are a consistent user interface, graphical input facilities, great flexibility in setting up new problems and a fluid work-flow during the whole optimization process from initial guess generation to optimization up to simulation and viewing results.

2 Initial Guess Generation

One of the core problems in trajectory optimization is to obtain an initial guess to get the optimizer started. That means, the initial guess has to be within the convergence radius of the applied optimization method.

Guidance laws have been implemented which allow generation of an initial guess for all kinds of launcher and reentry missions. The guidance laws are tuned by specifying parameters which can easily be estimated by an engineer. Moreover, since the guidance laws are compatible to the optimization programs, the control discretization can be reduced to very few parameters, a fact that speeds up the optimization process considerably while hardly reducing the performance index. The mission can even consist of a mixture of fully optimizable phases and guidance phases. Fig. 1 shows an Ariane 5 ascent trajectory to an escape orbit. Controls in the atmospheric phase are fully optimizable while the upper stage is using a bilinear tangent control law. For more details, see [4].

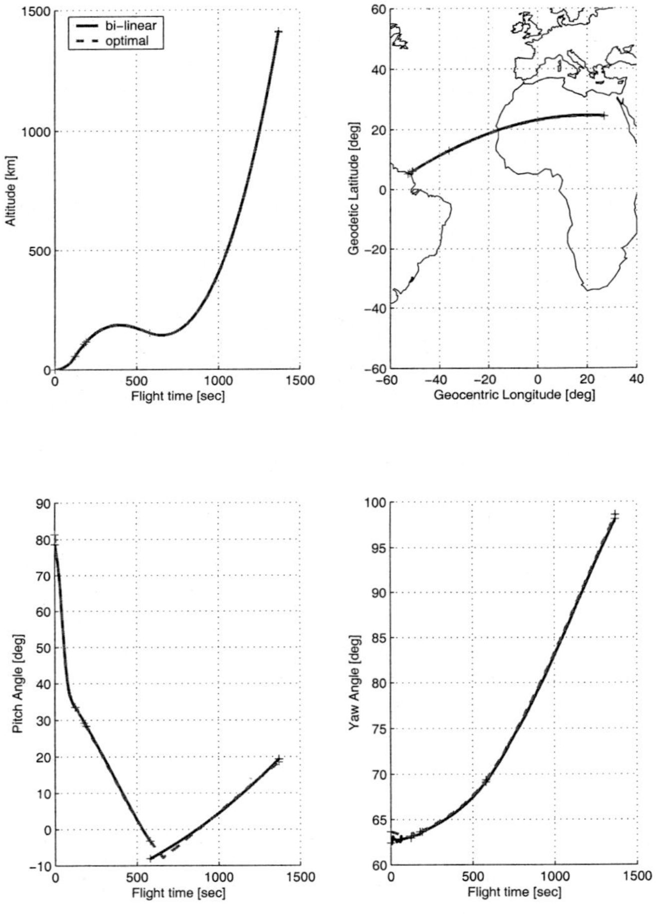

Fig. 1. Ariane 5 ascent to escape orbit with guidance law in upper stage

3 Branched Trajectories

Many problems in advanced launchers and reentry missions involve branched trajectories. An optimal solution must ensure, that the lower stage can return to its launch site (Fig. 2) or another destination, e.g. a reentry vehicle like the CRV drops its deorbit propulsion stage and this part must not crash into inhabited areas.

Branching has been included in the ASTOS Model Library in a fairly generic way. The software allows for multiple branches which can even be divided into sub-branches again. Every branch can contain its own set of path and boundary constraints (Fig. 3).

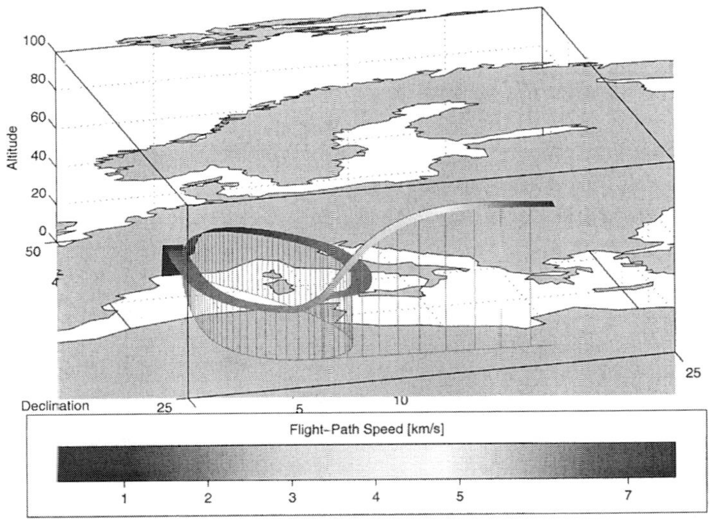

Fig. 2. TSTO ascent with optimal return. In order to save fuel mass of the return leg the ascent starts slightly to the west instead of straight to the south

The branched trajectories are computed sequentially, that is by processing the phases in the order of their numbering. In this way, the real flight time is replaced by an ”´integrated”´ time. Offset parameters p_{BO} serve the purpose to relate the integrated time to the real flight time. The initial times, states, controls, and design parameters of each branch are defined by the introduction of branching parameters p_b

$$p_b^j = [t_f^i, x_f^i, u_f^i, p^i]^T, \tag{1}$$

$$[t_0^k, x_0^k, u_0^k, p^k]^T = p_b^j. \tag{2}$$

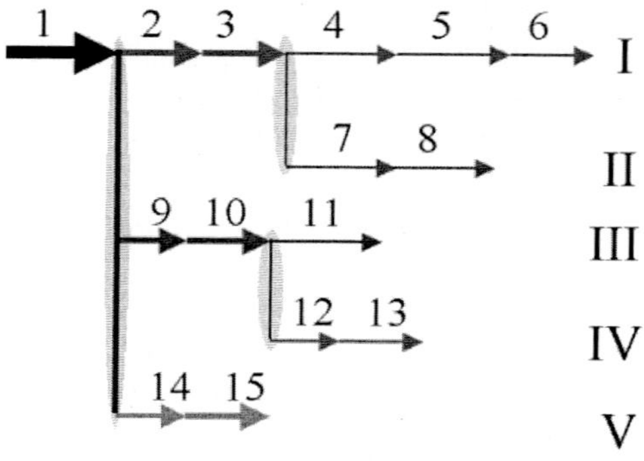

Fig. 3. Branched phase structure with 15 phases, divided into 5 branches by 3 branching points

Phase i (equation (3) in Fig. 3) ends with a branching point and phases j (eg. 4-6) with $i < j < k$ are the phases within the first branch. Phase k (eg. 7) is the first phase of the second branch. The connect conditions for additional branches are given in (2).

The offset parameter to compute the real flight time t_r from the integrated time t (3) is computed by

$$t_r = t - p^k_{BO}. \tag{3}$$

with

$$p^k_{BO} = t^j_f - t^i_f, \tag{4}$$

$$p^{k+1}_{BO} = p^k_{BO}. \tag{5}$$

Computationally this can be managed through a further set of parameters: a last offset parameter p_{LO} and a current offset parameter p_{CO}. They are necessary in all phases beginning with the phase after the first branching point until the last phase of the last but one branch. Their purpose is to compute the time difference between the integrated time and the real time.

$$p^{j+1}_{LO} = p^j_{CO}, \tag{6}$$

$$p^j_{CO} - p^j_{LO} - t^j + t^{j-1} = 0, \tag{7}$$

$$p^{j+1}_{BO} = p^j_{CO}. \tag{8}$$

The parameter p_{LO} is connected to p_{CO} of the previous phase (6) and represents the time difference at the beginning of phase j. p_{CO} is computed through a parameter constraint, adding together p_{LO} and the current phase duration (7). At the beginning of each branch, p_{BO} is now only connected to the current offset parameter of the previous phase (8).

4 New Optimization Methods

So far, the two major optimization methods for ASTOS have been PROMIS, which is a direct multiple shooting method, and TROPIC, which is a direct collocation method [5]. Lately, the software CAMTOS (Collocation And Multiple shooting Trajectory Optimization Software) has been developed which is a combination of PROMIS and TROPIC with several extensions. Within CAMTOS, the transcription method for every phase can be different. Currently, Trapezoidal and Hermite-Simpson transcription are supported for collocation phases and several different Runge-Kutta schemes with and without step-size control are implemented for multiple shooting phases. Since CAMTOS has been developed in an object-oriented approach, using Ada95, it can be easily extended to use additional integration methods, transcription schemes or even other SQP solvers. Currently, the SQP solver SNOPT 5.3-5 (see [6]) is used to solve the transcribed problem.

By applying some modifications to the discretization of a collocation phase, significant improvements in the number of required iterations were achieved compared to TROPIC (see Table 1). The major change in comparison to TROPIC is the addition of a support for piecewise constant and discontinuous control. In order to achieve this, each collocation node contains two control values, one for the collocation intervall to the left side, the other one for the collocation intervall to the right side. In case of a continuous control, additional control continuity constraints are added to the NLP problem.

Table 1. Performance of TROPIC vs. CAMTOS

Method	Iterations	Payload
TROPIC	189	6346 kg
CAMTOS Collocation	101	6344 kg
CAMTOS Shooting + Collocation	108	6346 kg

The combination of direct multiple shooting with direct collocation also simplifies the process of selecting appropriate grids for specific probems. In addition, higher-order integration methods can be used where appropriate (e.g. during coasting phases or for solving boundary value problems), while other phases of a problem can use lower-order integration methods or collocation. Provisions have been made to include special integrators for stiff systems in a future version. More details can be found in [7].

In addition to the direct optimization method, an indirect method has been developed for rocket launch vehicles (see [8] and [9]). This method can also be applied to any finite thrust maneuver which involves only time-dependent massflow and thrust time histories. The convergence radius of this method is very large and arbitrary values for the initial costates can be used as initial guesses. Since this method does not allow for complex path and boundary constraints, it is not able to meet industry requirements for a complete optimization. However, it is very useful for generating an initial guess and analyzing the applied sequence of burn-coast-burn phases for its optimality. Analysis of an Ariane 5 dual-payload mission with this method has lead to a solution with an increase of 40% in the deliverable payload to GTO (final trajectory is shown in Fig. 4).

Fig. 4. Ariane 5 dual payload mission

A future enhancement will be the integration of the indirect optimization method into the ASTOS Model Library. This will allow its use not only for initial guess generation, but also during optimization for the upper stages, where only a few boundary constraints have to be met (e.g. specific orbital conditions). The combination with direct optimization will allow for problems with several cost functions at once, e.g. maximizing the payload delivered to an orbit while minimizing heatload for a reentry vehicle during the same mission.

5 New Applications of the ASTOS Model Library

The ASTOS model library offers an extensive set of more than 80 different types of constraints which can be applied as path and/or boundary constraints in any phase. About 15 different cost functions are currently available and control smoothing can be applied if necessary. Another important enhancement is the capability of importing external data files for any item of the user setup (e.g. propulsion, aerodynamics, constraints etc.).

In order to improve the accuracy of long-duration coast arcs, two additional dynamics have been implemented. One set is using osculating Keplerian elements, the other one employs so-called equinoctial elements [10]. Especially the latter set has been succesfully used for low-thrust missions in the literature and similar missions have been analyzed using ASTOS.

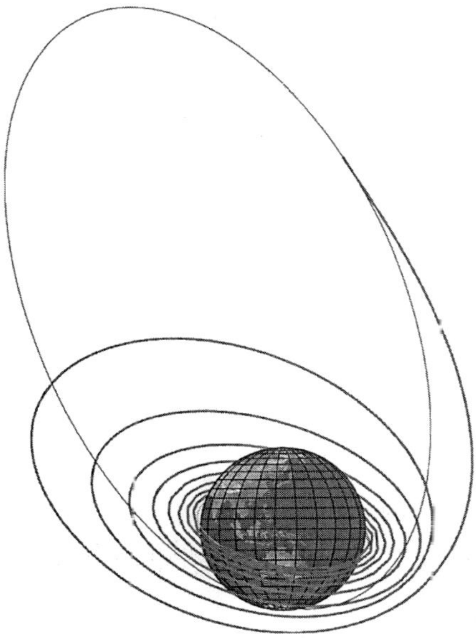

Fig. 5. Low thrust orbit transfer mission

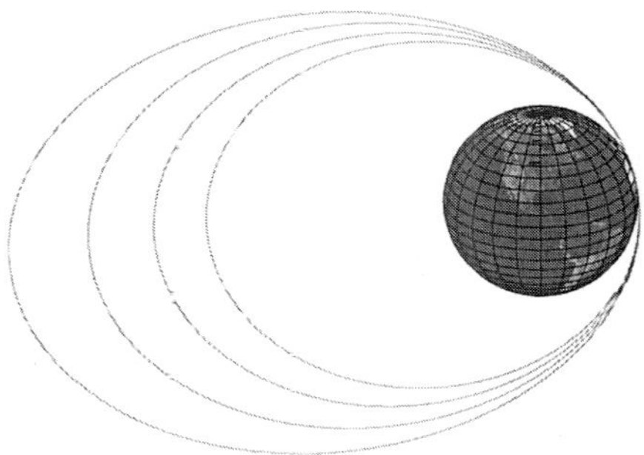

Fig. 6. Atmospheric transfer mission

The flexibility of ASTOS also allows for the investigation of reentry missions to other planets. A reentry into Marsian atmosphere using a low L/D capsule was investigated in the framework of vehicle design. Mass of the vehicle was maximized for a set of entry conditions in order to calculate sensitivities. Another recent application is the investigation of atmospheric orbit transfers, employing several atmospheric passes (see Fig. 6). This investigation was based on a vehicle presented in [11].

References

1. ASTOS Model Library Reference Manual. (2001) ASTOS MLR 4.3.0, Institute of Fligth Mechanics and Control, University of Stuttgart, Germany, March 2001
2. Wiegand, A., Markl, A., Well, K. H., Mehlem, K., Ortega, G., Steinkopf, M. (1999) ALTOS - ESA's Trajectory Optimization Tool Applied to Reentry Vehicle Trajectory Design. IAF-99-A6.09, 50th International Astronautical Congress, Amsterdam, The Netherlands, 4-8 October 1999
3. Well, K. H., Markl, A., Mehlem, K. (1997) ALTOS - A Software Packages for Simulation and Optimization of Trajectories of Launch- and Reentry Vehicles. IAF-97-V4.04, 48th International Astronautical Congress, Turin, Italy, 6-10 October 1997
4. Markl A., (to be published 2001) An Initial Guess Generator for Launch and Reentry Vehicle Trajectory Optimization. Dissertation, Institute of Fligth Mechanics and Control, University of Stuttgart, Germany
5. Jänsch, C., Schnepper, K.,Well, K. H. (1994) Multiphase Trajectory Optimization Methods with Application to Hypersonic Vehicles. Applied Mathematics in Aerospace Science and Engineering, Edited by A. Miele and A. Salvetti, Plenum Press New York, New York, chapter 8
6. Gill, P. E., Murray, W., Saunders, M. A. (1997) SNOPT: An SQP Algorithm for Large-Scale Constrained Optimization. Report NA 97-2, University of California, San Diego
7. Gath, P. F., Well, K. H. (to be published) Trajectory Optimization Using a Combination of Direct Multiple Shooting and Collocation. AIAA 2001-4047, AIAA Guidance, Navigation and Control Conference, Montreal, Quebec, 6-9 August 2001
8. Gath, P. F., Well, K. H., Mehlem, K. (2000) Automatic Initial Guess Generation for Ariane 5 Dual Payload Ascent Trajectory Optimization. AIAA 2000-4589, AIAA Guidance, Navigation and Control Conference, Denver, CO, 14-17 August 2000
9. Gath, P. F., Calise, A. J. (2001) Optimization of Launch Vehicle Ascent Trajectories with Path Constraints and Coast Arcs. Journal of Guidance, Control and Dynamics, Vol. **24**, No. 2, March-April 2001, 296–304
10. Betts, J. T. (1994) Optimal Interplanetary Orbit Transfer by Direct Transcription. The Journal of the Astronautical Sciences, Vol. **42**, No. 3, July-September 1994, 247–268
11. Miele, A., Wang, T. (1993) Nominal Trajectories for the Aeroassisted Flight Experiment. The Journal of Astronautical Sciences, Vol. **41**, No. 2, 139–163

Optimal Design of the Power Train of Vehicles: Modelling, Simulation and Optimization

D. Tscharnuter[1,2,3]

[1] reports and works done at Lehrstuhl für Höhere Mathematik und Numerische Mathematik — FORTWIHR, Zentrum Mathematik, TU München, Deutschland and

[2] Abt. EB-24, BMW AG Knorrstr. 147, 80788 München, Deutschland

[3] recent affiliation AMATEC Robotics GmbH, Landsbergerstr. 63a, 82110 Germering, Deutschland

Abstract. In the early stage of development of a vehicle, the design of the power train is encouraged by numerical simulations. Virtual testing of different variants and configurations of a car can be well-investigated without constructing any prototype. The mathematical models used to calculate the fuel consumption and driving power differ. Therefore, there exist different program packages to calculate these qualifying criteria of a vehicle. A standard interface between these packages and the optimization program has to be developed. Different, partly antagonistic objective functions are presented to optimize the design of the power train. The most important of the criteria are driving power and fuel consumption. The aim of development of new vehicles is to reduce the fuel consumption without loss of driving power. To fulfil these requirements it is not enough to investigate on each part of the power train separately. It is more efficient to optimize the whole system, using a mathematical model of the power train. The aim of the investigations is to find optimal characteristics of the torque converter, automatic gear transmissions, and the transmission of the rear axle differential, using mathematical methods for systematic optimization. In this paper methods are developed which can be used to optimize both model parameters and whole characteristics. The investigations are split up into the topics: mathematical description of objective functions, appropriate parameterization of characteristics, sensitivity analyses of the objective function. Some results are presented.

1 Motivation and Modelling

In the early stage of development of a vehicle the design of the power train is encouraged by numerical simulations. Virtual testing of different variants and configurations of a car, especially of the power train, can be well-investigated without constructing any prototype. Thus, the costs and the development time can be reduced. Even car industry suppliers can be involved in the early stage of development.

The mathematical models used to calculate the fuel consumption and driving power differ. The control of the converter bridge clutch is realised differently in the various models:

- backwards calculation: The power train is modelled as a quasi-static, longitudinal dynamic model. Giving time-speed-diagrams (see Fig. 1) the torque and revolutions per minutes of the motor is iteratively calculated, depending on the states of

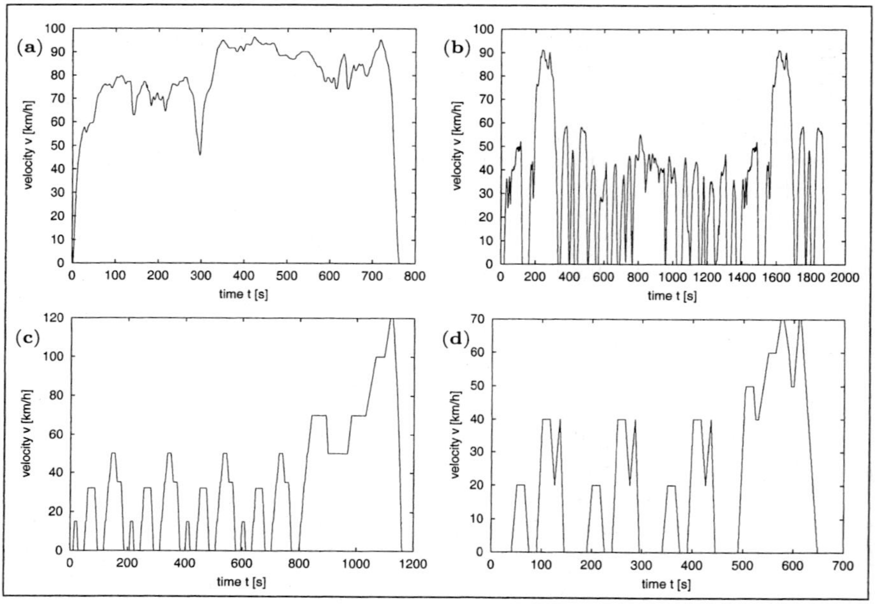

cycle name	country	remarks	
Highway	USA	interurban	(hot test)
FTP 75	USA	city	(cold test)
MVEG (KV96)	EU	city and interurban	(cold test)
ECE	EU	city (tolerance is $\pm 2\,\mathrm{km/h}$)	(hot test)
10•15-Mode	Japan	city and interurban	(hot test)

Fig. 1. Typical velocity cycles to calculate the fuel consumption of a vehicle: (**a**) US Highway cycle, (**b**) Federal Test Procedure or FTP 75, (**c**) ECE/EG cycle, (**d**) 10•15-Mode

all power train components: the gear G, the state of the converter bridge clutch W, etc. The backwards calculation is done iteratively, since the choice of the gear and the state of the converter bridge clutch depend on the motor load, given by a controller, and the speed number of revolutions of the main drive pinion. Then the fuel consumption is given as a function of motor torque and motor speed number of revolutions, and can be interpolated using the data of the motor characteristic fields. Starting up the torque flow at the gearbox, its partition is not yet fixed on the side of the turbine of the converter. There is no information about the motor load α. The gear $G = G_1$ is arbitrarily set. Both, α and G, have to be fixed iteratively. On the side of the turbine of the converter, the torque is partitioned in a mechanic and a hydrodynamic part. Thus, the slippage in the converter, which is a control variable, is defined. In each time step $W = W_1 = 1$ is initialized. This means that the converter bridge clutch is open. Having this partitioning of the torque, the state of the pump and the motor load signal, which is given by the motor characteristics depending on motor speed number of revolutions, and motor torque are calculated, using the converter characteristics. Now the acceptances of the chosen gear and state of the converter bridge clutch can be verified. They are iteratively fixed if necessary until the conditions

$$G_1 = G_2 \quad \text{for the gear,} \tag{1}$$

$$W_1 = W_2 \quad \text{for the converter bridge clutch, and} \tag{2}$$

$$S_1 = S_2 \quad \text{for the slippage} \tag{3}$$

are fulfilled. The condition for the slippage is taken into account only if the clutch is controlled.

- forwards calculation: The dynamic model of the power train is especially developed to calculate the fuel consumption of a vehicle. The mechanical part of the power train is modelled as a one-dimensional, longitudinal, dynamic model using differential equations. It is a torsion dynamic, two masses, spring-damper system developed by Hülsmann [9] and Andritzky [2]. To guarantee the availability of the model under real-time conditions, e.g. hardware in the loop tests, the non-linear characteristics of some parts of the system are given by data tables those are bi-linearly interpolated. There are different controllers in the model.
  The model is represented by a system of differential equations [18]:

$$\boldsymbol{M}(\boldsymbol{q})\,\ddot{\boldsymbol{q}} = \boldsymbol{f}(\boldsymbol{q}, \dot{\boldsymbol{q}}, \boldsymbol{\lambda}, \boldsymbol{u}, \boldsymbol{k}, t) - \left(\frac{\partial \boldsymbol{g}}{\partial \boldsymbol{q}}\right)^T (\boldsymbol{q})\,\boldsymbol{\lambda} \tag{4}$$

$$\dot{\boldsymbol{u}} = \boldsymbol{d}(\boldsymbol{q}, \dot{\boldsymbol{q}}, \boldsymbol{u}, \boldsymbol{p}_u, t) \tag{5}$$

$$\boldsymbol{k} = \boldsymbol{s}(\boldsymbol{q}, \dot{\boldsymbol{q}}, \boldsymbol{u}, \boldsymbol{p}_s, t) \tag{6}$$

$$\boldsymbol{0} = \boldsymbol{g}(\boldsymbol{q}) \tag{7}$$

The motion of the bodies is described by n_q generalized position coordinates $\boldsymbol{q}$ and n_q generalized velocity coordinates $\dot{\boldsymbol{q}}$. The vector $\boldsymbol{u}$ denotes the control variables and the vector $\boldsymbol{k}$ all characteristics of the system. Lagrange multipliers $\boldsymbol{\lambda}$ are introduced into the mechanical principle, yielding these equations of constrained mechanical motion or Lagrange equations of first kind. The vector $\boldsymbol{\lambda}$ stands for the generalized constraint forces.

The system of algebraic equations $\mathbf{0} = \boldsymbol{g}(\boldsymbol{q})$ expresses the connections like joints. $\boldsymbol{M}(\boldsymbol{q})$ denotes the positive definite $n_q \times n_q$ mass matrix and the vector $\boldsymbol{f}$ the applied and outer forces. The controllers are described by the vector $\boldsymbol{d}$ depending on the control parameters $\boldsymbol{p}_u$. The spline functions $\boldsymbol{s}$ with the de Boor points $\boldsymbol{p}_s$ denotes all characteristics of the system. The parameters $\boldsymbol{p} = (\boldsymbol{p}_u, \boldsymbol{p}_s)$ are the degrees of freedom when optimizing the dynamic behavior of the system. When integrating these differential equations, all generalized variables and the torques are known at each time step. For describing the physical model of an idealization of a power train, a lot of detailed examples are given in the literature [12,13]. In order to use the mathematical model under real-time conditions, complicated physical behaviors are approximated by one- and two-dimensional characteristics. Examples are the states of the automatic gearbox and of the converter bridge clutch. The given data are linearly or bi-linearly interpolated in the standard models. In this paper however the characteristics of the torque converter are described by cubic B-spline functions.
Iterations that are necessary to identify the gear and torque partition in case of backward calculation are not needed here. The controller of the converter bridge clutch is realized by a classic PID-controller.

2 Optimal Design of the Power Train

For designing the power train there are various, partly antagonistic, criteria to optimize the system. The most important of them are driving power and fuel consumption. The aim of development of new vehicles is to reduce the fuel consumption without loss of driving power. To fulfil these requirements, it is not enough to investigate each part of the power train separately. It is more efficient to optimize the whole system of the power train using a mathematical model.

Optimization criteria and optimality conditions for the design parameters and the state variables are often given in words. They have to be formulated as mathematical equations and inequalities. Lower and upper limits for the model parameters, e.g. transmissions of the gearbox, must be given.

As it was mentioned above, it is necessary to look at the whole system when designing the power train of a vehicle with an automatic gearbox and optimizing the transmissions and torque converter characteristics. The dynamic behavior of the torque converter is approximated by two characteristics. When optimizing these degrees of freedom, using various values of the design parameters $\boldsymbol{p}_s$, one has to guarantee the quality of these characteristics when varying the parameters of the model which interpolates the given data, because there are only embedding shapes accepted.

The primary object of investigations in this paper is the optimal design of the torque converter characteristics, and of the transmissions of the automatic gearbox and rear axle differential, using a systematic mathematical optimization method. For that purpose, an optimization method has to be developed that allows to vary both, model parameters, and one- and two-dimensional characteristics. The investigations are split up into 3 topics:

1. finding objective functions that are numerical stable to minimize the fuel consumption and to maximize the driving power simultaneously,
2. appropriate parameterizations of characteristics to get form keeping variations during the optimization,
3. sensitivity analyses of the objective function depending on design parameters variations.

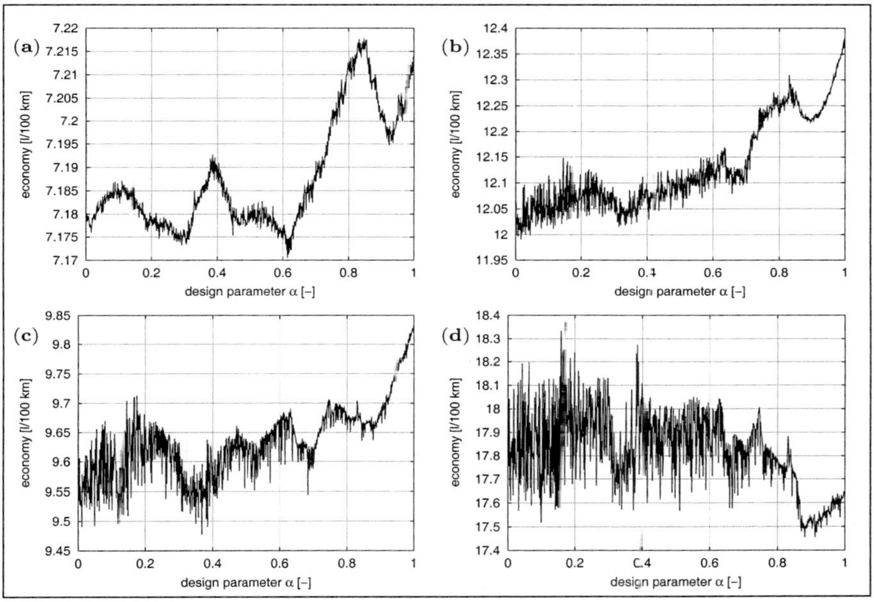

Fig. 2. Sensitivity investigations. The fuel consumption as a function of the torque converter. The fuel consumption per 100 km is calculated based on different time-speed-diagrams: (**a**) US Highway cycle, (**b**) Federal Test Procedure, (**c**) ECE/EG Test cycle and (**d**) 10•15-Mode. For drawing these plots the design parameter varies with the step size 10^{-3}

Several investigations of the objective function behavior have shown that the performance index used to optimize the fuel consumption is a smooth function superposed by high frequency, noisy functions, which can be interpreted as perturbations with low amplitude, see Fig. 2. In the present case, the high frequent superpose of low amplitude is not even differentiable. Therefore, a robust optimization method has to be applied that only use information of the objective function values. The results, presented in this paper, are calculated by the implicit filtering method [3,4,7,8]. Further methods of optimization that can be used are:

- bundle-trust-methods [1],
- multidirectional pattern search algorithm [15–17],
- simulated annealing [5,10,14,19],
- genetic algorithm [11].

Design Criteria to Optimize Driving Power and Fuel Consumption

As it was mentioned above, the dynamic behavior of the torque converter is approximated by two characteristics. These curves are varied during the calculation of the optimal torque converter. The following points have to be considered:

- The characteristics of the torque converter have special shapes which have to be maintained during the optimization.
- To guarantee the construction of the new optimal torque converter only small variations are allowed.
- Ideal converter characteristics are generated, embedding the measured characteristics that belong to real torque converters. Thus, the problem of the optimization of the characteristic is reduced to a parameter optimization problem. Embedding the converter characteristics into two-dimensional surfaces, there are two possibilities for choosing the design parameter:
  1. The sequence of N existing, measured, torque converters is arbitrary, since the embedding parameters α_i $(i = 1, \ldots, N)$ can be freely chosen within the interval $[0, 1]$. Thus, a surface is built by interpolation. During the optimization, the embedding parameter α is one degree of freedom, and the actual ideal torque converter is presented when α is constant. Criteria to choose the sequence are [18]:
     - ascending pump acceptance torque,
     - ascending torque super elevation,
     - ascending fuel consumption,
     - ascending driving power,
     - an especially developed embedding algorithm.
  2. The characteristics of the converter in the model are always given by setting the embedding parameter $\alpha = 0.5$. The degrees of freedom of the optimization problem are the arbitrary parameters α_i. Then the embedding surface varies during the optimization using different sequences of the arbitrary parameters. It has to be guaranteed that all α_i differ. Otherwise, the surface is not continuous in α.
- The problem can be treated as an optimal control problem with the torque converter characteristics as control functions. In this case a lot of conditions on the state and control variables have to be fulfilled to guarantee the quality of the characteristics. But these conditions are completely unknown.

The torque converter is now optimized using the developed mathematical dynamic model to calculate the fuel consumption and driving power of a vehicle. The objective function is a linear combination of the results (8) – (12) of the dynamic simulations. Because of bi-linearly interpolating the motor characteristics the objective function is a non-differentiable continuous function. It was shown that the fuel consumption has to be related to the distance covered, thus L/100 km is calculated. Parameters are encouraged for those the vehicle covers a longer distance burning the same volume of fuel V_{fuel}:

$$\min_{\boldsymbol{p}\in\mathbb{R}^{n_p}} F_1(\boldsymbol{p}) = \min_{\boldsymbol{p}\in\mathbb{R}^{n_p}} 100 \frac{V_{\text{fuel}}(t_f)}{s(t_f)}. \tag{8}$$

Following criteria characterize the driving power:

1. Maximum distance covered in the first 4 s

$$\max_{\boldsymbol{p}\in\mathbb{R}^{n_p}} F_2(\boldsymbol{p}) = \max_{\boldsymbol{p}\in\mathbb{R}^{n_p}} s(t = 4\,\text{s}). \tag{9}$$

2. Minimum time for the first 400 m

$$\min_{\boldsymbol{p}\in\mathbb{R}^{n_p}} F_3(\boldsymbol{p}) = \min_{\boldsymbol{p}\in\mathbb{R}^{n_p}} t(s = 400\,\text{m}). \tag{10}$$

3. Time minimal acceleration 0–100 km/h

$$\min_{\boldsymbol{p}\in\mathbb{R}^{n_p}} F_4(\boldsymbol{p}) = \min_{\boldsymbol{p}\in\mathbb{R}^{n_p}} t(v = 100\,\text{km/h}). \tag{11}$$

4. Time minimal acceleration 80–120 km/h in the 5th gear, the so-call elasticity

$$\min_{\boldsymbol{p}\in\mathbb{R}^{n_p}} F_5(\boldsymbol{p}) = \min_{\boldsymbol{p}\in\mathbb{R}^{n_p}} \left(t(v = 120\,\text{km/h}) - t(v = 80\,\text{km/h})\right). \tag{12}$$

To reach the maximum of reduction of the fuel consumption further degrees of freedom are set. The transmission of the rear axle differential is a new design parameter that is varying within the box condition

$$p_l = 0.5 \le p = i_{\text{AG}_\text{h}} \le p_u = 5.0. \tag{13}$$

Once again the fuel consumption per 100 km (8) has to be minimized. Further conditions are not necessary. The variation of this transmission results in scaling the gear switch characteristics because the vehicle makes a higher maximum speed. Thus, the number of revolution of the main drive pinion at the exit of the gearbox is greater.

The driving power of a vehicle can be maximized identifying the optimal transmissions of the gearbox. Therefore there are four parameters to be optimized while the 4th gear is constant $i_{\text{G}_4} = 1$. The transmissions vary within the box conditions

$$p = i_{\text{G}_k} \pm 20\% \qquad \text{with} \quad k = 1, 2, 3, 5. \tag{14}$$

Objective functions of this optimization problem are (9) – (12). Further conditions are not necessary.

The box conditions for these parameters are chosen so that the sequence of the gears keeps in order. Some simulations have shown that the transmission of the 5th gear should be very small. Then the maximum speed of 260 km/h is attained in the 4th gear and falls down to 245 km/h in the 5th gear. The reduction of fuel consumption, when driving very fast, is attained by losing driving power. To maximize the driving power in the 5th gear, the distance covered is a term

$$\max_{\boldsymbol{p}\in\mathbb{R}^{n_p}} F_6(\boldsymbol{p}) = \max_{\boldsymbol{p}\in\mathbb{R}^{n_p}} s(t_f) \tag{15}$$

of the weighted objective function with the condition

$$v(t_f) = v_{\max} \quad (= 260\,\mathrm{km/h}), \tag{16}$$

which is formulated as a penalty term of the objective function

$$F_s = c_s \left(1 - \exp\left(-\left\{\ln\left[\left(\frac{v(t_f)}{v_{\max}}\right)^{d_s}\right]\right\}^2\right)\right) \tag{17}$$

with the two parameters c_s and d_s. The convex penalty term (17) is set in the following way: If the condition is fulfilled, the penalty term vanishes. Variations of the form $v(t_f) < v_{\max}$ are stronger penalized than variations of the form $v(t_f) > v_{\max}$, when the maximum speed is higher than the given limit. The singularity of the logarithm near $v(t_f) = 0$ will never be received, because the velocity of the vehicle is positive at the end of the acceleration phase.

Without the parameter c_s the term (17) locates within the interval $[0, 1]$. Here, this parameter is initialized so that the penalty term rests inside the domain of attraction, if the maximum speed varies caused by the variation of the design parameters. In this case here, the weight factor can be set to $c_s = 1$.

The arbitrary parameter d_s indicates how fast the penalty term increases, if the equality condition is hurt. Using the parameter d_s the domain can be varied where the penalty term is sensitive when varying the velocity $v(t_f)$. In this example, the parameter is set to $d_s = 10$ so that the term (17) is near to 1, if $v = 200\,\mathrm{km/h}$. This means: Outside the domain $190 < v(t_f) < 340$ the variations of F_s are very small varying $v(t_f)$.

The weight factor of the term of the distance covered $F_6(t_f)$ is very small, because the distance is measured in meter so that its value is 10^3 times greater than the other terms. Thus, the objective function has the following form:

$$\min_{\boldsymbol{p}\in\mathbb{R}^{n_p}} F(\boldsymbol{p}) = \min_{\boldsymbol{p}\in\mathbb{R}^{n_p}} \frac{1}{3} \{c_1 F_1(\boldsymbol{p}) - c_2 F_2(\boldsymbol{p}) + c_3 F_3(\boldsymbol{p}) + c_4 F_4(\boldsymbol{p}) + c_5 F_5(\boldsymbol{p}) + c_6 F_6(\boldsymbol{p}) + F_\varepsilon(\boldsymbol{p})\} \tag{18}$$

with the coefficients $c_1 = 0.139$, $c_2 = \text{-}0.024$, $c_3 = 0.067$, $c_4 = 0.152$, $c_5 = 0.058$, $c_6 = \text{-}0.125 \cdot 10^{-3}$.

3 Results

By simultaneous variation of all transmissions and the characteristics of the torque converter, using a method of the non-smooth optimization, the power train is designed so that the driving power is maximized and the fuel consumption is minimized. This is shown by two examples, see Table 1 and 2.

Table 1. Data sheet of simulations after the optimization using ECE cycle

objective function	unit	improvement
driving power	[s]	
0-100 km/h		6.68 %
400 m standing start		1.98 %
elasticity (5th gear)		1.29 %
fuel consumption	[l/100 km]	
US Highway		1.95 %
ECE		0.70 %
FTP 75		2.07 %
10•15-Mode		-14.06 %
Improvement of the weighted objective function: 8.9 %		
Keeping the conditions: 99.9 %		

A PID-controller is used to keep the velocity so that the gap to the given velocity is minimal. Its parameters are initialized so that the fuel consumption per 100 km is minimal. In the second step the characteristics are optimized. The start up configuration is a weak torque converter. The result of the optimization confirm the knowledge got by trial. Since the effort of the reduction was not satisfying enough, the transmission of the rear axle was optimized in the next step too. The fuel consumption was reduced about 4.58 %.

Table 2. Data sheet of simulations after the optimization using FTP cycle.

objective function	unit	improvement
driving power	[s]	
0-100 km/h		7.58 %
400 m standing start		2.61 %
elasticity (5th gear)		1.63 %
fuel consumption	[l/100 km]	
US Highway		2.60 %
ECE		0.82 %
FTP 75		2.64 %
10•15-Mode		-7.62 %
Improvement of the weighted objective function: 17.3 %		
Keeping the conditions: 99.9 %		

Using the US Highway cycle to calculate the fuel consumption, the objective function (8) is more sensitive, when varying the characteristics of the converter than varying the transmission of the rear axle differential. The reason is the following one: The torque converter is bridged, because the converter bridge clutch is always controlled. In this example the equality condition on the final velocity in the 5th gear is fulfilled with 99.9 %. The improvement of the driving power is 9 % with a simultaneous reduction about 1.49 % of the fuel consumption calculated with the US Highway cycle. For all other cycles, however, the virtual vehicle initialized with the optimized configuration needs more fuel. This fact will be avoided, using the ECE city cycle, which is a composition of some stop and go sequences. Above all, additional criteria are taken into account to the objective function to qualify the driving power: It is the acceleration from 0 km/h up to 100 km/h (11), the first 400 m with standing start (10) and the distance covered in 4 s (9). All these criteria can be calculated by simulating the acceleration up to maximal speed under full motor load. A separate simulation is necessary to calculate the elasticity. The virtual vehicle has to be accelerated from 80 km/h up to 120 km/h (12).

Optimizing the transmissions of the gearbox too, driving power and fuel consumption can become much better. That is why the transmissions of the 1st, 2nd, 3rd and 5th gear are the additional degrees of freedom to be optimally identified.

A better configuration of the power train concerning the fuel consumption is the result of optimization using the ECE or FTP velocity cycle, e.g. Table 1 and 2. The interplay between the motor and the automatic gearbox with a torque converter and controlled converter bridge clutch take the best effect here. Using 3 of 4 cycles to optimize the configuration of a vehicle the fuel consumption decreases about 0.7 % and 2.64 %. The exception is the Japanese city cycle 10•15-Mode. With this cycle one gains the best improvement of the fuel consumption: 3 L per 100 km. The optimized configuration however worked out badly with the other velocity cycles. That is why the Japanese cycle is not a good chose looking for the optimal design of the power train. Using the ECE or FTP cycle, the calculation of the fuel consumption is independent of the velocity cycle as far as possible, because in both cycles there are some stops an goes, simulating the city traffic and intercity sequences too. Looking at the acceleration 0–100 km/h the improvement is about 6.68 % up to 7.58 %.

4 Conclusion

In this paper some aspects of modelling a power train of a vehicle are shown, when simulating its fuel consumption and driving power. Design criteria to optimize transmissions and the torque converter characteristics are developed and investigated. Further on methods for calculating embedding characteristics are developed. The results, which are presented in this paper, show that both fuel consumption and driving power can be optimized simultaneously, when optimizing the characteristics of the torque converter and the transmissions of the gearbox and rear axle. Configurations of the power train were found so that the vehicle will be better in economics being more agile.

References

1. Achtziger, W., Zowe, J. (1996) Nichtglatte Optimierung. Vorlesungsskript, Institut f. Angewandte Mathematik, Friedrich-Alexander-Universität Erlangen-Nürnberg.
2. Andritzky, B. (1996) Verifikation und Erweiterung des konfigurierbaren Berechnungskerns zur Simulation der Längsdynamik und zur Antriebsstrangauslegung von PKW mit Automatikgetrieben, Diplomarbeit, Technische Universität München, Lehrstuhl B für Mechanik.
3. Choi, T., Gilmore, P., Eslinger, O. J., Keller, C., Patrick, A., Gablonsky, J. (1999) IFFCO, Implicit Filtering for Constrained Optimization, Version 2, Techn. Rep., Center for Research in Scientific Computation, North Carolina State University, Raleigh.
4. Choi, T., Kelley, C. (1999) Superlinear Convergence and Implicit Filtering, Techn. Rep. CRSC-TR99-14, Center for Research in Scientific Computation, North Carolina State University, Raleigh. To appear in SIAM J. on Optimization.
5. Cohn, H., Fielding, M. (1999) Simulated Annealing: Searching for an optimal Temperature Schedule, SIAM J. Optimization.
6. Gill, P. E., Murray, W., Wright, M. H. (1981) Practical Optimization, Academic Press, Inc., San Diego, CA.

7. Gilmore, P. (1993) IFFCO, Implicit Filtering for Constrained Optimization, Techn. Rep. CRSC-TR93-7, Center for Research in Scientific Computation, North Carolina State University, Raleigh.
8. Gilmore, P., Kelley, C. (1995) An implicit filtering algorithm for optimization of functions with many local minima, SIAM J. Optimization **5**.
9. Hülsmann, W. (1995) Entwicklung eines konfigurierbaren Berechnungskerns zur Simulation der Fahrzeuglängsdynamik und Antriebsstrangauslegung, Diplomarbeit, Technische Universität München, Lehrstuhl B für Mechanik.
10. Ingber, L. (1989) Very fast simulated re-annealing, Mathl. Comput. Modelling **12**.
11. Ingber, L., Rosen, B. (1992) Genetic algorithms and very fast simulated reannealing: A comparison, Mathematical and Computer Modelling **16**.
12. Mitschke, M. (1995) Dynamik der Kraftfahrzeuge Band A: Antrieb und Bremsung, Springer, 3. Auflage.
13. Mitschke, M. (1990) Dynamik der Kraftfahrzeuge Band C: Fahrverhalten, Springer, 2. Auflage.
14. Papageorgiou, M. (1996) Optimierung: Statische, dynamische und stochastische Verfahren für die Anwendung, R. Oldenbourg Verlag München Wien, 2. Auflage.
15. Lewis, R. M., Torczon, V. (1999) Pattern Search Algorithm for Bound Constrained Minimization, SIAM J. Optimization **9**.
16. Torczon, V. (1991) On the Convergence of the Multidirectional Search Algorithm, SIAM J. Optimization **1**.
17. Torczon, V. (1997) On the Convergence of Pattern Search Algorithms, SIAM J. Optimization **7**.
18. Tscharnuter, D. (2000) Optimale Auslegung des Antriebsstrangs von Kraftfahrzeugen — Modellbildung, Simulation und Optimierung, Dissertation, Technische Universität München, Zentrum Mathematik. To be published.
19. Wah, B. W. and Chen, Y. X. (2000) Optimal Anytime Constrained Simulated Annealing for Constrained Global Optimization, Proc. 6th Int. Conf. on Principles and Practice of Constraint Prop., Springer-Verlag.

Unsteady Heat Load Simulation for Hypersonic Cruise Optimization

M. Wächter and G. Sachs

Institute of Flight Mechanics and Flight Control, Technische Universität München, Boltzmannstr. 15, 85747 Garching, Germany
e-mail: waechter@lfm.mw.tum.de

Abstract. Unsteady heat loads during the range cruise of a hypersonic vehicle propelled by a turbo/ram jet engines combination are considered. The unsteady heat load effects are simulated using a a realistic mathematical model. This model is coupled to the equations of motion of the vehicle. A two step technique for generating a solution is applied. First, an efficient optimization technique is used and, second, the obtained results are recalculated with an ODE-Solver. It is shown that the heat input can be significantly reduced, with only a small increase in fuel consumption.

1 Introduction

A hypersonic flight sytem equipped with wings and airbreathing engines can be considered a promising concept for an economic and safe access to space. It provides the capability of a range cruise so that the orbital stage can be released at a location distant from the takeoff site.

A main problem at hypersonic speed concerns the high temperatures to which the vehicle is exposed. This is illustrated in Fig. 1, which shows areas where heat loading is particularly significant. An efficient thermal protection system is required to protect the structure of the flight system. It will be shown in this paper that the heat flux into the vehicle can be significantly reduced using optimal trajectory control.

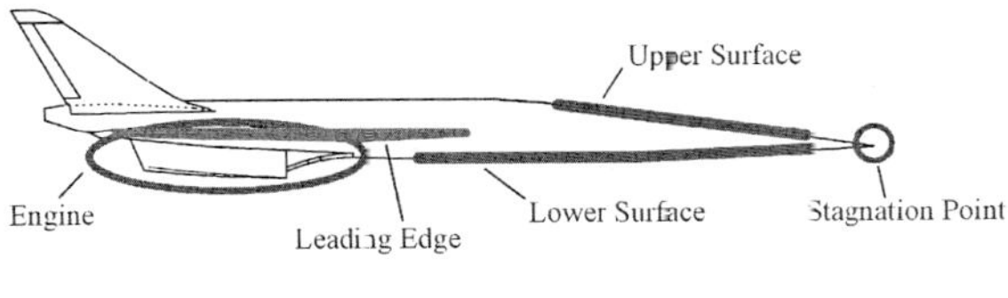

Fig. 1. Areas with significant heat loading

In recent years, important results have been achieved in the trajectory optimization of hypersonic vehicles [2, 4, 6], including effects of heat load and heat flux [5, 11]. This paper presents results for an optimized range cruise, with special emphasis placed on a realistic simulation of the unsteady heat flux.

2 Modelling

2.1 Flight System Dynamics Modelling

Modelling of the vehicle for trajectory optimization is based on point mass dynamics. The equations of motion read, with reference to a rotating spherical earth model [10], Fig. 2:

$$\begin{aligned}
\dot V &= \frac{1}{m}\Big[T(V,h;\alpha,\delta_T)\,\cos\alpha - D(V,h;\alpha)\Big] - m\,g(h)\,\sin\gamma + \\
&\quad + \omega_E^2\, r(h)\,\sin\gamma\,, \\
\dot\gamma &= \frac{1}{m\,V}\Big[T(V,h;\alpha,\delta_T)\,\sin\alpha + L(V,h;\alpha)\Big] + \\
&\quad + \cos\gamma\left[\frac{V}{r(h)} - \frac{g(h)}{V} + \frac{\omega_E^2\, r(h)}{V}\right] + 2\,\omega_E\,, \\
\dot h &= V\sin\gamma\,, \\
\dot x &= V\cos\gamma\,, \\
\dot m &= -\dot m_{fuel}(V,h;\alpha,\delta_T)\,.
\end{aligned} \tag{1}$$

State variables are speed V, flight path angle γ, altitude h, coordinate x and mass m. Control variables are angle of attack α and throttle setting δ_T.

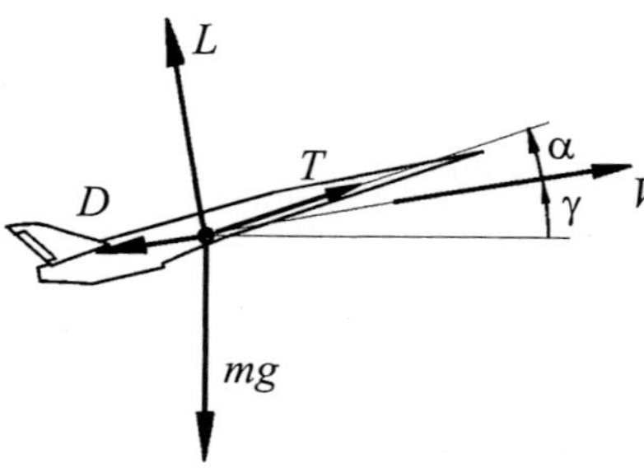

Fig. 2. Forces acting on the vehicle

To describe the aerodynamics and powerplant characteristics a complex mathematical model is applied, including multifunctional dependencies for lift L, drag D and thrust T.

2.2 Heat Flux Modelling

A complex model for realistic simulation of the unsteady heat flux into the vehicle was developed [4]. A region at the lower side of the vehicle at half the distance between the nose and the liquid hydrogen tank is considered ($x_{le} = 20$ m).

The wall structure of the thermal protection system is shown in Fig. 3. The thermal protection system consists of several layers of different material and thickness.

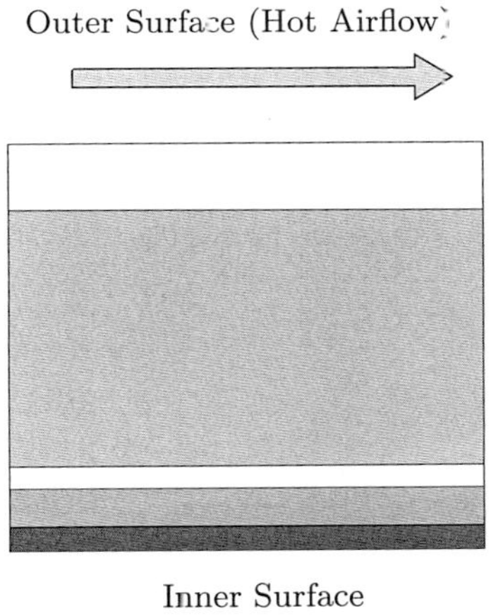

Fig. 3. Wall structure with heat protection (5-layer model)

To model the heat flux, the wall is split up into n layers, Fig. 4. A one-dimensional knot model is used to describe the heat flux from one layer to the other. The heat flux into the first layer can be described by

$$q_1 = q_{air} - q_{rad} = q_{air}(V, h, T_1; \alpha) - \varepsilon\, \sigma \left[T_1^4 - T_\infty^4\right] \tag{2}$$

where q_{air} and T_1 depend on the flight condition.

The heat fluxes for the remaining layers are given by

$$q_i = C_i(T_i)\left[T_{i-1} - T_i\right] + \sigma\, R_i(T_i)\left[T_{i-1}^4 - T_i^4\right], \qquad i = 2, \cdots, n\,. \tag{3}$$

With (2) and (3), a system of n differential equations for the temperatures T_i of each layer is obtained

$$\dot{T}_i = \frac{q_i(T_{i-1}, T_i) - q_{i+1}(T_i, T_{i+1})}{C_{P_i}(T_i)}, \qquad i = 1, \cdots, n\,. \tag{4}$$

The quantity ε denotes the emissivity of the material and σ is the Boltzmann constant. C_i, R_i and C_{P_i} are model functions describing the properties of the different materials.

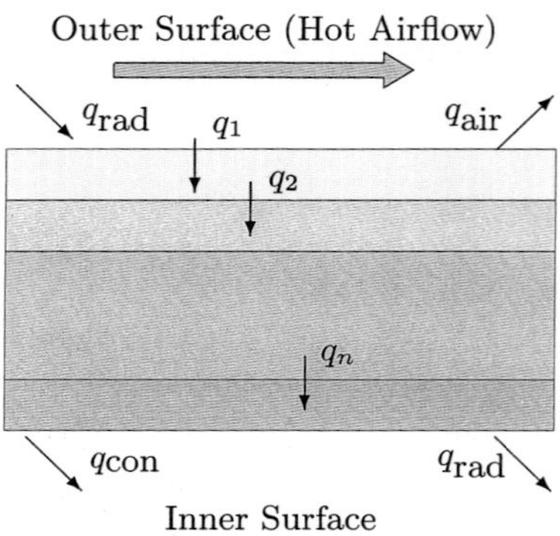

Fig. 4. Heat flux model with thermal knots

3 Optimization Problem

The performance index to be minimized is the fuel consumption for a range cruise on a great circle, subject to (1) and (4) as well as to control and state constraints. Applying the Minimum Principle [3] the cost functional reads

$$I[\mathbf{u}] = \frac{1}{m|_{t_f}} \rightarrow \text{Min} \tag{5}$$

where **u** denotes the control vector. The control and state constraints are given by

$$\begin{array}{llll} \alpha_{min} & \leq & \alpha & \leq \alpha_{max} \, , \\ 0 & \leq & \delta_T & \leq \delta_{Tmax} \, , \\ \bar{q}_{min} & \leq & \bar{q}(V,h) & \leq \bar{q}_{max} \, , \\ n_{min} & \leq & n(V,h,m;\alpha) & \leq n_{max} \end{array} \tag{6}$$

where $\bar{q}$ is the dynamic pressure and n the load factor.

The integrated heat flux at the most interior wall layer is also treated as a constraint

$$\int_0^{t_f} q_n \, dt \leq \Gamma_{n_{limit}} \, . \tag{7}$$

This relation couples the differential equations for the dynamics of the vehicle and for the heat flux.

Boundary conditions are specified for the initial and final states. They are given in Table 1.

The resulting optimal control problem is solved by using the efficient optimization technique *DIRCOL*, based on a direct collocation approach [9]. To reduce the dimension of the optimal control problem the number of layers of the thermal protection system is set to five. The obtained results are then recalculated with the advanced ODE-Solver *LSODA* which is capable of dealing with stiff and non-stiff systems [7, 8]. For this case the number of layers is increased to ten to improve the model.

Table 1. Boundary conditions

	h [m]	V [m/s]	γ [°]	x [km]	m [kg]	T_i [K]	Γ_n [kJ/m^2]
$t=0$	500	150	0	0	244000	300	0
$t=t_f$	500	150	0	9000	–	–	–

4 Results

Results for the optimization problem are presented in Figs. 5 and 6. In Fig. 5, the time histories for state and control variables are depicted. Figure 6 shows the temperatures of the different layers for the 10-layer model. The case without a heat flux constraint is used as a reference. Table 2 presents the most relevant quantities for the constrained and unconstrained cases.

The results depicted in Fig. 6 show that there is a significant reduction of the temperature level during the entire flight. The decrease of the maximum temperature at the first layer is presented in Table 2.

The heat flux reduction is due to conducting the range cruise at significantly lower levels of speed and altitude. This leads to an increase of the overall flight time of almost 20%. The fuel penalty is comparatively small, showing an increase of about 2%.

Table 2. Optimization results

	Γ_{10} [kJ/m^2]	$T_{1_{max}}$ [K]	m_{fuel} [kg]	t_f [s]	I [kg^{-1}]
Constrained	1935.21	702.2	63950.3	8113.4	$5.554 \cdot 10^{-6}$
Unconstrained	2129.82	816.97	62659.9	6800.0	$5.515 \cdot 10^{-6}$

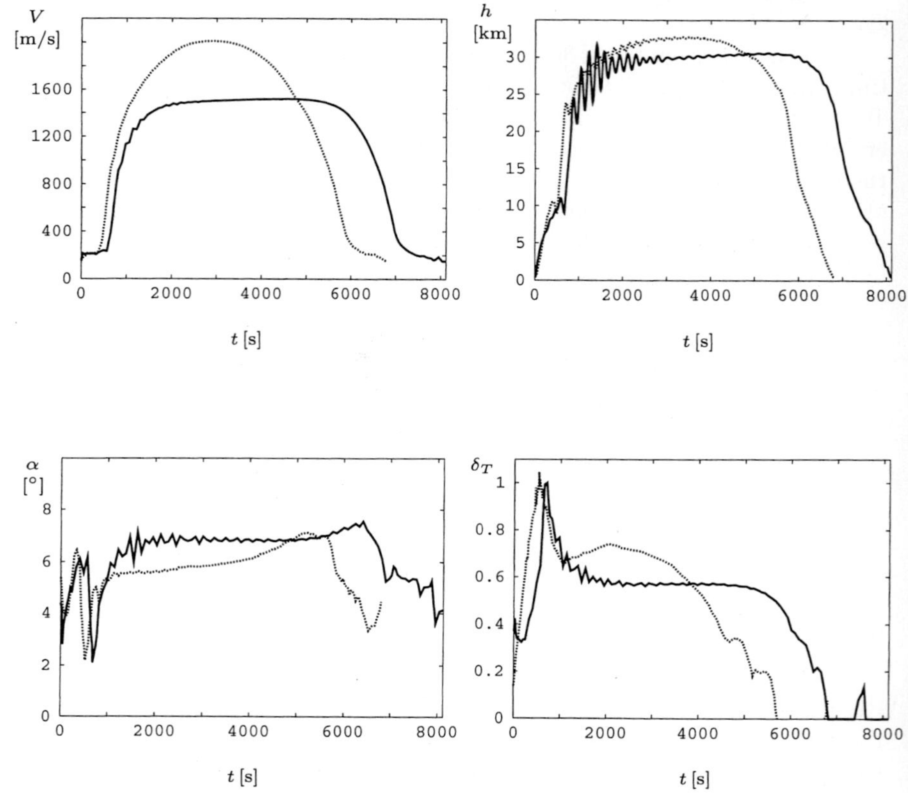

Fig. 5. State and control variables during the range cruise for the unconstrained (dotted lines) and the constrained cases (solid lines)

In addition to the described optimization, an investigation of the heat load at the lower surface of the vehicle was conducted for the region between the nose and the tank area, applying a 10-layer model. The ODE-Solver *LSODA* was used for the computation. The results presented in Fig. 7 show that there are two parts with a different behavior. This can be attributed to the transition from laminar to turbulent flow characteristics, using a model according to [1, 4].

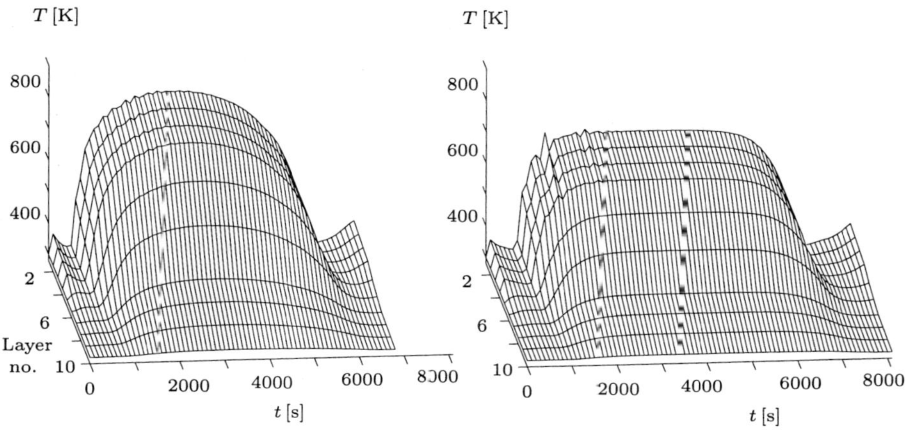

Fig. 6. Temperatures in the different layers during the range cruise for the unconstrained (left) and the constrained (right) cases

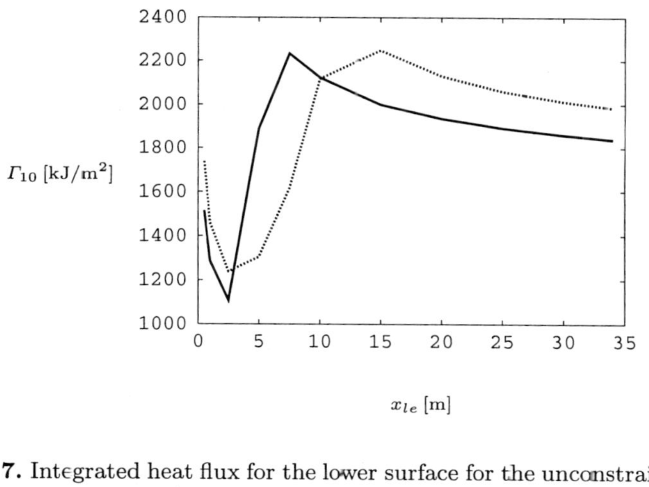

Fig. 7. Integrated heat flux for the lower surface for the unconstrained (dotted line) and the constrained (solid line) cases

5 Conclusions

Heat load reduction by means of optimal trajectory control is considered for a winged hypersonic vehicle equipped with a turbo/ram jet engines combination. A complex mathematical model is used to simulate the heat flux through the thermal protection system, with particular emphasis placed on unsteady effects. Furthermore, effects concerning transition from laminar to turbulent flow are accounted for. The systems of the differential equations for describing the motion of the vehicle and the unsteady heat load are coupled, using a constraint for the heat flux. Solutions are constructed with an efficient optimization technique and an advanced ODE-Solver. It is shown that the heat input can be significantly reduced, with only a small increase in fuel consumption.

References

1. Anderson, J. D. (1989) Hypersonic and High Temperature Gas Dynamics. McGraw-Hill Book Company, New York.
2. Bayer, R., Sachs, G. (1995) Optimal Return-to-Base Cruise of Hypersonic Carrier Vehicles. Zeitschrift für Flugwissenschaften und Weltraumforschung **19**(1), 47-54.
3. Bryson A. E., Ho Y. C. (1975) Applied Optimal Control, Hemisphere Publishing Corporation, New York.
4. Dinkelmann, M. (1997) Reduzierung der thermischen Belastung eines Hyperschallflugzeugs durch optimale Bahnsteuerung. Ph.D. Thesis, Technische Universität München, München.
5. Dinkelmann, M., Wächter, M., Sachs, G. (2000) Modelling and Simulation of Unsteady Heat Transfer for Aerospacecraft Trajectory Optimization. Mathematics and Computers in Simulation **53**, 389-394.
6. Grallert, H., Vollmer, K. (1993) Conceptual Design of the Reference Concept SÄNGER by Means of Advanced Methods. In AIAA/DGLR 5th International Aerospace Planes and Hypersonics Technologies Conference, München, AIAA-93-5085.
7. Hairer, E., Wanner, G. (1991) Solving Ordinary Differential Equations II, Springer Verlag, Berlin.
8. Strehmel, K., Weiner, R. (1995) Numerik gewöhnlicher Differentialgleichungen, Teubner, Stuttgart.
9. Stryk, O. von (1999) User's Guide for DIRCOL. A Direct Collocation Method for the Numerical Solution of Optimal Control Problems. User Manual, Technische Universität München, München.
10. Vinh, N. X., Busemann, A., Culp, R. D. (1980) Hypersonic and Planetary Entry Flight Mechanics, The University of Michigan Press, Ann Arbor.
11. Windhorst, R., Ardema, M. D., Bowles, J. V. (1997) Minimum Heating Reentry Trajectories for Advanced Hypersonic Launch Vehicles. In AIAA Guidance, Navigation, and Control Conference, New Orleans, LA, AIAA-97-3535.

Modeling Techniques and Parameter Estimation for the Simulation of Complex Vehicle Structures

T. Butz[1,2], O. von Stryk[2], C. Chucholowski[1], St. Truskawa[1], and T.-M. Wolter[1]

[1] TESIS *DYNAware* GmbH, Implerstraße 26, D-81371 München, www.tesis.de
[2] FG Simulation und Systemoptimierung, Technische Universität Darmstadt, Alexanderstr. 10, D-64283 Darmstadt, www.sim.informatik.tu-darmstadt.de

Abstract. The numerical simulation of complex vehicle structures requires dynamic models for passenger cars as well as for trucks and vehicles with trailers. Tailored numerical modeling and integration techniques must be employed to achieve real-time capability of the considered vehicle dynamics program which is vital for its use within hardware-in-the-loop test-benches. To efficiently calibrate the vehicle model a parameter estimation tool was developed which relies on observations obtained from driving tests. Combining robust nonlinear optimization algorithms and careful numerical differentiation it is well suited for low-cost parallel computing platforms, such as heterogeneous PC clusters, which are usually available for automotive suppliers and industries employing vehicle dynamics simulations.

1 Introduction

Simulations of the full vehicle dynamics play a key role in automotive development, since they enable the road performance and the handling properties of a new car to be investigated in advance. Besides reducing the need for physical prototyping, numerical simulations may be used within software- and hardware-in-the-loop test-benches, which allow control units, such as anti-lock braking systems and electronic stability programs, to be tested without danger for test driver and vehicle.

The development of intricate control devices and strategies requires the virtual car to reproduce the complex behavior of the real vehicle in detail. Therefore, we employ a sophisticated vehicle model which comprises a suitable multi-body system as well as a realistic tire model. In this paper, the model is extended to trucks and vehicles with trailers. The use of tailored modeling and numerical integration techniques enables the entire vehicle dynamics to be described by a large system of ordinary differential equations which can be solved in real time.

The calibration of the model substantially relies on the application of robust nonlinear optimization algorithms and careful numerical differentiation. The resulting parameter estimation scheme allows model coefficients which are not available from physical measurements to be determined efficiently. The developed program tool is suitable for the parallel use on low-cost computing platforms such as heterogeneous PC networks. It was successfully applied to adjust selected vehicle model parameters for a passenger car.

2 Simulation of Full Vehicle Dynamics

The vehicle dynamics program *veDYNA* [10] which forms the basis of the following investigations is developed and commercially distributed by TESIS *DYNAware*, München. The integration of the program core into a Matlab/Simulink [12,13] environment enables versatile application and easy handling through a graphical user interface.

The vehicle model in *veDYNA* consists of a system of rigid bodies which comprise the vehicle body, the axle suspensions and the wheels. Additional partial models depict the characteristics of the drive train, the steering mechanism and the tires [9]. While general purpose methods for modeling multibody dynamics would yield a differential-algebraic system of index 3, we make use of appropriate minimum coordinates which avoid algebraic constraints in the equations of motion

$$
\begin{aligned}
M_{BV}\,\dot{z}_{BV} &= Q_{BV}(y_{BV}, z_{BV}, y_{ST}, z_{ST}, y_{DT}, z_{DT}) && (1)\\
\dot{y}_{BV} &= K_{BV}^{-1}(y_{BV})\,z_{BV} && (2)\\
M_{DT}\,\dot{z}_{DT} &= Q_{DT}(y_{DT}, z_{DT}) && (3)\\
\dot{y}_{DT} &= V_{DT}\,z_{DT} && (4)\\
M_{ST}\,\dot{z}_{ST} &= Q_{ST}(y_{ST}, z_{ST}) && (5)\\
\dot{y}_{ST} &= V_{ST}\,z_{ST} && (6)\\
D\,\dot{y}_{T} &= F_{stat} - C\,y_{T}. && (7)
\end{aligned}
$$

Thus, the vehicle dynamics in *veDYNA* is fully described by a system of first-order differential equations comprising 24 ODEs (1), (2) for the basic vehicle, 19 equations (3), (4) governing the dynamics of the drive train, five ODEs (5), (6) for the steering system as well as eight additional ODEs (7) which depict the deviations of the tires. Couplings between the separate systems occur by way of the generalized forces and torques Q_{BV}.

The basic vehicle model is now extended to vehicles with two rear axles and vehicles with trailers. Thus, it is possible to analyze the dynamics of single trucks as well as of trucks and passenger cars with trailers. Besides investigating stability issues in truck and trailer design, this feature allows the development of anti-roll control strategies for semi-trailers. Moreover, *veDYNA* may be employed in a hardware-in-the-loop setup to investigate anti-lock braking systems for vehicles with caravans (cf. Fig. 1).

In *veDYNA* the trailer is treated as a separate vehicle which lacks a drive train and is coupled mechanically to the leading vehicle. The transmission of

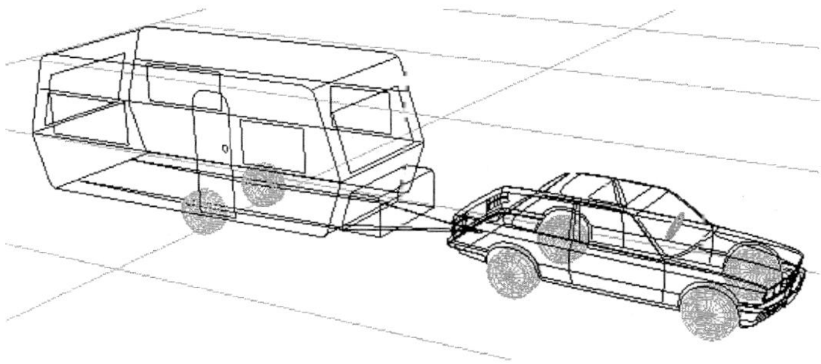

Fig. 1. *veDYNA* model for vehicles with trailers

the driving moment to the trailer is modeled by elastic and frictional forces whose strengths depend on the geometry and the rigidity of the coupling. To compute the magnitude of these forces a set of local coordinates and angles is introduced which allow the relative position and orientation of vehicle and trailer to be determined.

For a realistic implementation of virtual driving tests on the computer also a nonlinear driver model and a program module for the simulation of arbitrary road geometries and conditions were developed [3,10].

The equations of motion for vehicle, trailer and driver result in a system of stiff ordinary differential equations which need to be integrated with a specifically tailored algorithm. The use of a semi-implicit Euler scheme [9] enables a stable numerical solution for integration steps of several milliseconds. Due to the absence of drive train and driver in the model for the trailer the computational complexity of the coupled vehicle models is reduced proportionately. Thus, step sizes in the range of few milliseconds allow real-time simulations of complex vehicle structures on current PC hardware.

3 Estimation of Vehicle Model Parameters

The system of differential equations describing the dynamics of vehicle, trailer and driver in *veDYNA* may be summarized by

$$\dot{x}(t) = g(x(t), u(t), p, t) \tag{8}$$

with suitable initial values $x(t_0) = x_0$. In addition to the vehicle's state variables $x(t) \in \mathbb{R}^{n_x}$ comprising the generalized coordinates and velocities, the full-vehicle performance is also governed by a number of control variables $u(t) \in \mathbb{R}^{n_u}$ which determine the respective driving maneuver. The unknown model parameters $p \in \mathbb{R}^{n_p}$ are constant for all times t.

The calibration of the *veDYNA* vehicle model aims to determine model coefficients of the virtual car which are not directly available from physical measurements. Suitable parameter values shall be estimated such that the numerical integration results closely fit the observed behavior of a physical prototype. Consequently, we obtain the nonlinear least-squares problem

$$\underset{p \in \mathbb{R}^{n_p}}{\text{minimize}}\ r(p) := \frac{1}{2}\|f(p)\|_2^2 := \frac{1}{2}\sum_{j=1}^{n_t}\sum_{i \in I_j}\left(\eta_{ij} - x_i(t_j, p)\right)^2 \tag{9}$$

where the η_{ij}, $i \in I_j$, are measurements of selected vehicle state variables recorded at the times t_j during a driving test, and $x(t,p)$ denotes the corresponding numerical solution for a specific parameter set $p \in \mathbb{R}^{n_p}$. Usually, additional box constraints

$$l_i \leq p_i \leq u_i, \quad i = 1, ..., n_p, \tag{10}$$

on the parameter range have to be considered which shall ensure optimization results compatible with the real vehicle properties.

For the solution of (9), (10) a parameter estimation tool was developed which allows the application of robust mathematical optimization algorithms [1,2]. In addition, a Matlab implementation is now provided which enables comfortable pre-processing of the experimental data and easy handling of the numerical optimization codes through a graphical user interface. Moreover, it supports the visualization of the identification results as well as a mathematical sensitivity analysis, which can be used to determine the impact of single parameters on (9) and the reliability of their numerical estimates.

Currently, the employed optimization algorithms consist of the implementation LMDER of the Levenberg-Marquardt method [7], which is based on the iterative solution of approximate linear least-squares problems, and the implicit filtering code IFFCO [6], a quasi-Newton algorithm for noisy optimization problems. Both algorithms are available from the World Wide Web and were furnished with an application program interface for Matlab which allows them to be called from the Matlab prompt.

Moreover, the general design of the user interface also enables a variety of other methods, as for instance from the Matlab Optimization Toolbox [11], to be incorporated easily. As an example, two gradient-free search techniques were employed for the parameter estimation. The Matlab implementation FMINS of the Nelder-Mead simplex algorithm was extended to handle parallel objective evaluations and problems with simple bounds (10). Further, the evolutionary strategy EVOCLASS due to [8] was implemented whereupon a parallel interface and the treatment of box constraints were introduced.

For the optimization with gradient-based methods the Jacobian matrix $\partial f / \partial p$ of the objective (9) must be supplied. However, the complexity of the underlying vehicle model does not support analytic or internal numerical differentiation techniques for this purpose. Rather the required first-order derivatives need to be determined by means of finite difference approximations where the increments must be chosen carefully such as to account for truncation, condition, and rounding errors [5].

For LMDER the partial derivatives $\partial_i f(p)$ with respect to the i-th parameter are obtained from the one-sided differences

$$(\partial_i f(p))_{\pm h} = \frac{f(p \pm h e_i) - f(p)}{\pm h}, \tag{11}$$

depending on the feasibility of $p + he_i$ or $p - he_i$. Further, the variable finite difference strategy in IFFCO is based on central differences

$$(\partial_i r(p))_{2h} = \frac{r(p + he_i) - r(p - he_i)}{2h} \tag{12}$$

provided that both $p + he_i$ and $p - he_i$ are feasible; otherwise (11) is used as well. In both cases, $e_i \in \mathbb{R}^{n_p}$ denotes the i-th canonical unit vector, and $h > 0$ is a suitable finite difference increment.

Obviously, this strategy entails a large number of additional evaluations of the objective function depending on the derivative approximation being used. But also the application of gradient-free methods often requires multiple function evaluations to be performed at the same time, when for instance the Nelder-Mead simplex or new populations for the evolutionary strategy are generated. Accordingly, the computational time for the optimization is significantly reduced by distributing the simultaneous objective evaluations among several processors.

As to the parallel platform a heterogeneous PC cluster with MS Windows operating systems was presumed which is well-priced and usually available for small and medium-sized automotive suppliers and industries. The communication of data across the network was realized by means of the ONC RPC library from Sun Microsystems, ported to MS Windows systems [4]. Matlab application program interfaces for the relevant routines were developed making the parallel setup available from the Matlab prompt. Thus, the parallel distribution of simultaneous function evaluations and the management of the RPC network is enabled through a graphical user interface for Matlab.

4 Results and Conclusions

In [1,2] the numerical parameter estimation scheme was successfully employed to determine selected coefficients for the vehicle model of a passenger car; though, the parameter estimation had been carried out with a stand-alone console application of *veDYNA* and the associated optimization frame. Here, we present results which were obtained from the newly developed parameter estimation tool for the Matlab/Simulink emulation of *veDYNA*.

As an example, we consider the calibration of the brake friction coefficients of the vehicle model which indicate the ratio for the generated braking moments transmitted to the wheels. Since the brakes in *veDYNA* are uniformly modeled as disc brakes, the adjustment of these parameters to the actual brake mechanism represents a common problem in practice. The remaining parameters of the *veDYNA* vehicle model as well as the underlying maneuver data were provided by an automotive manufacturer for testing purposes.

The recorded measurements consisted of the brake pressures and the rotational wheel speeds throughout the full braking of a passenger car equipped with an anti-lock braking system. The actual braking was preceded by a speed-up phase where the vehicle was accelerated to the experimentally observed initial speed of approximately 106.8 km/h. Thus, the replication of the driving maneuver by simulation resulted in a total maneuver time of 20.25 s where 17.1 s and 3.15 s were required for the acceleration and the braking phase respectively. For the purpose of a realistic simulation the observed brake pressures which were measured every 0.001 s were used as input control variables. The nonlinear least-squares criterion (9) consisted of the experimental wheel speeds, also recorded at regular intervals of 0.001 s, set off against the corresponding numerical integration results, which yields a total number of 12600 addends.

Starting with initial values

$$p^0 = (0.45, 0.45, 0.45, 0.45)^T \tag{13}$$

which were given by the default values from the provided vehicle database, our parameter identification tool was used to determine suitable estimates for the four unknown coefficients. Here, the initial guesses (13) refer to the friction coefficients at the left and right front wheels and the left and right rear wheels respectively. The associated least-squares residual (9) was given by $r(p^0) = 6.374 \cdot 10^6$.

Due to the optimization the objective value was reduced to $r(p^*) = 7.713 \cdot 10^4$ which is approximately one percent of its initial size. The minimum residual was assumed for the parameter values

$$p^* = (0.2505, 0.2670, 0.4253, 0.3662)^T \; . \tag{14}$$

A numerical sensitivity analysis revealed small confidence intervals indicating that reliable estimates were obtained. The deviations between the coefficients (14) at the rear wheels allow for the different wear of the respective braking devices and tires as well as for possible measurement errors.

A comparison between the measured wheel speeds and the corresponding simulation results for the optimal solution (14) is depicted in Fig. 2. Obviously, good agreement was achieved for the respective characteristics. The remaining deviations between the computed and the experimentally observed values must be attributed to the uneven surface of the test track which is not depicted in detail by the road model of *veDYNA*.

The quoted computations were carried out on a homogeneous PC cluster consisting of eight Intel Pentium 500 MHz processors with Windows NT 4.0 operating systems. Each evaluation of the least-squares residual (9), which comprised a full vehicle dynamics simulation with *veDYNA* and a record of the simulation data, required a computational time of about 45 seconds. The

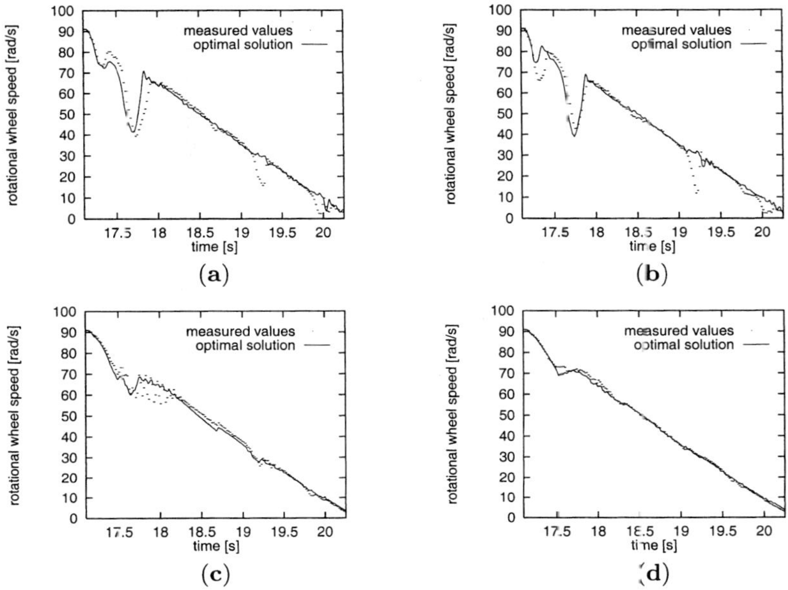

Fig. 2. Comparison between observed and computed rotational speeds for the left (**a**) and right (**b**) front wheels and the left (**c**) and right (**d**) rear wheels

software packages Matlab 5.3 and Simulink Toolbox 3.0 as well as the present version *veDYNA* 3.3.2 were used.

For the numerical solution of the parameter estimation problem we employed the Matlab ports of LMDER und IFFCO as well as the extended simplex algorithm FMINSB and the newly developed evolutionary strategy EVOCLASS. The latter pursued a (3,8)-strategy where three current parameter sets served to generate eight new estimates in each iteration.

For comparison, the numerical optimization was carried out sequentially as well as in parallel. The parallel optimization with LMDER and FMINSB required four processors treating the additional function evaluations for the one-sided differences (11) and the geometrical transformations of the simplex. For the application of IFFCO and EVOCLASS eight processors were supplied which served to compute the symmetric differences (12) and the respective new generations. Dynamic load sharing was implemented such as to achieve minimum CPU times.

Table 1 compares the different optimization codes, giving the least-squares residual $r(p^*)$ for the respective solutions, and the computational times t_{seq} and t_{par} for the sequential and parallel executions of the parameter estimation. The ratios of these CPU times, i. e., the achieved parallel speed-ups, are shown in the last column. Also listed are the numbers of objective evaluations n_{seq} and n_{par} performed during the entire optimization and by the client process in the parallel framework respectively.

Table 1. Computational results for the employed optimization algorithms

Optimization code	$r(p^*)$	n_{seq}	t_{seq} [s]	n_{par}	t_{par} [s]	$n_{\text{seq}}/n_{\text{par}}$	$t_{\text{seq}}/t_{\text{par}}$
LMDER	$7.730 \cdot 10^4$	43	1946.1	19	897.0	2.26	2.17
IFFCO	$7.726 \cdot 10^4$	288	12070.0	103	4410.3	2.80	2.74
FMINSB	$7.733 \cdot 10^4$	234	9987.9	134	5961.8	1.75	1.68
EVOCLASS	$7.716 \cdot 10^4$	1201	52225.8	151	6715.6	7.95	7.78

For this problem, all investigated algorithms produced reasonably small residuals. The parallel treatment of the finite difference computations reduced the CPU time for both gradient-based codes by more than a factor of two. The maximum parallel speed-up is achieved by the evolutionary strategy where each new generation requires a corresponding number of simultaneous evaluations of the objective function. Accordingly, the parallel approach achieves competitive computational times also for the gradient-free methods.

References

1. T. Butz, O. von Stryk, M. Vögel, T.-M. Wolter, C. Chucholowski: *Parallel parameter estimation in full motor vehicle dynamics.* SIAM News 33, 4 (2000)
2. T. Butz, O. von Stryk, T.-M. Wolter: *A parallel optimization scheme for parameter estimation in motor vehicle dynamics.* In: A. Bode et al. (eds.): Euro-Par 2000 - Parallel Processing. Lecture Notes in Computer Science 1900, Springer, Berlin (2000) 829–834
3. C. Chucholowski, M. Vögel, O. von Stryk, T.-M. Wolter: *Real time simulation and online control for virtual test drives of cars.* In: H.-J. Bungartz et al. (eds.): High Performance Scientific and Engineering Computing. Lecture Notes in Computational Science and Engineering 8, Springer, Berlin (1999) 157–166
4. M. Gergeleit: *ONC RPC for Windows NT Homepage.* World Wide Web, http://www.dcs.qmw.ac.uk/~williams/nisgina-current/src/rpc110/oncrpc.htm (1996)
5. P. E. Gill, W. Murray, M. H. Wright: *Practical Optimization.* Academic Press, London New York (1981)
6. P. Gilmore: *IFFCO: Implicit Filtering for Constrained Optimization, User's Guide.* Technical Report CRSC-TR93-7, Center for Research in Scientific Computation, North Carolina State University, Raleigh (1993)
7. J. J. Moré: *The Levenberg-Marquardt Algorithm: Implementation and Theory.* In: A. Dold, B. Eckmann (eds.): Numerical Analysis. Lecture Notes in Mathematics 630. Springer, Berlin Heidelberg (1978) 105–116
8. I. Rechenberg: Evolutionsstrategie. Frommann-Holzboog, Stuttgart (1994)
9. G. Rill: *Simulation von Kraftfahrzeugen.* Vieweg, Braunschweig (1994)
10. TESIS *DYNAware: veDYNA User's Guide.* München (1997)
11. The MathWorks Inc.: *Optimization Toolbox User's Guide.* Nattick (1999)
12. The MathWorks Inc.: *Using MATLAB.* Nattick (1999)
13. The MathWorks Inc.: *Using Simulink.* Nattick (1999)

Part V

Optimization of Electronic Circuits

Numerical Techniques for Different Time Scales in Electric Circuit Simulation

A. Bartel, M. Günther, R. Pulch, and P. Rentrop

Centre for Scientific Computing and Mathematical Modelling, University of Karlsruhe, Engesser Straße 6, D-76128 Karlsruhe, Germany.

Abstract. The simulation of integrated circuits demands an increasing amount of computational resources, since systems become more and more complex and parasitic effects are included to a larger extent. However, the number of active elements keeps relatively small: less than 10%. In other settings digital and analog circuits are coupled, which yields systems with largely differing time scales. We will discuss two techniques, which are adapted to those cases. One approach is based on multirate Rosenbrock-Wanner schemes, the other leads to a PDE-model for driven oscillators.

1 Introduction

Today, electronic design rests upon numerical simulation. Due to an ongoing development, the integration rate and complexity increases extensively. Therefore computation time is one decisive factor and algorithms have to exploit as much structure as possible. Since latency accompanies complexity, multirate algorithms, which are adapted to latent-active settings, will be more and more considerable. Here, only active parts are closely tracked, while the remainder is computed on a larger scale.
Another current application field is the coupling of analog and digital circuits. Signals with widely separated times rates meet, e.g. analog signals in 10^{-6} sec and digital signals in 10^{-9} sec. In radio frequency applications, a multidimensional signal model allows to decouple time scales of mixed signals. This leads to a PDE description for the circuit, which avoids the disadvantageous time behaviour of the ODE model.

First, we outline the usual modelling procedure in electric circuit simulation. Chapter 3 concentrates on the numerical time integration. The standard multistep approach BDF is compared to the one-step method CHORAL. Then the latent-active multirate scheme is discussed and applied to the inverter chain benchmark. In Chapter 5, the multidimensional signal model is presented and its ODE-PDE relations are stated. Finally an outlook on future work is given.

2 Basics of Electric Circuit Simulation

Electric circuits are modelled as lumped networks. Therefore, the setup splits in a sole description of basic elements, which specifies the local electrical behaviour, and topological rules, which take interconnections into account. This represents a typical network approach.

2.1 Setup of Network Equations: MNA

Circuit simulation analyses the transient behaviour of node-voltages and branch-currents. First, current-voltage relations characterise each element of the circuit. Table 1 summarises the basic elements. Second, the topology of the circuit specifies the connections of the elements. Both together comprise the network equations.

The common strategy to derive the network equations is modified nodal analysis (MNA) [4,14]. This technique applies Kirchhoff's laws for the mathematical formulation of the topology part. Kirchhoff's current law (KCL) states that the sum of currents traversing any node is equal to zero. Thus KCL describes the topology of the electrical network. Kirchhoff's voltage law (KVL) implies that the sum of voltage drops of any loop is equal to zero. This allows us the introduction of node potentials with one reference node, e.g. the ground.

The MNA approach works as follows. First, we apply KCL in each node. Then all branch-currents, which possess an element relation depending on node voltages, are substituted. Finally, we add the equations of current controlled elements. As an example, we assemble the network equations for a tunnel-diode-oscillator, sketched in Fig. 1. Here, the voltage-current relation of the diode is given by a nonlinear function g.

$$
\begin{aligned}
\text{node 1}: &\quad 0 = I_S - I_R = I_S - R^{-1}(U_1 - U_2)\\
\text{node 2}: &\quad 0 = I_R - I_L = R^{-1}(U_1 - U_2) - I_L\\
\text{node 3}: &\quad 0 = I_L - I_C - I_D = I_L - C(\dot{U}_3 - \dot{U}_0) - g(U_3 - U_0)\\
\text{voltage source}: &\quad U_1 = U_{op}(t)\\
\text{inductor}: &\quad L\dot{I}_L = U_2 - U_3
\end{aligned}
$$

Table 1. Basic elements

element	characteristic equation	symbol
linear resistor	$I = U/R$	
linear capacitor	$I = C\dot{U}$	
linear inductor	$U = L\dot{I}$	
voltage source	$U = U(t)$	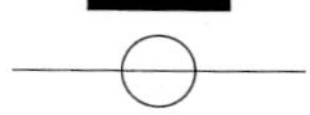
current source	$I = I(t)$	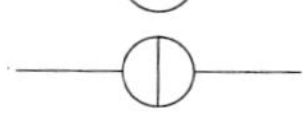

We rewrite these network equations in matrix notation

$$\begin{pmatrix} 0 & 0 & 0 & 0 & 0 \\ 0 & 0 & 0 & 0 & 0 \\ 0 & 0 & -C & 0 & 0 \\ 0 & 0 & 0 & 0 & 0 \\ 0 & 0 & 0 & 0 & L \end{pmatrix} \begin{pmatrix} \dot{U}_1 \\ \dot{U}_2 \\ \dot{U}_3 \\ \dot{I}_S \\ \dot{I}_L \end{pmatrix} = \begin{pmatrix} R^{-1}U_1 - R^{-1}U_2 - I_S \\ R^{-1}U_2 - R^{-1}U_1 - I_L \\ g(U_3) - I_L \\ U_{op}(t) - U_1 \\ U_3 - U_2 \end{pmatrix} \tag{1}$$

and we immediately realize that the 5×5 capacitance matrix has only rank 2 here.

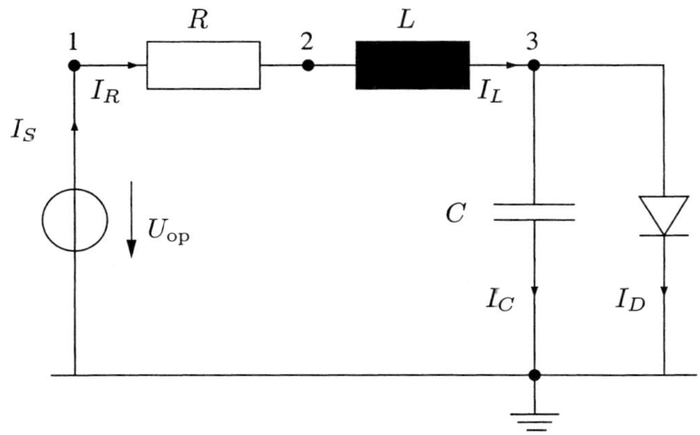

Fig. 1. Tunnel-diode-oscillator

In general, the MNA yields large systems of differential equations of the form

$$C(\mathbf{x})\dot{\mathbf{x}} = \mathbf{f}(t, \mathbf{x}), \tag{2}$$

where the matrix $C(\mathbf{x})$ is singular. Hence (2) is referred to as a singular quasi-linear implicit ordinary differential equation (ODE). This type of differential equations demands a special numerical treatment. Next section will focus on this aspect. The advantage of the MNA is that it enables an automatic generation using the element and topology information only.

2.2 Substitute Circuits for Devices

Semi-conductor devices are more complex, since their electric effects are described by partial differential equations. To be able to handle those elements in the MNA, too, the device is replaced by a substitute circuit, which includes only basic elements. Fig. 2 illustrates a circuit of a single inverter, which is used for inverting the logical values 0 or 1. It contains one semi-conductor device, namely a MOSFET (metal oxide substratum field effect transistor). The four terminals of the MOSFET are called gate (G), drain (D), source (S) and bulk (B). Now the MOSFET is substituted by one of the shown networks. The simplest model consists in replacing the MOSFET by a switch between source and drain, which is controlled by the voltages of gate and bulk. A better approach is to introduce a current source between source and drain, which is also voltage controlled by gate and bulk. This current results in

$$\begin{aligned} i_{DS} &= \eta f(u_G, u_D, u_S, u_B) \\ f(u_G, u_D, u_S, u_B) &= \max\{(u_G - u_B - u_S - U_{thres}), 0\}^2 \\ &\quad - \max\{(u_G - u_B - u_D - U_{thres}), 0\}^2 \end{aligned} \tag{3}$$

with a coefficient η and a threshold voltage of 1V, for example. The model by Shichman-Hodges substitutes the MOSFET by a larger alternative network, which is illustrated in the right of Fig. 2. More sophisticated substitute circuits for the MOSFET also exist, which include up to 150 fitting parameters. Thus we are able to apply the MNA technique to the substitute model.

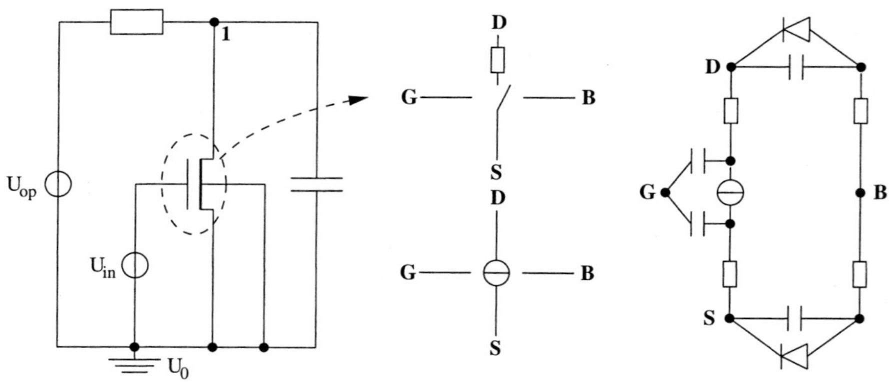

Fig. 2. Single inverter with MOSFET substitute circuits

2.3 TITAN

Now numerical simulation consists in the time integration of the network equations. A typical simulation tool in industry is the TITAN package, which has been developed by Infineon. The code consists of more than 700.000 lines basing mainly on a FORTRAN source and more recently C and C++ sections. TITAN generates the network equations by MNA. Therefore the topology of the network and its elements have to be specified in a netlist. This input data is of the same form as for the simulator SPICE [18]. Typically, the equations do not exist in symbolic mode. For more details, we refer to [7].

3 Numerical Discretisation Techniques

The automatic modelling process based on MNA preserves the network topology, while it allows redundancy in the variables. Thus the network equations are generally formulated in descriptor form, which combines implicit differential equations and nonlinear equations:

$$\mathcal{F}(\dot{\mathbf{q}}, \mathbf{x}, t) = A \cdot \dot{\mathbf{q}}(\mathbf{x}, t) + \mathbf{f}(\mathbf{x}) + \mathbf{s}(t) = \mathbf{0} \quad \text{on} \quad t \in [0, T], \tag{4}$$

where A denotes the capacitive and inductive part of the network incidence matrix and in general fails to have full row rank. Therefore consistent initial values $\mathbf{x}(0) = \mathbf{x}_0$ have to be provided; the notation involves $\mathbf{x}$: vector of all node potentials and branch currents of voltage defining elements, $\mathbf{q}(\mathbf{x}, t)$: terminal charges and branch-fluxes, $\mathbf{f}(\mathbf{x})$: currents through static elements and voltage drops of voltage controlling elements and $\mathbf{s}(t)$: independent sources. It has to be noted, that (4) is in *charge-oriented* form, which is standard in circuit simulation and provides charge conservation in the devices, cf. [10].

The first questions to be answered are, when do solutions of systems (4) exist, when is uniqueness achieved. Furthermore, the matter of well-posedness has to be viewed. This leads to the *index* concept.

3.1 Index for DAEs

We investigate a linearisation of (4) along a solution, namely

$$\mathcal{A}\dot{\mathbf{x}} = \mathcal{B}\mathbf{x} + \mathbf{s}(t). \tag{5}$$

Then the conductance matrix $\mathcal{B}$ is regular, since loops, which consist of independent voltage sources and inductors only, and cutsets, which just comprise current sources and capacitors, are forbidden (otherwise the circuit would be inconsistent or there would be islands of undetermined variables). Thus we can left-multiply (5) by $\mathcal{B}^{-1}$ and decompose $\mathcal{B}^{-1}\mathcal{A} = \mathcal{T}^{-1}\mathcal{J}\mathcal{T}$, where the Jordan form $\mathcal{J} = \mathrm{diag}(\mathcal{C}, \mathcal{N})$ comprises the regular $\mathcal{C}$ and nilpotent $\mathcal{N}$. Now, we split up the vector of unknowns: according to the Jordan form we have $\mathcal{T}\mathbf{x} = (\mathbf{y}, \mathbf{z})^\top$, such that $\mathbf{z}$ describes the coordinates of $\mathbf{x}$ belonging to eigenvalue zero. Therefore (5) becomes

$$\begin{pmatrix} \mathcal{C} & \mathbf{0} \\ \mathbf{0} & \mathcal{N} \end{pmatrix} \begin{pmatrix} \dot{\mathbf{y}} \\ \dot{\mathbf{z}} \end{pmatrix} = \begin{pmatrix} \mathbf{y} + \mathbf{f}(t) \\ \mathbf{z} + \mathbf{g}(t) \end{pmatrix} \quad \text{with} \quad \begin{pmatrix} \mathbf{f}(t) \\ \mathbf{g}(t) \end{pmatrix} = \mathcal{T}\mathcal{B}^{-1}\mathbf{s}(t), \tag{6}$$

that is, our system splits in an ODE for $\mathbf{y}$ and the nilpotent part

$$\mathcal{N}\dot{\mathbf{z}} = \mathbf{z} + \mathbf{g}(t). \tag{7}$$

Assuming nilpotency k, then by a chain of $k-1$ differentiations of (7) one obtains an explicit expression for $\mathbf{z}$

$$\mathbf{0} = \mathcal{N}^k \mathbf{z}^{(k)} = \mathbf{z} + \sum_{j=0}^{k-1} \mathbf{g}^{(j)}(t), \tag{8}$$

which fixes $\mathbf{z}$ completely. Therefore these components are referred to as the algebraic part. Moreover, we need to impose on (5) an additional hypothesis: $\mathbf{s} \in C^{k-1}[0,T]$ [1]. From (8) we encounter, that derivatives of inputs may enter the solution; thus the solution does not depend continuously differentiable on input data. Since the integer k indicates the degree of ill-posedness it is referred to as the index of the DAE.

Various generalisations to nonlinear settings exist. The special structure of the network equations in circuit simulation guarantees the index to be at most two by excluding arbitrary controlled sources [6,23]. In addition, the index just depends on the topology under that assumption.

3.2 Standard Approach: BDF

Starting with SPICE2 [18] solvers in circuit simulation tools are based on either *trapezoidal rule* (TR) or *backward differentiation formulae* (BDF). These have a quite natural application to the network equations (4): Let us assume the numerical approximation in computed for $t = t_{k-l}, \ldots, t_{k-1}$, then at time $t_k = t_{k-1} + h$ the derivative $\dot{\mathbf{q}}(\mathbf{x}(t_k), t_k)$ in (4) is replaced by a backward difference (BDF) operator $\rho\, \mathbf{q}_k$, which is given by

$$\rho\, \mathbf{q}_k := \frac{\alpha_0}{h} \mathbf{q}(\mathbf{x}_k, t_k) + \frac{1}{h} \sum_{i=1}^{l} \alpha_i \mathbf{q}(\mathbf{x}_{k-i}, t_{k-i})$$

involving real coefficients α_i. Thus to obtain a numerical approximation at t_k, a set of nonlinear equations

$$\mathcal{F}(\rho \mathbf{q}_k, \mathbf{x}_k, t_k) = 0$$

has to be solved. Imposing the structure of (4) the iteration matrix in Newton's method is found to be

$$\mathcal{M} = \frac{\alpha_0}{h} \cdot \mathcal{F}_{\dot{\mathbf{x}}} + \mathcal{F}_{\mathbf{x}}\,, \text{ where } \mathcal{F}_{\dot{\mathbf{x}}} = A \cdot \frac{\partial \mathbf{q}(\mathbf{x},t)}{\partial \mathbf{x}}\Big|_{(\mathbf{x},t):=(\mathbf{x}_k,t_k)},\ \mathcal{F}_{\mathbf{x}} = \frac{\partial \mathbf{f}(\mathbf{x})}{\partial \mathbf{x}}\Big|_{\mathbf{x}:=\mathbf{x}_k}.$$

From the previous section we know, a unique solution for any index can be guaranteed, if the matrix pencil $\lambda(\mathcal{F}_{\dot{x}}, \mathcal{F}_x) = \frac{\alpha_0}{h} \cdot \mathcal{F}_{\dot{\mathbf{x}}} + \mathcal{F}_{\mathbf{x}}$ is regular,

[1] Actually, smoothness conditions are only required for the algebraic components. The number of differentiations involved in the determination of a given component z_i depends on the structure of $\mathcal{N}$.

sufficient smoothness provided. However, convergence to the analytic solution is a different matter for index larger than 1.

A set of various modifications to obtain efficient and robust algorithms have been developed over the years for BDF methods: e.g. Index-2 adaptations by März and Tischendorf [17] or TR-BDF strategies [7], such that physical oscillations are preserved, but numerical noise damped. However, integration is performed on charges and fluxes, therefore these variables are primarily controlled and only additional heuristics transfer errors to the desired quantities of node potentials, see [22,5].

3.3 A Charge-Oriented Algorithm: CHORAL

Even so the above described conventional methods obtained a large degree of maturity, that is, are reliable, robust and efficient for a large class of problems, seeking alternatives can be fruitful in many respects: for instance, result verification, extension to further problem classes and direct error control on node-voltages and branch-currents.

For the charge-oriented network equations, an one-step scheme based on the embedded *Rosenbrock-Wanner* (ROW) method [20] has been developed [9], which meet the above criteria, for low accuracy demands and is suitable for an index ≤ 2. Here, the numerical approximation $\mathbf{x}_k$ for $t_k = t_{k-1} + h$ equipped with an embedded approximation $\widehat{\mathbf{x}}_k$ of lower order is computed by

$$\mathbf{x}_k = \mathbf{x}_{k-1} + \sum_{i=1}^{s} d_i \kappa_i, \quad \widehat{\mathbf{x}}_k = \mathbf{x}_{k-1} + \sum_{i=1}^{s} \widehat{d}_i \kappa_i$$

where increments κ_i in the circuit setup (4) are determined by

$$\begin{aligned}\mathcal{M}\kappa_i = & \frac{1}{\gamma h} A\left(\mathbf{q}(\mathbf{x}_{k-1}) - \mathbf{q}(a_i)\right) - \sum_{j=1}^{i} \widetilde{\beta}_{ij}\left[\mathbf{f}(\mathbf{a}_j) - \mathbf{s}(t_{k-1} + h\alpha_j)\right] \\ & - \sum_{j=1}^{i-1} \widetilde{\beta}_{ij} \frac{\partial \mathbf{f}}{\partial \mathbf{x}}(\mathbf{x}_{k-1})\kappa_j + h\,\widetilde{\tau}_i\,\dot{\mathbf{s}}(t_0)\end{aligned}$$

using internal stages

$$\mathbf{a}_i := \mathbf{x}_{k-1} + \sum_{j=1}^{i-1} \sigma_{ij}\kappa_j$$

and $\mathcal{M}$ taken from the previous section (BDF). The mathematical program involves the study of consistency (normally via Butcher-Theory) and stability analysis (via Laguerre polynomials) of the underlying abstract ROW-method. Both together guarantee convergence and define sets of coefficients, which fulfil the order conditions and stability requirements, cf. [9].

Taking low smoothness properties of transistor models and moderate user accuracy demands into account, CHORAL (CHarge ORiented ALgorithm) was developed as an embedded method of order (2)3. It comprises four stages using three function evaluations per times step. A choice of both embedded and parent scheme to be stiffly accurate, provides a reliable error estimator even for inconsistent initial values. The corresponding coefficient set can be found in [9] (using: $\widetilde{\beta}_{ij} := \beta_{ij}/\gamma$ and $\widetilde{\tau}_i := \tau_i/\gamma$ and $\alpha_0 = 1/\gamma$).

Summing up, CHORAL is directly adapted to the circuit equations (4), not to the general implicit nonlinear equation as BDF. Error control and time step prediction is naturally computed on errors in node-potentials and branch-currents. And furthermore the order (2)3 scheme can handle index-2 problems.

3.4 CHORAL in Industrial Simulation

The above algorithm is already part of TITAN's analysis kernel, which shows its compatibility to circuit simulation packages and data structures. Obviously, the coincides of the system matrices for BDF and CHORAL helps to integrate the ROW-method in a multistep based system. For details see [15]. Moreover, it enables the usage of one-step methods in industrial circuit simulation, which is unique at this time.

CHORAL has been tested within TITAN using standard industrial benchmarks and Table 2 gives an excerpt from numerical results.

Table 2. Numerical results: CHORAL versus BDF within TITAN

Circuit	# transistors	# equations	CPU time	
			CHORAL	BDF
LC oscillator	0	3	0.57s	0.33s
MOS ring oscillator	134	73	30.13s	27.61s
16-Bit adder	544	283	2m41.32s	2m30.1s
arithmetic logical unit	13005	32639	97m31.64	82m21.03s

Of course tolerances were adapted, such that CHORAL waveforms for a ring oscillator matched the waveforms produced by the conventional method using standard tolerances. The results are very encouraging, since first tests almost agree in CPU-time to BDF runs, which were tuned over ten years.

It seems that one reason for CHORAL being competitive is a very reliable error control and step size selection algorithm, which produces large time steps, but only a few rejections. Fig. 3 shows that CHORAL needs less timesteps than the standard integrator, but reproduces the signal with the same accuracy.

One last feature, which we mention, is CHORAL's damping behaviour for oscillating circuits. CHORAL's parameters are tuned, such that physical

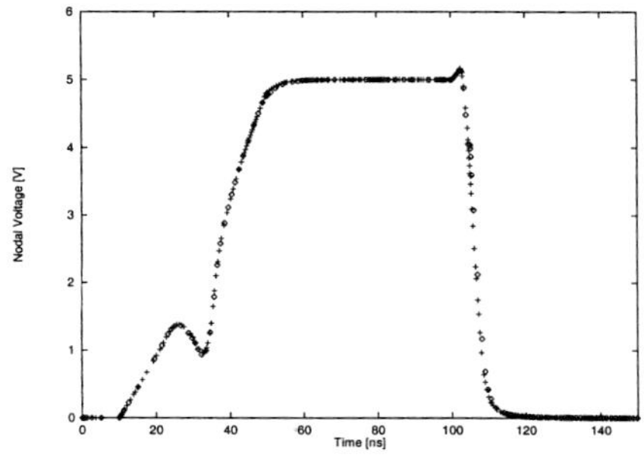

Fig. 3. Output voltage, 16-Bit adder: Integration steps CHORAL (◇) vs. BDF (+)

oscillations and their excitations are preserved on the scale of usual sampling rates, whereas BDF2 involves already a fair amount of damping, see Fig. 4. One notices a phase shift and amplitude damping for BDF2 [13].

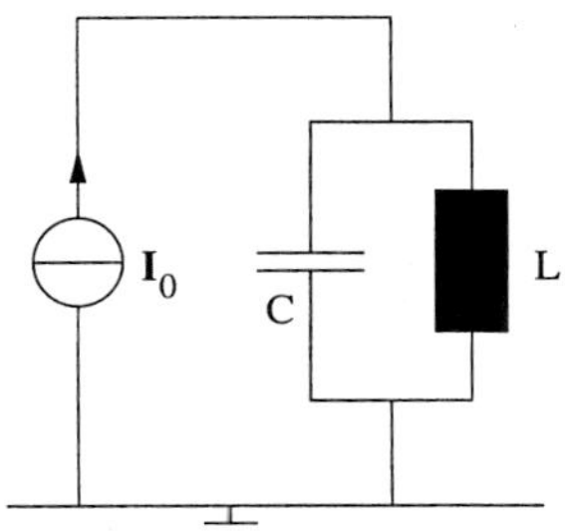

(a)

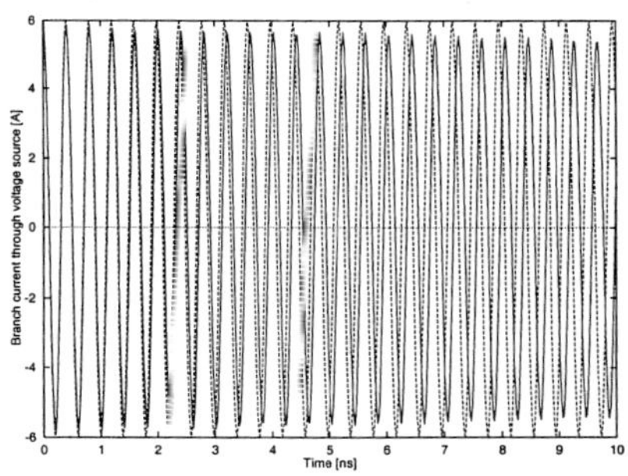

(b)

Fig. 4. LC oscillator (**a**); simulation results (**b**): BDF2 (—) vs. CHORAL (- - -)

4 Multirate

High integrated circuits show the phenomenon of latency. That is, large parts of a circuit keep almost constant, while others change rapidly. One may imagine signal processing, where an input, fed to some terminals, is traced through a circuit causing some part to be active (see Fig. 5). Thus activity is time dependent. Usually about 98% are latent, while just 2% are involved in the processing.

Therefore we have a partitioning in two subsets of nodes, which behave on differing time scales similar to the coupling of analog and digital circuits. For any transient analysis using ad-hoc numerics, the time discretisation will be governed by the fastest changing component and thus integration will resolve a very huge part of the circuit with unnecessary high accuracy.

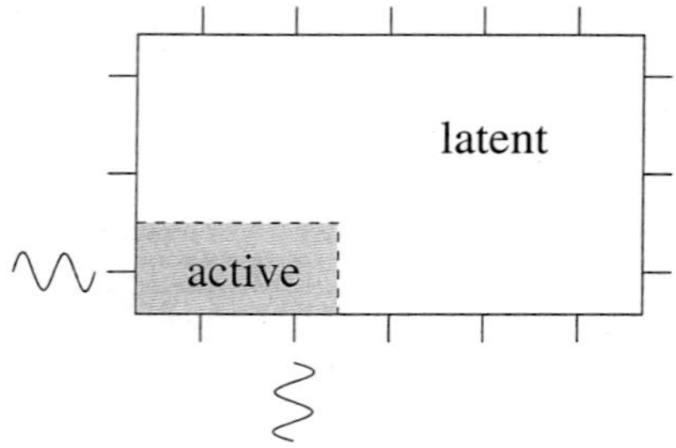

Fig. 5. Schematic Circuit.

4.1 One-Step Based Multirate Algorithms

Motivated by the above behaviour we subdivide the vector of unknowns into latent and active variables: $\mathbf{x} = (\mathbf{x}_L, \mathbf{x}_A)^\top$. Plugging this into an ODE founded, autonomous initial value problem $\dot{\mathbf{x}} = \mathbf{f}(\mathbf{x})$, $\mathbf{x}(t_0) = \mathbf{x}_0$, we obtain a *partitioned system*

$$\begin{array}{l} \text{latent} \\ \text{active} \end{array} \quad \begin{pmatrix} \dot{\mathbf{x}}_L \\ \dot{\mathbf{x}}_A \end{pmatrix} = \begin{pmatrix} \mathbf{f}_L(\ \mathbf{x}_L\ ,\boxed{\mathbf{x}_A}) \\ \mathbf{f}_A(\boxed{\mathbf{x}_L},\ \mathbf{x}_A\ ) \end{pmatrix} \qquad \begin{array}{l} \mathbf{x}_L(t_0) = \mathbf{x}_{L,0} \\ \mathbf{x}_A(t_0) = \mathbf{x}_{A,0}\ . \end{array} \tag{9}$$

Now, one idea of multirate is to impose an inherent step size for each subsystem (introduced by Gear and Wells [8]), in other words use a *micro step* of size h for the active part and a *macro step* of size H for the latent part. The relation

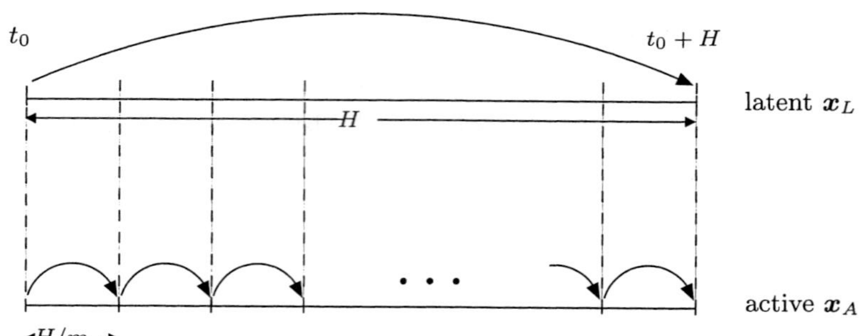

Fig. 6. Macro step vs. micro steps

$H = m_i h$ ($m_i \in \mathbb{N}$) is fixed for each macro step, such that we have synchronisations on the macro step grid (Fig. 6). In the decoupled case this approach is evident, so various types of multirate step sizes differ in the computation of the coupling terms, which are boxed in (9).

Before we consider the computation of couplings, we need to discuss the order of integration. Assuming small couplings from active to latent, one may start integration with the micro steps (fastest first). Then wake-up of latent variables will not be detected before performing the macro step, which means an update of all the computation is likely. Therefore, we look at the opposite strategy (slowest first): the integration of latent components over a large macro step will be only successful, if all components keep latent. On the other hand it does not matter, if some component goes latent[2].

Investigating Fig. 6, the first idea for the coupling terms is inter- and extrapolation from already known values. With a restriction to extrapolation, for simplicity and efficiency, one arrives at the mode of the first ROW approach: first macro step for the latent component [12]

$$\mathbf{x}_L(t_0 + H) = \mathbf{x}_{L,0} + \sum_{i=1}^{s} b_i^L \mathbf{k}_{L,i}$$

$$\left(\mathcal{I}_L - H\gamma^L \left.\frac{\partial \mathbf{f}_L}{\partial \mathbf{x}_L}\right|_{\binom{\mathbf{x}_{L,0}}{\mathbf{x}_{A,0}}}\right)\mathbf{k}_{L,i} = H\mathbf{f}_L\left(\mathbf{x}_{L,0} + \sum_{j=1}^{i-1} \alpha_{ij}^L \mathbf{k}_{L,j}, \boxed{\hat{\mathbf{x}}_A(t_0 + \alpha_i^L H)}\right)$$
$$+ H \left.\frac{\partial \mathbf{f}_L}{\partial \mathbf{x}_L}\right|_{\binom{\mathbf{x}_{L,0}}{\mathbf{x}_{A,0}}} \sum_{j=1}^{i-1} \gamma_{ij}^L \mathbf{k}_{L,j}$$

second the micro steps for $\chi = 0, 1, \ldots, m-1$

$$\mathbf{x}_{A,\chi+1} = \mathbf{x}_A(t_0 + (\chi+1)h) = \mathbf{x}_{A,\chi} + \sum_{i=1}^{s} b_i^A \mathbf{k}_{A\,i}^{\chi}$$

$$\left(\mathcal{I}_A - h\gamma \left.\frac{\partial \mathbf{f}_A}{\partial \mathbf{x}_A}\right|_{\left(\begin{smallmatrix}\mathbf{x}_{A,\chi} \\ \boxed{\hat{\mathbf{x}}_L(t_0+\chi h)}\end{smallmatrix}\right)}\right)\mathbf{k}_{A,i} = h \left.\frac{\partial \mathbf{f}_A}{\partial \mathbf{x}_A}\right|_{\left(\begin{smallmatrix}\mathbf{x}_{A,\chi} \\ \boxed{\hat{\mathbf{x}}_L(t_0+\chi h)}\end{smallmatrix}\right)} \sum_{j=1}^{i-1} \gamma_{ij}^A \mathbf{k}_{A,j}$$
$$+ h\mathbf{f}_A\left(\boxed{\hat{\mathbf{x}}_L(t_0 + \chi h + \alpha_i^A h)}, \mathbf{x}_A(t_0 + \chi h) + \sum_{j=1}^{i-1} \alpha_{ij}^A \mathbf{k}_{A,j}\right).$$

Additionally, we have to provide extrapolation formulae for $\hat{\mathbf{x}}$. Therefore, we loose the one-step nature, since this is a two step procedure and rather a two-step method with all stability issues: externally the extrapolation is computed, afterwards the increments for the one-step based method can be done.

To have the underlying one-step method reflected by the multirate algorithm, the concept of generalised multirate was introduced [11]: Here, integration starts with a combined macro and first micro step (so-called *compound*

[2] Note, if signals are pulsed with a high frequency, then it can be possible, that one misses something at all. Of course, this is true for any to large integration step.

step), before the remaining micro steps are worked out. Furthermore, couplings are based on internal stages, such that their computation resembles an 'own' RK-like methods

$$\boxed{\hat{\mathbf{x}}_A(t_0 + \alpha_i^L H)} = \mathbf{x}_{A,0} + m \sum_{j=1}^{s} \delta_{ij}^L \mathbf{k}_{A,j}^0,$$

$$\boxed{\hat{\mathbf{x}}_L(t_0 + \chi h + \alpha_i^A h)} = \mathbf{x}_{L,0} + 1/m \sum_{j=1}^{s} (\delta_{ij}^A + \eta_j(\chi)) \mathbf{k}_{L,j},$$

($\eta(.)$ denotes a polynomial, which takes the number of the currently computed micro steps into account). This allows to view the multirate method as a one-step method on the macro step scale.
A further specialisation is *mixed strategy* [1], which employs the compound step, but calculates the later micro steps via a dense output scheme. This allows a step size control for the later micro steps.

4.2 Test Example

One usual benchmark in electric circuit simulation is the inverter chain, which mainly delays a traversing signal. In addition, it is Infineon Technologies' record benchmark, for the largest simulated circuit in the world, which comprised 3 Mill. MOSFETs.

In order to obtain a system with latency, we concatenate n inverters (cf. Fig. 2),

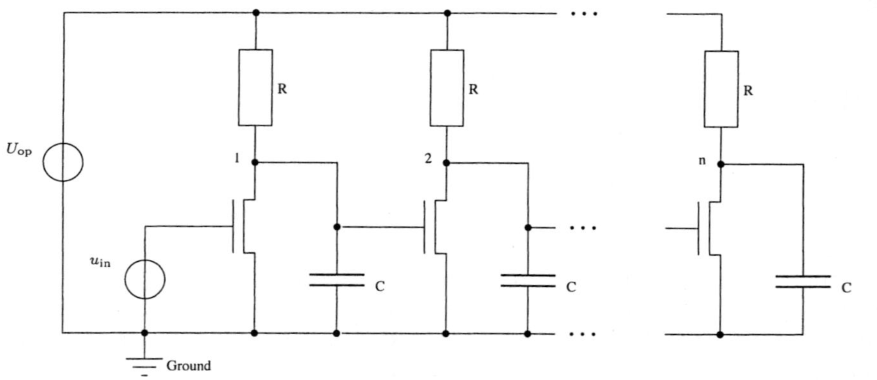

Fig. 7. The inverter chain

where the output of an inverter serves as input of the succeeding one (see Fig. 7). Thus, the equations for n inverters are given by

$$\dot{u}_1 = (U_{op} - u_1)/(RC) - i_{ds}(u_{in}, u_1, 0, 0)/C$$
$$\dot{u}_k = (U_{op} - u_k)/(RC) - i_{ds}(u_{k-1}, u_k, 0, 0)/C \quad \text{for } k = 2, \ldots, n$$

with i_{ds} from Chapter 2.2, formula (3). The equations are completed by the initial value: $\mathbf{u} = (5, 0, 5, \ldots, 0)^\top$.

Now, we look at some test results for mixed multirate (cf. [1]). Therefore, we take a stiff setting ($\eta = 100$ in eq. (3)) for $n = 200$ inverters and we assign the single pulse signal u_{in}

$$(0\text{ns}, 0\text{V}), (5\text{ns}, 0\text{V}), (10\text{ns}, 5\text{V}), (15\text{ns}, 5\text{V}), (17\text{ns}, 0\text{V}), (70\text{ns}, 0\text{V}).$$

A partition is obtained by inspecting the right-hand side excitation (u_i is active, if $|\text{rhs}_i| \geq \varepsilon_{\text{actlevel}}$) and as reference we compare the multirate results with single rate computations produced by MATLAB's build-in routine `odes23s`. Results are summarised in Table 3.

Table 3. Stiff inverter chain tests with mixed multirate

mode	err scal	$\varepsilon_{\text{actlevel}}$	macro steps	micro steps	# G ops	RHS-eval	ratio eval	remark
sr (ode23s)	-	-	-	2049	35.21	819 600	-	(+rej stp)
mr (mxrow)	off	0.5	248(16)	3289(111)	4.35	327 418	2.50	slight delay
	on	0.7	135(4)	2573(8)	2.36	178 114	4.60	(delay)
	on	0.05	289(11)	3364(47)	5.60	527 054	1.56	

Speed-ups are measured in scalar function evaluations, since the load part is most costly in circuit simulation tools (due to modelling reasons). A fair speed-up of 2.50 is obtained, but the number of micro steps has increased compared to the single rate algorithm. Tests with an error scaling of the form $\mathbf{err} = 2\,(\mathcal{I} - h\gamma \mathbf{f_x})^{-1}(\mathbf{err}_{\text{embedded}})$yield larger speed-ups at the cost of a delay. However, the number of micro steps mirrors much better the number of steps in single rate. – The effect of the delay can be decreased by tuning the activation level $\varepsilon_{\text{actlevel}}$ (see Table 3).

Roughly speaking, the speed-up, which is eventually reached depends on the signal width (active parts), so one has to expect less than 10. Additionally, the inevitable overhead has to be taken into account.

For 500 inverters the results are summarised in Table 4 ($\varepsilon_{actlevel} = 0.7$, 140 ns).

The mixed multirate method, here with step size control for the interior micro steps, uses about twice as many micro steps, which indicates that the employed step size selection scheme is somewhat more conservative than the scheme used in `ode23s`. However, a speed-up of 10 is achieved.

Table 4. Test results for the stiff inverter chain problem (500)

	Macro/micro steps	Rejected macro/ micro steps	Scalar function evaluations	flops
`ode23s`	5445	1625	$9.80 \cdot 10^6$	$5.1 \cdot 10^8$
Mixed multirate	287/11175	6/127	$0.93 \cdot 10^6$	$2.0 \cdot 10^8$

5 PDE Modelling for Source Driven Oscillators

We consider again an implicit ODE, which describes the time behaviour of a circuit

$$\dot{\mathbf{q}}(\mathbf{x}(t)) = \mathbf{f}(\mathbf{x}(t)) + \mathbf{s}(t), \tag{10}$$

where $\mathbf{x}(t) \in \mathbb{R}^n$ represents the unknown node-voltages and branch-currents and $\mathbf{s}(t) \in \mathbb{R}^n$ independent input signals.

In radio frequency application, widely separated time scales arise in signals. An example is given by a fast varying oscillation, which is modulated by a slow varying oscillation, like

$$s(t) = \sin\left(\tfrac{2\pi}{T_1}t\right) \sin\left(\tfrac{2\pi}{T_2}t\right), \tag{11}$$

where $T1 \gg T2$. Fig. 8 illustrates the qualitative behaviour of this signal. Such an input signal causes a forced oscillation in the circuit, that is, the output signals inherit its time performance. Thus numerical simulation becomes costly, since one has to restrict the integration step size to the fastest rate. Brachtendorf [2] and Roychowdhury [21] propose a multidimensional approach to avoid this problem. We are able to change from the time-dependent function s to a representation $\hat{s}$, where each separate time scale complies with an own variable. We call $\hat{s}$ the *multivariate function* (MVF) of the signal s. In our example, the MVF results to

$$\hat{s}(t_1, t_2) = \sin\left(\tfrac{2\pi}{T_1}t_1\right) \sin\left(\tfrac{2\pi}{T_2}t_2\right), \tag{12}$$

which is biperiodic and therefore just fixed by its values in the rectangle $[0, T_1[\times[0, T_2[$. As shown in Fig. 8, the MVF does not have high frequency oscillations, since time scales are decoupled. Hence we need relatively few points to represent the MVF accurately. Nevertheless, the MVF includes the original signal, because $s(t) = \hat{s}(t, t)$ holds. Consequently, the MVFs yield a compact and efficient description of such signals.

The transition to multidimensional functions change the ODE model of a circuit into a PDE model. If a multirate behaviour with m separate time scales arises, then let $\hat{\mathbf{x}}(t_1, \ldots, t_m) \in \mathbb{R}^n$ and $\hat{\mathbf{s}}(t_1, \ldots, t_m) \in \mathbb{R}^n$ be the MVFs of $\mathbf{x}$ and $\mathbf{s}$ in (10). Accordingly, we define the *multirate partial differential equation* (MPDE) corresponding to (10) to be

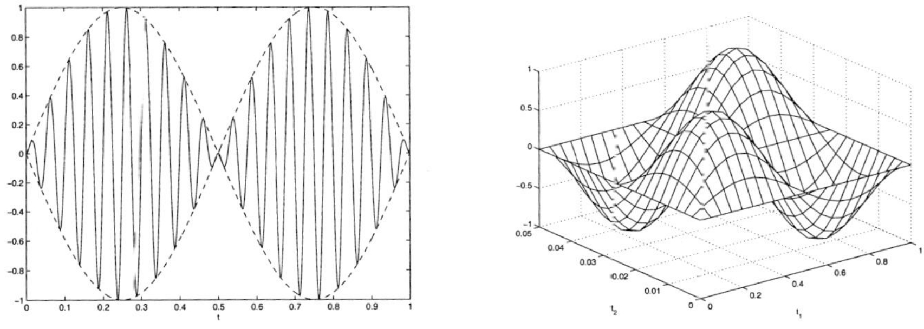

Fig. 8. Signal s and its MVF $\hat{s}$

$$\frac{\partial \mathbf{q}(\hat{\mathbf{x}})}{\partial t_1} + \cdots + \frac{\partial \mathbf{q}(\hat{\mathbf{x}})}{\partial t_m} = \mathbf{f}(\hat{\mathbf{x}}) + \hat{\mathbf{s}}(t_1, \ldots, t_m) \tag{13}$$

with $\mathbf{q}$ and $\mathbf{f}$ from (10). Now the key relation is that solutions of the MPDE (13) include solutions of the corresponding ODE (10) on diagonals.

Theorem 1. *If $\hat{\mathbf{x}}(t_1, \ldots, t_m)$ and $\hat{\mathbf{s}}(t_1, \ldots, t_m)$ solve the MPDE (13), then $\mathbf{x}(t) = \hat{\mathbf{x}}(t + c_1, \ldots, t + c_m)$ and $\mathbf{s}(t) = \hat{\mathbf{s}}(t + c_1, \ldots, t + c_m)$ satisfy the ODE (10) for any fixed $c_1, \ldots, c_m \in \mathbb{R}$.*

The proof of this statement is straightforward. Therefore $x(t) = \hat{x}(t,t)$ solves the ODE with $s(t) = \hat{s}(t,t)$ in our two-rate example. Moreover, we obtain strong connections between m-tone quasi-periodic solutions of the ODE and m-periodic solutions of its MPDE. The following theorems give sufficient and necessary criteria.

Theorem 2. *If $\hat{\mathbf{s}}$ in (13) is m-periodic and $\hat{\mathbf{x}}$ an m-periodic solution, then $\mathbf{x}(t) = \hat{\mathbf{x}}(t+c_1, \ldots, t+c_m)$ is an m-tone quasi-periodic solution of (10) under the m-tone quasi-periodic function $\mathbf{s}(t) = \hat{\mathbf{s}}(t + c_1, \ldots, t + c_m)$ for any fixed $c_1, \ldots, c_m \in \mathbb{R}$.*

Theorem 3. *If $\mathbf{s}$ in (10) is an m-tone quasi-periodic function and $\mathbf{x}$ an m-tone quasi-periodic solution, then for any given $c_1, \ldots, c_m \in \mathbb{R}$, m-periodic functions $\hat{\mathbf{s}}$ and $\hat{\mathbf{x}}$ exist satisfying (13) such that $\mathbf{x}(t) = \hat{\mathbf{x}}(t + c_1, \ldots, t + c_m)$ and $\mathbf{s}(t) = \hat{\mathbf{s}}(t + c_1, \ldots, t + c_m)$.*

Quasi-periodic functions represent an important signal class in radio frequency application. Thus if the ODE (10) has m-tone quasi-periodic input signals, then we change to the corresponding MPDE (13) and determine an m-periodic solution, since it also results in an m-tone quasi-periodic solution of the ODE. The numerical handling of the MPDE saves time and memory compared to computations using the corresponding ODE due to the behaviour of the according functions. Moreover, the MPDE model is also suitable for envelope-modulated signals. In both cases, periodicities in the MPDE solutions play an important role.

The MPDE (13) represents a hyperbolic system. However, conservative methods are not suitable for this PDE, since the physical meaning differs from conservation laws here. Furthermore, the periodic boundary conditions do not fit for such techniques. Finite difference methods, that is discretisation of the PDE in time domain, are qualified to compute solutions of the system. The harmonic balance approach can also be generalised to solve the MPDE in frequency domain [3]. Numerical computation of a periodic solution using a method of characteristics in time domain is described in [19].

6 Electro-Thermal Coupling

Due to miniaturisation and increasing density of devices on circuit boards, heat production and convection will be more and more important for the time behaviour of circuits. Currently, Infineon Technologies speaks of power losses of about 100 Watt/cm^2 for a CPU in the near future.

Therefore, we will investigate the coupling of network equations and PDEs, in the future. One interesting topic will be existence and uniqueness of solutions for the arising PDAEs, another the realisation and justification of simulator coupling.

As first object of investigation, we take a simple circuit, cf. Fig. 9, which comprises all heat effects, that is: heat generation (amplifier), heat conduction (wire as a resistor) and a temperature dependent characteristic (diode).

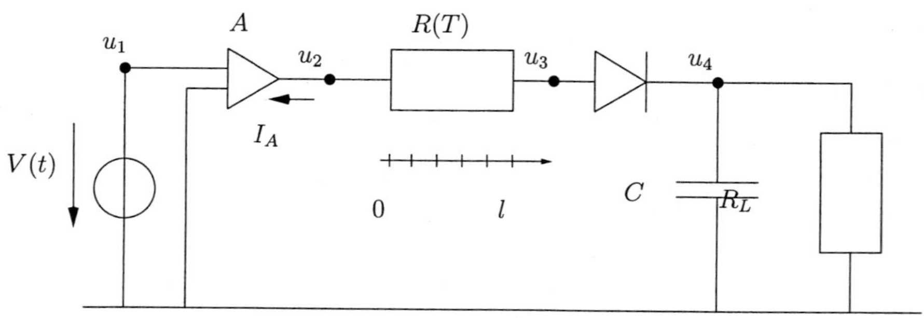

Fig. 9. Benchmark circuit: electro-thermal coupling

Its mathematical model is obtained by MNA, which gives the network equations, these are completed by the parabolic heat convection for the wire:

$$\begin{aligned} I_A + \frac{u_2 - u_3}{R(T)} &= 0 \\ -\frac{u_2 - u_3}{R(T)} + I_D(u_3 - u_4) &= 0 \\ -I_D(u_3 - u_4) + C\dot{u}_4 + \frac{u_4}{R_L} &= 0 \\ u_1 &= V(t) \\ u_2 - Au_1 &= 0 \end{aligned} \qquad \begin{aligned} R(T) &= \int_0^l \frac{\rho(x,T)}{a_0} dx \\ I_D(T) &= \hat{I}_D(T) \\ &\cdot(\exp((u_3 - u_4)/v_T) - 1) \end{aligned}$$

$$\rho_d c \frac{\partial T}{\partial t} = \lambda \frac{\partial^2 T}{(\partial x)^2} + \left(\frac{u_2 - u_3}{Ra_0}\right)^2 \rho(x,T) - \mathrm{cool}(T - T_{\mathrm{env}}) + \gamma(u_2 I_A)\,\delta_0(x)$$

$$\text{initial values: } T(0,t) = T_0(t)$$
$$\text{boundary conditions:} \quad T_x(0,t) = T_x(l,t) = 0\,.$$

The heat source (amplifier) is modelled as delta function, which give us a local source for the heat equation and enables us to supply von-Neumann conditions at the boundary of the wire. Hence we just couple the network with the heat equation via source terms.

Acknowledgements

This work is part of the project "Numerical Simulation of Electric Networks" within the federal BMBF program, as well as a part of the WiR BaWü project "A3: Simulation Tools for Ordinary Differential Equations".
For continuous discussions and many fruitful contributions the authors are indebted to Dr. U. Feldmann, Infineon Technologies.

References

1. Bartel, A.: Multirate ROW Methods of Mixed Type for Circuit Simulation. To appear in [24].
2. Brachtendorf, H. G.: Simulation des eingeschwungenen Verhaltens elektronischer Schaltungen. Shaker, Aachen, 1994.
3. Brachtendorf, H. G., Welsch, G., Laur, R., Bunse-Gerstner, A.: Numerical steady state analysis of electronic circuits driven by multi-tone signals. Electrical Engineering **79** (1996) 103–112.
4. McCalla, W. J.: Fundamentals of Computer Aided Circuit Simulation. Acad. Publ. Group, Dordrecht (1988) Kluwer.
5. Denk, G.: An improved numerical integration method in the circuit simulator SPICE2-S. In: Bank, R.E., Bulirsch, R., Merten, K.: Mathematical modelling and simulation of electrical circuits and semiconductor devices. ISNM 93. Basel: Birkhäuser-Verlag (1990) 84–90.

6. Estévez Schwarz, D., Tischendorf, C.: Structural analysis of electric circuits and consequences for MNA. Int. J. Circ. Theor. Appl. **28** (2000) 131–162.
7. Feldmann, U., Wever, U., Zheng, Q., Schultz, R., Wriedt, H.: Algorithms for modern circuit simulation. AEÜ **46** (1992) 274–285.
8. Gear, C.W., Wells, D.R. Multirate Linear Multistep Methods. BIT **24** (1984) 484–502.
9. Günther, M.: Simulating digital circuits numerically – a charge-oriented ROW approach. Num. Math. **79** (1998) 203–212.
10. Günther, M., Feldmann, U.: CAD based electric circuit modeling I: mathematical structure and index of network equations. Surv. Math. Ind. **8** (1999) 97–129.
11. Günther, M., Kværnø, A., Rentrop, P.: Multirate Partitioned Runge-Kutta Methods. To appear in BIT.
12. Günther, M., Rentrop, P.: Multirate ROW methods and latency of electric circuits. Applied Numerical Mathematics **13** (1993) 83–102.
13. Günther, M., Rentrop, P., Feldmann, U.: CHORAL — a one step method as numerical low pass filters. To appear in: [24].
14. Ho, C. W., Ruehli, A. E., Brennan, P. A.: The modified nodal approach to network analysis. IEEE Trans. Circuits and Systems, CAS **22** (1975) 505–509.
15. Hoschek, M.: Einschrittverfahren zur numerischen Simulation elektrischer Schaltungen. VDI Verlag, Düsseldorf, 1999.
16. Kampowsky, W., Rentrop, P., Schmitt, W.: Classification and numerical simulation of electric circuits. Surv. Math. Ind. **2** (1992) 23–65.
17. März, R., Tischendorf, C.: Recent results in solving index 2 differential algebraic equations in circuit simulation. SIAM. J. Sci. Comp. **18** (1997) 139–159.
18. Nagel, W.: SPICE 2 – a computer program to simulate semiconductor circuits. Dissertation. Berkeley, CA: UC Berkeley, 1975.
19. Pulch, R., Günther, M.: A Method of Characteristics for Solving Multirate Partial Differential Equations in Radio Frequency Application. Preprint 00/07, IWRMM, University Karlsruhe (2000).
20. Rentrop, P., Roche, M., Steinebach, G.: The application of Rosenbrock–Wanner type methods with stepsize control in differential–algebraic equations. Numer. Math. **55** (1989) 545–563.
21. Roychowdhury, J.: Efficient Methods for Simulating Highly Nonlinear Multi-Rate Circuits. Proc. Design Automation Conference, Anaheim, CA (1997).
22. Sieber, E.-R., Feldmann, U., Schultz, R., Wriedt, H.: Timestep control for charge conserving integration in circuit simulation. In: Bank, R.E., Bulirsch, R., Gajewski, H., Merten, K.: Mathematical modelling and simulation of electrical circuits and semiconductor devices. ISNM 117. Basel: Birkhäuser-Verlag (1994) 103–113.
23. Tischendorf, C.: Topological Index Calculation of DAEs in circuit simulation. Surv. Math. Ind **8**(3-4) (1999) 187–199.
24. van Rienen, U., Günther, M., Hecht, D.: Scientific Computing in Electrical Engineering — Proceedings of the 3rd International Workshop — 20 - 23. August 2000, Warnemünde, Germany. Springer (2001).

Transient Noise Analysis in Circuit Simulation

G. Denk

Infineon Technologies, Memory Products, Balanstr. 73, D-81541 München

Abstract. Circuit simulation is one of the most important steps during the development of electronic circuits. A standard task for circuit simulation is the transient analysis, the simulation of the chip on transistor level in the time domain. Normally, this simulation does not include noise source. As the supply voltage decreases, the signal-to-noise ratio also decreases and noise effects become more and more important. In this paper the modeling of noise in the time-domain will be described. Especially for flicker noise, new and efficient approaches are presented. For a numerical treatment of the resulting stochastic differential-algebraic equations, a new numerical scheme is given for the charge-oriented formulation of the circuit equations. Finally, the results of numerical experiments are presented.

1 Introduction

The simulation of the design steps plays an important role for the development of integrated circuits. Simulation helps to reduce the time necessary for the design as well as the production costs. A crucial point is the circuit simulation, this is the simulation of the chip on transistor level. Circuit simulation helps to prove the physical behavior of the circuit before producing test chips and allows for early corrections. Several simulators are available for this task allowing different analyses (operation point analysis, small-signal analysis, non-linear frequency analysis, transient analysis, noise analysis, ...).

Transient analysis is usually computed without taking noise effects into account. But due to the decreasing supply voltages and due to the decreasing size of the elements, this is no longer possible. The signal-to-noise ratio is getting so small that the noise effects have to be simulated, too. In most simulators this is done within the small-signal analysis, which means that only a linearization of the circuit is handled. Another approach is the non-linear frequency analysis for oscillatory circuits, which restricts the application to that special class of chips. To overcome this restriction, we will present a non-linear noise analysis combined with the transient simulation of the circuit. This analysis computes the so-called path of the noisy signals and allows the computation of the moments of the signals in a post-processing step.

There are three main sources for noise in electronic circuits: Thermal noise, shot noise, and flicker noise [3]. While the simulation of thermal and shot noise in the time domain is well understood [1], the numerical treatment of flicker noise in the time domain requires some new approaches which will be given in the next section.

From an engineer's point of view, the consideration of noise during the transient simulation means that appropriate noise sources are shunt in parallel to the ideal noiseless circuit elements. This gives an additional term in the underlying system of equations. For an efficient numerical treatment of the modified system, we need a new integration scheme specialized on the structure of the charge-oriented formulation of the equations. This will be given in Sect. 3. Numerical experiments are presented at the end of this paper.

2 Modeling of Flicker Noise in the Time Domain

The numerical treatment of thermal noise and shot noise is done with the help of the white noise stochastic process (cf. [7]), which leads in the case of the charge-oriented formulation [4] to the stochastic differential-algebraic equation (SDAE)

$$A \cdot q(X_s)\Big|_{t_0}^{t} + \int_{t_0}^{t} f(s, X_s)\, ds + \int_{t_0}^{t} g(s, X_s)\, dB_s = 0 \,, \tag{1}$$

with the initial value $X_{t_0} = x_0$. $t \in [t_0, t_{\mathrm{end}}]$ describes the simulation time, $X_t(\omega) : \mathbb{R} \times \Omega \to \mathbb{R}^d$ is the stochastic vector process of the nodal voltages and branch currents through voltage-defined branches with $(\Omega, \mathcal{S}, \mathcal{P})$ an appropriate probability space, $A \in \mathbb{R}^{d \times d_q}$ is the generally singular incidence matrix describing the topology of the circuit, $q(X_t) : \mathbb{R}^d \to \mathbb{R}^{d_q}$ is the vector of the terminal charges of the dynamic circuit elements, $f(t, X_t) : \mathbb{R} \times \mathbb{R}^d \to \mathbb{R}^d$ describes the static parts of the elements and the independent sources, $g(t, X_t) : \mathbb{R} \times \mathbb{R}^d \to \mathbb{R}^{d,m}$ is the diffusion matrix which contains the spectral power densities of the noise sources, and $B_t(\omega) : \mathbb{R} \times \Omega \to \mathbb{R}^m$ is a multi-dimensional Brownian motion. Without loss of generality we set $t_0 = 0$ for the following.

Equation (1) shows that the integral $\int g(s, X_s) W_s ds$ over the generalized Wiener process W_t describing the white noise can be replaced with the integral $\int g(s, X_s) dB_s$, where the integration is performed over the ordinary stochastic process of the Brownian motion. A similar approach has been developed for flicker noise [2]: The corresponding integral for flicker noise is given by $\int g(s, X_s) \eta_{1/f} ds$ where $\eta_{1/f}$ is an unknown generalized process which is proportional to $1/f^{\beta}$ with $\beta \in (0, 1)$. Again, this integral can be transformed into $\int g(s, X_s) dB_{\mathrm{FBM}}$, where B_{FBM} is a yet unknown ordinary process which (generalized) derivative has a spectrum proportional to $1/f^{\beta}$. Such a process has been described by Mandelbrot and van Ness [6] and is called *fractional Brownian motion*. It is a Gaussian process determined uniquely by its expectation and covariance matrix:

$$\begin{aligned} E(B_{\mathrm{FBM}}(s)) &= 0 \quad \forall s \in \mathbb{R}_0^+ \,, \\ Cov(B_{\mathrm{FBM}}(s), B_{\mathrm{FBM}}(t)) &= \mathrm{const} \cdot \left(|s|^{\beta+1} + |t|^{\beta+1} - |t - s|^{\beta+1}\right) \quad \forall s, t \in \mathbb{R}_0^+ \,. \end{aligned}$$

The numerical approximation can also be computed similar to the case of white noise as

$$\int_0^t g(s, X_s) dB_{\text{FBM}} \approx \sum_{i=1}^{n} g(t_i, X_{t_i}) \cdot \Delta B_{\text{FBM}}(i)$$

with $0 = t_0 < \ldots < t_n = t$ a discretization of $[0, t]$ and the increments $\Delta B_{\text{FBM}}(i) = B_{\text{FBM}}(t_i) - B_{\text{FBM}}(t_{i-1})$. The increments of the fractional Brownian motion describe a Gaussian process and are given by

$$\begin{aligned} E(\Delta B_{\text{FBM}}(i)) &= 0\,, \\ Cov(\Delta B_{\text{FBM}}(i), \Delta B_{\text{FBM}}(j)) &= \text{const} \cdot (-|t_j - t_i|^{\beta+1} + |t_{j-1} - t_i|^{\beta+1} \\ &\quad + |t_j - t_{i-1}|^{\beta+1} - |t_{j-1} - t_{i-1}|^{\beta+1})\,. \end{aligned} \tag{2}$$

This means that the flicker noise within the circuit elements can be approximated by random numbers fulfilling (2). In opposite to the approximation of the Brownian motion, the increments $\Delta B_{\text{FBM}}(i)$ are not stochastic independent, all time steps of the simulation enter into the covariance matrix. As these time steps are not known in advance due to the adaptive step-size control, an stepwise computation of the random variables has to be developed. This can be done with conditional densities. Let $x_{(n-1)} = (\Delta B_{FBM}(1), \ldots, \Delta B_{FBM}(n-1))^T$. The conditional density for $\Delta B_{\text{FBM}}(n)$ is given by

$$f(\Delta B_{\text{FBM}}(n)|x_{(n-1)}) = \frac{1}{\sqrt{2\pi\sigma^2}} \exp\left\{-\frac{1}{2\sigma^2}(\Delta B_{\text{FBM}}(n) - \mu)^2\right\}$$

with

$$\mu = -\frac{{x_{(n-1)}}^T \cdot \Sigma^{-1}_{\bullet,n}}{\Sigma^{-1}_{n,n}} \quad \text{and} \quad \sigma^2 = \frac{1}{\Sigma^{-1}_{n,n}}\,.$$

$\Sigma^{-1}_{\bullet,n} \in \mathbb{R}^{n-1}$ and $\Sigma^{-1}_{n,n} \in \mathbb{R}$ are given by the partitioned inverse of the covariance matrix

$$\Sigma^{-1}_{(n)} = \left(\begin{array}{c|c} \Sigma^{-1}_{(n-1)} & \Sigma^{-1}_{\bullet,n} \\ \hline (\Sigma^{-1}_{\bullet,n})^T & \Sigma^{-1}_{n,n} \end{array}\right)\,. \tag{3}$$

$\Sigma^{-1}_{(n-1)} \in \mathbb{R}^{n \times n}$ describes the inverse of the covariance matrix of the step before. The computation of (3) can be done efficiently with Schur complement techniques.

3 Numerical Scheme for Transient Noise Simulation

As described in the previous section, all relevant noise sources in electronic circuits can be modeled correctly for transient noise analysis. The remaining task is the efficient numerical solution of (1). In order to make notation simpler, we restrict ourself to the index-1 case, as most real-world circuits fall into this class. Chr. Penski [8] has constructed the following scheme for the charge-oriented formulation of the circuit equations:

$$
\begin{aligned}
&A\big(q_{n+1} + \lambda(\dot{q}_{n+1} - h\ddot{q}_{n+1})\big) + (\lambda + h) f_{n+1} - \lambda h \dot{f}_{n+1} - \\
&\quad - A\big(q_n + \lambda \dot{q}_n\big) - \lambda f_n + g_n \begin{pmatrix} I_{(1)} \\ \vdots \\ I_{(m)} \end{pmatrix} - \\
&\quad - \sum_{j=1}^{m} g_x^{\bullet,j} \Big(A(q_x + \lambda \dot{q}_x) + \lambda f_x\Big)^{-1} g\Big|_{(t_n, X_n)} \begin{pmatrix} I_{(1,j)} \\ \vdots \\ I_{(m,j)} \end{pmatrix} = 0\,.
\end{aligned} \tag{4}
$$

Here $h = (t_{\text{end}} - t_0)/N$, $N \in \mathbb{N}$, denotes the step size, $t_n = t_0 + n \cdot h$, $n = 0, 1, \ldots, N$. $I_{(j)} = I_{(j),t_n,t_{n+1}}$, $j = 1, \ldots, m$, are the Wiener increments, $I_{(i,j)} = I_{(i,j),t_n,t_{n+1}}$, $i, j = 1, \ldots, m$, are double Itô integrals. X_n is the numerical approximation of X_{t_n}, the numerical evaluations of F, g, Q, $\dot{Q}$ at X_n and (t_n, X_n), resp., are indicated by the index $_n$ and $|_{(t_n,X_n)}$, resp. The j-th column of the matrix g is denoted by $g^{\bullet,j}$ for $j = 1, \ldots, m$. The parameter $\lambda \in \mathbb{R}$ of the scheme can be chosen almost arbitrarily. The numerical approximation given by (4) converges with strong order 1 to the exact solution of (1). For details see [8].

The idea behind this scheme is to add

$$
\lambda(A\dot{q}(X_s) + f(X_s, s))\Big|_{t_0}^{t}
$$

with some λ to both sides of (1):

$$
\begin{aligned}
&A \cdot q(X_s) + \lambda(A\dot{q}(X_s) + f(X_s, s))\Big|_{t_0}^{t} \\
&= \lambda(A\dot{q}(X_s) + f(X_s, s))\Big|_{t_0}^{t} - \int_{t_0}^{t} f(X_s, s)\, ds - \int_{t_0}^{t} g(X_s, s)\, dB_s
\end{aligned}
$$

Due to the index-1 assumption, the partial derivative of $A \cdot q(X_t) + \lambda(A\dot{q}(X_t) + f(X_t, t))$ given on the left-hand side with respect to x is regular for almost all λ. Together with the truncated Itô expansion, this gives the formula (4).

In the context of circuit simulation, the derivatives $\ddot{q}$, $\dot{f}$ present in (4) are not available. In order to get a numerical scheme applicable for circuit simulation, these derivatives are replaced by some numerical difference approximations yielding

$$A\big(q_{n+1}-q_n\big)+hf_{n+1}+g_n\begin{pmatrix}I_{(1)}\\ \vdots\\ I_{(m)}\end{pmatrix}-$$

$$-\sum_{j=1}^{m} g_x^{\bullet,j}\Big(A(q_x+\lambda\dot{q}_x)+\lambda f_x\Big)^{-1} g\Big|_{(t_n,X_n)}\begin{pmatrix}I_{(1,j)}\\ \vdots\\ I_{(m,j)}\end{pmatrix}=0\,. \qquad (5)$$

The analysis of this numerical schemes shows that the strong convergence order is 1, it is stable in the mean and A-stable. For the parameter λ, the choice $\lambda = h$ gives reasonable results. Details are given in [8]. These properties indicate the usefulness of the formula for the application in the area of circuit simulation.

Remark For linear SDAEs with additive noise ($g(t, X_t) \equiv g(t)$), method (5) coincides with the numerical scheme given by Schein in [9].

4 Numerical Experiments

The first numerical test has been carried out using the benchmark circuit of a ring oscillator shown in Fig. 1. It confirms the theoretical results for method (5) and helps to do comparisons with other numerical schemes. A highly accurate solution of the nodal voltage X_t^5 for a ring oscillator consisting of $d = 5$ inverters is given in Fig. 2, where the mean value of 100 paths is plotted.

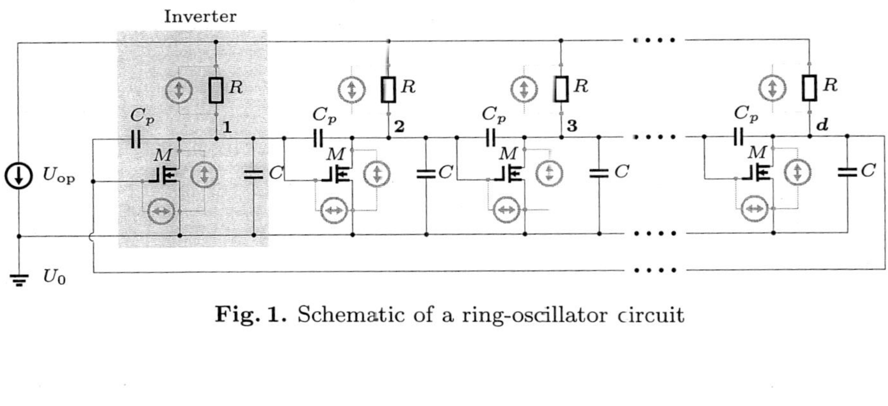

Fig. 1. Schematic of a ring-oscillator circuit

In order to investigate the convergence order of (5), the benchmark circuit was simulated with different constant step sizes and compared with the results

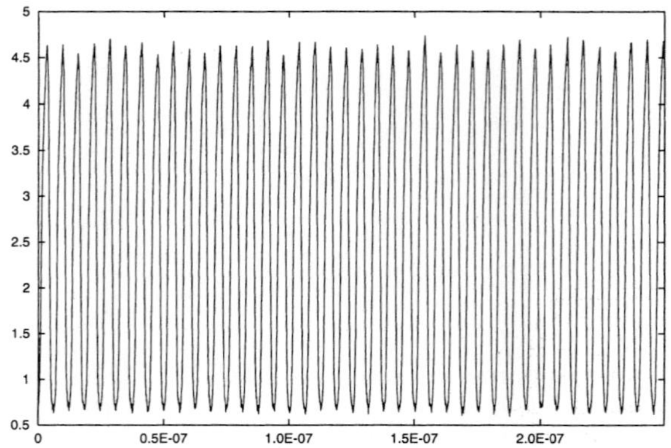

Fig. 2. Highly accurate solution for nodal voltage X_t^5

using a well-known method for stochastic differential equations (SDEs) given in [5, p. 376]. To simulate the charge-oriented equations for the ring oscillator with a scheme for SDEs, the stochastic differential-algebraic equations have to be transformed into a system without algebraic constraints. Though this is always possible for index-1 systems, it is not feasible for the computer-aided design as the index reduction cannot be performed automatically. In Fig. 3, the error of the solution compared to the results of the standard scheme is given for $t \in [0 : 19\,\text{ns}]$. It can be seen clearly that the error is reduced by one magnitude if the step size is reduced by a factor of 10. This proves numerically the strong order of convergence 1 of the numerical scheme. It also shows that the formula (5) treats the algebraic constraints correctly.

For the transient noise simulation of real-world circuits, a modified version of the numerical scheme (5) was implemented in Infineon's in-house simulator TITAN. The transient noise analysis was performed for a cross-section of a 256 Mb DRAM chip which contains approximately 37,000 elements and 150,000 noise sources leading to a system with nearly 17,000 equations. In Fig. 4 the path of nodal voltage is given. The simulation with noise sources gives a zig-zag line around the straight line resulting from the noiseless simulation. The peaks arising from some switching events within the circuits are quite similar between both simulations. For this circuit the transient noise simulation shows that no problems due to noise are to be expected.

5 Conclusion

The modeling of noise sources in the time domain and the numerical integration scheme for stochastic differential-algebraic equations presented here give a good starting point for the transient noise analysis of electric circuits. It

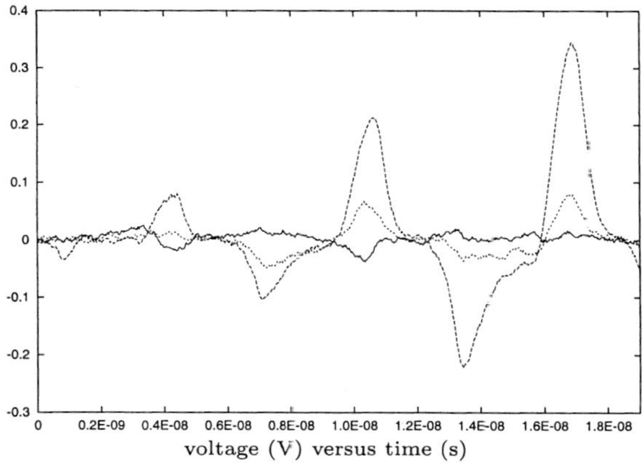

Fig. 3. Difference of the numerical solutions for X_t^5 computed with a standard scheme for SDEs and the new scheme for SDAEs: step size $h = 10^{-13}$ (*solid line*), step size $h = 10^{-12}$ (*dotted line*), step size $h = 10^{-11}$ (*dashed line*)

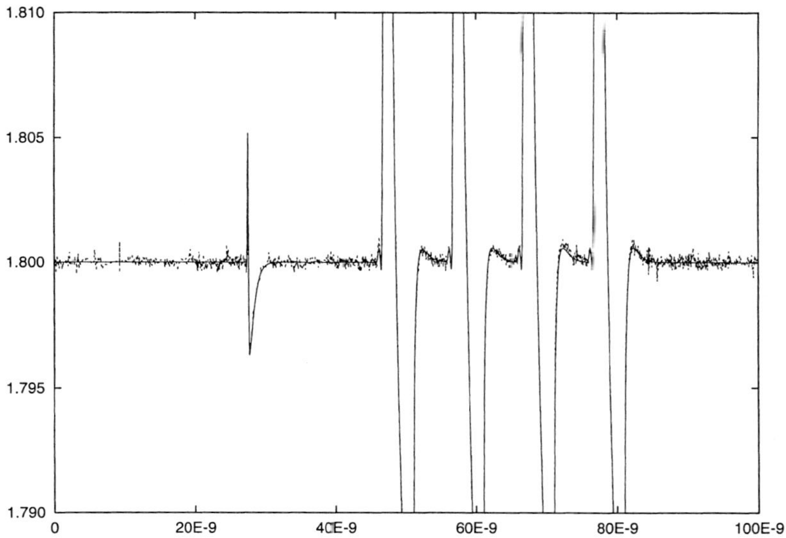

Fig. 4. Simulation result for a cross section of a 256 Mb DRAM: without noise (*solid line*), with noise (*dotted line*)

was shown that the numerical formula is applicable to circuit simulation. For a productive application of transient analysis analysis, further investigations have to be done, especially in the context of step-size control.

Acknowledgment

The author thanks the Bayerische Forschungsstiftung for the support by FORTWIHR (Bavarian Consortium for High Performance Scientific Computing).

References

1. Demir A., Sangiovanni-Vincentelli A. L. (1998) Analysis and Simulation of Noise in Nonlinear Electronic Circuits and Systems. Kluwer Academic Publishers
2. Denk G., Hillermeier C., Schäffler St. (2000) Ein Online-Verfahren zur numerischen Simulation von Systemen mit $1/f^{\beta}$-Rauschquellen. Internal report, Infineon Technologies
3. Engberg J., Larsen T. (1995) Noise theory of linear and nonlinear circuits. John Wiley & Sons
4. Günther M., Feldmann U. (1999) CAD-based electric-circuit modeling in industry I, II. Surv Math Ind 8:97–157
5. Kloeden P., Platen E. (1992) Numerical Solution of Stochastic Differential Equations. Applications of Mathematics 23, Springer Verlag
6. Mandelbrot B., van Ness J. (1968) Fractional Brownian motions, fractional noises and applications. SIAM Review 10:422–437
7. Øksendal B. (1995) Stochastic Differential Equations. Springer Verlag
8. Penski C. (2001) Numerische Integration stochastischer differential-algebraischer Gleichungen in elektrischen Schaltungen. Ph.D. thesis, Zentrum Mathematik, TU München
9. Schein O. Denk G. (1998) Numerical solution of stochastic differential-algebraic equations with applications to transient noise simulation of microelectronic circuits. J. Comput. Appl. Math. 100:77–92

Realistic Step Flow Model for Orientation-Dependent Wet Etching

A. Horn and G. Wachutka

Institute for Physics of Electrotechnology, Munich University of Technology, Arcisstrasse 21, D-80290 München, Germany

Abstract. We present a new simulation tool for orientation-dependent etching of silicon. The implemented algorithm is based on a model proposed by Schröder [1], which can explain the convex corner undercutting in pure aqueous KOH solutions. Essential is the experimental observation that the so called fast etching planes, which hitherto were assumed to cause the characteristic shape of underetched convex etchmask corners, are not really crystallographic planes. Referring to some basic examples we demonstrate that our simulation approach using this "step flow model of 3D structuring" is able to reproduce the detailed morphology of the etched structures.

1 Introduction

Orientation-dependent wet chemical etching is one of the basic techniques in silicon bulk micromachining. However, the etching mechanisms and several phenomena observed in orientation-dependent etching are still not completely understood. Consequently, in the simulation of the etching process there are still unsolved problems preventing the accurate calculation of shape and morphology of the etch front. In this paper we present a method and a simulation tool for three-dimensional orientation-dependent wet etching of silicon exhibiting {100}-surface orientation[1]. The implemented algorithm is based on the step flow model proposed by Schröder [1], which can explain the shape and morphology of singular parts of the etch front such as the convex corner undercutting (Fig. 1) in pure aqueous KOH solutions. The essential experimental finding is that the so-called "fast etching planes", which are commonly invoked as the cause of the characteristic shape of underetched convex etchmask corners, are not really crystallographic planes. Instead, these areas (denoted as "B" in Fig. 1) are the envelope surfaces of <110> oriented, bunched steplines originating from kink sites on the intersecting {111} planes. During the etching process these steplines move in $<11\bar{2}>$ direction, which may be conceived as "peeling off" stable {111} planes on the surface in lateral direction. The coarse "A" areas are residues of this process. Further details are given in [2].

[1] The notation {...} means a set of symmetrically equivalent crystallographic planes, while crystallographic directions are indicated by <...>

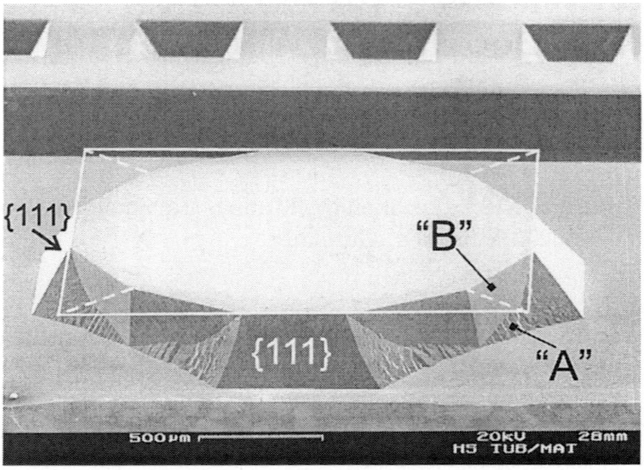

Fig. 1. SEM micrograph of a mesa structure with square etch mask (indicated by the white lines) and four convex corners

2 Numerical Modeling

The mathematical method used in our simulation approach was originally developed for digital image processing and later adapted to the purpose of efficient topography simulation [3] [4]. The basic idea is to represent the etch body and its exterior as a black-white image (black = material, white = no material), which is altered by the action of certain set operations. Each time step of the etching process is modeled as so-called "erosion operation", where a properly chosen "structuring element" acts on the etch body to mark all material that has to be removed next. The structuring element is a three-dimensional body such as a sphere or an ellipsoid, the shape of which has to be adjusted to the specific etching mechanism under consideration. The geometrical configurations during the etch process are described by a cellular representation in order to ensure an efficient numerical implementation. The simulated region is discretized by a partition of equally shaped volume elements ("cells") which are labeled by a material index indicating their location inside or outside the etch body. The time evolution of the etch front is represented by the temporal change of the material indices of the cells which, in turn, is controlled by the structuring element. We adapted this method for the implementation of the step flow model in an already existing simulation tool [3]. To this end, we had to extend the original erosion algorithm in the way that the structuring element, which is determined by the local etch rate vector, is not moved along the whole surface of the simulation area. Instead we have to trace the locations where kink sites are generated as well as the positions of the already existing steps. This means that we have to set up and solve the equations of motion for kinks and steps (see Fig. 2, left part) and then to apply the next erosion step accordingly. This basic idea of the step flow model is illustrated in Fig. 2 in a two dimensional schematic example. We introduce a coordinate system at the convex corner of the mask to determine the location of steps $\boldsymbol{r}_{step}$ and kinks $\boldsymbol{r}_{kink}$ at which the steplines have their origin. The existing steps (drawn in dark gray) are moved in lateral direction with an etch rate $\boldsymbol{R}_{step}$, as indicated by the arrows in the left part of Fig. 2, while the etching of kink sites (at the same time) generates further steps on the new etch front. On the right part of Fig. 2 these new steps are marked, and the removed cells are indicated by the dashed lines.

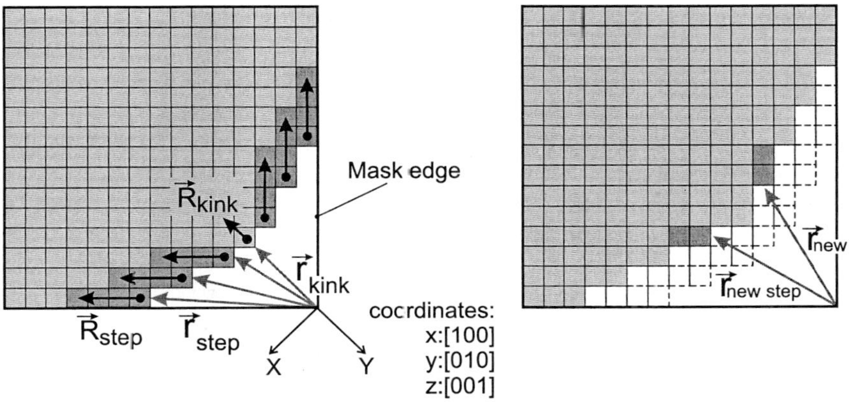

Fig. 2. Left: Top view on the simulation area of a convex corner showing locations and etch velocities of kinks and steps. Right: New etch front with two new generated steps

The orientation and magnitude of the etch rate vectors of kinks and steps ($\boldsymbol{R}_{kink}$, $\boldsymbol{R}_{step}$) at a given cell position ($\boldsymbol{r}_{kink}$, $\boldsymbol{r}_{step}$) on the etch front is derived from the specific properties of the respective cell which, in turn, result from the equations of motion of kinks and steps. In this way the new position of a step of the etch body after a time step Δt is given by the following equation of motion:

$$\boldsymbol{r}_{step}(t + \Delta t) = \boldsymbol{r}_{step}(t) + \boldsymbol{R}_{step} \cdot \Delta t \tag{1}$$

The step etch rate vector $\boldsymbol{R}_{step}$, which is regarded as velocity of the moving step, is assumed to be constant.

$$\boldsymbol{R}_{step} = \frac{\partial \boldsymbol{x}_{step}}{\partial t} \tag{2}$$

Actually the vector $\frac{\partial \boldsymbol{x}_{step}}{\partial t}$ indicating the motion of the steps is oriented along the <11$\overline{2}$> direction, i.e. the steps move in <11$\overline{2}$> direction. For the simulation we split this vector into its components parallel to the material grid. Therefore, in this two-dimensional example, the unit vector $\boldsymbol{e}_{step}$ which determines the projection R_{step} of $\boldsymbol{R}_{step}$ has only a component in <110> direction:

$$R_{step} = \frac{\partial \boldsymbol{x}_{step}}{\partial t} \cdot \boldsymbol{e}_{step} = const. \tag{3}$$

In the same way the kinetic equation describing the motion for the kink sites can be derived. Proceeding this way the progressive erosion of material is described as a layer-by-layer peel-off process.

3 Simulation Results

Fig. 3 shows the simulation result of the mesa structure of Fig. 1 (silicon-{100}, KOH/water solution with a concentration of 33 weight% at a temperature of 80°C, the square etch mask is not displayed). Comparing the SEM micrograph in Fig. 1 with the simulated structure clearly demonstrates that the typical morphology observed along underetched convex corners is correctly reproduced by our simulation method.

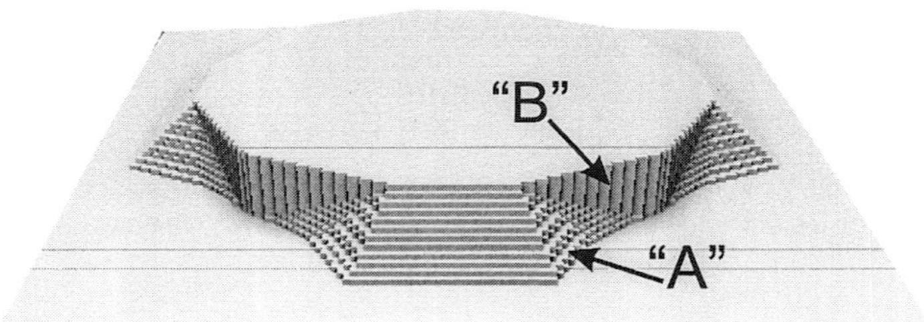

Fig. 3. Simulation of a mesa with four convex corners etched in pure KOH

The discretization of the simulated region is chosen coarser than the real dimensions of the fine steps in area B in consequence of the limited computational resources available. Furthermore, the material discretization is performed on a simple cubic grid oriented along <110> direction to ensure compatibility with another simulation tool. Therefore, the real shape and orientation of the fine steps visible in SEM micrographs is not fully resolved, but the enfolding envelope surface is calculated accurately. The etch rate vector of the real steps is decomposed into components along the <110>- and <100> direction to obtain the correct surface orientation of area B. The coarse A-type areas also appear regularly shaped in the figures, since the statistical mechanisms that cause the irregular morphology are currently not included in our simulation model. It is notable that the interaction between two B-type areas coming from different convex corners and contacting each other is accurately described. This is demonstrated in Fig. 4, which displays the etch mask, the SEM micrograph and the simulated structure of three <110> oriented beams which have the same length but differ in their width. Structures of this kind can serve as basic elements for the design of corner compensation structures.

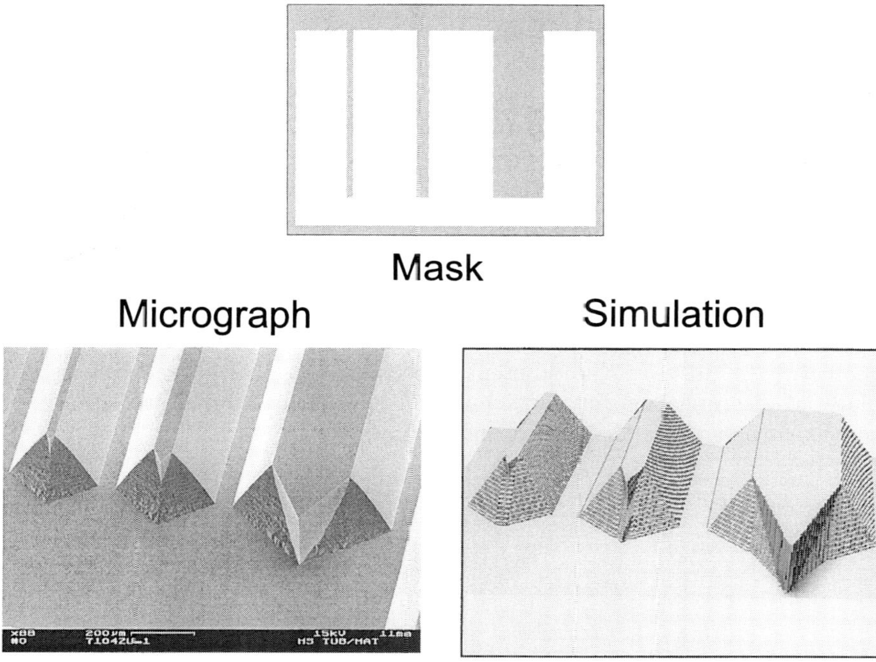

Fig. 4. Mask, SEM micrograph and simulated structure of three beams oriented along the <110> direction

The morphology of these structures is the result of the interaction of the two convex corners on the mask edges of each beam. Therefore, the structures exhibit a symmetry different from that of a single convex corner, but we can observe also smooth areas of type B and coarse A-type areas. It is remarkable that the shape of the narrow beams is more determined by the A-type areas than by the smooth B-areas. This effect is even more pronounced for narrow beams because of their smaller lateral extension compared to the etch depth. After sufficiently long etching, a change of the characteristic shape would also occur with the broader beam. Hence, etching a test structure featuring an array of several beams of different width for a fixed period of time is equivalent to the evolution of the shape of one broad beam observed at different subsequent time steps. The characteristic shape of the beams is correctly reproduced by the simulation, as it is evident by comparing the micrograph and the simulation result displayed in Fig. 4. Please note that the simulation was visualized by a ray-tracing tool, so the perspective view is slightly different from that in the micrograph. Evidently the accurate reproduction of the etch front requires a large number of cells. In this case the simulation area was discretized in $4.5 \cdot 10^6$ cells ($450 \cdot 250 \cdot 40$), and the simulation time was about 10 minutes on a standard workstation.

A quite complex interaction of convex corners appears if an etchmask exhibits edges with arbitrary angles or even curved edges. According to the step flow model, every mask edge which is not aligned in <110> direction exhibits unstable points, where kink sites are induced, and the characteristic shape of a convex corner emerges right at the beginning of the etching. Therefore, the 3D structure of the etch body resulting from an arbitrary oriented mask edge is caused by the interaction of those convex corners. Consequently, in our simulation approach any curved or bevelled mask edge can be composed of an array of convex corners in the way that the edge under consideration is approximated by line segments oriented in <110> direction. As an example, Fig. 5 shows the approximation of a circle which, in turn, represents the mapping of the circle to the cellular material discretization of the mask geometry. In this way the inherent approximation of a curved line by the material discretization for our simulation approach automatically produces convex corners along the curved mask edge.

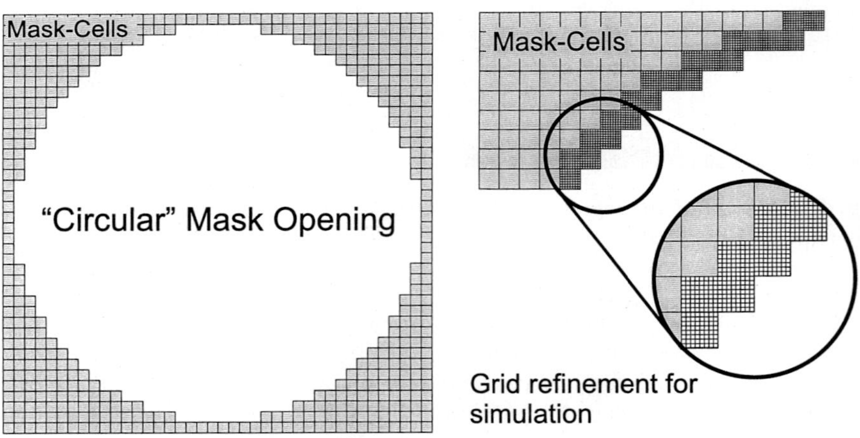

Fig. 5. Top view on a circular mask opening. The curved line of a real circle is approximated by the cellular material representation creating a large number of convex corners. A grid refinement is required for the evolution of these convex corners at the beginning of the simulation

For the etch simulation underneath the circular mask in Fig. 5, the material discretization of the mask has to be refined. Therefore every mask cell is subdivided into 64 cells again. Then all convex corners created for approximating the curved mask are treated like single convex corners at the very beginning of the etching. With progressing etch depth, the interaction of kinks and steps which originate from all these convex corners determine the further evolution of the etch body. The right part of Fig. 5 displays a zoom out of a part of the circular etch mask with the refined grid on which the etching simulation is carried out.

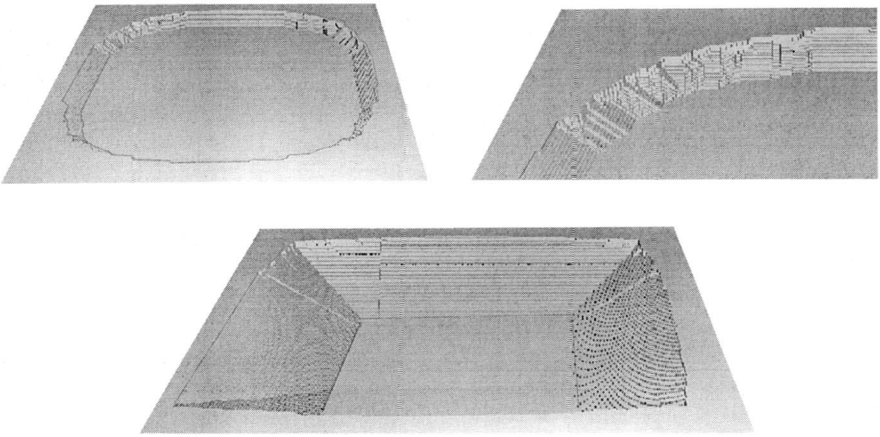

Fig. 6. Etch simulation of a circular mask. After sufficiently long etch time, only the stable {111} planes remain as part of the etch front

Fig. 6 displays the simulation results obtained with the circular etch mask of Fig. 5 (the mask is not drawn for better visualization). After sufficiently long time of etching only the stable {111} planes remain and form a pyramid-shaped hole.

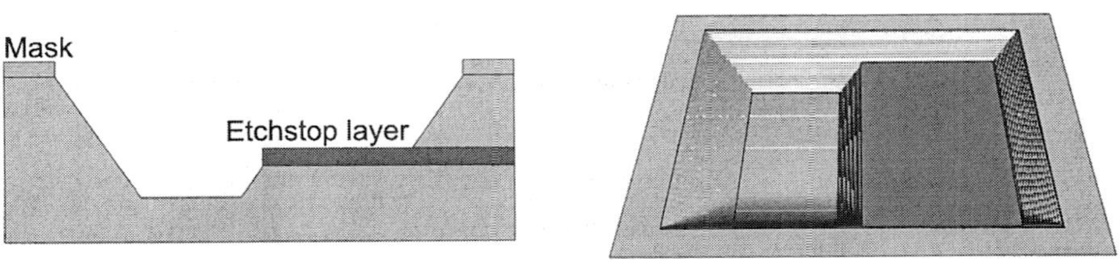

Fig. 7. Incorporation of etch stop layers. Schematic cross-section and three-dimensional simulation

The material index that characterizes the particular etching behavior of each individual cell is multi-valued. In this way, the simulation of substrates containing layers of different materials can be carried out efficiently. As an example, Fig. 7 displays the incorporation of a buried etch stop layer. The cells forming this layer are marked by a material index that is not attacked by the etchant (e.g. silicon nitride). The method is very flexible and allows for the simulation of realistic three-dimensional etch bodies. It also offers the possibility of simulating the etching of pre-structured wafers obtained, e.g., by laser micromachining or other etching processes (RIE, plasma etching), as well as double-sided wafer etching.

4 Conclusion

We conclude that the simulation of orientation-dependent wet etching on the basis of the step flow model as proposed in [1] is able to correctly reproduce the detailed morphology of the etch front. Thus it provides a solid basis for the predictive simulation of progressively complex MEMS structures. The compatibility with a professional TCAD environment allows for the efficient design of complicated etch mask compensation structures, potentially in combination with other materials and structuring techniques, since the use of a realistic three-dimensional model offers the possibility of simulating the etching of pre-structured wafers, obtained by laser micromachining or other etching processes (RIE, plasma etching), as well as double-sided wafer etching. This flexibility is the prerequisite for embedding the etching simulator in a modular TCAD platform which comprises all process steps encountered in microtechnology. Moreover, employing methods adapted from digital image processing, we can keep the simulation times short and the computational effort affordable, even for three-dimensional process simulation.

Acknowledgement

The authors are very grateful to Prof. Dr. S. Selberherr and Dr. W. Pyka from the Institute for Microelectronics, Technical University of Vienna, and to Dr. H. Schröder and Prof. Dr. E. Obermeier from the Technical University of Berlin, MAT, for providing valuable assistance and support.

References

1. H. Schröder, E. Obermeier, Workshop on Physical Chemistry of Wet Chemical Etching of Silicon, Holten, 1998, Book of Abstracts, pp 31-32.
2. H. Schröder, PhD Thesis, Technical University of Berlin, 2000.
3. Strasser, E., Selberherr, S., IEEE Trans. on CAD of Integrated Circuits and Systems, Vol. 14, 9, (1995), pp 1104-1114.
4. Strasser, E., Selberherr, S., Proc. of SISDEP-93, Vienna, Austria, in Simulation of Semiconductor Devices and Processes, Eds.: S. Selberherr, H. Stippel, E. Strasser, (Springer Verlag, Wien, 1993), pp. 357-360, 1993.

Modeling of Ion-Induced Charge Generation in High Voltage Diodes

W. Kaindl[1], G. Sölkner[2], and G. Wachutka[1]

[1] Institute for Physics of Electrotechnology, Munich University of Technology, D-80333 Munich, Germany
[2] Infineon Technologies AG, Power Semiconductors, D-81730 Munich, Germany

Abstract. Additional terms have to be added to the right-hand sides of the carrier balance equations for electrons and holes in order to describe the initial injection of charge caused by the loss of kinetic energy of a single ion penetrating a semiconductor device. Two-dimensional simulations of a reverse biased power diode yield the temporal and spatial distribution of the device-internal electric field initiated by an intruding ion. For small reverse biases the charge generated within the device corresponds to the total absorption of the ion's kinetic energy. Applying a sufficiently high reverse bias a steep field peak forms which is able to propagate through the whole device with an undiminished peak height. Due to the corresponding strong avalanche multiplication a large amount of additional charge can be generated. The results obtained from our device simulations conform well to recent experimental findings.

1 Introduction

Meanwhile, there is undisputed evidence that cosmic radiation is a potential source for the failure of high voltage power electronic devices effected by single event burnout (SEB). This has been proven by numerous tests in low and high altitudes as well as by irradiation experiments [1–3]. Incident neutrons have been identified as the decisive particles which are able to cause SEB even at sea level [4]. A widely accepted model of the failure mechanism assumes that light ions are produced by a nuclear reaction between an intruding neutron and a silicon nucleus of the semiconductor substrate material. The ions initiate a highly localized burst of charge which is amplified in the high-field region of a reverse-biased power semiconductor device. Eventually, this leads to the thermal breakdown of the device.

In order to gain more insight into this mechanism, irradiation experiments with ions were performed [3,5]. An advantage of ion irradiation experiments is that, in contrast to neutrons, ions have a well-defined penetration depth after entering the semiconductor material, whereas nuclear reactions between neutrons and silicon nuclei could occur anywhere within the device due to the highly statistical nature of the reaction process.

In this work, we report the modeling and simulation of an ion irradiation experiment, which was discussed in detail in [3]. In this experiment, a high power diode with a reverse blocking voltage of about 3.5 kV was irradiated by several types of ions using different ion beam energies. The flux of the ion beam was reduced, so that the charge generated by one single ion could be measured. The fundamental experimental findings reported in [3] are resumed in Fig. 1. Applying small reverse voltage the diode behaves like a surface barrier counter. The charge generated in the device corresponds to the total absorption of the ion's kinetic energy. For sufficiently high reverse bias charge carrier multiplication sets on. As a consequence, up to four orders of magnitude of additional charge carriers can be generated. For still higher reverse bias solely the multiplication events occur. The onset voltage of charge multiplication as well as the multiplication factor depend on the type of the ion and on the kinetic energy.

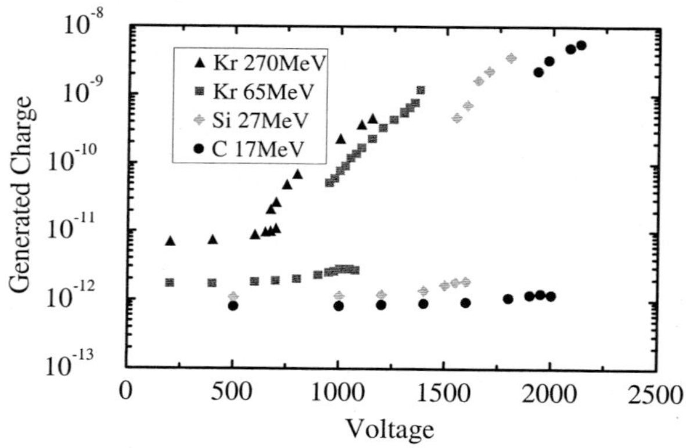

Fig. 1. Charge generated by one single ion within the diode for different types of ions and kinetic energies

2 Modeling

2.1 Extended Drift-Diffusion Model

In order to enable a detailed numerical analysis of the experimental processes with conventional device simulators, additional generation terms G_{ion} have to be added to the right-hand sides of the particle balance equations for electrons

$$\frac{\partial n}{\partial t} - \frac{1}{q}\nabla \boldsymbol{j}_n = -R + G_{ava} + G_{ion}, \tag{1}$$

and holes

$$\frac{\partial p}{\partial t} + \frac{1}{q}\nabla \boldsymbol{j}_p = -R + G_{ava} + G_{ion}, \tag{2}$$

respectively. Here, n and p denote the densities of electrons and holes, respectively, while $\boldsymbol{j_n}$ and $\boldsymbol{j_p}$ are the respective current densities. q denotes the elementary charge. Usually, generation/recombination includes the sum of Shockley-Read-Hall (SRH) recombination R and impact ionization (also known as avalanche generation)

$$G_{ava} = \frac{1}{q}\left(\alpha_n|\boldsymbol{j_n}| + \alpha_p|\boldsymbol{j_p}|\right). \tag{3}$$

Impact Ionization strongly depends on the electric field F in the interior of the device according to the relation

$$\alpha_{n,p} = a_{n,p} \exp\left(-\frac{b_{n,p}}{F}\right). \tag{4}$$

Here, $a_{n,p}$ and $b_{n,p}$ are material parameters. Equations (3) and (4) state that the higher the electric field, the higher is the avalanche generation of additional free electrons and holes.

Together with Poisson's equation, and the constitutive current relations, the particle balance equations (1) and (2) describe the transient transport behaviour of power semiconductor devices. A more detailed description of the drift-diffusion model can be found elsewhere, e.g. [6,7].

2.2 Modeling of Initial Charge Generation

For a detailed numerical analysis of the experimental processes and the visualization of the dynamical behavior in the interior of the device, we developed the following model describing the initial generation of charge induced by the loss of kinetic energy of the ion. First, the track of an ion penetrating into the semiconductor material is computed using the simulator TRIM (Transport of Ions in Matter) [8]. TRIM calculates the energy loss function $\frac{dE}{dz}$ of the ion along its path through the semiconductor material on the basis of a full quantum mechanical treatment of the ion-atom collisions. The resulting initial charge carrier generation rate

$$G_{ion}(\boldsymbol{x}, t) = \frac{A}{2\pi w s} \cdot \frac{dE}{dz} \cdot \exp\left(-\frac{r^2}{2w^2}\right) \cdot \exp\left(-\frac{1}{2}\left(\frac{t - t_0}{s}\right)^2\right) \tag{5}$$

is added to the generation/recombination terms on the right-hand sides of the particle balance equations for electrons and holes, respectively (see (1) and (2)). In (5), z is the coordinate parallel to the trajectory of the ion within the device. t represents the time, whereas r denotes the radial distance from the ion track. From the sum rule

$$\int\int G_{ion}(\boldsymbol{x}, t) d^3x \, dt = \frac{E}{E_p} \tag{6}$$

the normalization constant A can be calculated. $E_p = 3.6\,\mathrm{eV}$ denotes the energy that is necessary to generate one electron-hole pair [9]. Therefore, (6) ensures that the ratio of the kinetic energy E of the ion and E_p determines the total initial amount of charge injected into the device.

Solving the equation of motion for the ion during the stopping process within the device,

$$\frac{dz}{dt} = \sqrt{\frac{2}{m_{ion}} \int \frac{dE}{dz} dz} \tag{7}$$

we obtain estimates of the parameters w and s. Here, m_{ion} denotes the mass of the ion. Fig. 2 shows the calculated velocity and depth of penetrating ^{12}C and ^{86}Kr ions within the semiconductor material. As the velocity of the ^{12}C and ^{86}Kr ions is well below 10% of the velocity of light c, a non-relativistic treatment of the equation of motion is justified. The ^{12}C (^{86}Kr) ions loose all their kinetic energy within approximately 2 ps (4 ps). Within this time, all the initial charge is generated by the ions. Half of this time is therefore a reasonable approximation of the temporal Gaussian width s. Consequently, s is 1 ps for the ^{12}C ions with an kinetic energy of 17 MeV and 2 ps for ^{86}Kr with 270 MeV.

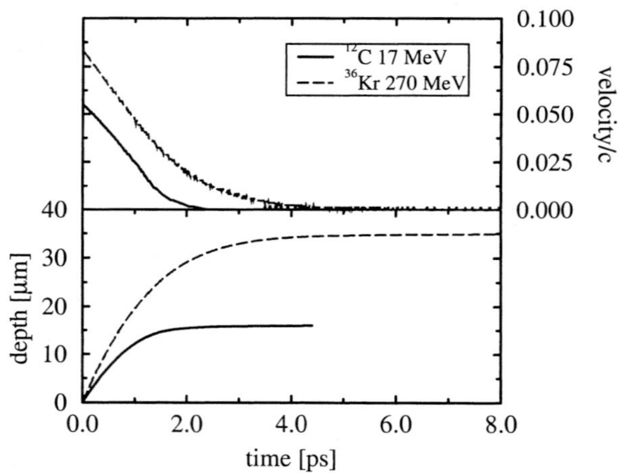

Fig. 2. Velocity and penetration depth of ^{12}C (solid lines) and ^{86}Kr (dashed lines) ions within the semiconductor material

Assuming that carrier diffusion starts immediately in the moment of the penetration of the ion, the width of the diffusion profile is about 0.2 μm after 2 ps. Consequently, the radial Gaussian width w is 0.1μm for the ^{12}C ions and 0.2μm for ^{86}Kr.

Of course, these estimates are quite coarse but, nevertheless, they provide a valueable guess of the order of magnitude of the parameters w and s.

The stopping time s is very short on the time scale relevant to the electric dynamics of the device. This justifies the approximation made in (5) by assuming that a static generation profile develops quasi-instantaneously along the ion track.

Implementing the described model in the multidimensional device simulator DESSIS$_{-\text{ISE}}$ [7] allowed us to perform numerical simulations of the ion irradiation experiment.

3 Simulation

Among others we performed a numerical analysis of a p^+n-diode of 450 μm length, where a ^{12}C ion intrudes perpendicularly into the device at the anode contact (z=0 μm). With a kinetic energy of 17 MeV, the penetration depth of the ^{12}C ions is about 16 μm. The pn-junction of the diode was modelled using a 1.5 μm deep Gaussian p-diffusion in a lowly doped n-base. The whole structure was discretized using about 10000 finite elements in 2D. The width of the simulated structure was 100 μm around the track of the ion.

Fig. 3 and 4 show the temporal and spatial evolution of the electric field in the plane of incidence of the ion. The respective applied reverse biases are 1800 V (Fig. 3) and 2200 V (Fig. 4). The density of the initial charge generated by the intruding ion exceeds the background charge of the dopants. Therefore, the initial triangular electric field (thick lines in Fig. 3 and 4, respectively) is disturbed. A steep field peak forms (t=25 ps) and starts propagating through the device towards the cathode. On the anode region left-sided to the peak, the electric field is almost completely suppressed due to charge neutrality occuring under plasma conditions. While propagating through the device, the

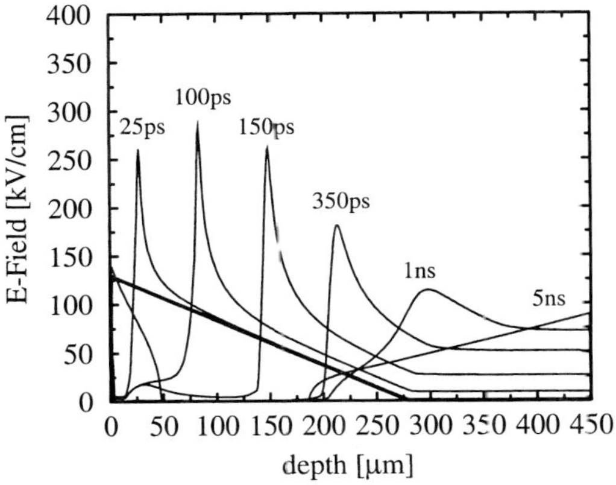

Fig. 3. Temporal and spatial evolution of the electric field inside the device at a reverse bias of 1800 V

field peak smoothes out for a reverse bias of 1800 V (Fig. 3). The period of strong avalanche multiplication caused by the high electric field peak is too short to initiate the generation of a large amount of additional free charge carriers. The charge flowing out of the device thus corresponds to the total absorption of the ion's kinetic energy. Consequently, the resulting flow of current remains relatively low (≈ 0.1 mA, see Fig. 5). As the free carrier concentration near the anode falls below the background doping concentration ($t = 5$ ns), the initial triangular field starts building up again. After $t = 50$ ns the initial electric field is completely restored.

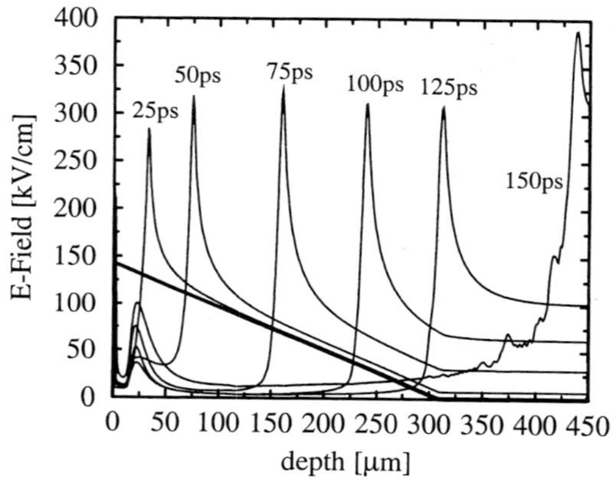

Fig. 4. Temporal and spatial evolution of the electric field inside the device at a reverse bias of 2200 V

A completely different situation occurs for a reverse bias of 2200 V. The maximum value of the field peak remains almost unchanged while it propagates through the device (Fig. 4). Therefore, the whole diode is flooded with free charge carriers along the plane of incidence due to the corresponding strong avalanche multiplication. As a consequence, a highly conductive filament can form which initiates a high flow of current (≈ 20 mA) that stays on for serveral 10 ns, until the initial situation is restored after about 150 ns. In Fig. 5, we recognize that the high current is initiated just at the moment when the field peak and, therefore, the conductive filament reaches the cathode contact at about 150 ps. In this case, a large amount of additional free electrons and holes is generated.

An interesting detail of the propagation mechanism of the electric field is that the velocity of the field peak is faster than the saturation velocity of the free carriers. This clearly indicates that the propagation of the electric field is driven by the continuous generation of charge and not by the ambipolar transport of charge. The higher the avalanche generation rate, the higher is the velocity of the field peak. The propagation and the distribution of the

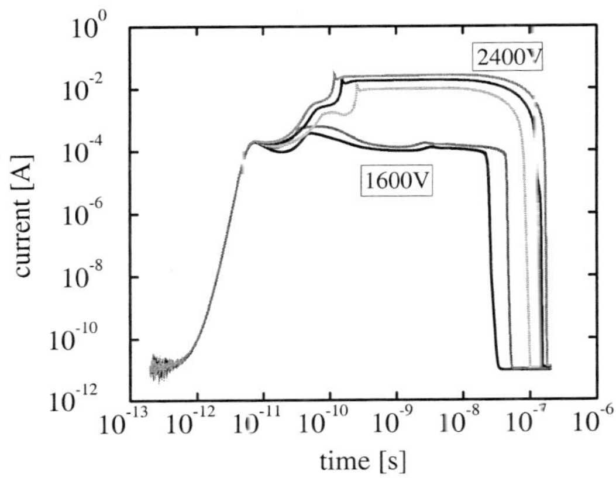

Fig. 5. Current transients for applied reverse voltages between 1600 V and 2400 V (in steps of 200 V)

electric field is very similar to a phenomenon that occurs in gas discharges and has become well-known as so-called streamer [10].

4 Discussion and Conclusion

In Fig. 6 the simulated generated charge is shown and compared with the experimental data from [3] for ^{12}C (17 MeV) and ^{86}Kr (270 MeV) ions.

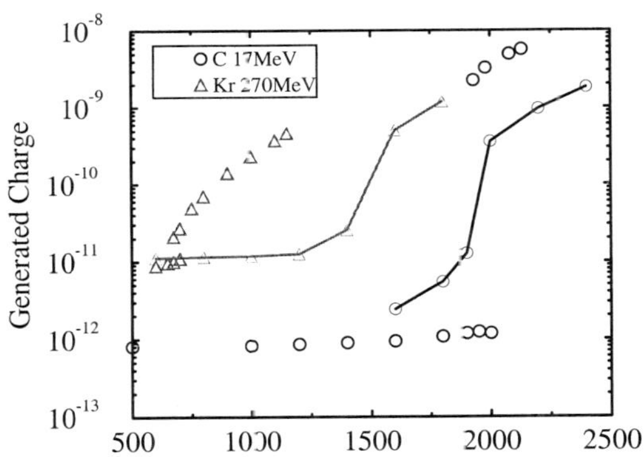

Fig. 6. Generated charge within the device: comparison of experiment (symbols) and simulation (solid lines) for the case of irradiation with ^{12}C (17 MeV) ions and ^{86}Kr (270 MeV) ions

In the case of the ^{12}C ions, we find good agreement between simulation and experiment. Furthermore, the abrupt onset of charge multiplication as well as the onset voltage are satisfactorily reproduced in the simulation. But for the ^{86}Kr ions, the simulated onset voltage is shifted to much higher voltages compared to the experimental findings (see also Fig. 1). Here, the limitations of the 2D simulation become noticeable. In 2D the field peak is not a real peak, but a field front. This means that the third dimension, in which the electric field has a finite extension is missing which finally leads to a lower value of the electric field peak and an accordingly lower avalanche generation of free charge carriers. Using cylindrical coordinates in the simulation is likewise problematic, since the gradients of the state variables become singular on the axis of symmetry which is just the line of incidence in our case. The nature of the problem would require a full 3D simulation, but with the computational resources currently available to us a 3D simulation is not (yet) possible.

Nevertheless, already with 2D simulations we were able to get a qualitative visualization of the mechanisms that occur during charge multiplication events. The basic features of the device behaviour resulting from the irradiation by ions or cosmic rays can be simulated and explained. This knowledge is very useful for the design of new devices to make them more robust against cosmic radiation.

References

1. Kabza H., Schulze H.-J. et al. (1994) Cosmic Radiation as a Cause for Power Device Failure and Possible Countermeasures. Proc. 6th IEEE int. Symp. on Power Semiconductor Devices and ICs (ISPSD), Davos, Switzerland.
2. Findeisen C., Herr E. et al. (1998) Extrapolation of cosmic ray induced failures from test to field conditions for IGBT moduls. Microelectron. Reliab. **38**, 1335–1339.
3. Maier K.H., Denker A. et al. (1998) Single event burnout of high-power diodes. Nucl. Instr. and Methods in Phys. Res. **B146**, 596–600.
4. Tang H.H.K. (1996) Nuclear Physics of cosmic ray interation with semiconductor materials. IBM J. Res. Develop. **40(1)**.
5. Voss P., Maier K.H. et al. (1997) Irradiation experiments with high-voltage power devices as a possible means to predict failure rates due to cosmic rays. Proc. 9th IEEE int. Symp. on Power Semiconductor Devices and ICs (ISPSD), Weimar, Germany.
6. Selberherr S., (1984) Analysis and Simulation of Semiconductor Devices. Springer-Verlag, Wien.
7. ISE Integrated Systems Engineering AG (1996) DESSIS user manual 5.0.
8. Ziegler J.F., Biersack J.P., Littmark U., (1985) The Stopping and Range of Ions in Solids. Pergamon Press, New York.
9. McKay K.G., McAfee K.B. (1953) Electron Multiplication in Silicon and Germanium. Physical Review **91**, 1079–1084.
10. Morrow R., Lowke J.J. (1997) Streamer Propagation in Air. Jour. Appl. Phys. D **30**, 614–627.

Modelling and Simulation of the Transient Electromagnetic Behavior of High Power Bus Bars

P. Böhm[1], G. Wachutka[1], and R.H.W. Hoppe[2]

[1] Institute for Physics of Electrotechnology, Munich University of Technology, D-80290 Munich, Germany
[2] Institute of Mathematics, University of Augsburg, D-86159 Augsburg, Germany

Abstract. An extended concept for the characterization of the electromagnetic behavior of high power bus bars in high power semiconductor modules is presented. Based on a three-dimensional transient finite element analysis of the electromagnetic field under realistic switching conditions, the eigendynamics of interconnects is evaluated. It is demonstrated that the investigation of distributed parasitic effects caused by short switching times and steep current or voltage ramps complements the common circuit analysis approach. The different modelling and simulation techniques in use are discussed focusing on specific numerical requirements. The capabilities of the method are demonstrated by some illustrative numerical examples.

1 Introduction

The optimization of high power components based on fast switching power semiconductor devices requires an integral approach that takes into account the overall performance of a module including the system-specific parasitic effects. With increasing switching frequencies and ever shorter pulse width, the eigendynamics inherent in the system becomes relevant and, hence, has to be included in the system analysis. Along with electrothermal aspects, the transient electromagnetic behavior of the interconnects can no longer be neglected. Therefore, the characterization of a complex, multiply contacted bus bar, or a system of bus bars, respectively, has to include the concept of time-dependent inductance. Induced overvoltage peaks, not only resulting from the time-derivative of the terminal current but also reflecting the time-dependence of the inductance, have to be considered as they can damage the attached semiconductor devices. Inhomogeneous current distributions in the interconnects do not only cause non-negligible electrothermal heating, but they can also lead to malfunction of large-scaled semiconductor switches which require a most homogeneous current distribution along the contacts. Moreover, the switching behavior may no longer be governed by the turn-on time of the semiconductor switches in use, but rather by the turn-on delay caused by the inductances of the bus bars.

2 Conventional Approach

Commonly the characterization and description of the electromagnetic behavior of high power bus bar structures is restricted to the determination of the integral quantities resistance, terminal currents and terminal voltages, and the extraction of the static inductance matrix of self- and mutual inductance coefficients. Apart from the widely employed heuristic approaches [1], there exists a well-established numerical method for inductance extraction, namely the Partial-Element-Equivalent-Circuit-Method (PEEC) [2]. This method is based on the calculation of the partial inductances by evaluating Neumann's formula.

$$L_{f_{ij}} = \frac{\mu}{4\pi} \oint_{C_i} \oint_{C_j} \frac{d\boldsymbol{l}_i d\boldsymbol{l}_j}{r_{ij}} \tag{1}$$

$L_{f_{ij}}$ denotes an inductance matrix element, the index f indicates that current filaments are considered, $r_{ij} = |\boldsymbol{r}_i - \boldsymbol{r}_j|$ and $d\boldsymbol{l}_i$ is an element of conductor i. For the evaluation of (1), the interconnects are partitioned into brick-shaped parts where each brick is associated with a current filament [3]. Therefore the PEEC-method is quite restricted with respect to the geometrical shape.

Besides other integration techniques are used to evaluate this formula, for example by multipole expansion [4] or Monte Carlo integration [5].

3 Extended FEM Based Approach

However, all above-mentioned methods are restricted to the quasi-magneto–static or the time-harmonic case. But for the analysis of the fast transient behavior of interconnects and in view of the wide variety of associated target functionals, a detailed understanding of the electromagnetic fields inside and outside the interconnects is essential. For this purpose it is more appropriate to stay in the time domain and not to change to the frequency domain by Fourier analysis.

We have already presented a methodology for the analysis of the transient electromagnetic behavior of multiply contacted interconnects based on the "field diffusion approximation“ [6,7]. This methodology requires an efficient solver for the associated electric and magnetic potentials. To this end, we extended the finite element simulator SESES [8] by an electromagnetic kernel based on the $\boldsymbol{A}, V$-$\boldsymbol{A}$ formalism [9] which stands for solving the magnetic vector and electric scalar potentials $(\boldsymbol{A}, \varphi)$ inside the conducting region(s) and $\boldsymbol{A}$ in the nonconducting region(s).

The finite element simulator SESES provides a number of features essential for the simulation of the transient electromagnetic behavior of interconnects. Besides a mesh generator based on hexahedral isoparametric elements, which are very well suited to the shape of multiply banded and cutted plates, the simulator features effective refinement techniques in space and time, e. g. a recursive adaptation of the mesh [10]. This is indispensable for economizing the number of elements, in order to save computing time while keeping the discretization error small at the same time.

These capabilities are demonstrated by a simple two-dimensional example, namely a rectangular conductor embedded in a non-conducting environment (Fig. 1). A current ramp is used as bias condition. As one of the features of the simulator, the efficiency of a global refinement strategy is demonstrated in Fig. 2. We find that the mesh is refined at the corners of the conductive rectangle, where the current density has its maximum inhomogeneity due to the skin effect, while it remains nearly unchanged elsewhere. As we want to solve the problem in the time domain, the number of time steps of transient simulations has to be minimized by an efficient time-stepping scheme as illustrated in Fig. 3. Starting with a time step of $1ns$, for instance, the program soon reaches a default value of $100ns$. The reduced time step at about $t = 3.0 \cdot 10^{-7}s$ is caused by a coincidence of the prescribed, user defined maximum time step and the automatically chosen time step.

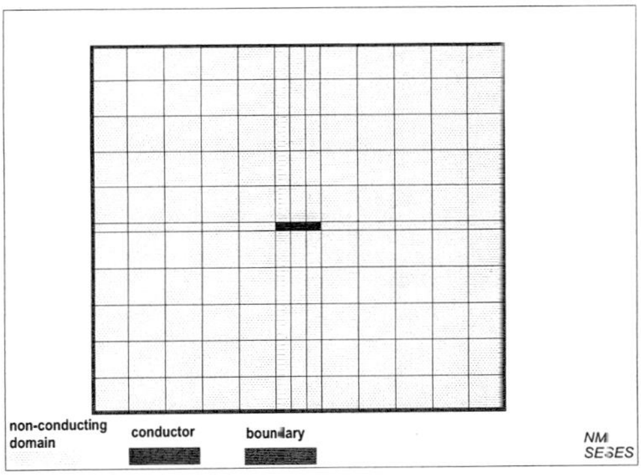

Fig. 1. Exemplary problem: Simulation domain and initial mesh of a two-dimensional rectangular conductor embedded in a non-conducting environment

4 Examples

In an illustrative example, the above-described computational functionality of the simulator is demonstrated by the simulation of a three-dimensional skin effect problem. Fig. 4 shows the finite element grid for the computation of a conducting bar in a non-conducting environment. For symmetry reasons, only one fourth of the bus bar is simulated. According to most of the cases encountered in practice, the skin depth is much less than the dimension of

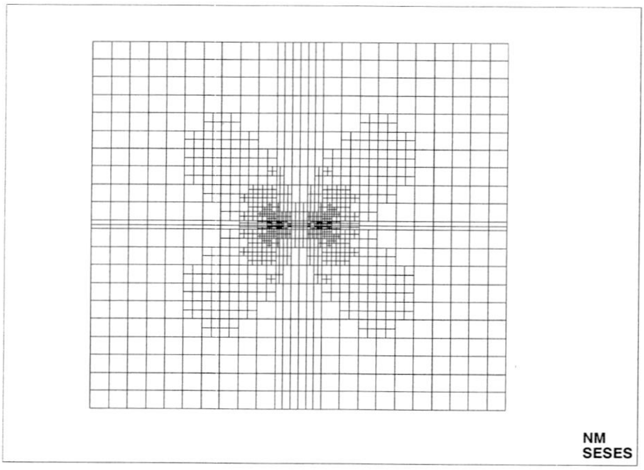

Fig. 2. Final mesh as obtained by global refinement strategy (number of elements: 1400)

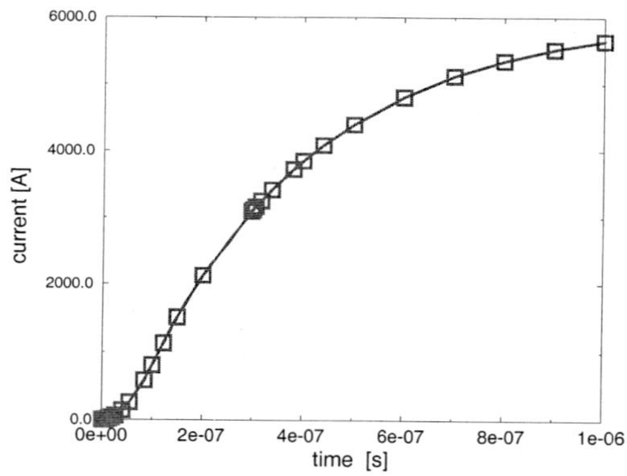

Fig. 3. Simulated time-dependent current flow demonstrating the capabilities of SESES with respect to time adaptivity

the investigated bus bar structure. Therefore, we start with a coarse grid for both the conducting and the non-conducting region. Then a user defined thin "surface layer" of elements with a thickness in the range of the skin depths is added underneath the conductor surface. This layer is subsequently refined and adapted to the surrounding grid by automatic adaptive mesh refinement. Elements with a bad aspect ratio in the non-conducting area are of minor influence, whereas a non-adapted grid in the conducting region, which would lead to an ill-conditioned discretized problem, is avoided by the automatic mesh refinement. The mesh is flexible enough to resolve the skin effect, but also to reduce the number of elements to a minimum.

Fig. 4 and Fig. 5 show the current distribution inside the conducting bar at different time steps. A voltage ramp with a pulse rise-time of $0.1\mu s$ is used as bias condition. At $t = 10\mu s$ the current distribution is still dominated by the skin effect, which is clearly resolved. The current density in the bulk is by some orders of magnitude smaller than the current density in the skin layer. Even at $t = 1000\mu s$, the skin effect is still noticeable and the current distribution reaches a steady state not before $t = 10ms$.

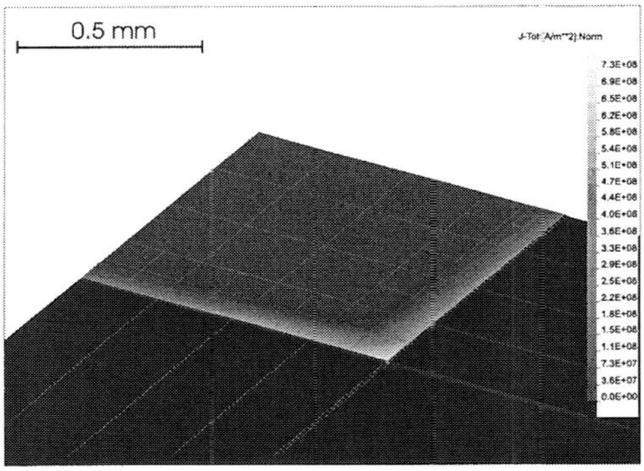

Fig. 4. Simulated current distribution at $t = 10\mu s$ for a 3D skin effect problem of a conducting bar in a non-conducting environment

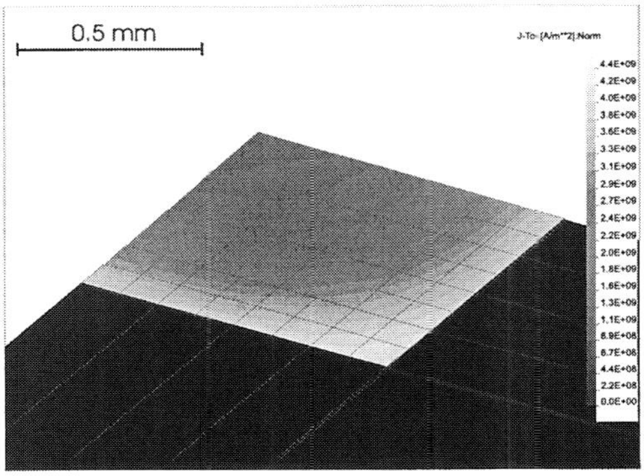

Fig. 5. Simulated current distribution at $t = 1000\mu s$ for a 3D skin effect problem

As a further example, a simplified but typical high power bus bar structure was analyzed for different geometrical modifications (Fig. 6). This bus bar is made of copper plates with a thickness of $1\,mm$ and a length of $14\,cm$. The plate is connected at the left and right side faces to a voltage source with a rise-time of $100ns$ and a constant value of $U = 0.01V$ afterwards. Special attention is paid to the transient states, where the skin effect is the dominant factor, by comparing them with the static state.

At $t = 1\mu s$, ten times later than the end of turn-on, the current distribution is still dominated by the skin effect which displaces the current to the outer corners of the structure (Fig. 6). For comparison, Fig. 7 shows the quasistatic current distribution which is attained at $t \approx 1000\mu s$ and plotted for $t = 1.11s$. The terminal current passing the contact electrodes shows, as a function of time, a behavior similar to that of a simple RL network element.

In Fig. 8, the time-dependent inductance is plotted. As the bus bars considered here have just two contacts, the terminal behavior of the structure is characterized by its self-inductance, which increases for $0 \leq t \leq 1000\mu s$ until the so called inner self-inductance has fully built-up. A structure possessing a L-shaped structure (Fig. 6) causes a partial compensation of the magnetic field between the L-structure, which results in a reduced inductance in comparison with a plain structure (dotted line in Fig. 8).

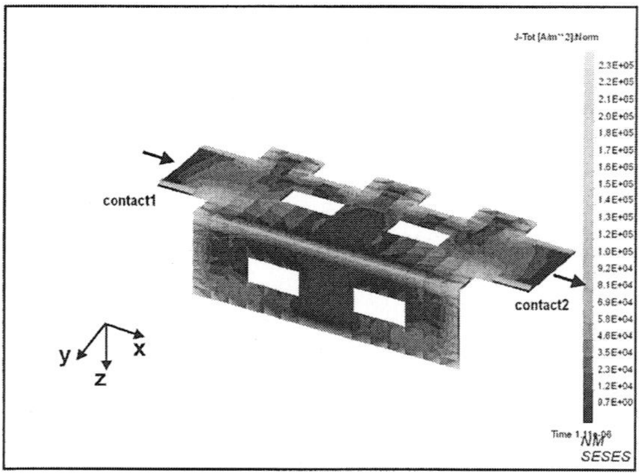

Fig. 6. Simulated current distribution at $t = 1.11\mu s$ for a simplified bus bar structure

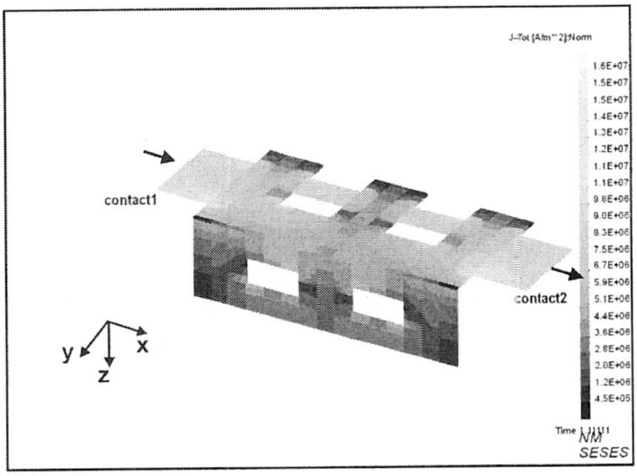

Fig. 7. Stationary current distribution at $t = 1.11s$

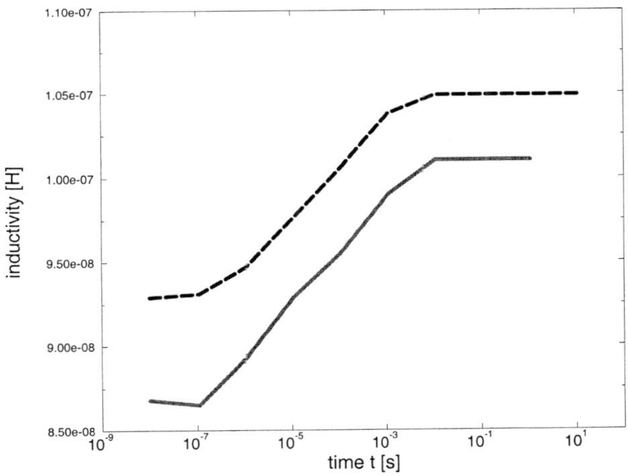

Fig. 8. Simulated time-dependent inductance of the interconnect for different geometries

5 Conclusion

The finite element simulator SESES allows the numerical analysis of the transient electromagnetic behavior of interconnects under realistic switching conditions. Equipped with the terminal characteristics and time-dependent impedence coefficients extracted from the transient electric and magnetic fields and with a view to shape optimization, we are able to define target functionals to assess the quality of a given interconnect set-up with respect to the uniformity of current flow, switching time delay, overvoltage and overheating limitations and related quantities of interest. In addition, our methodological concept provides a future perspective of extracting a set of time-dependent inductance coefficients required as input for circuit-level full system simulations of high power modules.

References

1. Skibinski, G.L., Divan, D.M.: Design Methodology & Modeling of Low Inductance Planar Bus Structures. Proc. of EPE, Brigthon, Great Britain, (1993) 98–105
2. Ruehli, A.E.: Inductance Calculation in a Complex Integrated Circuit Environment. IBM-Journal of Research and Development (1972) 470–481
3. Clavel, E., Schanen, J.L., Roudet, J.: Electromagnetic Modeling of a Power Module Case. IEEE EUROEM, (1994)
4. Kamon, M., Tsuk, M.J., White, J.: Fasthenry: A multipole-accelerated 3-d inductance extraction program. IEEE Trans. On Microwave Theory and Techniques **42** (1994) 1750–1758
5. Leonhardt, G., Fichtner, W.: Acceleration of Inductance Extraction by Means of the Monte Carlo Method. Proc. of MSM, San Juan, Puerto Rico (1999) 147–150
6. Böhm, P., Falck, E., Sigg, J., Wachutka, G.: Continuous Field Analysis of Distributed Parasitic Effects by Interconnects in High Power Semiconductor Modules. Proc. of SISPAD, Leuven, Belgium (1998) 340–343
7. Böhm, P., Wachutka, G.: Transient Electromagnetic Behavior of Multiply Contacted Interconnects. Proc. of MSM, San Juan, Puerto Rico (1999) 301–304
8. NM-SESES User Manual 5.0. Published by NM Numerical Modelling GmbH, Winterthur, Switzerland (1998)
9. Bíró, O., Preis, K.: On the Use of the Magnetic Vector Potential in the Finite Element Analysis of Three-Dimensional Eddy Currents. IEEE Trans. Mag. **25** (1989) 3145–3159
10. Korvink, J. G.: An Implementation of the Adaptive Finite Element Method for Semiconductor Sensor Simulation. Dissertation 10143, ETH Zürich (1993)

Modeling and Simulation of Electrothermomechanical Coupling Phenomena in High Power Electronics

P. Böhm[1], Y.C. Gerstenmaier[1], R.H.W. Hoppe[2], Y. Iliash[2], G. Mazurkevitch[2], and G. Wachutka[1]

[1] Physics of Electrotechnology, Munich University of Technology, D-80290 Munich, Germany
[2] Institute of Mathematics , University of Augsburg, D-86159 Augsburg, Germany

Abstract. High power electronic devices based on innovative semiconductor technology play a significant role in technical applications. A robust operating behavior of such devices can be achieved by an optimal design taking into account the presence of coupled physical effects.
In this contribution, we focus on electrothermomechanical coupling phenomena in Integrated-High-Voltage Modules with housing and cooling mechanisms that are used as electric drives for high power electromotors. For the numerical solution of the underlying coupled systems of partial differential equations we consider efficient algorithmic tools such as domain decomposition techniques.

1 Introduction

The development and realization of innovative technologies for contact systems in high power electronics are currently subject of various activities both in basic research as well as in industrial practice. Particular interest is devoted to semiconductor based power devices as they are used in converter modules for energy transmission or as energy saving and low noise electric drives for high power electromotors in public transportation systems like trams or high speed trains. Significant aspects that have to be taken into account are reliability and safety of the used components. In order to guarantee a long term durability, the specific semiconductor devices have to be packed into appropriately designed modules that are integrated in housings filled with hard- and/or soft-cast.
Figure 1 displays a typical Integrated High-Voltage- (IHV-) module as developed by Siemens AG and eupec GmbH & Co KG (without the housing). In operational mode, the semiconductor devices produce Joule heat which results in a considerable self-heating of the entire module causing mechanical deformations due to heat stresses that can even lead to mechanical damage without appropriate cooling mechanisms. Therefore, the optimal design of the modules requires a subtle understanding of the electrothermomechanical coupling phenomena. This cannot be achieved by experimental means, since

during operation the most interesting parts in the module are not accessible to measurements. Therefore, efficient numerical simulation tools based on appropriate models have to be developed and implemented.

Concerning the applications oriented treatment of electrothermomechanical

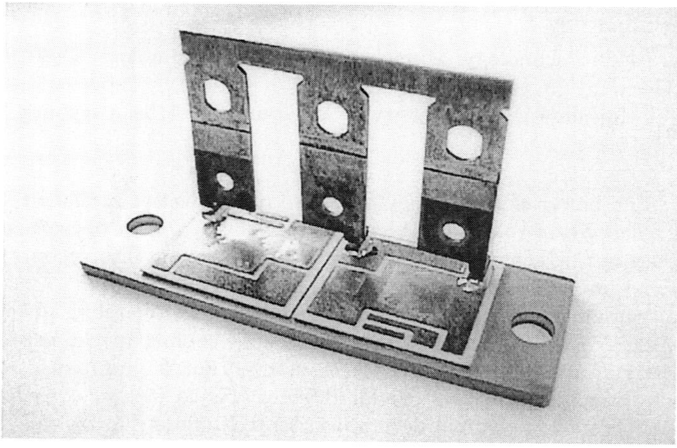

Fig. 1. IHV-module (without housing and semiconductor chips)

couplings in semiconductor device simulation, physically consistent and transparent models have been derived and validated on the basis of phenomenological thermodynamics [8,13–15], and numerical simulation techniques have been addressed in [3,7,9–12].

In this paper, suggested both by the discontinuous behavior of the material parameters across material interfaces and by the geometric structure of the module (cf. Fig. 2), we will use domain decomposition techniques on nonmatching grids also known as mortar element methods based on individual triangulations and discretizations of the subdomains and subdomain problems. This allows a great amount of flexibility at the expense of imposing weak continuity constraints across the interfaces to ensure consistency of the overall discretization (cf., e.g., [1,2,4–6,16]).

We discuss both the numerical solution of the resulting saddle point problems by multilevel block-preconditioned Lanczos iterations and adaptive grid refinement by efficient and reliable residual-type a posteriori error estimators. Numerical results are given illustrating the temperature distrubution, mechanical deformations, and stresses on selected clipping planes of real IHV-modules.

2 Electrothermomechanical Couplings in IHV-Modules

IHV-modules consist of specific semiconductor devices such as Insulated Gate Bipolar Transistors (IGBTs) and power diodes serving as switches for electric currents that are connected to the current contacts by wire bonds and integrated in copper housings (see Fig. 2). To prevent mechanical damage due to self-heating, the housings are mounted on several layers of different materials attached to each other by thin soldered joints. The copper ground plate is fixed on a cooling device (not explicitly shown in Fig. 2).

The operating behavior of the IGBTs and power diodes can be described by

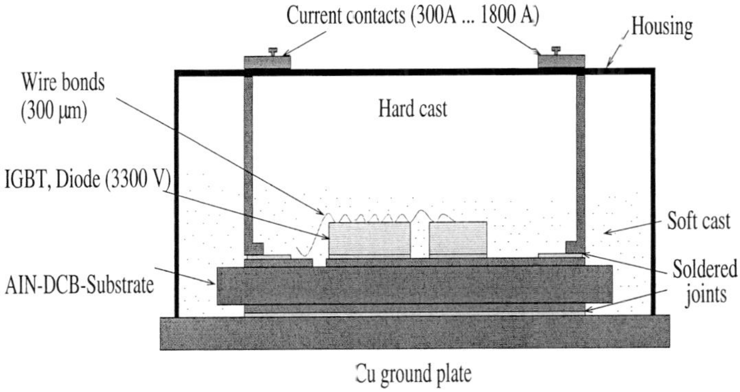

Fig. 2. Schematic representation of an IHV-module

the drift diffusion model (cf., e.g., [13–15]). Neglecting Thomson/Peltier and Nernst heat effects, the dominant heat source is the electric energy converted to Joule heat as given by $H_{Joule} := \mathbf{J}_n \cdot \mathbf{J}_n/(q\,\mu_n n) + \mathbf{J}_p \cdot \mathbf{J}_p/(q\,\mu_p p)$ where n and p are the carrier concentrations, $\mathbf{J}_n$ and $\mathbf{J}_p$ refer to the current densities, μ_n and μ_p stand for the mobilities, and q denotes the elementary charge.

The heat source is modelled by a heat flux $J(t)$ through the upper boundary Γ_0 of the module. At the lower boundary Γ_1 a heat exchange with the cooling device is assumed whereas the lateral boundary Γ_2 is supposed to be thermally insulated. Then, the temporal and spatial temperature distribution within the module is given by the following initial-boundary value problem

$$\rho c \frac{\partial T}{\partial t} = \nabla \cdot (\kappa \nabla T) \quad \text{in } Q := \Omega \times (t_0, t_1) \ , \tag{1}$$

$$\mathbf{n} \cdot \kappa \nabla T = J(t) \quad \text{on } \Gamma_0 \times (t_0, t_1) \ , \tag{2}$$

$$\mathbf{n} \cdot \kappa \nabla T = h\,(T^* - T) \quad \text{on } \Gamma_1 \times (t_0, t_1) \ , \tag{3}$$

$$\mathbf{n} \cdot \kappa \nabla T = 0 \quad \text{on } \Gamma_2 \times (t_0, t_1) \ , \tag{4}$$

$$T(x, t_0) = T_0(x) \quad \text{in } \Omega \ . \tag{5}$$

Here, the functions ρ, c and κ refer to the density, heat capacity, and thermal conductivity of the different materials. Moreover, h is the heat transition coefficient and T^* stands for the temperature of the ambient medium.
The self-heating of the module causes heat stresses leading to mechanical deformations $\mathbf{u}$ that can be described by the equilibrium equations

$$\mathrm{div}\,\sigma(\mathbf{u}) \;=\; \frac{\alpha(1-\nu)E}{(1+\nu)(1-2\nu)}\,\nabla(T-T_0) \quad \text{in } \Omega \subset \mathbf{R}^3 \;, \tag{6}$$

$$\mathbf{u} \;=\; \mathbf{0} \quad \text{on } \Gamma_0 \quad , \quad \mathbf{n}\cdot\sigma(\mathbf{u}) \;=\; \mathbf{0} \quad \text{elsewhere} \;. \tag{7}$$

where the stress tensor $\sigma(\mathbf{u})$ is related to the deformation rate tensor $\mathcal{D}(\mathbf{u}) := \frac{1}{2}(\nabla\mathbf{u} + (\nabla\mathbf{u})^T)$ by the constitutive equation

$$\sigma(\mathbf{u}) \;=\; \frac{E}{1+\nu}\,[\,\mathcal{D}(\mathbf{u}) \;+\; \frac{\nu}{1-2\nu}\,\mathrm{tr}\,\mathcal{D}(\mathbf{u})\,\mathbf{I}\,]\,.$$

Here, E and ν stand for Young's modulus and Poisson's ratio whereas α in (6) refers to the heat expansion coefficient.

3 Domain Decomposition Methods on Nonmatching Grids

The implicitly in time discretized heat equation (1)-(5) and the equilibrium equation (6),(7) give rise to boundary value problems for PDEs of the form $Lu = f$ with a second order elliptic differential operator L. Their weak formulations lead to variational problems

$$a(u,v) \;=\; \ell(v) \quad , \; v \in V(\Omega) \;.$$

Here, $V(\Omega) \subset H^1(\Omega)$ resp. $V(\Omega) \subset (H^1(\Omega))^3$, $a : V(\Omega) \times V(\Omega) \to \mathbf{R}$ is a continuous $V(\Omega)$-elliptic bilinear form, and $\ell \in (V(\Omega))'$.
The computational domain $\Omega \subset \mathbf{R}^3$ is the bounded domain occupied by the IHV-module without housing. According to the sandwich-type structure of the module (cf. Fig. 2), we subdivide Ω into n mutually disjoint hexahedral subdomains $\bar{\Omega} := \bigcup_{i=1}^n \bar{\Omega}_i\,,\ \Omega_i \cap \Omega_j = \emptyset\,,\ 1 \le i \ne j \le n$. We assume the partition to be geometrically conforming and refer to the union of the interfaces between adjacent subdomains $S := \bigcup_{i \ne j} \Gamma_{ij}\,,\ \Gamma_{ij} := \bar{\Omega}_i \cap \bar{\Omega}_j\,,\ |\,\Gamma_{ij}\,| > 0$ as the skeleton of the decomposition.
We choose individual simplicial triangulations $\mathcal{T}_i$ of $\Omega_i\,,\ 1 \le i \le n$, allowing nonconforming nodal points on the interfaces $\Gamma_{ij} \subset S$. We note that Γ_{ij} inherits a triangulation $\mathcal{T}_{ij}$ from $\mathcal{T}_i$ of Ω_i and a triangulation $\mathcal{T}_{ji}$ from $\mathcal{T}_j$ of Ω_j. We refer to Ω_j resp. Ω_i as the mortar resp. nonmortar side of Γ_{ij}.
The subdomain problems are discretized by continuous, piecewise linear finite elements. We denote by $V_h(\Omega_i;\mathcal{T}_i)$ the associated finite element spaces and consider the product space $V_h(\Omega) := \prod_{i=1}^n V_h(\Omega_i;\mathcal{T}_i)$. Clearly, functions

$v_h \in V_h(\Omega)$ may exhibit jumps $[v_h]_J$ across the interfaces $\Gamma_{ij} \subset S$. Therefore, in order to ensure consistency of the approach, we have to impose continuity constraints by means of appropriately chosen Lagrange multipliers which is done using the traces inherited from the nonmortar side. In particular, denoting by $S_{1,0}(\Gamma_{ij}; \mathcal{T}_{ij})$ the finite element space of continuous, piecewise linear finite elements with respect to $\mathcal{T}_{ij}$ vanishing on $\partial\Gamma_{ij}$, we construct $M_h(\Gamma_{ij}; \mathcal{T}_{ij})$ by a modification of those nodal basis functions in Γ_{ij} that have neighboring nodal points on $\partial\Gamma_{ij}$ such that $S_{1,0}(\Gamma_{ij}; \mathcal{T}_{ij}) \subset M_h(\Gamma_{ij}; \mathcal{T}_{ij})$ and $\dim M_h(\Gamma_{ij}; \mathcal{T}_{ij}) = \dim S_{1,0}(\Gamma_{ij}; \mathcal{T}_{ij})$ (for details see, e.g., [2]). The multiplier space is then given by $M_h(S) := \prod_{\Gamma_{ij} \subset S} M_h(\Gamma_{ij}; \mathcal{T}_{ij})$, and weak continuity constraints are realized by the bilinear form $b_h : V_h(\Omega) \times M_h(S) \to \mathbf{R}$ with $b_h(v_h, \mu_h) := -\sum_{\Gamma_{ij} \subset S} \int_{\Gamma_{ij}} \mu_h [v_h]_J \, d\sigma$, $v_h \in V_h(\Omega)$, $\mu_h \in M_h(S)$. Moreover, setting $a_h : V_h(\Omega) \times V_h(\Omega) \to \mathbf{R}$ with $a_h(u_h, v_h) := \sum_{i=1}^n a_{\Omega_i}(u_h, v_h)$ where $a_{\Omega_i} := a\,|_{\Omega_i}$, the mortar finite element approximation is given by

$$
\begin{aligned}
a_h(u_h, v_h) \; + \; b_h(v_h, \lambda_h) &= \ell(v_h) \quad , \quad v_h \in V_h(\Omega) \; , &(8)\\
b_h(u_h, \mu_h) \qquad\qquad\quad &= \quad 0 \quad\;\; , \quad \mu_h \in M_h(S) \; . &(9)
\end{aligned}
$$

It can be shown that the bilinear form a_h is elliptic on the kernel of the operator associated with b_h and that b_h satisfies an LBB-condition (cf., e.g., [2]). In algebraic form the mortar element approximation (8),(9) gives rise to the saddle point problem

$$
\begin{pmatrix} A & B^T \\ B & 0 \end{pmatrix} \begin{pmatrix} u \\ \lambda \end{pmatrix} = \begin{pmatrix} b \\ 0 \end{pmatrix} \tag{10}
$$

where $A := \operatorname{diag}(A_1, ..., A_n)$, $B^T := (B_1^T, ..., B_n^T)^T$. For its numerical solution we use a preconditioned Lanczos-type iteration with a block-diagonal preconditioner of the form $\begin{pmatrix} R_u & 0 \\ 0 & R_\lambda \end{pmatrix}$. Here, $R_u := \operatorname{diag}(R_1, ..., R_n)$ with subdomain preconditioners R_i being spectrally equivalent to the subdomain stiffness matrices A_i, $1 \le i \le n$, and R_λ represents an interface preconditioner being spectrally equivalent to the Schur complement $S_\lambda := B\,A^{-1}\,B^T$. We use multilevel preconditioners with respect to hierarchies of simplicial triangulations. In particular, the preconditioners R_i are chosen as the subdomain diagonal blocks of the BPX preconditioner. On the other hand, the construction of R_λ is realized by inner preconditioned Chebyshev iterations using a decomposition of S_λ into its subdomainwise counterparts and the construction of preconditioners for the corresponding subdomain Schur complements (for details we refer to [6]).

Grid adaptation is performed on the basis of an efficient and reliable residual-type a posteriori error estimator (cf. [4–6]). Assuming $h_i \le C\,h_j$ where $h_i := \{\operatorname{diam}(T) \mid T \in \mathcal{T}_i\,,\, T \cap \Gamma_{ij} \neq \emptyset\}$ and $h_j := \{\operatorname{diam}(T) \mid T \in \mathcal{T}_j\,,\, T \cap \Gamma_{ij} \neq \emptyset\}$, it can be shown that there exist constants $0 < \gamma_R \le \Gamma_R$ independent of the refinement level such that

$$
\gamma_R\,\eta_R^2 \;\le\; \|u - u_M\|^2_{V(\Omega)} \;\le\; \Gamma_R\,\eta_R^2
$$

where in case the principal part of L is of the form $-\operatorname{div} a \operatorname{grad}$

$$\eta_R^2 := \sum_{i=1}^{n} \sum_{T \in \mathcal{T}_i} \eta_{R;T}^2 ,$$

$$\eta_{R;T}^2 := \underbrace{h_T^2 \, \|f - L\, u_M\|_{0;T}^2 + \sum_{F \subset \partial T \cap \Omega_i} h_F \, \|[\mathbf{n} \cdot a \, \nabla \, u_M]_J\|_{0;F}^2}_{\text{standard residual based error estimate w.r.t. } \mathcal{T}_i}$$

$$+ \underbrace{\sum_{F \subset T \cap S} h_F \, \|\mathbf{n} \cdot a \, \nabla \, u_M - \lambda_M\|_{0;F}^2 + \sum_{\ell=1}^{L} \sum_{F \subset \partial T \cap \Gamma_\ell} h_F^{-1} \, \|[u_M]_J\|_{0;F}^2}_{\text{additional part on the interfaces}} .$$

4 Numerical Results

For the design of IHV-modules one is interested in the distribution of the temperature T, the mechanical deformations $\mathbf{u}$, and the equivalence stresses $\sigma_E := \|\sigma^{(D)}\|_F$ where $\sigma^{(D)} := \sigma - \frac{1}{3}\operatorname{tr}(\sigma)\,\mathbf{I}$ is the deviatoric stress tensor and $\|\cdot\|_F$ refers to the Frobenius norm, i.e., $\|\sigma^{(D)}\|_F := (\sum_{i,j=1}^{3}(\sigma_{ij}^{(D)})^2)^{1/2}$. In particular, large equivalence stresses indicate possible mechanical failure. They are most likely to occur at the soldered joints. Figs. 3, 4, and 5 display the distribution of the computed values of $T, \mathbf{u}$, and σ_E in the clipping plane corresponding to the lower soldered joint between the copper ground plate and the copper plate above (cf. Fig. 2). We observe temperature peaks in the left and right part of the plane at locations just below the copper blocks housing the IGBTs and power diodes (Fig. 3). The mechanical deformations caused by the associated heat stresses are radially distributed towards the boundaries of the joint (Fig. 4). Moreover, we see that high equivalence stresses occur at the left and right lateral boundary of the joint which is in accordance with experimental findings.

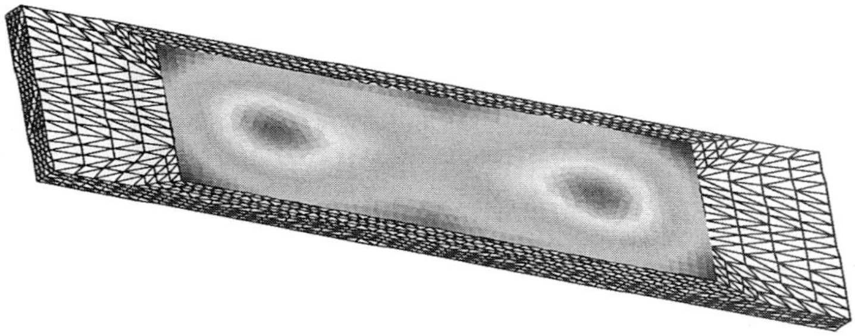

Fig. 3. Temperature distribution (cross section of lower soldered joint)

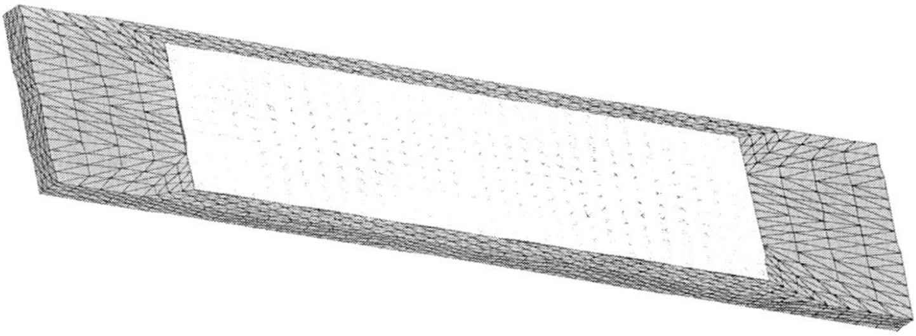

Fig. 4. Distribution of mechanical deformations (lower soldered joint)

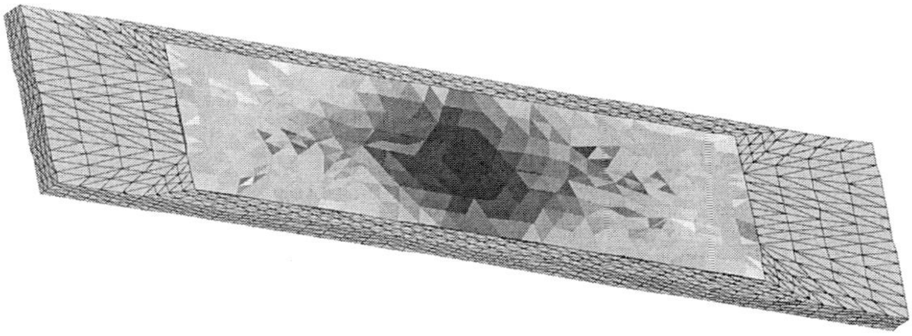

Fig. 5. Distribution of equivalence stresses (lower soldered joint)

Acknowledgments

The work has been supported by grants from the Federal Ministry for Education and Research (BMBF) under Grant No. 03HOM3A1 and from the German National Science Foundation (DFG) under Grant No. HO877/4-1.

References

1. F. Ben Belgacem and Y. Maday, "The mortar element method for three dimensional finite elements", RAIRO M^{2}AN **31**, 289-302, (1997).
2. D. Braess and W. Dahmen, "Stability estimates of the mortar finite element method for 3-dimensional problems", East-West J. Numer. Math. **6**, 249-263, (1998).
3. St. Dürndorfer, V. Gradinaru, R.H.W. Hoppe, E.-R. König, G. Schrag, and G. Wachutka, "Numerical simulation of microstructured semiconductor devices, transducers, and systems", In: High Performance Scientific and Engineering Computing. Proc. "Int. FORTWIHR-Symposium", Munich, March 1998 (H. Bungartz, F. Durst, and Chr. Zenger; eds.), pp. 309–323, Lecture Notes in Computational Science and Engineering, Vol. **8**, Springer, Berlin-Heidelberg-New York, (1999).
4. B. Engelmann, R.H.W. Hoppe, Y. Iliash, Y. Kuznetsov, Y. Vassilevski, and B. Wohlmuth, "Adaptive macro-hybrid finite element methods", In: Proc. 2nd European Conference on Numerical Methods (ENUMATH 97), Heidelberg, Sept.

29 - Oct. 3, 1997 (Bock,H.G.,Brezzi,F.,Glowinski,R.,Kanschat,G.Kuznetsov,Y., Périaux,J., and Rannacher,R.;eds.), p. 294-302, World Scientific, Singapore, (1998).

5. B.Engelmann, R.H.W.Hoppe, Y.Iliash, Y.Kuznetsov, Y.Vassilevski, and B. Wohlmuth, "Adaptive finite element methods for domain decomposition on nonmatching grids", In: "Parallel Solution of PDEs", IMA Volume in Mathematics and its Applications, Vol. 120 (P. Bjørstad and M. Luskin; eds.), pp. 57-83, Springer, Berlin-Heidelberg-New York, (2000).
6. R.H.W. Hoppe, Y. Iliash, Y. Kuznetsov, Y. Vassilevski, and B. Wohlmuth, "Analysis and parallel implementation of adaptive mortar element methods", East-West J. Numer. Math. **6**, 223-248, (1998).
7. R.H.W. Hoppe, Y. Iliash, and G. Mazurkevitch, "Domain decomposition methods in the design of high power electronic devices", In: "Multifield Problems. State of the Art" (M. Sändig, W. Schiehlen, and W. Wendland; eds.), pp. 169-182, Springer, Berlin-Heidelberg-New York, (2000).
8. K.D. Kells, "General electrothermal semiconductor device simulation", Hartung-Gorre, Konstanz, (1994).
9. K. Kells, K. Lilja, G. Wachutka, and W. Fichtner, "Detailed study of electrothermal effects in semiconductor power devices using multidimensional numerical simulation", In: Verhandlg. DPG(VI) **29**, HL-28.12, (1994).
10. K. Kells, S. Müller, W. Fichtner, and G. Wachutka, "Simulation of self-heating effects in a power p-i-n diode", In: Simulation of Semiconductor Devices and Processes (SISDEP-93), Vol. 5 (S. Selberherr, H. Stippel, and E. Strasser; eds.), p. 41-48, (1993).
11. S. Ramminger, G. Mitic, P. Türkes, and G. Wachutka, "Thermomechanical simulation of wire bonding joints in power modules", In: Proc. 2nd Conf. on Modeling and Simulation of Microsystems, Sensors, and Actuators (MSM-99), San Juan, Puerto Rico, pp. 483-486, (1999).
12. R. Thalhammer, G. Deboy, W. Keilitz, U. Müller, and G. Wachutka, "Electrothermal effects in semiconductor power devices analyzed by numerical simulation and internal laser deflection measurements", In: Proc. Int. Semiconductor Device Research Symp. (ISDRS-95), Charlottesville, U.S.A., pp. 51-54, (1995).
13. G.Wachutka, "Rigorous thermodynamic treatment of heat generation and conduction in semiconductor device modeling", IEEE Trans. CAD Integr. Circuits and Syst. **9**, 1141-1149, (1990).
14. G.Wachutka, "Problem-oriented modeling of coupled physical effects in microtransducers and electronic devices", In: Proc. 20th Int. Conf. on Microelectronics (MIEL-95), Nis, Serbia, Vol. 2, p. 539-547, (1995).
15. G. Wachutka, "The art of modeling coupled field effects in microdevices and microsystems", In: Proc. 2nd Int. Conf. on Modeling and Simulation of Microsystems, Sensors, and Actuators (MSM-99), San Juan, Puerto Rico, p. 14-19, (1999).
16. B. Wohlmuth, "Discretization Methods and Iterative Solvers Based on Domain Decomposition", Lecture Notes in Computational Science and Engineering, Vol. 17, Springer, Berlin-Heidelberg-New York, (2001).

Heat Conduction as Eigenvalue Problem

Y.C. Gerstenmaier[1] and G. Wachutka[2]

[1] Siemens AG, Corporate Technology, Otto-Hahn-Ring 6, D-81730 München
[2] Institute for Physics of Electrotechnology, Munich University of Technology Arcisstr.21, D-80290 München, Germany

Abstract. The time dependent solution of the heat conduction equation is represented by expanding T in terms of eigenfunctions and eigenvalues (= inverse time constants). Once these are known, the time evolution of T can be computed readily for arbitrary heat sources. A method is presented for computing eigenfunctions and eigenvalues for a multilayer structure, where all layers have the same lateral extension and differ vertically in thickness and material parameters. Those structures are encountered in modeling IC packages and multichip modules in power electronic applications. Examples will be given for the temperature evolution calculated for the time dependent heat distribution in power converter applications. However, computation time and storage demands will be high if high spatial and temporal resolution for the temperature field is needed. Using effective time constants and functions the numerical effort can be reduced considerably.

1 Introduction

The thermal evolution in electronic systems for arbitrary heat generation may be computed by finite element (FEM) [1], finite differences or finite volume (FVM) methods. However, for models with a large number of grid points (83,100 grid points for the set-up of Fig. 1), CPU-time and storage requirements become to excessive for the transient calculation in case of complicated power pulses and load cycles over large time intervals.

In recent work [3,4], the time-dependent solution $T(x,t)$ of the heat conduction equation

$$\rho(\boldsymbol{x})c(\boldsymbol{x})\frac{\partial T(\boldsymbol{x},t)}{\partial t} = H(\boldsymbol{x},t) + \nabla_x \cdot (\lambda(\boldsymbol{x})\nabla_x T(\boldsymbol{x},t)) \tag{1}$$

was represented by expanding T in terms of eigenfunctions and eigenvalues of (1) similar to the treatment of the quantum mechanical one particle Schroedinger equation. This is possible, provided that the mass density ρ, the specific heat c and the thermal conductivity λ of the structure depend only on the position x and not on the temperature T itself, i.e. (1) is linear in T. The letter x denotes the 3D location of the space point x,y,z .

Once the eigenfunctions and eigenvalues of (1) are known, which do not depend on the heat generation density $H(x,t)$ but only on the geometry of the set-up, the boundary conditions, and the material parameters, the temporal evolution of T can be computed very fast for arbitrary heat sources

$H(x,t)$. However, it is a non-trivial task to determine the eigenfunctions and eigenvalues for a complex three-dimensional structure with given boundary conditions. The problem could be solved by FEM- computation, similar to a mechanical modal analysis for determining the vibration characteristics (natural frequencies and mode shapes) of machine components. However, since the superposition of several thousand eigenfunctions is needed for the precise representation of the temperature field, again large computation times would result. This paper will review an algorithm for computing eigenfunctions and eigenvalues for a rectangular multilayer structure, where all layers have the same lateral extension and differ vertically by thickness and material parameters.

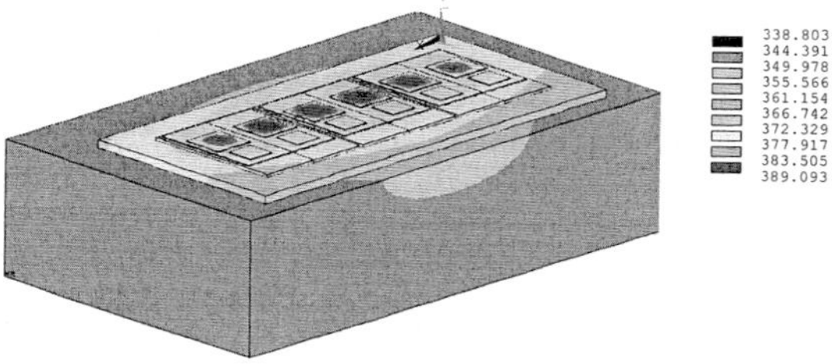

Fig. 1. FEM-simulation of the stationary temperature distribution in a 1700V IGBT and diode module mounted on a radiator. IGBT chips are heated with uniform power, diodes are not heated

2 Theoretical Background

In order to calculate the eigenfunctions and eigenvalues for (1) with zero heat generation H, a separation of variables is introduced. When $T(x,t) = \psi(x) \cdot \vartheta(t)$ equation (1) with heat $H = 0$ can be rewritten as:

$$\frac{1}{\vartheta(t)} \frac{d\vartheta(t)}{dt} = \frac{1}{\rho(x)c(x)\psi(x)} \nabla_x \cdot (\lambda(x) \nabla_x \psi(x)) \ .$$

Since both sides of the equation depend on different variables they must be constant, where the constant $v = 1/\tau$ is an eigenvalue of the problem. Therefore: $\vartheta(t) = \vartheta(0) exp(-t/\tau)$ and

$$\frac{1}{\tau} \psi(x) = \frac{-1}{\rho(x)c(x)} \nabla_x \cdot (\lambda(x) \nabla_x \psi(x)) \ . \tag{2}$$

The eigenvalue spectrum $v_i = 1/\tau_i$ of the last equation depends on the boundary conditions the eigenfunctions $\psi_i(x)$ are subjected to. As detailed in [3] the problem of determining the eigenvalue spectrum is similar to that of the quantum mechanical Schroedinger equation for one particle of mass m moving in a rectangular potential wall. In the same way as for this problem the eigenvalues for (1) with $H = 0$ are real, positive, discrete (lumped) and not bounded from above.

For the temperature problem usually other boundary conditions have to be assumed [5,6]. Some parts of the boundary ∂V may obey Neumann conditions $\partial T(x,t)/\partial n = 0$ (no heat flow along the outside normal of the set-up), while other parts obey Dirichlet conditions $T(x,t) = T_0$. The prescribed boundary ambient temperature T_0 (cooling medium temperature) can be chosen to be zero without loss of generality. Linear combinations of functions T satisfying these zero boundary conditions again satisfy these conditions.

The differential operators of the temperature problem $\Lambda_{op} = \frac{-1}{\rho(x)c(x)} \nabla_x \cdot \lambda(x)\nabla_x$ and of the Schroedinger wave problem $-\hbar^2/2m\,\Delta_x$ are symmetrical with respect to a properly chosen inner product for functions with zero boundary conditions on ∂V, thus ensuring that the eigenvalue spectrum is real. Usually the inner product of f and g is chosen as : $(f,g) = \int_V f(x)g(x)d^3x$. Λ_{op} would be symmetrical, if $(f, \Lambda_{op} g) = (\Lambda_{op} f, g)$, which obviously is not the case. But since the material functions $\rho(x)$ and $c(x)$ are always positive, another product can be defined, which obeys all mathematical conditions of an inner product:

$$(f,g) = \int_V \rho(x)c(x)f(x)g(x)d^3x \, . \tag{3}$$

It can be shown by partial integration that, with respect to this inner product, the operator Λ_{op} is symmetrical on the space of functions with the stated zero boundary conditions. Also mixed conditions of the 3rd type, $\alpha(x)T + \lambda(x)\partial T/\partial n = 0$, are allowed. This means the inclusion of a heat transfer coefficient $\alpha(x)$ (inverse contact resistance) between that part of the system surface ∂V and zero ambient temperature. No other values of the ambient temperature are allowed.

The eigenfunctions $\psi_i(x)$ of Λ_{op} can be normalised to $(\psi_i, \psi_i) = 1$ and constitute a complete set of orthogonal basis functions, provided that $\rho(x)c(x)$ vanishes nowhere in V. So any reasonable function with zero boundary conditions in the sense explained here can be expressed as a series of the $\psi_j(x)$. The general time-dependent solution of the heat conduction problem (1) can be written as $T(x,t) = \Sigma_j c_j(t)\psi_j(x)$. The modified heat generation $H(x,t)/(\rho(x)c(x))$ can also be written as $\Sigma_j h_j(t)\psi_j(x)$. With this (1) leads to the set of ordinary differential equations $dc_j(t)/dt = -c_j(t)/\tau_j + h_j(t)$ with the solution

$$c_j(t) = c_j(0)e^{-t/\tau_j} + \int_0^t e^{(\tau - t)/\tau_j} h_j(\tau)d\tau \, .$$

Since $(\psi_i, \psi_j) = \delta_{i,j}$:

$$h_j(t) = (\psi_j, \frac{H}{\rho c}) = \int_V \psi_j(x)H(x,t)d^3x \, .$$

The complete temperature field within V as solution of (1) for general $H(x,t)$ and with zero Dirichlet and Neumann boundary conditions is:

$$T(x,t) = \sum_{j=1}^{\infty} \psi_j(x) c_j(0) e^{-t/\tau_j} + \int_0^t \sum_{j=1}^{\infty} h_j(\tau) \psi_j(x) e^{(\tau-t)/\tau_j} d\tau \,. \tag{4}$$

The coefficients $c_j(0)$ determine the starting condition of the temperature field $T(x,0)$ at $t=0$. In case that the heat generation is switched off at $t=0$, $h_j(\tau)=0$ and only the first sum in (4) contributes, which describes the cool down of the system. The position-independent inverse eigenvalues τ_j are the time constants of the system and, in the same way as the eigenfunctions $\psi_j(x)$, only depend on the geometry of the set-up, its material parameters and its boundary conditions. Considering the heating case with the special assumption $H(x,t) = P(t) \cdot h(x)$ with $P(t) = \int H(x,t) d^3x$ being the overall heating, the time dependence of h_j is $h_j(t) = P(t)\, g_j$ with constant $g_j = \int_V h(x) \psi_j(x) d^3x$. Then

$$T(x,t) = \sum_{j=1}^{\infty} \psi_j(x) c_j(0) e^{-t/\tau_j} + \int_0^t P(\tau) \sum_{j=1}^{\infty} \psi_j(x) g_j e^{(\tau-t)/\tau_j} d\tau \,. \tag{5}$$

The sum under the integral can be identified with the time derivative of the thermal impedance for the position x , called $\dot{Z}th(x, t-\tau)$. $Zth(x,t)$ is the temperature rise at position x for constant dissipated power P of unit strength starting at $t=0$. Therefore $Zth(x,0)=0$, and the unit step response in terms of the eigenfunctions and eigenvalues can be written as [3,4]:

$$Zth(x,t) = \sum_{j=1}^{\infty} r_j(x)(1-e^{-t/\tau_j}) \quad ; \quad r_j(x) = \psi_j(x)\, g_j\, \tau_j \,. \tag{6}$$

The eigenvalues and eigenfunctions $1/\tau_j, \psi_j$ will be calculated for multilayer structures in the next section. It turns out that for an accurate description of the temperature field many thousand eigenfunctions have to be superposed. However, the measured or simulated $Zth(t)$ curve for a specific location x can be represented with very good accuracy by a sum of the form (6) with only $M=6$ to at most 16 effective pairs $reff_i, \tau eff_i$, also for systems with complex geometry [3]. The τeff_i can be chosen to be the same for all x:

$$Zth(x,t) = \sum_{i=1}^{M} reff_i(x)(1-e^{-t/\tau eff_i}) \,.$$

Once the τeff_i are properly chosen a linear combination of their exponentials gives: $e^{-t/\tau_j} = \sum_{i=1}^{M} a_{i,j} e^{-t/\tau eff_i}$ with $\sum_{i=1}^{M} a_{i,j} = 1$. With this the relation between $reff_i$ and r_j is obtained: $reff_i(x) = \sum_{j=1}^{\infty} a_{i,j} r_j(x)$. The temperature field for general time dependence of $P(t)$ can be represented according to (5) in the form of a convolution integral [6,3]: $T(x,t) =$

$\int_0^t P(\tau)\dot{Z}th(x, t-\tau)d\tau$. This is valid for a time dependent heat source whose location is fixed. For several heat sources, e.g. multichip modules with independent heating of the chips, the contribution of several thermal impedances has to be added to obtain the complete temperature field.

3 Multilayer Structures

3.1 Fundamental Equations

In [8] three-layer structures have been calculated analytically in the frequency domain and for the stationary case, where all layers have the same lateral extension and differ vertically by thickness and material parameters. In [2] this has been generalised to multilayer structures with arbitrary number of layers. In order to compute the time-dependent temperature field according to (4) or (5), in this paper the eigenfunctions and time constants will be calculated in a semianalytical way for multilayer structures with arbitrary number of layers. The structure consists of N layers starting at $z_0 = 0$ (bottom side). The top side of the structure is at z_N. z_i is the top side of the i-th layer. The layers are of rectangular shape and extend from 0 to Lx and Ly in x and y direction, respectively. The distribution of the heat generation H within the structure may be defined at will.

Since the material parameters are constant within each layer, (2) can be written with the 3D position x now expressed explicitly as x,y,z:

$$(1/\tau)\psi(x,y,z) = -(\lambda_i/\rho_i c_i)\,\Delta\psi(x,y,z) \tag{7}$$

for x,y,z in the i-th layer. The structure under consideration seems to be one of the most general for which the eigenfunctions can be represented by separation of variables in product form: $\psi(x,y,z) = a(x)\,b(y)\,f(z)$. ψ will be subjected to the boundary conditions $\psi(z=0) = 0$ at the bottom side and $\partial\psi/\partial n = 0$ on all other sides. Then the solutions for $a(x)$ and $b(y)$ have to be cosine-functions:

$$\psi_{mnl}(x,y,z) = \cos(m\pi x/L_x)\cdot\cos(n\pi y/L_y)\cdot f_l(z)\,. \tag{8}$$

The indices m, n, l denote integers running from zero to infinity. Inserting the ψ_{mnl} of (8) into (7) yields the governing equation for $f_l(z)$:

$$\left(\gamma_0^2 - \frac{\kappa_i}{\tau_{mnl}}\right) f_{lmn}(z) = \frac{d^2}{dz^2} f_{lmn}(z)\,;\; \gamma_0^2 = \left(\frac{m\pi}{L_x}\right)^2 + \left(\frac{n\pi}{L_y}\right)^2\,;\; \kappa_i = \frac{\rho_i c_i}{\lambda_i}\,. \tag{9}$$

Due to the definition of γ_0^2 the $f_l(z)$ generally also depend on m, n. The solution $f_l(z)$ for the layer i is:

$$f_{lmn}^i(z) = A_{mnl}^i e^{\gamma_i z} + B_{mnl}^i e^{-\gamma_i z} \quad ; \quad \gamma_i^2 = \left(\gamma_0^2 - \frac{\kappa_i}{\tau_{mnl}}\right)\,.$$

Since γ_i^2 may be positive or negative, γ_i may assume real or imaginary values. However $f_l(z)$ as component of the temperature field always has to be real. Therefore the following conditions result: γ_i imaginary $\Rightarrow A_i = B_i^*$; γ_i real $\Rightarrow A_i, B_i$ real. The asterix * denotes the complex conjugate.

If several layers are present, the individual f_l^i of the layers have to meet continuity conditions at the interfaces z_i , which result from the requirements of continuity for $f_l : f(z_i - \epsilon) = f(z_i + \epsilon)$ and for the heat flow at the interface z_i , $\lambda_i \partial f/\partial z(z_i - \epsilon) = \lambda_{i+1} \partial f/\partial z(z_i + \epsilon), (\epsilon \rightarrow 0)$:

$$
\begin{aligned}
A^i_{mnl} e^{\gamma_i z_i} + B^i_{mnl} e^{-\gamma_i z_i} &= A^{i+1}_{mnl} e^{\gamma_{i+1} z_i} + B^{i+1}_{mnl} e^{-\gamma_{i+1} z_i} , \\
\gamma_i \lambda_i \left(A^i_{mnl} e^{\gamma_i z_i} - B^i_{mnl} e^{-\gamma_i z_i} \right) &= \gamma_{i+1} \lambda_{i+1} \left(A^{i+1}_{mnl} e^{\gamma_{i+1} z_i} - B^{i+1}_{mnl} e^{-\gamma_{i+1} z_i} \right) .
\end{aligned}
\tag{10}
$$

The boundary conditions at the bottom side $f_l(z = 0) = 0$ and at the top side $\lambda_N \partial f_l(z_N)/\partial z = 0$ lead to:

$$
A^1_{mnl} + B^1_{mnl} = 0 ; \quad \gamma_N \lambda_N \left(A^N_{mnl} e^{\gamma_N z_N} - B^N_{mnl} e^{-\gamma_N z_N} \right) = 0 . \tag{11}
$$

This linear equation system for the A_i, B_i can be conceived as a linear system with a $2N \times 2N$ matrix $M(\tau_{mnl}, m, n, z_i, \kappa_i, i = 1, ...N)$ operating on the vector of the $(A_1, B_1, A_2, B_2, ..., A_N, B_N)$. For solving the eigenvalue problem, the matrix M is required to be singular in order that a non-trivial solution vector (A_i, B_i) different from zero exists. Therefore, for fixed m and n, the τ_{mnl} in M have to be chosen such that the determinant of M vanishes: det $M(\tau_{mnl}, m, n) = 0$. This constitutes a nonlinear equation for the unknown eigenvalues $1/\tau_{mnl}$ with several solutions, which are numerated by the index l for fixed m and n. Once the eigenvalues have been calculated the corresponding eigenfunctions are obtained by determining A_i, B_i for fixed m, n, l respectively.

The system (10) can be written as a recurrence relation to obtain the A^{i+1}, B^{i+1} values from the A^i, B^i values. Starting for general m, n, l with the first layer, $A^1 = -B^1$ gives one free variable A^1 , which at last serves to scale the eigenfunction so that its norm with respect to the inner product (3) becomes unity $(\psi_{mnl}, \psi_{mnl}) = 1$. The A^{i+1}, B^{i+1} may be obtained recursively from A^i, B^i for general variable τ_{mnl} . The top side boundary condition (11) gives the equation determining the numerical value of τ_{mnl}:

$$
A^N_{mnl}(\tau) e^{\gamma_N(\tau) z_N} - B^N_{mnl}(\tau) e^{-\gamma_N(\tau) z_N} = 0 . \tag{12}
$$

Since the functions in (12) have oscillatory behaviour, numerous roots τ are calculated for each fixed pair of numbers m,n. This step is the most time consuming in the procedure. To calculate the eigenfunction corresponding to each previously computed eigenvalue characterised by the triplet (m, n, l), the γ_i are computed and with the help of the recurrence relation (10) the coefficients are stored which determine $f_l(z)$. With this also the eigenfunctions (8) are determined. Next the norm of the eigenfunctions is calculated and the coefficients are rescaled in order that $(\psi_{mnl}, \psi_{mnl}) = 1$. Care must be taken in this process to prevent erroneous results due to the difference between large and nearly equal numbers.

3.2 Example: IGBT Multichip Module

As an example a structure similar to the power semiconductor module of Fig. 1 is treated. All layers now have the same lateral extension as the module base plate. The number of layers in the structure under investigation including the convection layer at the bottom side of the radiator is $N = 10$.

The spatial resolution of T in x-, y- and z- direction is determined by the summation limits for $(m, n, l) = j$ in (4). The time resolution of T is given by the quality of the spatial approximation of $H(x, y, z, t)/\rho c(xyz) = \Sigma_{mnl} h_{mnl}(t)\psi_{mnl}(x, y, z)$. If, for instance, this function has precisely the spatial waveform of one eigenfuntion $\psi_{mnl}(x, y, z)$, all other $h_{mnl}(t)$ are zero and the time evolution of T would be given precisely by only one time constant τ_{mnl} and one function $h_{mnl}(t)$. However, in cases of practical interest $H/\rho c$ always is a superposition of numerous eigenfunctions ψ_{mnl} .

The eigenfunctions are well adapted to the material parameter variations in the z-direction. To resolve spatial details in lateral direction to which the eigenfunctions are not adapted, e.g. boundaries of chip heat sources, the maximum values for m and n have to be chosen sufficiently large in order that the wavelength of the cosine functions in (8) are small enough. Figure 2 shows the topside of the heated IGBT module in stationary state. The max values for m and n are 80 and 40 respectively. The max value for l (depending on m, n) is 145 on the average. It took several hours to calculate the 446,745 eigenfunctions and values on a fast workstation, yet the program was not optimised with respect to speed. For an x-y plot with

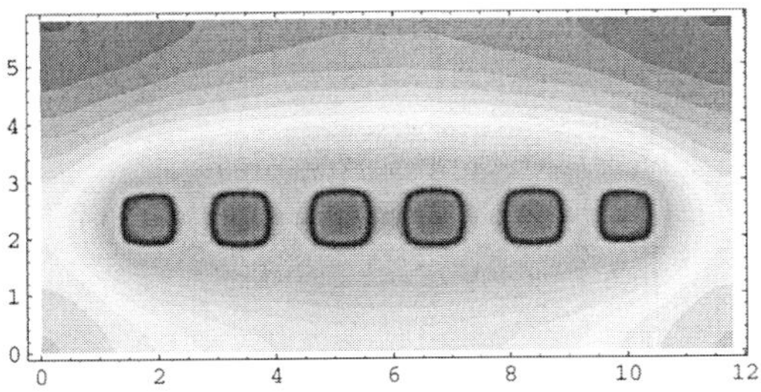

Fig. 2. Topside of the heated IGBT module in stationary state. Max temperature 80.79° C. Each IGBT is heated with 56.95W. The diodes in parallel to the IGBT are not heated and therefore do not appear. 446.745 eigenfunctions are superposed

20,000 stationary temperature calculations on the topside (Fig. 2) one minute is needed once the eigenfunctions and values have been calculated.

The computation time for the *time dependence* of T following (4) or (5) may be considerable for large number of eigenfunctions and complicated power cycles of the $h_j(t)$ over many periods. Thermal impedance matrices can be calculated for treating such problems and formerly developed procedures [3,6] for solving the convolution integral of impedances with power generation can be applied, also in the case of temperature dependent power sources. The advantage of the eigenfunction method compared to FEM computation is that the thermal impedances can be obtained very fast for any combination of heat sources once the eigenfunctions are known. On the other hand a new FEM computation has to be started for each individual heat source. A further advantage compared to FEM is that for our semi-analytical approach no grid has to be created.

To give some simple illustrations of the broad possibilities the thermal interaction of different chips in the module is studied. In power electronic (converter) applications often the power switch (IGBT) and diode are alternately switched on. In Fig. 3 only two chips are heated alternately. The temperatures in the centre of the chips (hot spot temperatures) are displayed. The temperature distribution was also calculated by the use of generalised

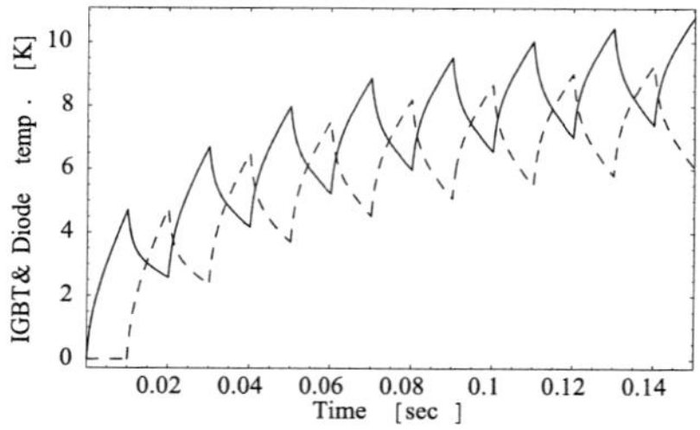

Fig. 3. 3rd IGBT and 5th Diode (dashed line) hot spot temperatures for alternating heating with 10msec period. Calculated using eigenfunctions from Fig. 3. IGBT heating 56.95W, diode 33.75W

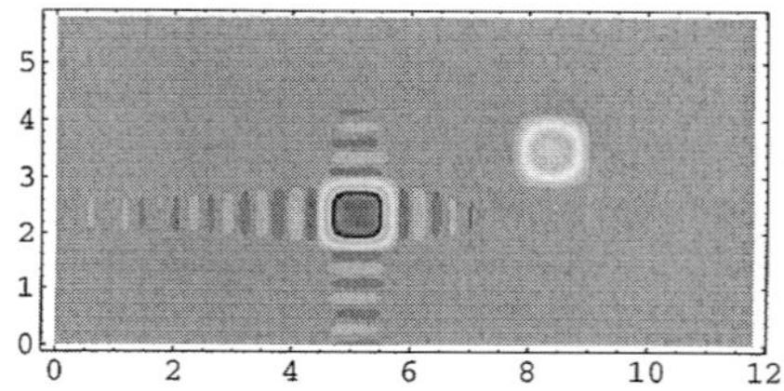

Fig. 4. Temperature distribution of the module top-side at t=0.05 sec for alternating heating of the 3rd IGBT and the 5th diode (intermediate stage of Fig. 3)

thermal impedances defined for every location on the topside. For 5,000 locations calculated in Fig. 4, the computation time amounts to some hours. Usually only the temperature evolution of the hot spots (chip centres) are of interest, which can be obtained much faster.

References

1. ANSYS Inc., Houston 1996.
2. A. Csendes, V. Székely, M. Rencz; Microelectro. Jour., 29, pp. 241-255, 1998.
3. Y.C. Gerstenmaier, G. Wachutka; Proc. IEEE SEMI-THERM XVI, San Jose, USA, pp.50-59, March 2000.
4. Y.C. Gerstenmaier, G. Wachutka; Proc. 6th THERMINIC Workshop, Budapest, Hungary, September 2000.
5. R.C. Bartels, R.V. Churchill; Bull. Amer. Math. Soc. 48, pp.276-282, 1942.
6. Y.C. Gerstenmaier, G. Wachutka; A new Procedure for the Calculation of the Temperature Development in Electronic Systems, EPE'99 conference, Lausanne, Switzerland, (CD) ISBN 90-75815-04-2, September 1999.
7. V. Székely; IEEE Trans. Circ. and Sys.-I, vol. CAS-45, no.3, pp.244-258, 1998.
8. A.G. Kokkas, IEEE Trans. on Electron Devices, ED-21, no.11, pp.674-681, 1974.

Editorial Policy

§1. Volumes in the following four categories will be published in LNCSE:

i) Research monographs
ii) Lecture and seminar notes
iii) Conference proceedings
iv) Textbooks

Those considering a book which might be suitable for the series are strongly advised to contact the publisher or the series editors at an early stage.

§2. Categories i) and ii). These categories will be emphasized by Lecture Notes in Computational Science and Engineering. **Submissions by interdisciplinary teams of authors are encouraged.** The goal is to report new developments – quickly, informally, and in a way that will make them accessible to non-specialists. In the evaluation of submissions timeliness of the work is an important criterion. Texts should be well-rounded, well-written and reasonably self-contained. In most cases the work will contain results of others as well as those of the author(s). In each case the author(s) should provide sufficient motivation, examples, and applications. In this respect, Ph.D. theses will usually be deemed unsuitable for the Lecture Notes series. Proposals for volumes in these categories should be submitted either to one of the series editors or to Springer-Verlag, Heidelberg, and will be refereed. A provisional judgment on the acceptability of a project can be based on partial information about the work: a detailed outline describing the contents of each chapter, the estimated length, a bibliography, and one or two sample chapters – or a first draft. A final decision whether to accept will rest on an evaluation of the completed work which should include

- at least 100 pages of text;
- a table of contents;
- an informative introduction perhaps with some historical remarks which should be accessible to readers unfamiliar with the topic treated;
- a subject index.

§3. Category iii). Conference proceedings will be considered for publication provided that they are both of exceptional interest and devoted to a single topic. One (or more) expert participants will act as the scientific editor(s) of the volume. They select the papers which are suitable for inclusion and have them individually refereed as for a journal. Papers not closely related to the central topic are to be excluded. Organizers should contact Lecture Notes in Computational Science and Engineering at the planning stage.

In exceptional cases some other multi-author-volumes may be considered in this category.

§4. Category iv) Textbooks on topics in the field of computational science and engineering will be considered. They should be written for courses in CSE education. Both graduate and undergraduate level are appropriate. Multidisciplinary topics are especially welcome.

§5. Format. Only works in English are considered. They should be submitted in camera-ready form according to Springer-Verlag's specifications. Electronic material can be included if appropriate. Please contact the publisher. Technical instructions and/or TEX macros are available via http://www.springer.de/author/tex/help-tex.html; the name of the macro package is "LNCSE – LaTEX2e class for Lecture Notes in Computational Science and Engineering". The macros can also be sent on request.

General Remarks

Lecture Notes are printed by photo-offset from the master-copy delivered in camera-ready form by the authors. For this purpose Springer-Verlag provides technical instructions for the preparation of manuscripts. See also *Editorial Policy*.

Careful preparation of manuscripts will help keep production time short and ensure a satisfactory appearance of the finished book. The actual production of a Lecture Notes volume normally takes approximately 12 weeks.

The following terms and conditions hold:

Categories i), ii), and iii):
Authors receive 50 free copies of their book. No royalty is paid. Commitment to publish is made by letter of intent rather than by signing a formal contract. Springer-Verlag secures the copyright for each volume.

For conference proceedings, editors receive a total of 50 free copies of their volume for distribution to the contributing authors.

Category iv):
Regarding free copies and royalties, the standard terms for Springer mathematics monographs and textbooks hold. Please write to Peters@springer.de for details. The standard contracts are used for publishing agreements.

All categories:
Authors are entitled to purchase further copies of their book and other Springer mathematics books for their personal use, at a discount of 33,3 % directly from Springer-Verlag.

Addresses:

Professor M. Griebel
Institut für Angewandte Mathematik
der Universität Bonn
Wegelerstr. 6
D-53115 Bonn, Germany
e-mail: griebel@iam.uni-bonn.de

Professor D. E. Keyes
Computer Science Department
Old Dominion University
Norfolk, VA 23529–0162, USA
e-mail: keyes@cs.odu.edu

Professor R. M. Nieminen
Laboratory of Physics
Helsinki University of Technology
02150 Espoo, Finland
e-mail: rni@fyslab.hut.fi

Professor D. Roose
Department of Computer Science
Katholieke Universiteit Leuven
Celestijnenlaan 200A
3001 Leuven-Heverlee, Belgium
e-mail: dirk.roose@cs.kuleuven.ac.be

Professor T. Schlick
Department of Chemistry and
Courant Institute of Mathematical
Sciences
New York University
and Howard Hughes Medical Institute
251 Mercer Street, Rm 509
New York, NY 10012-1548, USA
e-mail: schlick@nyu.edu

Springer-Verlag, Mathematics Editorial
Tiergartenstrasse 17
D-69121 Heidelberg, Germany
Tel.: *49 (6221) 487-185
e-mail: peters@springer.de
http://www.springer.de/math/
peters.html

Lecture Notes in Computational Science and Engineering

Vol. 1 D. Funaro, *Spectral Elements for Transport-Dominated Equations.* 1997. X, 211 pp. Softcover. ISBN 3-540-62649-2

Vol. 2 H. P. Langtangen, *Computational Partial Differential Equations.* Numerical Methods and Diffpack Programming. 1999. XXIII, 682 pp. Hardcover. ISBN 3-540-65274-4

Vol. 3 W. Hackbusch, G. Wittum (eds.), *Multigrid Methods V.* Proceedings of the Fifth European Multigrid Conference held in Stuttgart, Germany, October 1-4, 1996. 1998. VIII, 334 pp. Softcover. ISBN 3-540-63133-X

Vol. 4 P. Deuflhard, J. Hermans, B. Leimkuhler, A. E. Mark, S. Reich, R. D. Skeel (eds.), *Computational Molecular Dynamics: Challenges, Methods, Ideas.* Proceedings of the 2nd International Symposium on Algorithms for Macromolecular Modelling, Berlin, May 21-24, 1997. 1998. XI, 489 pp. Softcover. ISBN 3-540-63242-5

Vol. 5 D. Kröner, M. Ohlberger, C. Rohde (eds.), *An Introduction to Recent Developments in Theory and Numerics for Conservation Laws.* Proceedings of the International School on Theory and Numerics for Conservation Laws, Freiburg / Littenweiler, October 20-24, 1997. 1998. VII, 285 pp. Softcover. ISBN 3-540-65081-4

Vol. 6 S. Turek, *Efficient Solvers for Incompressible Flow Problems.* An Algorithmic and Computational Approach. 1999. XVII, 352 pp, with CD-ROM. Hardcover. ISBN 3-540-65433-X

Vol. 7 R. von Schwerin, *Multi Body System SIMulation.* Numerical Methods, Algorithms, and Software. 1999. XX, 338 pp. Softcover. ISBN 3-540-65662-6

Vol. 8 H.-J. Bungartz, F. Durst, C. Zenger (eds.), *High Performance Scientific and Engineering Computing.* Proceedings of the International FORTWIHR Conference on HPSEC, Munich, March 16-18, 1998. 1999. X, 471 pp. Softcover. 3-540-65730-4

Vol. 9 T. J. Barth, H. Deconinck (eds.), *High-Order Methods for Computational Physics.* 1999. VII, 582 pp. Hardcover. 3-540-65893-9

Vol. 10 H. P. Langtangen, A. M. Bruaset, E. Quak (eds.), *Advances in Software Tools for Scientific Computing.* 2000. X, 357 pp. Softcover. 3-540-66557-9

Vol. 11 B. Cockburn, G. E. Karniadakis, C.-W. Shu (eds.), *Discontinuous Galerkin Methods.* Theory, Computation and Applications. 2000. XI, 470 pp. Hardcover. 3-540-66787-3

Vol. 12 U. van Rienen, *Numerical Methods in Computational Electrodynamics.* Linear Systems in Practical Applications. 2000. XIII, 375 pp. Softcover. 3-540-67629-5

Vol. 13 B. Engquist, L. Johnsson, M. Hammill, F. Short (eds.), *Simulation and Visualization on the Grid.* Parallelldatorcentrum Seventh Annual Conference, Stockholm, December 1999, Proceedings. 2000. XIII, 301 pp. Softcover. 3-540-67264-8

Vol. 14 E. Dick, K. Riemslagh, J. Vierendeels (eds.), *Multigrid Methods VI.* Proceedings of the Sixth European Multigrid Conference Held in Gent, Belgium, September 27-30, 1999. 2000. IX, 293 pp. Softcover. 3-540-67157-9

Vol. 15 A. Frommer, T. Lippert, B. Medeke, K. Schilling (eds.), *Numerical Challenges in Lattice Quantum Chromodynamics.* Joint Interdisciplinary Workshop of John von Neumann Institute for Computing, Jülich and Institute of Applied Computer Science, Wuppertal University, August 1999. 2000. VIII, 184 pp. Softcover. 3-540-67732-1

Vol. 16 J. Lang, *Adaptive Multilevel Solution of Nonlinear Parabolic PDE Systems.* Theory, Algorithm, and Applications. 2001. XII, 157 pp. Softcover. 3-540-67900-6

Vol. 17 B. I. Wohlmuth, *Discretization Methods and Iterative Solvers Based on Domain Decomposition.* 2001. X, 197 pp. Softcover. 3-540-41083-X

Vol. 18 U. van Rienen, M. Günther, D. Hecht (eds.), *Scientific Computing in Electrical Engineering.* Proceedings of the 3rd International Workshop, August 20-23, 2000, Warnemünde, Germany. 2001. XII, 428 pp. Softcover. 3-540-42173-4

Vol. 19 I. Babuška, P. G. Ciarlet, T. Miyoshi (eds.), *Mathematical Modeling and Numerical Simulation in Continuum Mechanics.* Proceedings of the International Symposium on Mathematical Modeling and Numerical Simulation in Continuum Mechanics, September 29 - October 3, 2000, Yamaguchi, Japan. 2002. VIII, 301 pp. Softcover. 3-540-42399-0

Vol. 20 T. J. Barth, T. Chan, R. Haimes (eds.), *Multiscale and Multiresolution Methods.* Theory and Applications. 2002. X, 389 pp. Softcover. 3-540-42420-2

Vol. 21 M. Breuer, F. Durst, C. Zenger (eds.), *High Performance Scientific and Engineering Computing.* Proceedings of the 3rd International FORTWIHR Conference on HPSEC, Erlangen, March 12-14, 2001. 2002. XIII, 408 pp. Softcover. 3-540-42946-8

For further information on these books please have a look at our mathematics catalogue at the following URL: `http://www.springer.de/math/index.html`

Druck: Strauss Offsetdruck, Mörlenbach
Verarbeitung: Schäffer, Grünstadt